Next Generation Wireless Applications

Next Generation Wireless Applications

Creating Mobile Applications in a Web 2.0 and Mobile 2.0 World

Second Edition

Paul Golding
Independent Consultant, UK

John Wiley & Sons, Ltd

Other Wiley Editorial Offices

John Wiley & Sons Inc., 111 River Street, Hoboken, NJ 07030, USA

Jossey-Bass, 989 Market Street, San Francisco, CA 94103-1741, USA

Wiley-VCH Verlag GmbH, Boschstr. 12, D-69469 Weinheim, Germany

John Wiley & Sons Australia Ltd, 42 McDougall Street, Milton, Queensland 4064, Australia

John Wiley & Sons (Asia) Pte Ltd, 2 Clementi Loop #02-01, Jin Xing Distripark, Singapore 129809

John Wiley & Sons Canada Ltd, 6045 Freemont Blvd, Mississauga, Ontario L5R 4J3, Canada

Wiley also publishes its books in a variety of electronic formats. Some content that appears in print may not be
available in electronic books.

Library of Congress Cataloging-in-Publication Data

Golding, Paul, 1968–
    Next generation wireless applications : creating mobile applications in a Web 2.0
and Mobile 2.0 world / Paul Golding. – 2nd ed.
      p.   cm.
    Includes index.
    ISBN 978-0-470-72506-1 (cloth : alk. paper)
1. Wireless communication systems.   2. Mobile communication systems.   I. Title.
    TK5103.2.G63   2008
    621.384–dc22   2007045746

British Library Cataloguing in Publication Data

A catalogue record for this book is available from the British Library

ISBN 978-0-470-72506-1 (HB)

Typeset in 10/12pt Times by Aptara Inc., New Delhi, India
Printed and bound in Great Britain by Antony Rowe Ltd, Chippenham, England.
This book is printed on acid-free paper.

To my wife and kids

Contents

Acknowledgements

I would like to thank all those who have supported and assisted me in writing the second edition of this book, especially my wife and kids for their usual encouragement. Thanks to all those companies who have assisted me with images.

Preface

This book is intended to give wireless practitioners a means by which to navigate and think about the increasingly complex world of next generation mobile application technologies. It is an attempt to describe and contextualise a number of prevalent and emerging paradigms for providing services on mobile devices other than the familiar calling and texting. The availability of new wireless technologies increasingly coincides with many other exciting technologies that are now within the mobile solutions toolkit, such as accurate location finding, rich client solutions, smarter devices, efficient protocols and many other micro and macro components. Possibly the most enticing aspect of these technologies is their ability to integrate into the massively connected world of the Internet, allowing new levels of service integration not envisaged at the start of the mobile data era. This is thanks to the vast array of Internet software tools, methods and services that continue to develop rapidly, even after the dot.com collapse, including the so-called Web 2.0 platform.

The rapid rise of so many *anytime, anyplace* possibilities is proving to be a challenge in terms of how to understand the many ideas that can be utilised when providing new and exciting services. Perhaps a programmer working on an application for a Java-capable handset is unable to grasp how best to incorporate location-finding techniques, or how to think about the security implications of their design, or how to utilise SIP or other protocols. This 'think about' step is the aspect that I am mostly trying to aid with this book, introducing and connecting some of the major technologies and themes applicable to mobile applications.

Combining these technologies can lead us to a bristling array of possibilities. Revealing such potentials is the real purpose of this book, which is why the contents are not organised into a more conventional menu of subjects, but largely grouped into themes that seem to follow a useful mental model when thinking about mobile applications; hence, why I feel justified in talking about J2EE and CDMA in the same book. They are vastly different technologies, but also very much connected within the world of mobile applications.

I hope that you succeed with your wireless endeavours and that this book provides you with enough interesting material to 'think mobile' in your own way. Please email me your feedback, experiences, questions, or ideas to ngwa@paulgolding.com.

Regards
Paul Golding, 2007
www.paulgolding.com

Abbreviations and Acronyms

2G	Second Generation
3G	Third Generation
3GPP	Third Generation Partnership Project
AD	Analog–Digital
ADPCM	Adaptive Differential Pulse-Coded Modulation
A-GPS	Assisted GPS
AJAX	Asynchronous JavaScript and XML
AKS	Authentication Algorithm
AI	Application Interface
ALU	Arithmetic Logic Unit
AMR	Adaptive Multirate
AMS	Application Management System
API	Application Programming Interface
ARPU	Average Revenue Per User
AS	Application Server
ASCII	American Standard Code for Information Interchange
B2BUA	Back-to-Back User Agent
BCF	Border Control Function
BEEP	Blocks Extensible Exchange Protocol
BER	Bit Error Rate
BGCF	Border Gateway Control Function
BSC	Base Station Controller
BSS	Base Station Subsystem
BT	Bluetooth
BTS	Base Transceiver Station
CA	Certificate Authority
CAB	Cabinet
CAS	Conditional Access System
CB	Citizens' Band

CC/PP	Composite Capability/Preference Profile
CD	Client Domain
CDC	Connected Device Configuration
CDMA	Code Division Multiple Access
CEF	Common Executable Format
cHTML	compact HTML
CLDC	Connected Limited Device Configuration
CLR	Common Language Runtime
CMS	Content Management System
Codec	Coder–Decoder
COM	Communication Port
COPS	Common Open Policy Service
CORBA	Common Object Request Broker Architecture
CPC	Cost-Per-Click
CPI	Capability and Preference Information
CPIM	Common Presence and Instant Messaging
CPM	Cost-Per-Impression
CPU	Central Processing Unit
CRM	Customer Relationship Management
CS	Client–Server; Circuit-Switched
CSCF	Call Session Control Function
CSS	Cascading Style Sheets
DBMS	Database Management System
D2C	Direct-to-Consumer
DCT	Discrete Cosine Transform
DD	Download Descriptor
DDP	Packet Data Protocol
D-GPS	Differential GPS
DHCP	Dynamic Host Control Protocol
DIA	Device Independence Activity
DMA	Direct Memory Access
DNS	Domain Name System
DOM	Document Object Model
DoS	Denial of Service
DRM	Digital Rights Management
DSP	Digital Signal Processing; Digital Signal Processor
DTD	Document Type Definition
DVB-H	Digital Video Broadcasting Handheld
E911	Enhanced 911
ECC	Error Correction Correcting
EFI	External Functionality Interface
EFIAS	External Functionality Intraface Application Interface
EIS	Enterprise Information Services
EJB	Enterprise Java Bean
E-OTD	Enhanced OTD
ETSI	European Telecommunications Standards Institute
FCC	Federal Communications Commission

FDMA	Frequency Division Multiple Access
FMC	Fixed-Mobile Convergence
FML	Film XML
FOAF	Friend-Of-A-Friend
FTP	File Transfer Protocol
GA	General Access
GGSN	Gateway GPRS Service Node
GIF	Graphic Interchange Format
GIS	Global Information System
GMLC	Gateway Mobile Location Centre
GPRS	General Packet Radio Service
GPS	Global Positioning System
GSM	Global System for Mobile Communications
GUI	Graphical User Interface
H2C	Human-to-Content
H2H	Human-to-Human
H2M	Human-to-Machine
HDML	Handheld Device Markup Language
HLR	Home Location Register
HSDPA	High-Speed Downlink Packet Access
HSPA	High-Speed Packet Access
HSS	Home Subscriber Server
HSUPA	High-Speed Uplink Packet Access
HTML	HyperText Markup Language
HTTP	HyperText Transfer Protocol
HTTPS	HTTP Secured
ICMP	Internet Control Message Protocol
I-CSCF	Interrogating CSCF
ID	Identification
IDE	Integrated Development Environment
IDL	Interaction Data Language
IETF	Internet Engineering Task Force
IF	Intermediate Frequency
IIOP	Internet Inter-ORB Protocol
IM	Instant Messaging
IMAP4	Internet Mail Application Protocol 4
IMC	Inter-MIDlet Communications
IMM	Integrated Mobile Marketplace
IMP	Instant Messaging and Presence Protocol
IMS	IP Multimedia System
IMSI	International Mobile Subscriber Identity
IN	Intelligent Networking
IP	Internet Protocol
IPTV	Internet Protocol Television
ISDN	Integrated Services Digital Network
ISP	Internet Service Provider
ISUP	ISDN User Port

IVR	Interactive Voice Response
J2EE	Java 2 Enterprise Edition
J2ME	Java 2 Micro Edition
J2SE	Java 2 Standard Edition
JAAS	Java Authentication and Authorisation Service
JAD	Java Application Descriptor
JAR	Java Archive
JAXP	Java API for XML Processing
JDBC	Java Database Connector
JDE	Java Developer Environment
JEP	Jabber Enhancement Proposals
JIT	Just In Time
JMS	Java Messaging Service
JNDI	Java Naming and Directory Interface
JNI	Java Native Interface
JPEG	Joint Picture Experts Group
JRE	Java Runtime Evironment
JSF	Jabber Software Foundation
JSP	Java Server Page
JVM	Java Virtual Machine
JXTA	Juxtapose
KVM	Kilo Virtual Machine
LAN	Local Area Network
LBS	Location-Based Service
LCD	Liquid Crystal Display
LDAP	Lightweight Directory Access Protocol
LIF	Location Interoperability Forum
LMU	Location Measurement Unit
LOS	Line-of-Sight
LTN	Long Thin Networks
M2M	Machine-to-Machine
MBMS	Multimedia Broadcast Multicast Service
MDB	Message-Driven Bean
MD5	Message Digest (algorithm) 5
MGCF	Media Gateway Control Function
MGW	Media Gateway
MID	Mobile Information Device
MIDI	Musical Instrument Digital Interface
MIDP	Mobile Information Device Profile
MIME	Multipurpose Internet Mail Extension
MIPS	Millions of Instruction Per Second
MLG	Mobile Location Gateway
MLP	Mobile Location Protocol
MM	Multimedia Message
MMC	Multimedia Message Centre
MMI	Man–Machine Interface
MMS	Multimedia Messaging Service

MMSC	Multimedia Messaging Service Centre
MMX	Multimedia Extensions
MO	Mobile Originated
MOM	Message-Orientated Middleware
MNO	Mobile Network Operator
MP3	MPEG-1 Layer 3
MRSP	Messaging Relay Session Protocol
MS	Mobile Station
MSC	Mobile Switching Centre
MSISDN	Mobile Station Integrated Services Digital Network
MVC	Model View Controller
MVNO	Mobile Virtual Network Operator
NACK	Negative ACK
NAT	Network Address Translation
NGN	Next Generation Network
NTP	Network Time Protocol
O&M	Operations and Maintenance
OBEX	Object Exchange
ODBC	Open Database Connector
ODP	On-Device Portal
OFDM	Orthogonal Frequency Division Multiplexing
OMA	Open Mobile Alliance
OMC	Operations & Maintenance Centre
OOP	Object-Orientated Programming
ORB	Object Request Broker
OS	Operating System
OSA	Open Services Architecture
OSA-SCS	Open Service Access – Service Capability Server
OSI	Open System Interconnection
OTA	Over-The-Air
OTD	Observed Time Difference
P2P	Person-to-Person, Peer-to-Peer
PABX	Private Automatic Branch Exchange
PAM	Pluggable Authentication Module; Presence and Availability Management
PAN	Personal Area Network
PAP	Push Access Protocol
PCMCIA	Personal Computer Memory Card International Association
P-CSCF	Proxy CSCF
PCU	Packet Controller Unit
PD	Personal Domain
PDA	Personal Digital Assistant
PDF	Policy Decision Function
PDP	Packet Data Protocol
PDU	Packet Data Unit
PILC	Performance Implications of Link Characteristics
PIM	Personal Information Management
PKC	Public Key Cryptography

PKI	Public Key Infrastructure
PM	Program Management
PMG	Personal Mobile Gateway
PMP	Personal Media Player
PNG	Portable Network Graphics
PoC	Push-to-Talk over Cellular
POI	Points of Interest
POIX	Point of Interest Exchange Language
POP3	Post Office Protocol 3
POTS	Plain Old Telephone System
PPG	Push Proxy Gateway
PPP	Point-to-Point Protocol
PSAP	Public Safety Answering Point
PSMS	Premium-Rate Text-Messaging Service
PMPs	Personal Media Players
PS	Packet-Switch
PSTN	Public Switched Telephony Network
PT	Predictive Text
PTT	Push-to-Talk
PVR	Personal Video Recorder
QoS	Quality of Service
RADIUS	Remote Authentication Dial In User Service
RDF	Resource Description Framework
RF	Radio Frequency
RFC	Request For Comments
RIM	Research in Motion
RMI	Remote Method Invocation
RMI-IIOP	RMI Internet Inter-ORB Protocol
RNC	Radio Network Controller
ROM	Read-Only Memory
RPC	Remote Procedure Call
RSS	RDF Site Syndication
RTOS	Real-Time Operating System
RTP	Real-time Transport Protocol; Real-Time Protocol
RTSP	Real Time Streaming Protocol
RTTTL	Ringing Tones Text Transfer Language
SACK	Selective ACK
SAR	Segmentation and Reassembly
S-CSCF	Serving CSCF
SD	Social Domain
SDK	Software Developer's Kit
SDMA	Spatial Division Multiple Access
SDP	Service Delivery Platform; Session Description Protocol
SF	Service Functions
SGN	Signalling Gateway
SGSN	Serving GPRS Service Nodes
SI	Service Indication

SIM	Subscriber Identity Module
SIMD	Single Instruction Multiple Data
SIMPLE	SIP for Instant Messaging and Presence Leveraging Extensions
SIP	Session Initiation Protocol
SISD	Single Instruction Single Data
SLEE	Service Logic Execution Environment
SM	Spatial Messaging
SMIL	Synchronized Multimedia Integration Language
SMLC	Serving Mobile Location Centre
SMPP	Short Message Point-to-Point
SMS	Short Message Service
SMSC	Short Message Service Centre
SMTP	Simple Mail Transfer Protocol
SOA	Service-Orientated Architecture
SOAP	Simple Object Access Protocol
SP-MIDI	Scalable Polyphony MIDI
SQL	Structured Query Language
SRAM	Static Random Access Memory
SSE	Streaming SIMD Extensions
SSL	Secure Socket Layer
STB	Set-Top Box
SVG	Scalable Vector Graphics
Sync ML	Synchronization Markup Language
TCP	Transmission Control Protocol
TCP/IP	Transmission Control Protocol over Internet Protocol
TDD	Time-Division Duplexing
TDMA	Time Division Multiple Access
THIG	Topology Hiding Internetwork Gateway
TID	Transaction ID
TISPAN	Telecoms & Internet Converged Services & Protocols for Advanced Networks
TLD	Top-Level Domain
TLS	Transport Layer Security
TSTV	Time-Shifted TV
UAProf	User Agent Profile
UDP	User Datagram Protocol
UE	User Element
UGC	User-Generated Content
UI	User Interface
UMA	Universal Mobile Access
UML	Unified Modelling Language
UMTS	Universal Mobile Telecommunication System
URI	Uniform Resource Identifier
URL	Uniform Resource Locator
USB	Universal Serial Bus
UTF	Unicode Text Format
UTM	Universal Transverse Mercator

UTRAN	Universal Terrestrial Radio Access Network
VB	Visual Basic
VC	Visual C++
VGA	Videographics Array
VLR	Visitor Location Register
VOD	Video On Demand
VoIP	Voice over IP
VPN	Virtual Private Network
VR	Virtual Reality
W3C	World Wide Web Consortium
WAN	Wide Area Network
WAP	Wireless Application Protocol
WBMP	Wireless BitMap
WBXML	WAP Binary XML
WCDMA	Wideband CDMA
WCSS	WAP CSS
WDP	Wireless Datagram Protocol
WG	Working Group
W-HTTP	Wireless-profiled HTTP
WI	Work Item
WJMS	Wireless Java Messaging Service
WLS	WebLogic Server
WML	WAP Markup Language; Wireless Markup Language
WSDL	Web Services Description Language
WSP	Wireless Session Protocol
WTAI	Wireless Telephony Applications Interface
W-TCP	Wireless-profiled TCP
WTLS	Wireless TLS
WTP	Wireless Transport Protocol
WURFL	Wireless Universal Resource File
WYSIWYG	What You See Is What You Get
XHTML	eXtensible Hypertext Markup Language
XHTML-MP	XHTNL-Mobile Profile
XML	eXtensible Markup Language
XMPP	eXtensible Messaging and Presence Protocol
XSL	XML Stylesheet Language
XUL	eXtensible User Interface Language

1

Prelude – The Next Generation Experience

1.1 WHAT IS 'NEXT GENERATION' ANYHOW?

In a moment, we shall take a fictional tour in the day of a life of a mobile user. This will largely serve to illustrate what is meant by wireless applications within the context of this book. Before that, what do I mean by 'Next Generation', as used in the title of the book? In the first edition, I explained how a large number of technologies were going through significant 'upgrades' to newer versions. The most anticipated at the time was the evolution of mobile networks to their *Third Generation,* called 3G. Chapter 12 explains what 3G means, but Next Generation is more than just 3G. We can build mobile services using a mixture of technologies, many of which have also evolved. For example, we hear a lot about the so-called 'Web 2.0'. This book also talks a lot about these technologies, too.

What I have come to realise is that apart from versioning of specifications, when we need precise referencing and indexing of information, the concept of generations in mobile technology has become increasingly meaningless. This is because technological *solutions* use a variety of evolving technologies and it is difficult to talk about discrete versions in relation to the whole solution. The generational idea, like everything '2.0' these days, gets used a lot by marketers and publicists to convey 'newness' and thereby generate interest in new products. Given that some of these themes, like Web 2.0, have become important, I have included some of them in this book, all within the Next Generation theme.

When I first wrote this book, the evolution of many technologies and their confluence in the mobile world represented a significant departure from the conventional mobile experience; thus, I felt the need to document what this new and emerging landscape was like. This is still my intention.

What then, do I mean by 'Next Generation'? It is still a convenient way of saying 'latest' and is the intended meaning in the title. Hence, that's why in this second edition, I have

Next Generation Wireless Applications, Second Edition   Paul Golding
© 2008 Paul Golding

included some new technologies that didn't really figure in the mobile landscape a few years ago, such as IMS. We shall get to that in due course (in Chapter 14).

An unintended meaning, which I have grown to like in recent project experience, is the *next generation of mobile user,* as in 'the youth'. Mobile telephony clearly started out as a business tool without a youthful user in sight, unless we include the Yuppies of course. Over time, especially after the clever move to pre-pay accounts, younger users have become increasingly instrumental in influencing the thinking of innovators and strategists in the mobile world. In some circles, mobile service discussion is almost entirely about young people. Mobile puberty starts at an increasingly early age. Unofficially, one operator I worked with told me that their target market starts from the age of five! This is no longer shocking. Five-year-olds can and do use mobiles[1].

Today, we are still using mobiles mostly to talk and send messages, both of which are very communications-centric. We are used to a *mobile telephony* experience. Where we are headed is towards a *mobile computing* experience, and it will be strongly influenced by the next generation of users. Understanding the users is an important consideration in the mobile services landscape and we shall return to this theme throughout the book, starting with an exploration of the 'mobile mindset' in the next section.

1.2 THE MOBILE MINDSET

In the previous section, I discussed the meaning of 'Next Generation'. Now, I turn our attention towards the word 'application'. A *mindset* is a way of thinking, or a state of mind. Why is it important to talk about mindsets in a book about technology? Over time, technologists have slowly realised that technology isn't developed for its own sake, but to carry out tasks. Furthermore, the users intend certain results from carrying out those tasks. It is in achieving these results that technology has an application.

The issue of mindset is about how we perceive these applications. A small example is useful here. Jitterbug[2] is a brilliant mobile service in the United States. Essentially, it is aimed at users who are 'technically challenged'[3]. One category of such users might be the elderly or infirm who are not dexterous or confident with mobile telephones. A mobile service provider might think that *we sell phones*, *we send out bills* and *we support these activities*. In other words, *we are in the mobile phone business*. This is their mindset.

Not so Jitterbug. They appear to have a different mindset, which we might capture as *we are in the making-it-easy-to-complete-calls business*. There is a difference, perhaps subtle, but potentially significant than being in the *mobile phone business*. Making it easy to complete calls, with the target users in mind (i.e. technically challenged), leads to a particular set of service features. For example, the phones have simplified and easy-to-use interfaces. The phone presents a familiar dial tone, just like a landline. And – my favourite feature – the address book on the phone can be configured by the operator before and after the handset is shipped to the customer.

Clearly, mindset is important. It affects how mobile services are perceived and then designed. It is not just the job of a marketing person to develop these insights into the

---

[1] Although the operator was looking at how 5-year-olds play games on their parents' mobiles, not how they make calls.
[2] www.jitterbug.com
[3] This is my own wording, not the official Jitterbug marketing position.

application. Any technologist worth his or her salt should be able to think in this multidisciplinary way, too.

The mobile mindset is simply thinking about the *application of the mobile technology* with insight and then designing the application accordingly. Indeed, I tend to think that this approach has been largely ignored. Mobile operators and equipment providers have become institutionalised in their way of thinking about technology, which is largely equipment-centric – networks, protocols and devices dominates the thinking. The revolution on the Web has shown us a very different mindset. People dominate the Web. Yes, it is built on an internet of networks, protocols and devices (i.e. routers), but nobody in the Web applications business talks of networks and protocols. They talk of social networks, blogs and other *applications*.

Despite being what people of common sense call obvious, the mindset of 'think application' versus 'think technology' is not as commonplace as it ought to be. However, this is changing and we are slowly moving into a new generation of ideas, applications and technologies, which is why we can talk of *next generation wireless applications*. In truth, there probably needs to be a companion book, about next generation wireless businesses, but that's another story.

1.3 THE FUTURE'S BRIGHT, THE FUTURE'S UBIQUITY

Let's take a trip into an imagined future. It's a fine day for lunching outside of the sultry office. Although warm and slightly humid outside, the air stirs with a wonderfully cool breeze that could easily carry away the morning blues. Imagine you are standing in a bustling shopping plaza at one of the popular meeting points where you are soon going to join up with three of your closest friends for lunch. You have in mind a cool crisp salad, perhaps a *salade nicoise*[4], but as of yet, you don't know in which restaurant you are going to eat, just when, or thereabouts.

The meeting time was suggested after a flurry of interactive picture messages popped up on everyone's *communicators* when earlier you had sent a 'let's eat' invitation initiated by pressing a 'let's eat' hot button[5]. One of your friends received the invitation via their wrist-based 'sleek device' that connects using Bluetooth to their smartphone. This is why the word *communicator* has gradually replaced the word *phone*, it no longer being the best word to describe some of the new devices, often barely resembling phones at all. When exchanging information, sending a message or placing a call, you can never be sure what device is on the other end of 'the line'; it might be someone's watch, a pendant, a pen or even a fridge magnet! Kids on the street have taken to calling these devices 'commies', short for communicator (not communist[6]).

Whilst standing still for a moment, the shoppers swarming around you, your own wrist-pad commie alerts that you have an incoming call. It gently squeezes your wrist and vibrates, the outer ring glows and an agreeable glide of 'mood tones' emits forth. You notice hoards of other cohabitants of the meeting place, also seemingly engaged with their wrists, bracelets, pendants, eyewear and other forms of commies – all shapes, sizes and colours, by now a perfectly familiar sight. Being preoccupied with another world is no longer a novel sight;

---

[4] A French dish: tuna salad with olives and boiled egg, in case you didn't know ☺.
[5] A kind of widget that you can load onto your phone from an online library of hot buttons.
[6] They can't remember what commie used to mean, anyhow.

across all ages, there is no shyness. Your wrist pad is a fabric cuff that wraps neatly around your wrist and has a curved semi-flexible pop-up display. Like your friend's connected watch, it also breaths via a Bluetooth lifeline to your Personal Mobile Gateway (PMG). Unlike the old-fashioned world of binary call handling in Global Systems for Mobile Communications (GSM) – accept or busy – you have all kinds of options to process the inbound call. This is thanks to IP-Telephony (Internet Protocol Telephony) and the new world of IMS (IP Multimedia Subsystem).

With IMS, your device doesn't just alert you with the call and the caller-ID, you also get a tag line introducing the call, like the subject field in an email. Not just that, but up pops an athletic looking avatar for Kim. She's your personal trainer. From the tag line you know that it's your personal fitness-training instructor calling to 'fix our next spinning session' – words that tell you that this is a call you would like to take, words that your trainer probably didn't type, but were most likely translated from voice-to-text before she started the call.

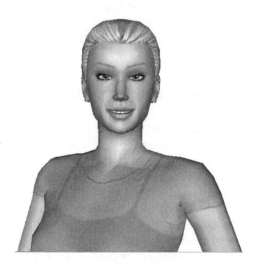

Figure 1.1 Kim's avatar. (Reproduced by permission of GoFigure mobile.)

There is no awkward conversation with your cuff as you have the latest-design Bluetooth earpiece snuggled discretely in your ear, much smaller than the earlier models. This one has been custom sculptured like a prosthetic for your ear, just like the earpieces that stage musicians such as Bono have used for years, but cheaply done using a laser-modelling scanner at a walk-in booth in one of the mobile phone shops. It is light and has a long battery life thanks to the latest generation Zinc-Air cell, which is convenient as you use your earpiece often. It also charges wirelessly whenever you take it out and place it on your induction-coupled *charger mat*. You make calls with it in the office via your tablet laptop that has Bluetooth built-in and can handle audio from the headset to a Session Initiation Protocol (SIP) client running on Mac OS X – total wireless freedom. It also has a noise-cancelling feature to blot out any background noise, which is great for calls in busy places and also good for use in aircraft.

Your visit to the office was necessary only to hold a face-to-face meeting. Otherwise, there's no need to come to work anymore. Most of the time you're talking, texting, messaging, chatting and emailing with people who are somewhere else, or you simply meet them online in a virtual world of one form or another, perhaps in *Second Life*. The detachment can sometimes be quite unbalancing. Time and space feels different than before, merging 'there' with 'here', 'private' with 'public' and other such fuzzifications.

Just prior to arrival, in the taxi ride to the plaza, you had received a call from a colleague who textually announced the call with 'wanting to discuss the latest sales figures', which you didn't have time enough to process as the taxi ride was coming to an end. You were tempted to hit the 'call back in 1 hour' response that would have sent the same words back to the caller's SIP client (probably on their desktop PC), but you diverted (forwarded) the call to a subordinate who you had previously primed with the task of looking over the figures. As it happened, your subordinate was in a meeting and they handled the call by converting it to an Instant Messaging (IM) session (still using SIP) for a brief and discrete chat with the caller, deferring the voice call to later. You then switched your *presence state* to 'out at lunch' to let callers know ahead of time that now might not be a good time to talk shop.

After forwarding the call from your colleague, you remembered that you had not yet reconciled the latest sales figures for European accounts. The easiest way to inform your colleague was to bring up their number on your wrist commie, simply by flicking your wrist back, and then saying 'Send voice message'. Speaking into your Bluetooth headset, you recorded a voice message that would go straight into their voicemail box without the need to chat. Your avatar would lip-sync the message at the other end.

Before setting out to lunch, a hot button labelled 'let's eat' initiated the upcoming gathering. You downloaded the hot button from a mobile mash-ups site. You pressed it, entered a start time and an end time and then selected invitees, which you did easily from your 'cool friends' list.

When browsing the list you also noticed that one of your friends had sent you an email, yet unread, and another had left you a voice message, whilst another had posted to their weblog a new entry that you ought to read. On the way through the list, you marked some of these tasks with the 'do next' command so that you didn't have to interrupt your current task flow. Later on, you may choose to resume these 'do next' tasks or receive a gentle reminder to look at them in your 'to do' list.

Your 'cool friends' folder is not confined to people you have met in person, far from it. There are people you have met online, via email, chat rooms and by tracking their daily weblog entries. There are people whose reputation you value, perhaps as particularly good soulmates on Amazon with a similar taste in books and films, or those with a similar taste in domestic products like washing machines – the most mundane information requirements are often the most valuable. These virtual soulmates will come in handy later when you drop by the bookstore whilst waiting for your friends to arrive.

Next, you clicked on the 'where?' option and you could elect to choose a place from your favourite places list; you could 'air write' using your finger in mid air, you could chose the 'nearby places' option or you could go for 'place-finder'. You went for nearby places and found Orbital Plaza, drilled down, went for 'meeting places', and chose 'Dancing Fountain' which is one of those synchronised water fountains that kids can run through. It's a popular meeting place and exists in the menu because other users of the 'let's meet' service use it often: social influence constantly shapes the user experience.

It didn't matter if your friends had never been to the fountain before, they'll be able to use the digital map to home in on this regular rendezvous point. On their commies they received a 'let's share food' invitation that popped up with your avatar in a tongue-licking hungry composure combined with a 'let's eat' speech bubble and a brief matching audio snippet. The avatar posed the question: 'Want to eat at 1:00 p.m. at Orbital Plaza?' which you didn't have to type, 1:00 p.m. being your suggested start time. Your friends could select 'available', 'not available' or 'change time' in response to the invite. If they changed the time, then it could only fall within the range you had suggested.

Once all the votes are in, the acceptances and the convergence time are notified to you for your confirmation. One-twenty is the agreed union time, which is why you arrived early as you already decided you would start your lunch break from 1:00 p.m. anyway, real-office fatigue having kicked-in early that day.

You have arrived early at the rendezvous point, having found where to go using the map and directions on your phone, not having been to Orbital Plaza before. Sitting on the seating next to a children's play area you check a few emails. In the play area, mums are pointing their camera phones, taking short video clips of little kids whooshing down slides. Moments later a clip pops up in daddy's email box at work, perhaps on the other side of the world. Dads everywhere are proud and happy.

You already know that an important client has sent you mail. Earlier, a notification pinged up on your phone accompanied by a soothing glide of musical notes with a softly announced 'you have mail from . . .' spoken in a natural voice, biometrically tuned to be assuring and comfortable to your psycho-acoustic appetite. It's the same voice used by your life-coach avatar. The notification occurs only because your phone automatically sensed that you are away from your desk where you would otherwise be checking your mail. You read the email and reply. Back at the office, or home, the reply is sitting in your 'sent items' folder and the email is marked as read: everything fully synchronised without the need to sync up.

Your commie detects that you are near a bookstore and brings up your Wish List on Amazon for a quick review. Via the world of Resource Description Framework (RDF) Site Syndication (known as RSS) feeds, your list neatly appears in your task manager, appropriately formatted for the device display – no messing around on Amazon's website. If you wanted to surf, you could bring the wrist display close to your eye and it would switch to pupil-projection mode and project a large-format screen into your field of vision.

Armed with your Wish List and enthusiasm for your next good read, you eagerly scuttle over to the nearby bookshop. You visually scan the bookshelf for the title you're after. It doesn't appear to be available but you notice something that seems similar. However, taking risks on buying books is something of the past. You consult your 'hive mind', namely those members of your online social network whose opinion you value, mostly people you have never met before but whose reputations figure highly in your estimation from past cyberencounters. They are most likely fellow reviewers on Amazon and elsewhere. Your phone is equipped with an optical bar code reader[7] and you scan the book's code and wait for a response.

Three opinions are available, one from 'Max32' who is a valued virtual soulmate on superfoods. She – or he – gives a five-star approval and even though you can see that Amazon is reporting a saving of 20% on the shop price, the book from them is on a 3–5 day lead. Your urge to read tells you that the extra amount is worth it: instant information, instant decision, and instant gratification. Thankfully, anxieties about buying goods are ephemeral these days, leaving plenty of time for you to worry about life's other uncertainties in the *post-mobile* age.

Before buying the book, you remember that you have a coupon from the store previously sent to your phone. You present the coupon as part of the checkout process using your

---

[7] The camera used for taking pictures also doubles as an optical scanner for 2D bar codes.

mobile wallet service. Swiping your phone over the NFC reader authorises the transaction and registers the coupon.

Leaving the bookstore, you notice a kid in the skate park using his commie to film his friend's skateboard trick. The clip is uploaded directly to YouTube with a tag 'Orbital Skate Crew' automatically appended based on the location of the trick. If you 'scanned for videos' in the vicinity, you would be able to pick it up, except that the skate crew are sharing the tag privately. You feel like having a go, but remember that you haven't been skateboarding for about ten years.

Time marches on and you notice that the lunch appointment is approaching. You would like to pop into the clothing shop next door, but you are not sure if there's enough time. You pull up your buddy-finder map on the screen and can see that your nearest friend, Joe, is showing on the map five minutes away from the rendezvous point. There's just enough time to have a quick perusal of the latest fashion lines, but just in case you get distracted you hit 'push to talk' on your buddy's icon and, just like with a walkie-talkie, your voice blurts out on their device (or ear piece):

'Hi Joe, I see you're nearly here. I'm just popping into EJ's. See you at the fountain or in EJ's.'

Joe replied, 'OK – see you in a minute.'

Neither of you had to dial and wait for a connection. Push-to-Talk[8] (PTT) happens instantly. For Joe the sound came out on their loud speaker built into their PMG. You heard the response in your headset.

You feel satisfied that you have this degree of control that let's you confidently decide that time can be used productively whilst waiting for the friendly come together.

You return to the Dancing Fountain next to the children's play area and the lunchtime gathering convenes. Next, your party talks about where to eat. There's a new bistro just opened along one of the polished walkways in the plaza. There is a consensus to try it out, but first you all check for ratings. Finding the bistro in the food listings is easy. It's immediately accessible via the 'nearby places' menu, your phone having sensed where you are. Looking at the entry for the bistro, one of your friends sees that a friend-of-a-friend (FOAF) in her social network has already tried it out. Unfortunately, it gets a low rating due to a scant vegetarian menu and this worries your friend. No problem. Her reviewer FOAF has posted an alternative venue that has a fantastic array of vegetarian options. You all head there with tummies rumbling.

In the eatery, the food is great and you all agree that it was worth a visit. It's time to pay the bill and get back to work. In the past, you would have spent the next five minutes working out how to split the bill. This time, thanks to mobile commerce, it's not a problem. The bill is 'swiped' from the waiter's Electronic Point of Sale (EPOS) terminal and then shared with the group. Each agrees to their share, including suggested gratuity, and the job is done.

Upon clearance of the bill, a courtesy message pops up on everyone's commie announcing a coupon offer on future meals. The coupon is only valid if within the next day *all of you* post a review of the restaurant. Communal offers like these are common. These reviews will be pinned in *space*[9] near the restaurant. Other friends who will pass by in the future might soon be reading them.

---

[8] Don't write PTT off just yet. Once IMS networks become more widely deployed and text messaging revenues become less previous, PTT shall most likely become a standard service.

[9] See the concluding section of the book for a discussion of *spatial messaging*.

For a while, you mingle outside the restaurant with the usual reluctance to part from each other and head back to stuffy cubes in the office. Two of you agree to meet later in the week for a book-reading event at the nearby bookstore. You 'beam' the appointment, along with its whereabouts and details, to your co-bookworm. Swapping electronic items in person has become a common social habit thanks to Bluetooth, but it doesn't have to be done face-to-face. Exchanging nuggets of information can take place just about anywhere and anytime, but often doesn't have the same exciting buzz generated in those irresistible spur-of-the-moment exchanges that you get when face-to-face.

You say your goodbyes and everyone melts back into the teeming milieu of shoppers. On your way back to the taxi rank, your wrist commie alerts you with a 'throbbing brain' graphic; a tiny animated cartoon brain pulsates below a glowing light bulb. This tells you that someone with a potentially similar interest profile is nearby. It could be someone potentially useful to 'mesh with' – the term used to indicate techno-assisted socialising purely for the sake of social networking. Meshing has become commonplace in this hypercompetitive world where the maxim 'network or die' isn't far off. As with many other pools of your personal digital information, you have assigned part of your contacts folder to be publicly accessible. Things have come a long way since the private and jealously guarded pride-and-joy address book. You have many public information trust zones in your digital vault, including much about your professional and social life.

Tapping the graphic reveals more information about the alert from your 'neighbour-hood watch' agent. Many times before, you had made interesting new acquaintances in your nomadic neighbourhood, a circle of interest that floats around like an imaginary X-ray beam illuminating the area all around you, virtually penetrating into the 'public persona' of willing networkers who happen to pass through your cloud with their mobile devices.

Upon inspection, you see that the nearby cohort is someone who you might want to partner with in your latest venture. It seems an interesting opportunity, so you call them without having to know their number or reveal yours. A *networking agent* automatically records the call for protection against malicious intent, just in case. The person takes the call, having already seen your profile pop up on their commie. You agree to meet for a coffee. The networking agent recognised the keywords 'meeting' and 'coffee'. It has already found the nearest coffee shop and popped it up on both commies. You both agree to meet there. It's also got WiFi[10] and will be a good chance to upload some pictures that you took earlier at the restaurant. You already selected 'send to gallery' when you took them earlier, but the device waited for a cheaper and faster network to complete the task automatically in the background.

You finish your meeting at the coffee bar. You speak 'taxi' into your commie and a request is made for the nearest taxi in your area to pick you up at your current location. The taxi driver gets the request and you are automatically connected via PTT to confirm details. The taxi soon arrives displaying your name on its roof-mounted electronic signboard. You enter the air-conditioned cab and sit back with the sun on your face and relax for the 20-minute journey back to work. You begin to watch the latest episode of a popular TV show that you noticed advertised inside the cab. You jumped straight to the channel by 'blinking' the ad's 2D bar code into your device. As the video player starts up, you see a message that one

---

[10] Just in case you didn't know, Wi-Fi is the abbreviation for Wireless Fidelity, a group of technical standards enabling the transmission of data over wireless networks.

of your friends has just uploaded their latest photo album and you click to view them as a streamed slide show, just before watching the TV episode.

You arrive back at the office and reflect on the events of that lunchtime with a sense of satisfaction that you seemed to get so much done in a lunch 'break'. You start wondering how you ever got by without the wireless umbilical of so many mobile services. You ask yourself 'how did I used to cope without my commie?' As hard as you try, you simply can't recall how you ever did.

1.4 OUR MULTITASKING MOBILE FUTURE

The point of the story is not to invent a new buzzword – commie. I admit, it does sound a rather unlikely name. I think the word 'mobile' will pervade, although the name isn't particularly important. The point is that new devices will continue to emerge that don't fit the traditional mobile phone form or function. Moreover, as the story tells, devices will be used for an increasing number of tasks besides talking, so the name 'phone' will become outdated.

The key message from the story is that the mobile experience will grow to include a number of tasks that can be easily carried out using a handheld device, or devices, and fully integrated into our daily routines. Perhaps your reaction to some of the mobile tasks in the story is a mixture of agreement and scepticism. You will see yourself doing some of those things and not others. Some you might think of as being just plain nuts. I agree with you. This is the normal reaction. However, there is one important consideration to keep in mind. It can best be summed up by McLuhan's famous aphorism – 'A fish doesn't know of the existence of water until beached.' In other words, we get used to services that we once didn't think we needed and only realise their value after they're removed, like losing a mobile phone. Remember life without the Web? When I worked on the very first GSM systems, nearly all my friends and associates clearly told me that they couldn't see themselves using a mobile phone. Here we are today with many people claiming that they simply couldn't imagine being without one. We all know this story.

The notion of doing so much 'on the move' – including watching TV – isn't so crazy any more. In fact, we might be better saying 'on the mobile device'. With the emergence of new technologies like Femtocells[11], it is possible that users shall use their mobile devices *in the home* – not on the move at all – for an increasingly wide range of tasks besides calling and texting.

Of course, we should be wary of hype. At the time of writing, the current fad is the '2.0' frenzy. Gosh, I should have called this second edition *Next Generation Wireless Applications 2.0*. There have been many false dawns in mobile services and I have seen most of them. I have participated in many of them. In the story just told, nearly all of the services could be implemented today using widely available technology. Some of them will pervade and some will fizzle out, or never see the light of day. We still don't know what the future mobile services landscape is going to look like, but this book charts most of the territory covered so far and tells us a good deal about where we are headed. Just for the record, my own favourites for the future are mobile TV and mobile commerce.

---

[11] *Femtocells* are very small mobile phone base stations for use inside the home.

2

Introduction

2.1 WHAT DOES 'NEXT GENERATION' MEAN?

Historically, most of the commercial activities of mobile operators have been dominated by voice and text-messaging (texting) services. No matter how these are delivered, whether by a GSM (i.e. 2G) or a Universal Mobile Telecommunications Systems (UMTS, i.e. 3G) network, this basic service portfolio represents the previous generation of mobile services. What we have been slowly (very slowly) moving towards is a next generation of services where the kinds of experience described in Chapter 1 become more commonplace and the mobile experience is no longer dominated by any one particular application. Essentially, the first three generations of mobile networks technology were largely about improving the mobilisation of voice (telephony), but the future evolution (next generation) of networks and related technologies is about the mobilisation of data. As a good deal of our data is increasingly bound to services on the Internet, we can obviously expect that the Internet will figure a great deal in the future of mobile. However, I wouldn't go so far as to characterise the future of mobile as being essentially the mobilised Internet. It is probably better to think of the future as the mobilisation of computing (including embedded computing), but likely to remain communications-centric rather than general purpose.

Another way to think about 'next generation' wireless applications is shown in Figure 2.1. The figure shows the explosion of technologies and related technological communities (e.g. standards) moving from the 2G world of voice-centric services to the Internet-connected 3G+ universe. There is massive convergence of mobile technology with other activities, particularly the Internet, which itself is converging with all other kinds of activities and constantly spawning new spin-off protocols and concepts, such as the Web 2.0 paradigm. What this diagram implies is that any notion of 'mobile services' has to be understood within the context of these related fields. Since first writing this book, the potential for convergence has only broadened to include totally new areas like Internet Protocol Television (IPTV). Therefore, when we talk about *next generation mobile services*, we are

Next Generation Wireless Applications, Second Edition   Paul Golding
© 2008 Paul Golding

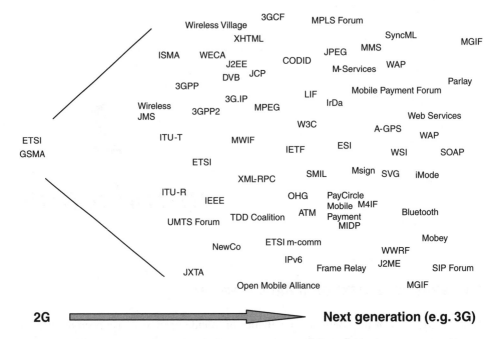

Figure 2.1 Explosion in technical activities relevant to mobile services.

really talking about a huge gamut of technologies, standards and industry bodies, such as those shown in Figure 2.1. For some, a mobile service is all about what can be done in the browser on the device, especially with newer programming possibilities like Asynchronous JavaScript Technology and XML (AJAX for short). For others, a mobile service is underpinned by an entire technology standard and related infrastructure, such as Digital Video Broadcasting – Handheld (DVB-H). Or, it might be related to what's possible within the evolving area of Java (e.g. J2ME) or the evolving area of SIP-based communications (e.g. IMS).

I have worked with many of these technologies and within various 'camps'. I have designed and built GSM networks and I have also run a software company on the 'bleeding' edge of Internet design. I have worked for operators, with operators and with their suppliers. What I have noticed is that the context of inventing, deploying and using technology does matter. It affects how you think. For example, an operator will have a totally different view of customer support than a small Web 2.0 start-up. They will also have totally different views of things like revenue generation, profit margins and operating costs, although the principles will be the same (i.e. business is *still* supposed to be about profit). Historical influences will be different, too. The history and operation of the Internet is wildly different to that of a telco network. Often, these different camps don't seem to understand each other. For example, advocates of the open Internet have difficulty coping with the relatively closed-network operation of a mobile operator. The plain fact is that most network operators have been highly successful in making a lot of money with voice and texting services and they mostly control the technology and networks used to deliver these services. In a nutshell, they have not been in a hurry to change.

Whilst we have been busy trying to connect some of the dots in Figure 2.1 to construct interesting and potentially profitable mobile services, many more dots have appeared on the map and continue to emerge at a pace that eludes us from keeping track. Most operators have understood that involving partners is the best way forward. Some operators have concluded that partnerships are the *only* way forward. Some are now questioning whether to be in the applications business at all. There is no easy answer. Web pundits vociferously argue for open and affordable access to the Internet as the model for delivery of mobile services, unhindered by the operators, much as broadband providers today don't really get in the way of which Internet service its customers want to connect with and enjoy. However, as we shall see, there is no simple answer here either, particularly if a service wishes to exploit aspects of the voice and messaging services completely under the control of the operators and not traditionally seen as 'Internet services', although this is changing of course in a post-VoIP (Voice over IP) world.

Throughout most of this book, I will try to present a context-independent view, so that it doesn't really matter where you are in the value chain or where you're coming from in terms of history and interests in mobile services. At some point it will help to have a model of the mobile services environment so that the discussion around applications can be more structured. I have therefore adopted a layered model that neatly divides the technologies into categories, mostly according to where they can be found in the network: device layer, network layer, services layer, etc. These layers often blur at the edges, but still serve as a useful way to think about the 'services stack'. The important thing to realise about my model, like all others, is that it is just a representation of what's out there and is a tool for understanding; so don't worry too much about the correctness of the model per se, but more about what it leads to in terms of understanding. The main point to keep in mind is that the model applies to operators or non-operators in terms of understanding the mobile applications universe.

2.2   WHAT IS A 'WIRELESS APPLICATION'?

In the previous section, I outlined what I had in mind when I used the words 'Next Generation' in the title of this book. The next idea for introduction and explanation is what I mean by 'Wireless Application'. It doesn't mean any application that has something to do with wireless communications. That might include, for example, billing software in a mobile operator's back office operation, or it might include a network accelerator application for speeding up access to the Internet from a mobile browser. I'm not going to discuss these types of application directly in this book. For the most part, I am interested in applications that have some kind of user interface on the handset and that are usually mobile in nature, meaning that they utilise the wireless communications capability of the device they run on. As we shall see later, most applications in this category either reside on the handset itself or are accessible via a player or browser on the device. Either way, the application on the handset might be pre-installed or downloadable *over the air* (OTA).

However, I am not concerned exclusively with describing only the technologies used to implement wireless applications directly. As I stated earlier, the *next generation* is underpinned by a huge variety of technologies, protocols and standards. It is important to understand at least some of these in order to implement mobile services more effectively. To enable us to navigate through the wider ecosystem of technologies, I have proposed a

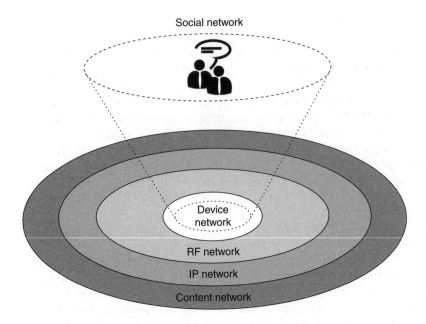

Figure 2.2 Mobile applications ecosystem.

concentric networks model of the ecosystem (see the next section) that allows us to categorise the various technologies that could be included in Figure 2.1 and to then understand broadly how they fit together.

2.3 A CONCENTRIC NETWORKS APPROACH

In order to introduce how we can proceed further in understanding next generation mobile applications, we need to find a useful starting point. To enable the delivery of mobile services to consumers, we need a network of some kind, plus some software applications that interact usefully across the network. Part of that network will obviously be wireless, but what might the entire network look like from end to end? When describing the big picture of mobile services, I have found it useful to use a model of concentric networks, as shown in Figure 2.2.

The entire ecosystem shown in Figure 2.2 is what I refer to throughout this book as the *mobile ecosystem*. This is in contradistinction to the actual radio frequency (RF) network (sometimes also referred to as the cellular or mobile network[1]). All of the various technologies found in Figure 2.1 could be plotted somewhere on Figure 2.2 (I dare not try).

There are five parts to the network, and these parts form the structure for discussing technologies throughout most of the book, including a discussion in the following chapter on how to make money in mobile services, the answer to which depends on where on the diagram you want to focus your business endeavours. All of our technological discussions will be relevant to activities that take place within these networks, so let's introduce the networks.

---

[1] Cellular network is too narrow a definition, as the RF component can sometimes be non-cellular. The notion of what is cellular will become clear later in the book when we look at the RF network in some depth.

2.3.1 Social Network

At the centre of the network is the *social network*. When I first wrote this book back in 2003, the importance of the user was very underestimated. It was common to find many participants in the wireless applications world who didn't really think much about the user beyond a set of eyes and fingers with which to operate the application. In the current era of 'communities' and user-generated content (UGC), our understanding of the user has shifted somewhat, so it is no longer odd to talk about the social network as part of the applications ecosystem.

Mobile networks offer some unique characteristics. A key feature is that users carry their interfaces with them everywhere they go. Indeed, this has led some industry experts, such as usability expert Barbara Ballard, to use this as a more meaningful way to categorise certain types of user behaviour and device, by what she calls the *carry principle*[2]. What this also means is that users can physically interact with each other as well as their applications at the same time. A group of kids can sit together in a coffee bar or gather on a street corner whilst carrying their devices at the same time. Rheingold noticed this in his observations of mobile users convening in physical meeting places whilst simultaneously engaged with their mobiles, either with each other or with others outside of the physically present social network. I shall explore this later when we look at the concept of *spatial messaging*[3] combined with messaging *semantics*[4] that enable some unique interactions between users based on their social networking habits within geographical constraints.

Users are at the centre of the wireless applications world and they almost always form networks. These might be formal or informal, geographical or virtual, business or pleasure and many other types. A better understanding of the social dynamics of these networks can lead to better applications for users, especially when combined with the carry principle.

There are some important issues to raise here about user networks:

- Users themselves can be, and often are, content providers. This trend has accelerated significantly since I first wrote this book with the likes of YouTube and various other popular sites built around UGC. This has impacted on the design of the physical networks used to carry data, which historically have been asymmetrically biased towards the delivery of content to the user, not the other way around. New wireless technologies like WiMAX and High-Speed Uplink Packet Access (HSUPA) are addressing the shift. Indeed, in some markets, the business case for these new technologies is being built around the very idea that UGC is a major characteristic of future services and that business models will emerge to generate revenue from it.

- Given the widespread availability of technologies that facilitate social exchange, users will form their own modes of communication and find new ways of socialising and forming social groups, such as LinkedIn and Facebook on the Internet. Rheingold identified the greater ability of social groups to form around technological interchange, which he referred to as 'swarming'. Such patterns of socialisation have become very evident

---

[2] http://www.littlespringsdesign.com/blog/2005/09/14/the-carry-principle/
[3] Spatial messaging is the ability to send messages that are tagged with positional information, so that we can target people at particular locations.
[4] Usually, messages are treated as black box entities by the infrastructure, which remains blind to the contents. By adding semantics to describe what's in the boxes, the infrastructure can handle the messages in new and interesting ways.

in the many outcomes of text messaging, as well documented mentioned in *Perpetual Contact – Mobile Communication, Private Talk, Public Performance*[5], which describes various academic studies in the usage patterns of 'texters'.

- Despite an initial assumption that mobile telephones were solely 'speech delivery devices', witness the trends in personalisation and emotional attachment to phones, much of which is driven by the power of social mechanisms like sharing, showing off and peer pressure.

The social impacts of mobile technology are important and have the potential not only to shape society, which they clearly already have, but also to influence how the technology itself evolves. With web applications, there is now a very tight loop between service features and user feedback. The hosted nature of web-based services allows a much quicker and fine-grain updating of service features than previously possible with applications more tightly bound to a device, such as traditional PC applications. Indeed, the idea of software release has been replaced by the 'perpetual beta' concept that is often associated with the Web 2.0 paradigm. However, to be fair, the traditional installed applications have also become more prone to frequent updates thanks to easier web distribution. Whilst this degree of agility is more difficult to implement on a mobile device, the trend is headed in that direction. If nothing else, users are gradually becoming more accustomed to participation in services, which in turn is influencing their expectations from digital services of all kinds, whether via the mobile or via the set-top box on their TV.

The carry principle means that users can interact *directly* with others using mobile devices, although this remains a surprisingly underutilised mode of interaction. For example, even though users are perfectly able to 'beam' content to one another using either infrared or Bluetooth, often users will merely exchange text messages in order to swap contacts.

In peer-to-peer computing networks (P2P), neighbouring devices may choose to swap information of their own accord *without* direct user intervention, which is likely to become commonplace in the future although this mode of computing generally remains a niche approach. No doubt, it will become more popular as bandwidth increases, probably with mobile WiMAX and beyond (wherever the uplink data path is fat).

We might ask the question 'can applications be run on the social network?' Clearly, the answer is 'yes' if we think of applications in the literal sense as an *application* of the technology, rather than a piece of software. Many uses of mobile technology require that the logic, or rules, reside in the social network. For example, kids who organise their social lives around texting are implementing their own rules about data origination and processing, such as who is included in a group, who gets to receive various messages and so on. The notion of 'running' an application on the social network is a useful aid to thinking about the social network, how it might work and its relationship to other networks in the ecosystem.

2.3.2 Device Network

The *device network* provides the primary *user interface* to the mobile services, although many services will be increasingly 'multichannel' in terms of the user interface, spanning a range of appliances (e.g. mobile, PC, set-top box, fridge!)

---

[5] Katz, J. and Aakhus, M. (eds), *Perpetual Contact – Mobile Communication, Private Talk, Public Performance,*. Cambridge University Press, UK (2002).

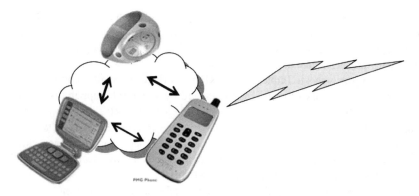

Figure 2.3 Personal network of communications devices. (Reproduced by permission of IXI Mobile.)

We are used to having only one type of mobile device, usually the mobile phone, but in the future, we may well end up having a cluster of devices that can talk to each other using Bluetooth technologies, or similar, and connect individually or collectively to the RF network, as shown in Figure 2.3. Out of a collection of devices in the device network, perhaps only one of them talks to the RF network and will form a PMG for the other devices, a concept that has been talked about for some time but is yet to gain momentum. It is useful to note that not all devices may be required at once and need carrying together all the time. Personal circumstances and the required mobile tasks at any given time will determine the device or devices needed for the occasion. Even fashion may influence which devices to take out with us. The consequence is that greater data fluidity will be required so that users can always access the data they need no matter which device they are carrying.

Again, because of the carry principle, devices can network directly with each other in personal area networks (PAN), or via the wide area networks (WAN). There are two modes of PAN connection: inter- and intra-personal. In the former case, devices might network directly using P2P techniques with or without user intervention. I have often used the example of a mobile device hunting its nearest neighbours for other devices belonging to an owner with a mutual interest, which might be a mutual friend. In this way, mobile devices are facilitating social introductions, which might sound odd, but is already underway with some niche services like Dodgeball[6].

2.3.3 Radio Frequency (RF – Wireless) Network

Radio frequency (RF) is the critical wireless link between the device network and the hardware infrastructure used to connect the device back to some content, person or machine somewhere on the planet. The job of the RF network is to enable the device network to interact with other networks irrespective of time and place. The only way to achieve freedom in these two dimensions is to bathe the landscape in RF signals that the mobile device can always pick up and reciprocate. In a wide area network, we achieve coverage by dividing the landscape into cells, each served by a cellular base station. A network of cells is necessary, as opposed to one big cell, because wireless communications is unique to each user, unlike

[6] http://www.dodgeball.com/

broadcast communications (e.g. TV stations). Each unique connection requires a portion of the limited RF spectrum and eventually it all gets used up. Therefore, the only way to increase capacity is to re-used the spectrum, but within a confined geography where it won't interfere with another area using the same spectrum.

There are various ways to implement an RF network, as we shall see in Chapter 12, but the net result is a network of cells, each managed by some radio apparatus called a base station. These stations mesh like a honeycomb and gather the RF signals, converting them into landline digital signals that ping to and from the Internet Protocol network that we introduce next. A key requirement of the RF network is that it has to manage mobility by keeping track of which cell each mobile device occupies at any given moment. *Mobility management* is a key function of any mobile network.

RF technologies are evolving all the time. Without detracting from the telecoms engineers who continue to innovate in radio communications, the principles of getting digital signals across the air as radio waves have been fairly well understood for some time. Much of the progress in RF communications that has fuelled, even enabled, the mobile revolution, has been in the area of low-cost and low-power silicon chip design. Some technologies are good for wide area communications (let's say hundreds of meters upwards) and some are good for local area communications (below 100 metres). Because of this and the various licensing issues to do with which parts of the RF spectrum can be used locally and nationally, a range of RF standards has emerged, such as GSM, UMTS, Bluetooth, WiFi, WiMAX and so on.

There is no one-size-fits-all solution and we shall need multiple RF technologies for some time to come. However, in terms of economics and efficiency, some of these standards are now competing for the same applications. For example, with the mobile version of WiMAX, it is possible to consider offering commercial services based on WiMAX to compete with UMTS (or other 3G technologies). As with all technologies, the inertia of investment, deployment and various commercial constraints will mean that there will always be overlapping services based on different technologies with no clear advantages to the consumer as to which technology to adopt. But in many markets this means increased competition with resulting better pricing and service choice for the consumer.

A question often asked in the cellular world is what has been the business case for moving to 3G? The answer is very often not clear, especially where an operator paid very high prices in an auction of the RF spectrum available to carry 3G signals. In many cases, perhaps most, it would be difficult to present a positive balance sheet for the investment in 3G versus the revenue as a direct result of 3G-specific services, if one can find them on many networks. Indeed, given the high proportion of revenue still coming from voice and texting, both of which don't require 3G networks, except perhaps for capacity reasons, it could be argued that some operators need not have adopted 3G at all. They have failed to deploy any services that require a 3G network and generate even a fraction of the revenue generated from voice and texting services. And in many cases, the fear of cannibalising voice and texting revenues has resulted in a very conservative approach to deploying new types of service over 3G, such as Instant Messaging. I will return to the revenue question in Chapter 3, but for now it is important to understand that for any particular kind of wireless application, there might be more than one way to offer it to consumers in terms of the underlying RF network type and that there is no clear answer indicating which to adopt. Sorry!

Some commentators have tried to argue that there is no such thing as value-added mobile services; that all we need is a wireless connection to the Internet; this being the 'operator-as-data-pipe-provider' view. In some ways, operators are beginning to submit to the inevitability that they may have to yield to this type of role. However, it is not as clear-cut as is sometimes made out. Operators do not run RF networks; they run mobile businesses. There is a big difference, with issues such as customer care, retailing, billing and providing a trusted brand being more important, complex and sophisticated than is perhaps appreciated by those not doing it. Managing and offering *reliable* mobility across a wide geographical area is an expensive and sophisticated business, not easily replaced by a DIY option on the Internet, although some will argue that it is. Moreover, we should be clear that providing the network itself is the value offered by an operator and therefore they are not likely to accelerate its commoditisation and hand it over at next to nothing to Internet-based services to reap the rewards.

2.3.4 Internet Protocol (IP) Network

Not surprisingly, the means by which we connect into the wired world of information and information-exchange is increasingly via the Internet or at least using its protocols, even when in a closed network. The rapidity at which the Internet has developed and its ability to support all kinds of interesting and scalable end-to-end services makes it the obvious choice for providing the basic plumbing for modern digital services. Most of the software world has already made this decision anyway. Indeed, we may argue that the Internet is the *'killer application'*[7] for 3G and it seems that operators are slowly coming around to this way of thinking and now offering open access to the Internet at an affordable fixed rate, now that they can do this using 3G (which wasn't previously possible with 2.5G without rapidly running into capacity and quality problems). Users' appetites for Internet services are such that demand for mobile access is gradually increasing, notwithstanding the major limitations of portable-device usability (which we shall discuss at length later the book). Indeed, usability remains a key issue and is probably the single limiting factor preventing any mass adoption of mobile data services to the same extent as voice and texting. One might ask whether or not there would be a sizeable market for mobile Internet devices if the devices weren't also capable of calling and texting. Even today, as the Internet matures into its so-called 2.0 phase, the answer is probably no.

The Internet is far from static. It often seems that we are constrained only by our ability to conceive of new applications, notwithstanding the communications and computer power needed to sustain them. However, there are clearly limitations to extending existing services into a mobile setting, most notably the relatively constrained user interface possibilities of current mobile devices. This is another reason why simply providing a wireless pipe to the Internet is perhaps not a useful approach. We shall talk about the mismatch between wired and wireless Internet later in the book when we look at mark-up languages and ways of presenting information on mobile devices. Whilst surfing the Web is possible on some devices, sites optimised for mobile are still a good idea and still widely used by operators to implement their portals.

Many flourishing Internet service concepts will work well in the mobile ecosystem and even provide particularly powerful and new opportunities specific to the mobile context. One such example is the ability to tap into *reputation networks* where we can gather trusted opinions about everyday products and services. This is a step towards imagining our mobile

---

[7] Can't really call the Internet an application in the conventional sense, although it can be viewed as 'an application' for UMTS, as indeed the UMTS Marketing Study apparently does.

device as a personal pocket advisor, engendering new modes of emotional attachment to our devices; attachment that I think will become popular and indispensable, especially when combined with other modes of communication like instant messaging, videophony and spatial messaging.

The golden protocol of the Internet is Hypertext Transfer Protocol (HTTP) and we shall examine this in some depth, giving special emphasis to its fulfilment and performance in a mobile context. The limitations of earlier mobile data solutions, before 3G, meant that HTTP wasn't always the ideal protocol to support mobile services. This limitation is partly what brought about the Wireless Access Protocol, or WAP, which is a set of specifications that now embraces HTTP with some optional features to optimise performance for bandwidth limited connections. We shall be examining the uses of WAP when we look at IP-based protocols later in the book.

2.3.5 Content Network

Many types of applications need go no further than the IP network. For example, Instant Messaging (IM) sits quite nicely in the IP layer. It is essentially an IP session used to pass packets of data between two end points. There is no access to content, unless one thinks of the messages themselves as a kind of content. This begs the question, *what is content?*

For the most part, when thinking and writing about content in this book, I am referring to any type of data that a user can somehow access and enjoy using a mobile device and which is generally accessed from a file store or database. The notion of a *content network* is useful – and real – because a substantial amount of different content and content types exist, sitting in a vast array of nodes and emanating from a wide variety of feeds and sources. There is a significant content ingestion and management problem in delivering content to mobile users; significant enough to warrant a range of technological and logistical solutions specifically to manage content. This problem only increases when we consider how to arrange content within meaningful retail packages across a wide variety of device types, as envisaged for converged services.

In the first edition, I made the case to focus mostly on the delivery of content to mobile devices, which was – and mostly still is – a significant challenge. As we all know, mobiles come in all types of shapes and sizes, so getting the content in the right format for the device is an important feature of the mobile ecosystem. A lot gets said about newer devices, like the Apple iPhone™, which have the ability to access any page hosted on the Web. This certainly isn't the case for the majority of mobile phones in the market. Moreover, it isn't really true for the iPhone either, which at least for the first edition couldn't render Flash files or Java contents (applets) in the browser. Additionally, to make services like YouTube work, deals had to be done to convert the content at its source into other video formats suitable for the iPhone and mobile generally. There is no doubt that content delivery still has many challenges for mobile.

Since the first edition, other challenges in the content layer have become more significant. Among them are digital rights management (DRM), advertisements, personalisation, retail bundling and converged services. These topics pose various additional challenges and add significant complexity to the content network, convergence in particular. In fact, with convergence, the notion of the network expands drastically, with whole new types of content network to absorb into a much larger ecosystem, often with very disparate and distinct workflows, data management and interface paradigms.

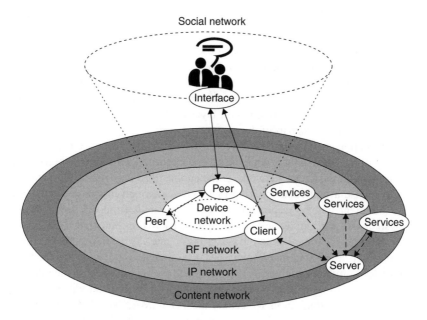

Figure 2.4 Application topologies in the mobile ecosystem.

2.4 APPLICATION TOPOLOGIES

In order to build services, we need to understand how applications might reside in the mobile ecosystem. There are two main planes of operation: client–server (CS) and peer-to-peer (P2P), and there is no reason why there can't be hybrids of the two – something we will look at later. Figure 2.4 shows the two planes of collaboration.

Notice that we have deliberately shown the user interface as a separate application component. The reasons for this will become clearer later, but it will help us to maintain a degree of focus on the user interface as an important and primary part of the applications domain. It also reminds us that the user interface and the actual bulk of the user-centred application could be running on separate devices within an interpersonal device network. Isolating the user interface in our model might also prompt us to consider more seriously the usability aspects of services and thereby give usability its due right on behalf of our end users. Regrettably, usability is often an afterthought.

If we focus on the CS plane for a moment, we notice that we are expecting the application to be in two parts. The server component of the application typically resides within the content network and we will appreciate that its essential role is to serve up content to the client on demand. However, we need to be aware that in mobile applications we have the possibility to push content to the client without demand. This latter scenario is what we call *asynchronous communication*, there being no deliberately sequenced and coordinated request and response for content – it can arrive at any time. It is still the case that out-of-band asynchronous communication is still unique to the mobile ecosystem and we shall examine enabling technologies like WAP Push and its powerful features and service possibilities.

You might be concerned, or potentially confused, to find the server component in the content network and not in the IP network. Don't forget that the concentric networks I have been proposing form a model, so it is a matter of preference how I partition the network, not forgetting that like all boundaries, the edges are fuzzy. However, ultimately content is managed by an application server of some description or another, so I prefer to think of this component as part of the content layer. We can then treat its attributes and parameters according to the language and paradigms of content management rather than the more abstract IP protocols and related concepts. If a more familiar view is important to you, then you can think of the Web presentation layer itself as part of the IP network and the application logic behind it as part of the content network.

The client component of the application clearly resides somewhere in our device network, and it is entirely possible that there could be several client applications either on each device or spread throughout the device network. Some applications may run on all devices acting in unison to realise a particular function or service, although this modality is still rare. Generally, the client will be responding to user commands from the user interface and converting some of these into requests for relaying back to the content network. In a device cluster, one of the devices could be acting as a proxy or gateway to relay requests to the server on behalf of other devices, perhaps solely because the gateway device has the necessary longer-range RF communications capability to reach the server whereas the other devices don't, as is shown in Figure 2.5. Perhaps even the gateway function will jump from one device to another depending on its RF capabilities. A WiMAX-enabled device may act as the gateway in a WiMAX area and a 3G device act as a gateway in a 3G area.

The server part of the application tends to be the workhorse that has to perform many actions to manage a meaningful content exchange with the client. We can see from Figure 2.5 that the server may have to call upon services in various layers of the mobile ecosystem. Such services may include access to location information for example, to enable tuning of content to end-user location without the user having to specify their whereabouts. Clearly, we shall need a means to access these services, this being part of the consideration of building a services platform for hosting mobile applications, and we shall be discussing this in detail in several parts of the book.

The availability of services in each layer of the concentric networks model serves as a reminder that each network is not monotonic. For example, as we discussed earlier, although the RF network is characterised by supporting the RF link between the device and the 'networked world', it contains a plethora of services to holistically support that function.

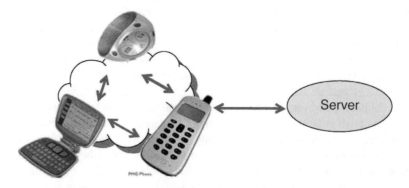

Figure 2.5 Application communication pathways in the device tier. (Reproduced by permission of IXI Mobile.)

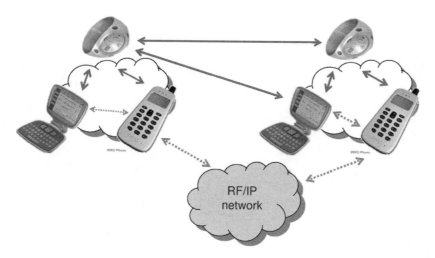

Figure 2.6 Peer-to-Peer (P2P) communication between devices. (Reproduced by permission of IXI Mobile.)

One example is billing, another is mobility management (often extended to include location-based services) and so on. In the IP network, a service might be a presence server, an identity management server (e.g. federated identity) and so on.

Later in the book, we shall focus heavily on the CS method of delivering services to the end users as it remains a key mechanism for enabling so many services. We should note that although the diagram shows a one-to-one relationship, implementations obviously have to support one-to-many – a single server-based application should cope with numerous clients. We shall review some key mechanisms for this in the next section as a necessary precursor to understanding the CS architecture and introducing and elaborating on the enabling technologies like Java 2 Enterprise Edition (J2EE).

It is not necessary to use servers, especially in a cloud of devices where users could potentially interchange information directly. This is the peer-to-peer (P2P) mode of application exchange and its simplest incarnation is 'beaming' of content between phones. Wireless gaming is also becoming popular. However, we need not confine our understanding of P2P to being contingent on physical proximity. It is perfectly feasible and reasonable that two devices could communicate directly with each other using any of the layers above the content network, without any server involvement. We can think of this as a shifting of the conventional CS roles whereby one of the participating end-user devices mimics a server role whilst the other becomes a client and vice versa. Such a configuration challenges many of the conventional understandings we may have about web-based computing. For example, imagine opening a web browser to view a page on your friend's device, such as their contact list. It is perfectly feasible, but not yet common within this context[8].

Figure 2.6 shows us two possible modes of P2P communication. The continuous lines indicate close proximity interdevice communication in the device network and will commonly be Bluetooth connections, but could also be infrared. These are both perfectly useable

---

[8] In other contexts, like embedded computing, the use of a browser to support localised communications is increasingly common; for example, a printer server or a vending machine.

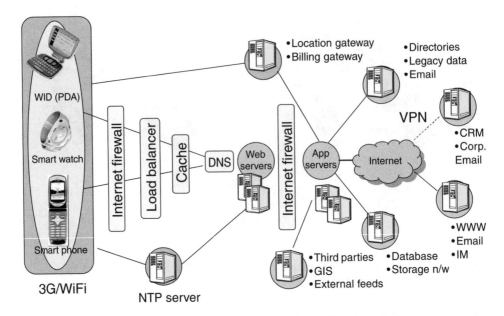

Figure 2.7 Simplified view of some important network elements. (Reproduced by permission of IXI Mobile.)

connection channels for P2P collaboration. The dotted arrows indicate a P2P session[9] taking place over a wider area network, possibly via the RF or the IP network layers. Note that text-messaging in existing 2G mobile phones is an example of a P2P application involving the RF network layer (although strictly speaking there is a store-and-forward server involved called the Short Message Service Centre (SMSC), but it is effectively transparent[10] as far as content exchange is concerned). In the figure, we show the P2P session taking place via personal mobile gateways that are enabling the wide area connection.

In the later sections of this book, we shall examine in detail the underlying mechanisms needed to enable both CS and P2P applications. Let us first introduce some of the physical elements or building blocks in the network layers throughout the mobile ecosystem.

2.5 PHYSICAL NETWORK ELEMENTS

For the rest of this book I shall be aiming to expand on the networks in the model just outlined. I shall do this by discussing either paradigms (essentialised ideas about what a particular network does) or technologies. Figure 2.7 offers an extremely simplified view of some of the more important building blocks that we might expect to encounter as we move across the networks, but I shall offer a much more detailed view later in the book, especially

---

[9] We can think of a session as being an entire communications exchange, but we shall formally define this term later as it has certain meanings that are assumed when discussing issues like transactions taking place via the Web (and HTTP).
[10] In as much as it is really a message queue and does not provide any transformative function.

as in my role as an applications architect I am often asked what does the applications ecosystem look like.

The rest of this book will explain how these elements work within the mobile ecosystem, focusing on various levels of detail depending on how relevant the particulars are to our wider concern with mobile applications and services. In some areas, we shall drill down several layers of detail as the technologies involved are very specific to the mobile framework, such as how Java-2 Micro Edition (J2ME) works on the devices and the dynamics and systems issues of loading J2ME programs, called MIDlets, over-the-air (OTA), which means across an IP connection supported by the RF network.

In other areas, like the IP network, we shall focus mainly on the key protocols and security aspects that are essential to understand in any mobile Internet application, but we shall not look into too much detail about the infrastructural workings of an IP network, like the use of routers and so on. The only time we shall delve into such discussions is when considering 'edge effects' associated with the implementation and workings of the boundary between the IP network and RF network, as the functional partition can begin to blur at this point, so some clarification is necessary, like IP address assignment to devices in a mobile network.

3

Becoming an Operator 2.0

3.1 INTRODUCTION

In the first edition of this book, I spent a lot of effort in the opening chapters trying to promote the idea of mobile applications. Since then, things have moved on. There are now many mobile services in the market, many companies producing mobile applications and there is significant interest in using mobiles to deliver services previously confined to the Internet. Indeed, many of the ideas that I introduced in the first edition, which were possibly novel or visionary at the time, have since been widely discussed or even implemented. Given the degree of progress since the first edition of the book, I shall spend some time in this chapter summarising what I see to be the major trends in next generation mobile applications. This will hopefully provide additional context for the rest of the book.

Of course, the critical question always arises – *where does the revenue come from?* In my experience, this question is often raised early and vigorously during presentations to operators. This is because they already run very successful businesses with clearly defined and measurable revenues. Indeed, the obsession with revenue in the mobile telecoms world is demonstrated by the invention of a metric to measure it, called ARPU, which stands for average revenue per user. Such a metric might cause an entrepreneur in Web 2.0 world to turn white, knowing that he or she doesn't yet know how to generate revenue from their prodigious efforts, never mind plotting it on a graph. With various web start-ups, there is often a notion of intrinsic value, like number of hits on the site, or some other measure of usage, but a difficulty converting this to extrinsic value, like revenue. For operators, the notion of intrinsic value is almost completely alien, except for the increasing realisation that they might have to offer some services just to retain users on the network, rather than generate additional revenue. I shall talk about this later in the chapter when I map out the things that an operator can do with mobile applications. For those of you still seeking to promote applications ideas to operators, I shall give some advice

Next Generation Wireless Applications, Second Edition   Paul Golding
© 2008 Paul Golding

here, but as you can imagine, most of it will centre on the revenue question. There is a change already underway in how operators think about applications, particularly within the Web 2.0 paradigm, so I shall offer some advice too on how to focus on value, not necessarily revenue.

3.2 WHAT APPLICATIONS CAN I SELL?

Let's tackle the relatively simple question of what applications you can sell. Let's imagine that you have an interest in making money from mobile applications. What are the types of product that you could create? The same question applies to an operator, except they would ask what type of applications they should *deploy*. The question ties back to our concentric networks ecosystem model described in Chapter 2. You could create an application to run anywhere in that model, as long as it adds value. Whom it adds value to and how it adds value will determine how you might make money. I tend to think of there being three types of mobile application:

1. User applications

2. Component applications

3. Infrastructure applications

These are explained as follows:

1. Create a complete application with an interface visible to the users, for which they might be willing to pay money to use it. I shall call these *user applications*. Examples include games, instant messaging, push email and probably the types of applications that many of you reading this book have in mind when thinking about mobile applications.

2. Create an application used by the operator as a building block in an applications ecosystem to support a user application, but that isn't necessarily used directly by the user and is almost never used without other components. I shall call these *component applications*. Examples include presence servers, identity servers, video codecs, streaming servers, on-device portals and so on.

3. Create an application used by the operator as a means to improve the efficiency of doing either of the two points mentioned above or to provide the various support services required to turn an application into a billable service. I shall call these *infrastructure applications*. Examples include service delivery platforms, authentication servers, billing platforms and so on.

Some of the principles in this book apply to all three of the applications areas mentioned. However, the focus is primarily on the first two: user and component applications. Infrastructure applications, such as billing platforms, are moving into a layer of software services in the 'back office' that, whilst important, is too divergent from the core themes covered by this book. You could make money in any of these areas. Additionally, you can make money by packaging services with these applications. It is important to keep this in mind if you are new to the business of selling applications to operators – *they seldom take an interest in buying only a piece of software.* They typically want any number of additional business support services in order to deploy your software application as part of a service they can monetise. For example, a common requirement, especially for smaller operators, is hosting. Also, be prepared to enter into trials of your application at your expense, although you should always ask the customer to pay you for it. This depends on the value of the trial to the operator, but you can always negotiate. Above all, *always be prepared to demo your application* and come prepared with *complete pricing* and model deployment plans at the ready. I shall return later to the topic of selling to operators.

3.3   WHERE DOES THE MONEY COME FROM?

If you are making and selling user applications, then you have two possible routes to market: direct or via the operator, or both. There is a market for direct-to-consumer (D2C) applications, but it is not an easy market to crack. Certain types of application can work well in this market, which is already the case. Games are a good example. If you are selling direct to the consumer then the business model is entirely up to you to decide, notwithstanding precedents set by the market. That said, the mobile content business is still young enough to welcome new pricing models. However, with D2C selling, it is also up to you to collect the money from the customers. This is where you will need to consider the charging model, which could be:

- *Free of charge* – you have indirect way of making money, such as advertising

- *One-time fee* – there is a one-off charge, usually upfront

- *Subscription* – money collected at regular intervals (e.g. monthly)

- *Usage based* – money collected according to some meter of usage (e.g. amount of data storage)

- *Transaction based* – money collected for a particular transaction (e.g. buying a piece of content)

- *Bundling* – some combination of all of the above

The charging model is often the key to success of a mobile service, so an additional consideration is ensuring enough flexibility and agility to change the model. Entire Mobile Virtual Network Operator (MVNO) businesses have failed because of their inability to change the charging model rapidly enough; thus, preventing the business from responding to market conditions in a timely manner (or at all). If you are selling D2C, then you will need to deploy all of the charging logic and infrastructure required to implement the chosen charging scheme, keeping in mind the need to be flexible. It might not be necessary to create the charging system itself, as charging platforms do exist that can be licensed, including open source solutions widely used in the ISP world. (As mentioned above, I shall not be covering these infrastructure applications in this book.) The important point is to know your charging model and to know that you will need a way of implementing it. This is a seemingly obvious point. Nonetheless, software vendors eager to provide a mobile application to the market frequently overlook it.

The discipline is to think of mobile *solutions*, not applications[1], unless you are working in partnership with another company who acts as the solution provider using your application as part of a wider solution. This might be one way to tackle the money issue – leave it to a more experienced and capable company to sell direct to the consumer. They will take care of the charging and pass on part of the revenue to you. This is the *revenue sharing model* and is a common approach to partnerships. If you think a solution provider will pay you up front for your software and then use it to build a solution, think again. This is an increasingly unlikely scenario. Therefore, be prepared to have revenue-share pricing available when working with a partner (and an operator). I would still suggest trying to squeeze some upfront or 'management' fee from a potential partner, as a means to test their intent and dedication towards the partnership.

3.4 DIRECT-TO-CONSUMER (D2C) RETAILING

If you are going for the D2C route, there are a large number of practical considerations specific to the mobile ecosystem, some of which you might not be used to if you are coming from an Internet background. There are four key considerations when trying to offer a mobile application D2C:

1. Discovery
2. Distribution
3. Service access
4. Charging mechanism

Let's look at each of these in more detail in the sections that follow.

3.4.1 Application Discovery

The discovery of the application is important to your business success. If no one finds the app, no one is going to use it. Like many other points made in this chapter, they are obvious, but frequently either overlooked or underestimated in terms of effort and difficulty. Don't spend 90% of your effort and budget in getting an application running and then realise that sales are incredibly slow because no one is able to find it. Ultimately, the mobile user has to be able to find the application and then activate it for their device. There are a number of options for the user to discover the application:

- Internet ('New Media')
- Retail outlets
- Traditional media (e.g. TV, magazines, etc.)
- Mobile

---

[1] Okay, so the title of my book says applications, not solutions, but that's a marketing decision. The word 'solutions' is too vague to use in a book title.

Let's consider each of these briefly:

- *Internet discovery* – This relies on your users accessing the Internet in order to find some kind of collateral describing the application, which might be a dedicated website, a partner website, a web advertisement, a blog, a news article, plus others. These are straightforward to figure out, although don't underestimate the effort required to successfully bring an application to your market's attention. Where possible, it might pay to use a site that is already generating a lot of traffic for applications similar or complimentary to yours, such as a gaming site for mobile games. The advantage of using an existing portal, like a gaming site, is that you can use their umbrella charging and customer care schemes and 'sit back' and take a revenue cut. If you are thinking about international distribution for your application, then it is important to understand the penetration of Internet in each territory. For example, in many parts of the Middle East and Asia, usage of the Internet might still be low whereas mobile penetration is high. Therefore, the market might still be there for your application, but you won't be able to use the Internet to enable users to discover your service.

- *Retail outlets* – such as the shops where users buy their phones. In the D2C market, there are always a number of retailers not controlled or owned by the operators. These independent dealers are usually free to distribute whatever additional services they like with the phones they sell. However, don't be surprised to find that this is often a poor channel to market. The business models of retailers are dominated by hardware and SIM-card sales. If your application, like most, only has the prospect of adding a tiny incremental revenue to these sales, then most retailers will not be interested. There are always exceptions. These are usually retailers with a particular motivation to sell applications, but they are rare. Some conventional software outlets will sell mobile applications, provided they are easily sold in a shrink-wrapped box.

- *Traditional media* – making use of conventional advertising and coverage is a possible option. It is not uncommon to find mobile applications advertised in magazines, newspapers or similar printed materials. The most common method is to provision a short code for a text message that will enable the user to send a message and then receive a link to access the application. This might be to download a software file to the device or to use a WAP site, or both. Traditional media advertising can be expensive, but for media companies wishing to extend their services to mobile, this is the obvious choice.

- *Mobile discovery* – is by far the most difficult discovery method in the independent distribution model because there is no clear way on a mobile to bring content to a user's attention. On an operator-retailed phone, there will almost certainly be a *WAP portal*, which US operators call a *deck*. However, for independently retailed phones, there is typically no default portal unless a retailer is big enough to provide their own (which isn't that often) or they have entered into some type of partnership with an independent portal provider, like Yahoo Mobile. If you can succeed in getting a deal with a device vendor directly, then it is possible to have something pre-installed on a range of devices. This might be the application itself, perhaps in a demo form, or a link in the browser, or an icon/link in the appropriate menu or folder to allow the user to download or access the application. Pre-installed applications will usually have to work out-of-the-box and therefore work initially free of charge. If the user likes the app and wants to keep it running, then you can charge them accordingly.

3.4.2 Application Distribution

The distribution problem for mobile applications is thinking about how you can get the application onto the device. There are several options:

- *Side-loading via a PC* – here the user has already discovered the application via a website where they are eventually led to a link that enables them to download the application to the PC where it can then be transferred to the device. This option sounds fine, but it is very clumsy and relatively unsuccessful, except for music content downloads. Many consumers will not know how to load the application to their device. Generally, you should avoid this method if you can.

- *Over-the-air (OTA) download* – here the user downloads the application via a WAP connection directly from their phone. This is probably the most common method and there are a number of ways of doing it. The most straightforward in terms of minimal effort on your part is to publish the Uniform Resource Locator (URL) where the application resides on the network. The user enters the URL into their WAP browser and proceeds to download the application. The pitfall here is avoiding complex URLs that the user has to enter accurately into their WAP browser, which is generally not an easy task to undertake in any case. A way around this problem is to publish a short code text address that the user sends a keyword to and then receives a reply that contains the URL (either via WAP Push or URL link in a standard text message). However, for international distribution, a short code has to be provided in each country and this isn't always easy. In some markets, it is very difficult. Whether via direct entry or short code, an additional challenge with OTA distribution via WAP is making sure that the application is suited to the device. If you have developed mobile applications already, then you will know that it is almost impossible to develop a one-size-fits-all version. You may need several versions of the application, each optimised for different devices or classes of device. Therefore, at the distribution stage you need to ensure this filtering mechanism is available. This can be done manually, such as offering different links for different devices (e.g. Basic Java link, Smartphone Java link, Symbian link, etc.) or by an automated process using an application vending server that can detect different device types and ensure that the right links are used accordingly. There are numerous vending solutions in the market, but you most likely want to find an existing site that is well visited and to somehow get your application included in their catalogue.

- *Pre-installation* – here the application is already installed on the device that the user buys from the retailer. How to arrange this is not always clear and depends on the market, which varies from region to region. It could be through striking a deal with the device vendor, the distributor (if one exists in between the vendor and the independent retailer, which it will for small retailers) or with the retailer directly, and sometimes with a combination. The economics of these deals aren't always clear. For example, a device vendor might not be that interested in making any profit from a pre-installed service. It might be a means to differentiate their devices from their competitors' devices. However, be warned! Don't assume that such a vendor isn't going to squeeze you financially. They want to minimise their costs like everybody else. Moreover, whilst it might not be their core business, that doesn't prevent them from trying to make some profit from the deal if they can. Pre-installation can also extend to memory devices, and even certain types of accessories, so don't overlook these options if you are keen to pursue this method of distribution.

3.4.3 Application Access

Your user has discovered the application and is now running it on their device, but do they have the required network resources to operate it? For example, if your application requires video streaming, then be aware that many networks in the world will not allow it to run because they will block video packets at the edge of their network. Don't think that there is an easy way around through the use of an unusual IP port assignment or protocol, or by trying to 'hijack' a conventional port/protocol pairing. Many firewalls are more than capable of detecting unusual traffic and then blocking it, especially on well-known ports. Operators usually run quite secure borders around their networks. Anyway, you wouldn't want to offer a commercial service that was subject to frequent disruption (i.e. playing cat and mouse with firewalls) or that you couldn't guarantee were accessible to your users in the first place. Also, keep in mind that none of the flat-rate mobile data packages offers truly unlimited bandwidth. They are subject to 'reasonable usage' policies, exceptions and other restrictions, which might include restricting what type of data traffic is allowed over the connection. Make sure that the network services that your users need are in place to the required level in each of the target markets. Consider different levels of service in different markets if you can't guarantee full service access in all regions. For example, if you are running a news service, then you might use video streaming in some markets, but not in others.

Additionally, if you are in a position to do so, you might want to negotiate directly with various network operators to ensure service access to the required level. For example, if you are offering a video streaming service, then you might be able to negotiate opening the firewall to your servers and then find additional business opportunities by allowing other applications providers to piggyback on your agreement and streaming resources. In many markets, such as much of the Middle East, operators will only deal with key partners to control their applications availability.

3.4.4 Charging Mechanism

Now that your service is in the user's hands, you will need a means to charge them. As a D2C provider, you may not have any financial relationship with your user, in which case you need to get one. Again, this is not always easy. By far the most common way to implement a financial transaction with the users is via the relationship that already exists with their network operator. This is possible for D2C channels via premium-rate text messaging service (PSMS). Operators can send or receive text messages on dedicated short codes that allow the messages to and from these codes to be charged at a specific rate that is typically much higher than standard text messaging rates. To limit fraud, these services are nearly always permission based. In other words, it ought not to be possible to send a PSMS message to a user without their permission. That's why the most common technique is to ensure that the message is mobile originated (MO) so that the provider is certain that the user is requesting to be charged. For periodic subscription models, this can be a bit cumbersome, in which case the user is first asked to request and then confirm their willingness to receive a PSMS on a frequent basis (e.g. monthly). Usually, operator (or operator watchdogs) will insist that the user can easily opt out of the subscription arrangement at any time, usually by sending some kind of termination message to the short code.

Outside of the PSMS charging mechanism, there aren't that many other mechanisms available via the mobile itself, except for premium-rate numbers for voice calls. Some of these can levy a minimum charge, which forms the bulk of the payment for the transaction. In terms of usability, the PSMS method is probably more convenient for users. The premium-rate number method exists mostly for services that require a call to be made (e.g. to access audible materials). However, some applications providers will use them for charging because they can get a better revenue-share deal from a particular provider who happens to have a number of premium-rate numbers available for this purpose. In some regions, micropayment mechanisms might be in place with the operator, perhaps using an m-commerce (mobile commerce) solution that involves a virtual wallet on the user's device. However, these are rare.

Outside of the mobile charging mechanisms, any other charging method is potentially viable, but most likely this would be Internet based, in which case the user will need access to an Internet account and a means of paying over the Internet, whether that be credit card or a micropayment account, like PayPal.

3.5   OPERATOR RETAILING

Much of what we have just discussed for D2C retailing applies generally to operator retailing, except to keep in mind that the operator always has a direct financial relationship with the customer, whether they are prepaid or post-paid users. Let us consider some of the key differences from D2C.

- *Discovery* – all of the means are the same except that the operator will usually have their own resources to hand, such as an existing Web and mobile portal. For pre-installation, operators can have a great deal of control or influence over device vendors, especially in markets where the operator subsidises the handset to the degree that they have a large amount of control over its configuration.

- *Distribution* – this is the same as for the D2C case, except that an operator will almost certainly have an existing vending service in place to manage the OTA download process.

- *Access* – an operator controls their own network, so they are able to make the appropriate network resources available to ensure that the service can be accessed. Moreover, they can set the charging rate for the resource, such as the price per message, price per megabyte and so on. However, never make any assumptions about an operator's willingness to do this. A golden rule seems to be that operators do not like to change the configuration of their network if they can avoid it. In some countries, this is cripplingly so. Worse still, if you have a service that requires fulfilling a particular optimisation of the network to reduce latency (e.g. for PTT services) then the business case for doing this needs to be solid. Even then, an operator might not want to do anything. In some cases, even if the operator agrees to configure resources in their network for your application to work, don't be surprised if it takes months, not weeks. This is more so in some countries, where a network change could take up to a year.

- *Charging mechanism* – this is perhaps the easiest bit. The operator has all the means necessary to extract money from their users. That's the good news. However, there is a caveat, which is similar to the problem with access. Even an operator, despite all their willingness to do so, might not have the means to bring the necessary flexibility and agility to the charging model for your application. I have seen this many times. There are two dimensions to the problem, both of which are worth keeping in mind. The first is a commercial issue, which is that the operator might have a problem with determining how to charge for your service and then take forever to make a decision. You might well come to the table with a particular charging model in mind, but you should expect the operator to have their own opinion, especially if the service could potentially compete in some way, even indirectly, with the usage and revenue of another service. This revenue cannibalisation problem seems to plague operators and prevent them from launching many new types of service. The second dimension to the problem of inflexibility is an operational issue. An operator's charging platform might not be up to the job of supporting flexibility and agility in the charging model, or at least for your application if it comes well down the list of priorities for change requests compared to a gamut of other more important services. Despite the various warnings just outlined, such as lack of charging model flexibility, don't make any assumptions about what the operator might expect from you. They might well insist that your application can support every conceivable charging model under the sun and be reconfigurable within hours, even if they can't utilise these models and even if they might take months to reconfigure them anyway. On this cautionary note, this seems a good point to discuss some tips for dealing with operators.

3.6 SELLING TO OPERATORS

You have read all of the above, built your app, configured and priced your services, done all your homework and are now ready to sell to operators, so you set up the first sales meeting and turn up in your best dress, ready to impress. Then the whole meeting goes sour, perhaps within the first five minutes. This is very common. I'll spare you a regurgitation of presentation basics, except to say that I've seen so many blunders here that it *always* pays to revise the basics, no matter how many years you've been doing presentations. Granted, many in the business have been doing presentations for years, but sadly, they've been doing it badly for all those years, too, and perhaps no one has ever told them. Make sure you're not one of these people. Oh yes, and don't forget to test your demo one hundred times and have at least two back-up plans. Whoops, I'm already giving you the tips, so let's get to them.

3.6.1 Top Ten Selling Tips

Based on many sales presentations and workshops given to operators, these are my top 10 tips for increasing your chances of having a productive meeting. It can take a lot of effort and time to get a meeting with an operator, so be prepared. Here's the top ten:

1. *Know your audience* – and their expectations and requirements as much as possible *before* the meeting. If your presentation is more of a 'workshop', make sure you know what this means – agree on an agenda and clarify expectations. So many meetings with operators are hit-and-miss with the audience and expectations. I don't know why.

2. *Be flexible with your pitch* – so be prepared to take your presentation in one of three directions (as hard as you try, you will seldom be able to control which way it's going to go):

   (a) A *top-down* presentation starting with what your product does (the user-experience), why users will need it and so on – probably the typical sales presentation that you had in mind to present.

   (b) A *get to the point* presentation, which is usually driven by a 'commercial' person who just wants to know what it does, for how much and what your business model is.

   (c) A *roller coaster* presentation, which is driven almost entirely by a random mix of technical and commercial questions coming from different types of questioners. Try to avoid combining technical meetings with commercial ones, although often you won't have that luxury.

3. *Know your product inside out* – and make sure that you are selling a 'whole product', which doesn't just mean calling it a 'solution'. Here's a tip for understanding what a whole product is: if you have to ask the operator for anything, other than their business, then the further you are from a whole product. Avoid statements like 'Oh, we thought that you would provide x, y or z.' Of course, there will be inevitable touch points with their network, so don't take this too literally.

4. *Understand your business model inside out* – try to avoid saying 'we'll get back to you.'

5. *Always start with the user experience* – ideally on a real handset, as nerve-wracking as demos can be in meetings. In many ways, this is your elevator pitch. 'Here's what my product does' and you show it to them, preferably in one slide.

6. *Run a pre-flight check on your demos* (see point 5) – make sure that they are going to work, especially in the hands of the customer (who will always press the 'wrong' button). Always have a backup plan, such as emulation on a PC. Always have a backup for the backup plan! You can always put screenshots in a presentation (amazing how many presentations don't do this).

7. *Avoid unnecessary lingo wherever possible* – your audience might not know all of it and it doesn't make you sound clever. There is no harm in saying something in full, as opposed to its acronym (except well-known ones of course). Most of all, avoid generic waffle. Don't say 'I've got a Web 2.0 ajaxian widget that exploits the long tail.' That might knock the socks off web-heads at a techno wizards Internet conference, but it tends to fall flat in a room full of operator types.

8. *Demonstrate operational competence–* be prepared to say how you can *deliver, deploy and support* your solution to the required level of robustness that an operator will want. If you are a small company, that's not a problem necessarily, but you might want to consider partnering with a bigger company who can cover the heavy stuff, like 24/7 support.

9. *Have references to hand* – be prepared to give proof that you have already deployed your product. If you haven't, then be ready to offer a trial of your product, which the operator may well expect to be at your cost.

10. *Be clear about which features are ready to deploy* – and be clear about which features are still an idea. Almost certainly, if you only have ideas, then your sales pitch isn't going to get very far, unless it's an idea based on existing product that you already ship. Even then, don't expect that an operator will pay you to implement your idea. This can happen, but it is very rare.

3.6.2 Selling Apps to Operators – Operator Perspective

Don't just take my word for it. I asked a well-placed person in a tier-1 operator for her tips on selling to operators. She sits on the other side of the table, which is the buyer's side, and was able to offer some very practical tips, including thoughts about the wider concerns that an operator might have about you and your company, not just your product. Unfortunately, we techies tend to be carried away with the bells and whistles of our product and can easily overlook the wider business concerns.

Most operators want to know the options for hosting versus in-house installation. Many projects require the application supplier to host under a managed service agreement due to time constraints and in order to reduce the risk if the application doesn't take off. This can lead to a big focus on the details of hosting. An operator will typically want a local (i.e. in-country) hosting option, preferably using a hosting center that the supplier has used before. However, be wary of hosted-only solutions. There will be some suspicion if the supplier doesn't want to give an option for the operator to license the application to put in-house – 'their software must be rubbish if they don't want us to touch it!'

Support is an equally important issue. Operators are used to very high levels of availability, 99.99% at least. They like to see a proposal showing a team of support staff, monitoring processes and local (in-country) 24/7 call centres. Many suppliers with otherwise good products have been rejected because of weak operational support.

Billing and provisioning is always a hot topic. Most applications should support self-provisioning. Some of the operator-provisioning systems might be old and difficult to change. It's better if customers are provisioned automatically. Billing integration, particularly on post-pay systems, which are usually the oldest, is always a challenge. Anything that requires tariff changes and event-based billing is difficult to do. In spite of this, or because of it, suppliers should have a range of flexible billing options that include subscription, all-you-can-eat, bundles, and the ability to support promotions.

The level of integration to the network should be as minimal as possible. Often, suppliers assume that they will have a supplier privilege of requesting system changes from the operator. Be aware that this is often not the case and that even relatively simple changes can cause major headaches to operators, to the point of becoming a show-stopper. The most likely solutions to get the green flag are those that affect operator systems the least. If possible, find out as much about the architecture of the operator network as you can before meeting with the operator. This will help to prepare a story that minimises system impact. In addition, this kind of knowledge and preparation always impresses. Knowing about wider operator issues, such as interoperability between other operators, is usually helpful.

Customer Relationship Management (CRM) is another important topic. Most operators want all the information they can get about their customers and expect to get fine-grain data (e.g. raw log files on a daily basis) to pull (or push) into their CRM systems for analysis.

Suppliers are often surprised by how much data is requested (even though operators often might not use it as intended). Providing low levels of detail with summary reports available on an extranet or via e-mail is a good approach.

Suppliers should provide demos wherever possible. It is still surprising that more suppliers don't do this, which can create suspicion about product readiness. Insight into market research and focus groups that were used to develop the app are also very helpful. Sometimes, a usability expert is expected to justify the design approach.

In terms of commercials, this is often a thorny area for suppliers. Operators don't always go for the cheapest! In fact, being too cheap sometimes gives the impression that the product isn't that good. Being flexible on price is helpful. Pricing should be clear with a detailed breakdown. Many products come with various options and often it isn't clear what is needed to do what, so the final pricing is difficult to determine. This should be avoided.

3.7 WHICH APPLICATIONS SHOULD AN OPERATOR DEPLOY?

Thus far, in this chapter, I have looked at ways in which software vendors can tackle the mobile applications market. A major channel for such vendors to consider is the operator market, as we have just discussed. Now I want to look at the question of which applications to deploy, which is an operator question. Of course, the answer is moving all the time, so this will just be a snapshot, although, as you will see, I shall attempt to break the problem down into categories of applications; thus, enabling some general principles to be established. In this section, I shall mostly be looking at the problem from the perspective of relatively mature operators. However, many of the issues covered still apply anyway to less mature operators.

3.7.1 The Market Challenges

What are the market conditions that operators face in mature markets? They can be summarised in three ways:

1. Declining subscriber growth (market saturation)

2. Falling ARPU (due to increased price pressure)

3. Evolving market (due to convergence, Internet threats, etc.)

Facing these market conditions, the major challenges facing operators are essentially four-fold:

1. Securing new revenue, or growing their share of the consumer wallet

2. Mitigating against churn, or creating service/brand stickiness

3. Increasing agility and innovation, or rapid deployment of new services and service variants

4. Developing new business models in the face of uncertainty and high risk

Given these challenges, many operators are looking towards mobile applications as part of their growth and competitive strategy. Traditionally, operators have built hugely successful businesses from voice and texting services. Applications have mostly been given very low priority, trying them out here and there, but not with any real enthusiasm, except some more visionary operators.

3.7.2 The User-Experience Focus

As Figure 3.1 shows, an operator can deploy platforms and networks as the base layer in their 'applications stack'. Platforms include things like Service Delivery Platforms (SDP), IP Multimedia Subsystem (IMS), billing solutions, mail servers and so on. Networks include 3G, WiMAX, DVB-H and so on. These are all technologies, not services. They are the foundation for building services. Running platforms and networks is a business that operators know well and do well.

The next layer on top of the platforms is the service enabler layer. This is mostly made from the *component applications* defined earlier, such as presence servers, IM servers (which can be used to deliver other services or be used standalone) on-device portals and so on. Operators are still relatively inexperienced at dealing with this layer in the stack. However, it is the next layer up that really matters, which I refer to as the *experiences layer*. How the users experience services on their device is important to mobile application adoption. It is not an exaggeration to *think of* the mobile applications business as being an *experience business*, not a technology business. The operator should always be thinking in terms of the end-user experience and this attitude should pervade all mobile applications projects. This includes dealing with suppliers, which is why I earlier mentioned the importance of the supplier presenting the user experience during the sales pitch. It is best to deal with companies who have a clear focus on user-centric thinking. This should be demonstrated by evidence of *formal usability testing* in their software testing.

What is a compelling experience for mobile users? There is no single answer, but I believe that certain attributes of a compelling experience are identifiable. Here are three attributes to think about:

1. *Seamlessness* – the user is able to enjoy a range of services across a range of delivery devices and channels (e.g. mobile, TV, Web) where their data and preferences are easily accessible across device and channel boundaries without the need to re-enter or reconfigure.

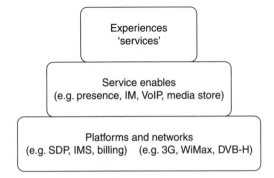

Figure 3.1 User-centric approach to applications.

2. *Connectedness* – the user should be able to connect easily to their 'digital world' including people and places, both real and virtual. The easier it is for users to connect with *their* blog, *their* social network, *their* train timetable, and *their* virtual life, the more compelling the experience.

3. *Personalised* – the users want and expect data *relevant* to them, tuned to their needs, their interests and their digital habits. Increasingly, the data should also be context-aware, which means that the application somehow seems to know the circumstances of the user in advance.

3.8 INTERPRETING USER-EXPERIENCE TRENDS INTO APPLICATIONS

Given the previous discussion about compelling user experience, how does an operator convert this into areas of activity within their mobile applications strategy? Here are five actions:

1. *Adopt a multichannel approach* – increasingly convert the platform and service enabling layer to allow the user experience to extend across multiple devices and channels, including mobile, Web, IPTV and the digital home network.

2. *Enhance real-time interpersonal connectivity* – increasingly enrich the 'call' experience by utilising new modalities (e.g. video share, presence, avatars, push-to-talk) and re-invigorating older ones (e.g. adding ring-back tones, creating single-click conferencing, visual voicemail). Here I believe it is important to maintain the 'call' concept as a metaphor for these services, mostly because the user is so familiar with the idea. This can be easily done via an active address book that allows a multitude of interpersonal services to be launched from within it.

3. *Extend connectivity modalities* – add new modalities to the phone experience, such as mobile TV (broadcast), video blogging, IM, avatars and 'remote control' for the home network.

4. *Drive content consumption* – utilising more on-device portal techniques, smart pre-loading of content over-the-air (or during charging) and moving towards an Amazon-like retail experience on the phone (and across converged channels) with minimal clicks to buy.

5. *Increase personalisation* – increasing the 'me' factor in all services wherever possible. Increasingly, personalisation is becoming a critical horizontal-service enabler across all channels. We can already see this with suites of web apps, like the various services provided by Google.

When thinking about new modalities of connectivity, three key trends should inform how you to take action to deploy new applications and services:

1. *Mobilisation of the Internet* – this can and is happening quicker than 'Internetisation' of mobile (i.e. mobile is already a hugely pervasive communications tool, but not a pervasive applications platform like the Internet).

2. *The rise of non-operator mobile services* – this is clearly already happening, with many operators making more revenue from off-deck (off-portal) content than on-deck content and with increasing trends towards flat-rate open Internet access.

3. *Key digital memes (themes)* – which are

   (a) Community and social networking
   (b) Advertising-funded services
   (c) Convergence, especially with the home network
   (d) Entertainment and increasingly *interactive* entertainment (e.g. extension of popular TV programmes, like *Lost*, into a rich suite of web applications)
   (e) Location and more generally context awareness
   (f) Web 2.0 paradigm (discussed below)

Your mobile applications strategy should be aligned with these major trends.

3.9 WIDER DIGITAL TRENDS INCLUDING WEB 2.0

We have thus far looked at some of the commercial trends and realities of dong business in the mobile applications world from within that world. However, the impact of the ongoing evolution of the Web is becoming increasingly important. We have even seen some operators, such as Three UK, evolve from a walled-garden operator (i.e. shut off from the Web) to an incredibly open operator who has fully embraced the Web. In my meetings with various operators, it is clear that they are very interested in how to understand and utilise emerging web trends, so this is the topic for the remainder of this chapter.

3.9.1 Web 2.0 and Mobile Web 2.0

Today's Web is a 'click-to-do' environment, having progressed from a 'click-to-read' one. It's just as easy to create a service to share calendars and photos as it is to create an online version of a marketing brochure. The Web has evolved from a publishing platform to a programming one. As we shall discuss in Chapter 4, the great power of the Web is the single universal client, the browser, which is not bound to any particular service or data source. The browser has grown up from an information-display client to a more generic user-interface client where the modes and level of interactivity are much richer.

With a web-based application, it is relatively easy to incorporate sophisticated components like blogs, email lists and forums without having to program them. These can be used on any site and have caused a gradual shift towards more participative websites where users share comments, views, ideas and their own content. The trends towards web-hosted applications, component-based assembly and greater user participation are now well established, which led some commentators, notably O'Reilly[2], to assert that we are in a new era of the Web, which he called Web 2.0.

Whilst O'Reilly's comments have sparked widespread use of the term Web 2.0, the term it not used consistently throughout the industry; part of the problem being that O'Reilly's

---

[2] Tim O'Reilly, founder of O'Reilly Media publishing company is thought to be the 'father' of Web 2.0 after the term was first coined at a conference brainstorming session organised by O'Reilly Media.

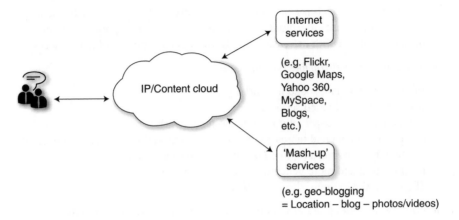

Figure 3.2 Web 2.0 trends.

original 'definition' included a large number of ideas. Web 2.0 is mentioned so often and in so many contexts, that it's meaning can vary, including: websites that use certain interface programming techniques (i.e. AJAX); websites that are built upon user-generated content (e.g. YouTube); websites that try to build communities; websites that are new and trendy; and so on. To be sure, there is much hype surrounding Web 2.0 and one could argue that it is such a vague catch-all description of web trends that it has become an almost meaningless term that doesn't serve any useful purpose.

As Figure 3.2 shows, an application or user can connect to the Internet to use web-based services. Moreover, increasingly they can use services that are a combination of other services, which are called *mash-ups*.

Within this type of environment, the possibilities for operators are as follows:

- *Offer '2.0' type websites to users via the desktop browser* – in other words, follow the trend of Web 2.0 and exploit the best of it to offer users compelling services in their web browsers. These don't need to be mobile related, nor does usage need to be restricted to operator customers only. One approach is to offer everyone a compelling service, but limit certain features to the operator's own customers.

- *Support third-party mash-ups using operator portal services* – provide APIs to allow 3rd parties to access features in the operator's portal, such as access to a user's online address book (back up from their phone).

- *Build mash-ups using third-party websites* – provide users with services that utilise other websites via open Application Programming Interfaces (APIs). For example, offer a 'nearest friends' service to allow users to find where their friends are, showing them on a Google Map that has been mashed into the service.

- *Utilise the '2.0' approach, but towards mobile web applications (i.e. web/WAP applications accessed via handset browsers)* – this is somewhat vague in many respects because the notion of '2.0' is also vague. Some commentators, such as Jaokar[3], have suggested that this means placing more emphasis on the service domains of social networking and

---

[3] http://www.futuretext.com/publications/mobileweb2/authors.html

user-generated content whilst using more advanced browser techniques like AJAX and widgets. Inevitably, some of the programming techniques used to support a richer user experience in today's desktop browsers will end up in mobile browsers. However, the ongoing problem of limited screen and keyboard size are the real restricting factors. My view is that a new approach towards mobile applications is more likely to emerge with the adoption of SIP-based networks (i.e. IMS) which I discuss in the next section and more fully Chapter 14.

No doubt, you will want to know which of the above approaches you should take and, more pertinently, how do they affect revenue. Having explored these themes many times with various solution providers, service providers and technologists, it is clear to me that there is no single or easy formula for exploiting Web 2.0 trends in the mobile world. The biggest mistake is to think that Web 2.0 is actually something concrete like a technology that comes with a set of defining specifications. Web 2.0 is an ill-defined and problematic term. There are new trends, techniques and technologies emerging on the Web all the time. What really matters is to develop the understanding that these all have the potential to enable new user experiences for mobile operator customers. Hence, given the previous discussion about the importance of a good user experience, the Web really ought to be on the strategy map. It is too important to be of peripheral interest. How it then forms part of the strategy is an issue of innovation. There is nothing magic about Web 2.0 that allows it to be sprinkled on like salt to turn any food into a good meal. Thinking is still required, about the ingredients, the recipe, the process, the serving and the taste. In other words, innovation matters!

3.9.2 Mobile Web 2.0 or Mobile 2.0?

Is there a Mobile Web 2.0? Again, the poor definition of the term 'Web 2.0; means that adding the word 'Mobile' in front of it is problematic. Jaokar has attempted to define what he thinks this means and has written a book about it. However, I remain sceptical that such a thing really exists. As stated in the last section, some of the technologies that get associated with the Web 2.0 term, like AJAX and widgets, are already making their way into mobiles. Mobile browsers follow their more mature and capable cousins from the desktop family, but major restrictions on device size and resources still exist. As far as I have seen in the last few years, since the first edition of this book, the best user experiences on mobile are usually on native applications, not browsers. Moreover, they are also on devices with a very rich set of device APIs accessible via the application, which is definitely not possible with most mobile browsers.

Since about 2005, in various workshops and activities in the mobile applications world, I have been trying to introduce the idea of Mobile 2.0. It is not an attempt to jump on the 2.0 bandwagon, although the use of the '2.0' moniker is clearly headed in that direction. The notion of a Mobile 2.0 came about whilst working on early IMS solutions and trying to articulate a strategy for IMS Applications. The reason for the use of a new term was to mirror the overall evolution of the Web from being one-dimensional (i.e. publishing) to multi-dimensional (i.e. programming). I have summarised this as a progression from 'click-to-read' to 'click-to-do'. In some sense, this marks a paradigm shift – users think of the Web as a place to do things, not just to read things. However, I tend to think of the dominant sense of the Web as a place to 'find things out', which is closely linked to the process of search. This sense of using a mobile phone doesn't really exist today.

The IMS architecture can take mobility from a 'dial-to-talk' (or 'type-to-text') experience towards something a bit more generic, like 'click-to-connect'. This is narrower than the Web's 'click-to-do' paradigm. I believe that the mobile will remain overwhelmingly a communications tool. This 'click-to-connect' paradigm is something different from the old mobile habit of dialling, which is what led me to describe it as Mobile 2.0. I describe it fully in Chapter 14.

This paradigm aligns very well with my previous action items for operators, one of which was to increase the modes of connectivity available to users. IMS provides an ideal architecture for this to happen. Whether or not it will herald a new phase in the evolution of mobile remains to be seen, but the analogy with Web 2.0 is irresistible. We shouldn't forget though that Web 2.0 was identified as something that was already happening with web trends. Mobile Web 2.0 and Mobile 2.0 are mostly speculative.

3.9.3 Content Trends

Content is still important in the world of digital services. For entertainment, content is still the dominant factor for success, even as content increasingly comes packaged with various interactive services, such as blogs, wikis and forums. As Figure 3.3 shows, consumers receive content over multiple channels. The broadcast channel, in its various forms, remains important and lucrative, although it has lost some of its shine, mostly due to increase competition from the web channel – surfing the Web has replaced channel hopping on the TV.

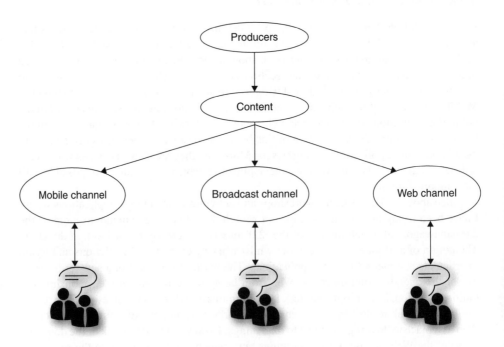

Figure 3.3 Content 1.0 – producer-fed channels.

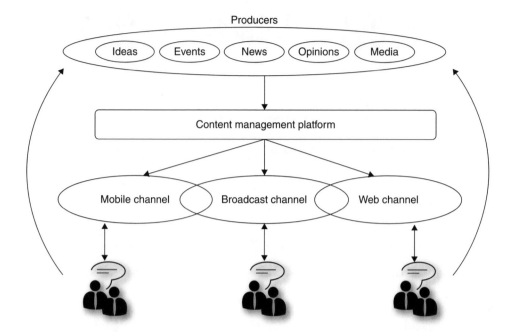

Figure 3.4 Content 2.0 – participative multichannel services.

As Figure 3.3 also shows, there is a general tendency for content to come from very few sources (at least in terms of distribution) and there is strong element of relatively few producers (in relation to users) dictating what content is produced. However, whilst the content might come from common sources, even across all distribution channels, the channels themselves remain relatively independent. There are very few services integrated across all the channels. An integrated multichannel model, as shown in Figure 3.4, is increasingly making more sense and is starting to emerge.

What Figure 3.4 shows is the trend towards multichannel services. This trend is often referred to as *convergence*, although the term is used to mean many things. Multichannel is a step further than triple play or quad play provision. Thus far, most of the multiplay service bundles are only integrated in terms of price bundling via a single bill. Our concern in this discussion is how content is used across multiple channels as part of the user experience. For example, when users buy access to a film on their set-top box, they are given access to an accompanying item of mobile content, such as a ring tone.

The additional trend is the content production itself, which is gradually extending to include content produced by the users, often called *user-generated content* (UGC). What typically springs to mind is photo or video sharing, such as services like YouTube. However, being able to share photos between users in a more automated fashion is desirable, particularly between extended family members in different households. Smart hubs in the home, such as routers and Femtocell access points can support peer-to-peer services between associated network storage devices (i.e. USB hard disks). This technology could be used to share media files between users.

With multichannel and UGC to think about, the need for common content management platforms becomes increasingly necessary. This trend is already underway.

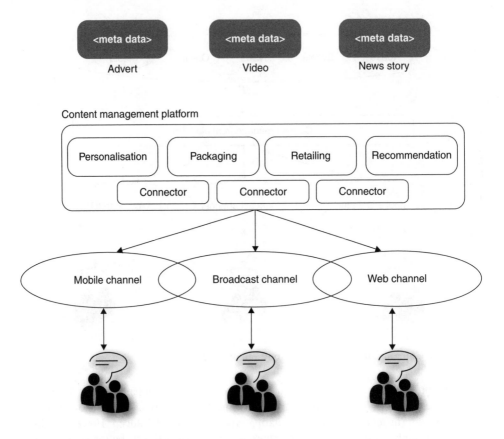

Figure 3.5 Common content platform.

As shown in Figure 3.5, platforms that can manage content across multiple channels and services can be extremely powerful in terms of supporting common service elements, like personalisation, packaging and recommendations. However, no single system exists that can provide every element of content management in the workflows for each channel. Therefore, the platforms must support the ability to connect with a variety of other systems that manage various parts of the channel workflows. The most important function of the common platform is the ability to understand all content flowing through the entire ecosystem, which means the ability to ingest, track and process huge amounts of metadata associated with the content. Of course, for UGC, there is a very real problem with the non-existence of metadata or the unreliability of user-generated metadata (e.g. tags). A user could well upload a picture of a set of false teeth, using the keyword 'Jaws'. Clearly, we don't want that content ending up being associated with *Jaws* the movie.

Various multichannel architectures are possible, especially with legacy services and systems to consider. However, the move towards IP in most services is providing a huge array of possibilities. The challenge for most operators is how to manage the complexity whilst still maintaining the agility goal identified earlier. This is not easy.

3.10   HARNESSING THE TRENDS

I have introduced many ideas so far in this chapter. Many man-years could be spent trying to construct a strategy and an action plan. We started with a very mobile-centric view, which is the natural starting point for this book. However, the view has blossomed out into a much bigger sphere of concern. If we consider the goal of increasing the modes of connectivity for mobile and combine this with the convergence trend, perhaps strongly influenced by the Web 2.0 paradigms, then we have a very different beast from what we have been used to historically in the mobile world. The trend away from talking and messaging continues. Let's summarise how operators can harness these trends:

- *Develop a 'Web 2.0' culture* – despite the vagueness of this term, let's consider it more as a 'way of doing things'. What seems to matter is that the Web is a place where new ideas can surface very quickly and attract a lot of user attention. Within the hive of activity on the Web, there are a lot of ideas, discussions, attitudes and even more tangible assets (like open-source code galore and web services APIs, e.g. Google Maps). Somewhere in this hive of activity, there are many ways to add value to operator services – you need to figure it out through innovation. If you wait for something on the Web to materialise in terms of a proven way to increase ARPU, then you might be too late. Think in terms of becoming 'Operator 2.0'. Go ahead; brainstorm it today throughout your organisation.

- *Adopt a service-orientated approach towards architecture* – the concept of a service delivery platform is also vague and can be very broad. Don't get bogged down with the terms. Think in terms of common architectures and distributed (i.e. message based) transactions. What does this mean? Think of everything the user does as an event and make sure that these events can be utilised to drive any process you might want to deploy across your entire multichannel ecosystem. For example, a user pushes a button on their mobile phone to cast a vote during a particular TV programme, which somehow tells us that the user likes female rock artists. Make sure that you can use this information to personalise, advertise and sell any service connected with female rock artists.

- *Focus on core themes* – you decide which themes you want to exploit, but I suggest basing the decision on competency rather than pure economics. Just because social networking looks good on paper in terms of revenue potential, doesn't mean you should do it, especially if you can't bring the right level of competency to the task. I would say that my first point about culture matters. If there are too few people in the organisation who really have their fingers on the pulse of a particular theme, then it would be a mistake to chase it. Culture change precedes a lot of other stuff you could do to become an Operator 2.0.

- *Accelerate the multichannel service experience* – this is simply inevitable, especially as certain services start to take hold, like Mobile TV. We should never forget that usage of consumer technology is essentially task-driven. A user doesn't want to *use* Mobile TV – they want to *watch* TV. As engagement with content becomes more interactive and more intimate, users will want to carry out the interaction and engagement tasks more frequently and efficiently, even when moving from one delivery channel to another.

3.11 CONCLUSION

In conclusion, the trends in digital services and technology are already combining across multiple channels to enable new modes of user experience. These modes in the mobile world are perhaps unknown or speculative, but they are still inevitable; thus, forcing operators to become an Operator 2.0. I don't have a concrete definition for what an Operator 2.0 looks like, but most of this chapter has outlined a set of trends and actions that can drive the creation of Operator 2.0 templates. There will be more than one of course. Some will fail and some will succeed. Innovation remains a key factor, which means the ability to understand the trends, understand the new technological possibilities, and understand new ways of thinking about value in an IP world and then converting this into compelling user experiences. The challenge is to do this in an agile fashion. It is not easy, but the types of services possible are tremendously exciting and undoubtedly lucrative for someone.

4

Introduction to Mobile Service Architectures and Paradigms

In this chapter, we examine the various ways in which to build a mobile service from end-to-end. We shall look at the main architectures and paradigms in preparation for the rest of the book where we shall elaborate on the various components.

4.1 POSSIBLE APPLICATION PARADIGMS FOR MOBILE SERVICES

Identifying what exactly constitutes a mobile application is increasingly difficult to say. Could it be defined as anything that runs on a mobile device? This definition isn't ideal because pages that run 'in' browsers don't really run 'on' a mobile device. Also, many of the service enablers, as discussed in the Chapter 3, don't run on mobile devices at all. How about any application that is accessible from a mobile device? Is this a good enough definition? This brings us back to the browser and accessing websites not designed with mobile in mind. Do these count as mobile apps? This is probably where I draw the line for his book. Here, I'm mostly interested in applications that are clearly intended for mobile – 'designed for mobile'. Given what we said earlier about the importance of a compelling end-user experience, this is really only possible by intending to design a service to exploit fully the technologies and potentials of mobile devices and their supporting ecosystem.

As Figure 4.1 shows, we have our roaming mobile user who wants a useful service (granted, they might not know what they want) and we have our service provider (in the

Figure 4.1 How do we connect the two ambitions?

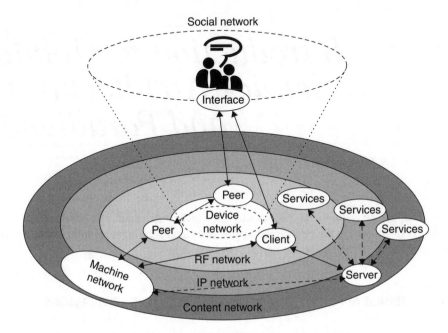

Figure 4.2 The possible network topologies for mobile services.

general sense[1]) who wants to provide a useful service? What, as the question mark is indicating, do we build in the middle to marry the intentions of these two parties? The first line of enquiry is to consider the essential nature of mobile services; so that we have an idea of what it is we might be trying to build. What are its parameters? What technologies are available? What architectures make sense? These are the kind of questions I want to start answering in this chapter, which I then go on to fill out the details in the rest of the book.

If we first look at the possible high-level architecture of any solution, Figure 4.2 reminds us of the basic nature of our mobile services ecosystem. Whatever service we build, we have to traverse some, or all, of the layers of networks moving out from the device network to the content network. However, the ways of traversing the layers in this networked model are evolving all the time. For example, the dominant means of implementing services has

---

[1] The use of the term *service provider* indicates any agent providing a mobile service. It is not restricted to the more traditional definition within the mobile telecoms markets.

been the classical client–server approach adopted from the world of networked computing[2]. The Web has taken this a step further with the advent of the browser *universal client*. This approach had been adopted by the WAP[3]. However, we now have new technologies to enable powerful software applications to run directly on the mobile devices (e.g. Java made available for small devices, called J2ME[4]) and we have new ideas concerning software architectures, like P2P. We also have protocols that are new to the mobile world, like SIP, which we discuss later in the book.

In this chapter, I aim to introduce the various possibilities for building services based on the network model shown in Figure 4.2. As indicated in the figure, there are three main paradigms for mobile software:

Mobile application paradigms:

1. Client–Server (CS)

2. Peer-to-peer (P2P)

3. Standalone

These are mostly distinguishable in terms of where the primary intelligence in the application resides. For CS, it is in the network and on the server. For P2P, it is on and shared across devices. For standalone, it is on the device. This doesn't mean that combinations can't exist. They often do. If we get too bogged down in how we define architectures, we shall get lost in a fuzz ball of semantics.

CS is the classical architecture for many services and there are good reasons for it. By the end of the book we shall have examined all its facets in detail, especially *within the mobile context*. Notably, this architecture usually assumes a continuously available connection between the device and a remote server in order for the service to work. In some geographical regions, this assumption is unwise, but the evolution of 'always on' packet mobile networks is the underlying reason for moving towards a CS approach.

P2P is an emerging contender for enabling mobile services, but it hasn't yet gained any significant attention in the mobile community, except for various prototypes and academic projects. When I wrote the first edition of this book I had more faith in P2P as a likely contender for mobile applications, but nothing much has happened. P2P continues to be a poorly exploited architecture in general, even on desktop PCs hooked to the Internet. I still believe in its potential because of some unique advantages in terms of privacy and trust, but most devices today don't readily support P2P architectures (due to lack of APIs).

Finally, the standalone paradigm is the most straightforward to understand, although not necessarily to implement because of significant porting and distribution problems. The application resides on the device and provides useful functions to the user without the need to be networked. A good example would be a game. Of course, standalone applications, also

---

[2] The client/server architecture arose out of the multi-user computing industry and has been adopted for many distributed computing solutions on The Internet, but it was not The Internet that led to the client/server approach – it had already emerged as an alternative to the mainframe or minicomputer approach.

[3] http://www.wapforum.org/

[4] J2ME, stands for Java 2 Micro Edition and we discuss it later in the book.

known as embedded applications, can also exploit the network and P2P communications, should that be useful.

These operating modes are the major paradigms available for implementing mobile services, but we are not limited to using only one of them. It is worth keeping this in mind. Of course, hybrids are possible, but it might also be the case that a particular service comes in different flavours, such as CS (i.e. browser-based) and embedded (i.e. J2ME-based) in order to ensure wide device coverage.

In terms of hybrid approaches, then perhaps a game operates most of its lifetime in standalone mode, but occasionally it networks with another user to enable collaborative playing or sharing of some game assets. Even in that sharing mode, it could be done via the content layer, where related game content resides, or it could be done P2P, with, or without, the need to cross the IP network to achieve the necessary peering infrastructure. This shall become apparent when we look at networking modes, such as infrastructural versus proximity (e.g. Bluetooth).

Because of the wide expanse of options, we should know about and understand the many different approaches to service implementation. This book addresses how to understand the wide range of possibilities. Many mobile services will involve a combination of programming techniques across all layers of the concentric networks model.

Even a standalone application that runs only on the device (e.g. a single-player game) and subsequently never needs to interact with a back-end server, or another peer, will probably be deployed using an OTA download mechanism (as discussed later in the book). In other words, the application itself would have originated from the content network and has to be sucked down to the device before it can be utilised, so knowledge of more than one paradigm is still necessary. However, the production of purely standalone applications seems more and more unlikely. If multimode design is made easy, then most developers and service creators will find an excuse to incorporate powerful mobile networking options wherever they can. For example, even an innocuous single-player game could become a shared network experience with a little bit of imagination. The incorporation of networked concepts in any mobile service will follow from the trends with Web 2.0, especially the mash-up approach (see later).

In addition to the networking paths between user and content (or other users, or machines), the network model in Figure 4.2 shows us how applications running in the network can also draw upon services from any layer, one or all of them. Of course, these services are themselves software applications and we very much need to examine their natures, so that we know what is available to empower our mobile service creation and how that empowerment takes place – how we plug our mobile application into these service-access points. As an example, we can draw upon telecoms services from the RF network layer, such as the ability to initiate a third-party call[5]. We could call upon services in the IP network layer, such as a Public Key Infrastructure (PKI) solution or some kind of payment gateway, or increasingly any of a growing number of web-based services with open interfaces. We could even call upon services offered by other devices, such as shared diary applications or shared contact books[6]. All these services are supplemental to our own service. We are not talking about having to develop a shared calendaring solution or third-party calling server

---

[5] A third-party call is placing a call between two (other) parties.
[6] See the concluding section of this book for a discussion on sharing diaries within a location-aware application context.

as part of our end-user service offering. We simply want to be able to access such facilities as if they were subcomponents in the application or service. All of this is possible with next generation mobile services and is made even more powerful by having more than one application paradigm at our disposal and the open-service trends that characterise Web 2.0.

4.2 MODES OF MOBILE INTERACTION

It is possible to identify several distinct modes of actor interaction in a mobile service. There may be others, but I suggest there are four principal ones that cover most scenarios for mobile services:

> **Possible modes of mobile interaction:**
>
> 1. Human to Human (H2H)
>
> 2. Human to Content (H2C)
>
> 3. Human to Machine (H2M)
>
> 4. Machine to Machine (M2M)

We can view these modes as being typed according to actors: human, content or machine. In all cases, at least one of the actors is *roaming* or *able to roam*. It is also possible that there are a multitude of actors involved, such as several humans interacting with the same machine in the case of shoppers using a mobile coupon fulfilment kiosk[7]. An example of a roaming machine is a telemetry device attached to a goods delivery vehicle. Other examples will be considered later on.

M2M is an important type to include in our list as it reminds us that networking, and therefore mobile services, is not restricted to humans alone. Anything that can 'talk' via a networking protocol that will work over the RF network can become an active participant (actor) in a mobile service. In many cases, the ability for a machine to roam and remain networked is a powerful enabler. They may want to report their activities to other machines that move goods or static machines that want to track the movement of goods. Even goods themselves could be enabled to talk in a mobile network by virtue of RF-enabled packaging[8].

In the case of H2H, this is clearly a social mode of interaction, but it does not have to be real-time, like making a voice call. The interaction can be asynchronous. This means that both parties need not be involved in the communications process at the same time. Leaving a voicemail message is a prime example of asynchronous interaction – we don't need the recipient to be available. Conducting an IM session is an example of real-time synchronous communication where all parties (two or more) must be engaged in the process at the same time. However, IM can be both H2H and H2M. The former is the familiar person-to-person chatting via IM. The latter might be using IM to talk to an *intelligent agent*, such as a bank. For example, we might ask the 'bank buddy' for our current balance.

---

[7] A mobile coupon kiosk is a coupon-printing machine in a supermarket that converts e-coupons held in a mobile device into laser-scan compatible coupons (i.e. printed paper coupons).
[8] This can be done by inserting an RF-device into the packing carton.

The fact that we have synchronous and asynchronous modes of H2H communication is important. It implies that we need our mobile ecosystem to support real-time and non real-time communications. Any synchronous communications requires a dedicated pathway with a minimum level of quality (Quality of Service, or QoS). The infrastructure required to do this can be found in the RF network as well as in the IP network, but they have different characteristics. Accordingly, they also have vastly different implications for how we build our service application and this will become abundantly clear when we discuss these parts of the network in the book.

As we shall see later in the book, when we look at location-based services (LBSs), the ability to cope with asynchronous events is a major challenge and a critical need for LBSs. Consequently, we require software processes and architectures suited to processing asynchronous events. We can use an architecture called a *messaging bus*. Otherwise, the design prospect of keeping all our software processes in lock-step (synchronised) across the entire mobile ecosystem (a 'network of networks') is too daunting. Fortunately, a similar challenge has existed for some time in financial processing ecosystems. This led to the development of a messaging-orientated (transaction-based) software architecture. You can think of this as lots of disparate software processes that communicate to each other using their own equivalent of email. We shall look at this later when we look at Java enterprise technologies.

H2C interaction involves a human wanting to gain access to any kind of content, the nature of which can be quite diverse. This could include anything from train timetables to pop-group pictures used to decorate the device display (wall paper). There are a number of architectures for implementing H2C interaction. The Web is the most obvious one and we shall examine this in depth. However, web browsers aren't necessarily the best way to handle content, so we shall look at other possibilities.

4.3 MAPPING THE INTERACTION TO THE NETWORK MODEL

In any of the identified modes of interaction, as our diagram in Figure 4.3 indicates, there are at least two end-points, sometimes more (as in group chat applications, videoconferencing and the like). As the diagram also shows, we can imagine that the end-points are participating sources of activity and information; they each have something to 'say' to each other, and so, in that sense, they can be viewed as actors in a *sequence of events* that takes place to complete meaningful tasks within the mobile context. The use of the term actors will be valuable in the following discussions, as an anthropomorphic[9] generalisation of entities as 'people' that can talk to each other is a useful analytical technique. *Actor* is also a term that crops up in formal software analysis techniques, like the increasingly popular UML[10] notation.

Importantly, the sequence of events and participating actors is not confined to the two end-points, which themselves may change throughout the life cycle of the service. As Figure 4.3 shows, quite elaborate interactions can arise during the life cycle of a single session of tasks

---

[9] Ascribing human characteristics to nonhuman things (except for our really human actors).
[10] UML stands for Universal Modelling Language and is a means of thinking about how to analyse a proposed system for the purposes of writing down its required behaviour for subsequent implementation using software technologies. See http://www.uml.org/.

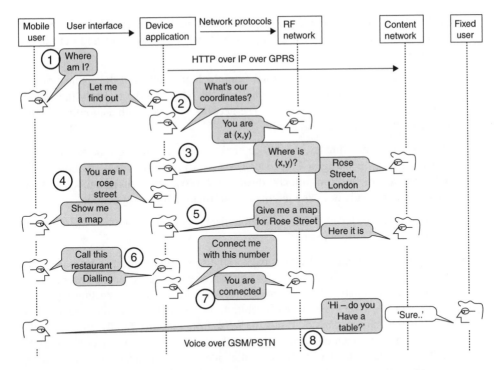

Figure 4.3 Sequence of events and participating actors in a sample mobile service.

carried out by the mobile user. It is worth dwelling on this example sequence chart[11] as it uncovers a lot of the issues that we need to address in building our mobile services.

The way to read the sequence chart is to follow it from top to bottom, which is tracking the flow of events in time. The basic intervals of the dialogue have been numbered to make it easier to follow. This sequence is a fictitious example of a user querying a location-services application on their device to find out where they are. They subsequently request a map and then notice a restaurant nearby. Clicking on the number displayed for the restaurant causes a phone call to be initiated to the restaurateur and a table is reserved. Sounds exciting and already has my mouth watering! Later in the book, when we look at the RF network and LBSs (Chapters 12 and 13, respectively), we shall see how we might implement such a system.

By the time you have finished reading this book, you should understand how every aspect of this sequence could be implemented. We can follow the number labels on Figure 4.3 to see what's going on:

1. The user makes a request to the device application via its user interface. The application could be an embedded application installed on the device (we shall look at this later in the book when we look at devices in Chapter 10) or it could be a browser. Let's assume that it is a Java application (implemented using J2ME). Within the Java environment on the device, we assume that there is a small program (an

---

[11] This is not a sequence chart in the formal sense in which they get used in UML notation, or any other notation, but illustrative of the concept.

API[12]) that is able to make location queries from the RF network, the network itself being more than capable of tracking device location, as we shall find out later in the book.

2. In response to the request for location from the higher layers of the application, the location services API on the device talks to the RF network, probably the Mobile Location Gateway (see Chapter 13). The network responds with a coordinate pair[13] to say where the mobile is situated.

3. Having got the coordinates, the device application then talks to a back-end server, so we are now operating in CS mode. The server is a location-information service delivery platform and can turn coordinates into geocoded[14] information, like street names, which it duly does and returns the information to the application.

4. The user interface layer of the application displays the information to the user, who subsequently requests a map.

5. The application on the device goes back to the server and requests a map. As in the previous client-server interaction, this connection relies on traversing several layers in the network. Firstly, the application talks directly to the server using Hypertext Transfer Protocol (HTTP), which we introduce in Chapter 5. An alternative would be to use the Wireless Session Protocol (WSP) from the WAP family, and this would have introduced yet another actor – the WAP gateway (proxy) to convert optimised[15] WSP messages to standard HTTP ones that our server can understand; HTTP being the lingua franca of most content servers deployed on IP networks (i.e. web servers). The HTTP messages are passed courtesy of a lower-level protocol called Internet Protocol (IP), which in turn is able to provide packet-based communication courtesy of the mobile data channels established by the RF network, in this case we assume GPRS (General Packet Radio System, which is an adjunct to GSM).

6. After fetching the map via HTTP from the location-information server, the user notices a nearby restaurant. Fancying a bite to eat, the user clicks on the phone number for the restaurant, which is hovering above the map, and the dialler application in the mobile device is initiated and starts to dial the requested number. This step assumes that numbers can be dialled programmatically from within an application. This is something we shall understand in more depth when we look at devices in Chapter 10.

7. The dialler calls the low-level call-handling program on the device, which in turn initiates communication over a signalling channel on the air interface, requesting a voice-call channel to the designated number. The dialler application gets a confirmation that the signalling was successful and a 'ringing' state event is passed back up

---

[12] There is an activity within the Java Community Process to define such an API, called 'JSR-179 Location Services'. See http://www.jcp.org/en/jsr/detail?id=179.

[13] As we shall find out, a coordinate pair by itself is unlikely or not that useful. We most likely also need, and would expect, a radius of uncertainty that we can rotate around the coordinate pair to define a zone that the device is probably located within. This is discussed later in the book when we look at location-based services in detail.

[14] Geocoding is turning any coordinate information into an alternative geographical reference that is more suitable for subsequent processing, such as postal codes.

[15] We shall look at the natures of the optimisations that WSP offers when we discuss IP-based protocols for mobile services later in the book, but WSP is essentially a compressed and more compact version of HTTP.

to the dialler, which interprets it graphically on the user interface and via an audio indication.

8. Finally, the restaurateur answers the destination phone and a voice call circuit is established and the user is now in a voice call and asks for a table to be booked.

As we have seen, for what appears to be a simple task from the user's perspective, we have used a lot of powerful resources in our mobile services network, at all layers in the network. There are interactions between layers, so this implies that we need defined interfaces. Furthermore, these interfaces need to be programmatically accessible so that our software can take advantage of them, the concept that we first showed in Figure 4.2 where we indicated that applications can call upon services from each layer in the network.

The availability of interfaces is a key area of understanding that we need to develop in our role as service creators. In the above sequence we can conceive of various enhancements to the mobile service. For example, we could have had the restaurant send a text message to the user to remind them later on that their table reservation time is approaching. This helps the customer to remember their booking and it helps the restaurant to ensure that people turn up on time, or at all, both being good for business. Let's say we wanted to implement such a feature, how would we do it?

The most crude and least automated way is for the restaurant to send a text message from a mobile phone at the appropriate time (manual method). Another way is to enable the restaurant to schedule a reminder at the time of taking the booking. This has the added advantage of being an activity that can be incorporated into the booking process. There are several ways we can tackle this problem. If the restaurant has a PC-based booking system, it could use an interface to send a text reminder from the diary, sending out reminders at the appropriate time. We could use a mobile device, like a phone, attached to the PC to send the message. Alternatively, we could submit a message via the Internet to a messaging gateway application, which either the RF network operator provides or a third party who knows how to link the gateway up to the text-messaging infrastructure in the operator network. Yet another option is to host an entire booking system on the Web and include a messaging alerts service.

This illustrates that there are often many ways to implement a mobile solution. The different methods will typically involve different actors and network services. When designing mobile services, it is crucial to understand which resources are available to us and how can we take advantage of them. Because there are many ways to build an application, it is important that we try to survey as many as possible. This book helps by trying to give as wide view as possible of the mobile services landscape. There are already many books that specialise in WAP, J2ME, and so on, but very few that tell us about the bigger picture. I hope this plugs the gaps for you.

4.4 MOBILE INTERACTION IN THE MOBILE ECOSYSTEM

Figure 4.4 shows our mobile ecosystem and the locations of each of the possible actor types: human, machine and content.

Let us now examine the layers in the ecosystem and how they support our primary modes of interaction: H2H, H2C, H2M and M2M.

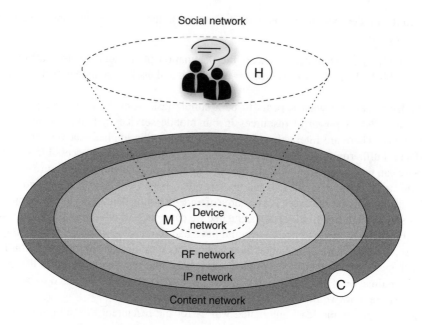

Figure 4.4 Positioning the actors in the mobile ecosystem.

4.4.1 Social Network

The social network is the people who use the mobile services. By explicitly including them in the model we are acknowledging and emphasising that people who use mobile services exhibit all forms of social interaction. A social network is a form of communications network in its own right. We often forget this.

The social network is the highest layer in our network, or its inner core, depending on how we view things. Either way, it is a sensible idea to give due reverence to the users! The inner core is a useful perspective to reinforce the notion that the users are at the centre of our mobile services cosmos. Users use (and abuse) the system and pay to do so. They are its lifeblood, so they deserve respect. In terms of software applications, it would be more usual not to include the social network as part of the model, but we are concerned with the higher abstraction of mobile services where we expect services to be more intimately tied to social habits given the intimate relationship between a user and a device that they carry all of the time, perhaps even sleep with[16].

The reason for including the social network is to underline that services are about people, except for the class of service M2M, where there are no people directly involved in the interaction. People will be the eventual beneficiaries, but not primary actors in an M2M service.

We can think of mobile services as social-networking tools, as facilitators, magnifiers and multipliers of social interaction. It is important to reiterate our discussion from the book's introduction (Chapter 2) wherein we posited that mobile technologies have

---

[16] Sleeping with a mobile next to the bed is a widespread habit. Some users will even wake up to answer text messages!

social-shaping powers[17]. New ways of communicating, interacting, socialising and living become possible and in ways that can become self-reinforcing. News ways of doing things will bring new, possibly unexpected demands for new technologies and services, which in turn will bring about new ways of doing things. This is what I refer to as the *virtuous circle of ubiquity*.

If we look at the text-messaging phenomenon that originally blossomed in GSM networks throughout Europe, particularly Scandinavian countries, then many observers recognise this as being a socially significant occurrence. Since the advent of the mechanical clock, time became a tool for social coordination, more so since the advent of portable time keeping (watches) and faster means of transport, which combine to make it possible to schedule social activities with a high degree of accuracy. The availability of mobile telephony, particularly text messaging, has made time more elastic again. As noted in their paper 'Hyper-coordination via mobile phones in Norway', by Ling and Yttri[18]:

> *Owing to the recent yet explosive growth of mobiles, it is quite noticeable as a cultural phenomenon. . . . [one] version is the 'softening' of time; for example, sitting in a traffic jam and [texting] ahead to the meeting to let them know that you will be late.*

The other reason that we need to include the social network is to understand that it has its own ability to move information sideward without the other layers being involved. This may be useful to study and understand. People meet and exchange information, such as reading out text messages, showing each other picture messages, swapping telephone numbers, and reporting locations where 'air graffiti' notes[19] are hanging in space, and so on. This P2P interaction is direct and we can think of it as the most basic method of achieving a H2H service, albeit limited. It is something that occurs naturally in any case and increasingly involves mobile technology. Some commentators have argued that there is something called Mobile Web 2.0, which is characterised by social networking. However, it seems clear to me that social networking has been a central feature of mobility for some time, ever since users started swapping texts and contact information, which happens in abundance. How many new social bonds must have been formed in this manner?

As we will consider in one of the following sections, the H2H interaction that occurs naturally can be assisted using devices, even without the other layers being involved (e.g. using Bluetooth).

4.4.2 Device Network

The device network comprises of all the devices that can attach and communicate via the RF network(s). They are the primary access points into the mobile service world and represent the main interface between the actors and the available services. However, as Figure 4.4 shows, it is entirely possible that a device is autonomous of any other actor and behaves

---

[17] It is interesting to note that even the possession of a mobile phone has social impact, such as their iconic status in some post Eastern-bloc countries and in places like Hong Kong and China.
[18] Ling, R. and Yttri, B. 'Hyper-coordination via mobile phones in Norway'. *Perpetual Contact – Mobile Communication, Private Talk, Public Performance,* pp. 139–169. Cambridge University Press, UK (2002).
[19] We shall examine the concept of spatial messaging later. 'Air graffiti' is a slang term that denotes how the idea of leaving a message pinned in mid-air (symbolically) is akin to daubing graffiti on a wall. Of course, this is a very narrow view of the potential of spatial messaging.

as a machine in the ecosystem, capable of communicating with a human or content source elsewhere in the network. The devices themselves have an intranetworking capability. They can network with each other, which is how the M2M mode of interaction is supported.

> **Device networking consists of three modes:**
>
> 1. *Client Domain (CD)* – between the device and the RF network to access other network
> 2. *Social Domain (SD)* – between devices belonging to different users
> 3. *Personal Domain (PD)* – between devices all belonging to the same user

The CD mode of networking will be commonplace in many mobile services. It is using the RF network to achieve nomadic or ubiquitous access to some other networked resource, such as another user (fixed or mobile), some content or a machine, all of which can be remote from the user.

Within both the SD and the PD, devices will probably interconnect using similar technologies, such as Bluetooth, infrared and WiFi. SD networking includes swapping contacts (e.g. phone numbers or other personal contact information). This used to be called 'beaming' in recognition of its light beam origins (infrared).

With the advent of low-power, cost-effective and short-range RF communications devices, it is no longer necessary to restrict interdevice communication to line-of-sight physics. This makes all kinds of exciting applications possible such a multiplayer gaming or proximity sensing[20]. Moreover, the ability to connect without regard for device orientation and positioning means that it is easy to connect personal devices to each other without requiring clumsy effort that puts users off. This means it is also easier to interface several devices within a PD, as there is no need to line them all up, nor be restricted to face-to-face alignment, which tends to preclude the use of more than one device pair. The introduction to this book talked about PD communications and applications, and the idea of personal mobile gateways. Later sections in this book will expound on all these topics in more detail when we look at devices in more depth (see Chapter 10).

The fact that we may have devices talking directly to each other, whether in a SD or PD context, will have implications for our software architecture approach and which software technologies and interaction mechanisms we need. If a user has a sleek wireless watch, a keyboarded smart communicator and a tablet PC, then some degree of coordination between the devices may be necessary. For example, if the user is accessing email via their tablet PC, there is no need to send email alerts to the watch. Similarly, all SIP-based calls could be routed through a SIP client on the tablet PC, perhaps itself on a WiFi link, with audio being handled via a Bluetooth headset. Another consideration is information synchronisation. Perhaps a contact is entered onto the tablet PIM application and this should be automatically available the next time a call is placed via the watch.

---

[20] For example, a Bluetooth network within a supermarket could be used to discover the presence of a mobile device.

In the SD networking scenario our software solutions need to include appropriate protocols to allow devices to talk to each other directly, possibly without any intervention at all from another networked resource. This may have implications for such considerations as security, making sure we have an appropriate mechanism to authenticate a user from one device to talk to a software service on another device. If this is going to use the same authentication principles used for network authentication onto the RF network (e.g. via a Subscriber Identity Module – SIM) then we need to ensure that we have the necessary provisions on the devices to access these authentication apparatus. Under normal circumstances, such authentication mechanisms may not be accessible for localised interdevice communication; they may ordinarily rely on networked resources, such as an authentication server. This has implications for our software architecture, as it means that we need a wide-area RF connection to achieve authentication.

4.4.3 RF Network

In this book, we shall consider two types of RF network: ad hoc and permanent. This is to enable us to distinguish CD device communications from the other domains.

> **Two types of RF network[21]:**
>
> 1. *Ad hoc* – exists for the duration of the communication only
>
> 2. *Permanent* – always exists, thus implying the need for permanently installed infrastructure

For devices that wish to talk to each other locally in the PD or SD configurations, I refer to this as an *ad hoc*[22] RF network. This alludes to the nature of the network being transitory and only coming together for the short time it is needed and without the need for any permanent communications infrastructure, as indicated in Figure 4.5. The network arises for a period of use by virtue of devices talking directly to each other in sufficiently close proximity[23], and then the network dissolves.

Ad hoc networks are continually set up and torn down as needed. In the case of social networks, devices cluster together to talk and then disperse again. With personal networks, the devices may well be within range for a lot of the time. The network is set up and negotiated between devices on an as-needed basis. We should not confuse this method of RF networking with the previously introduced method of software collaboration called P2P. This is a common confusion. As we shall see later in this chapter when the modes of communication across the network layers are discussed, P2P does not have to involve devices that are in physical proximity and devices talking directly[24] to each other.

---

[21] Note that I do mean network here, not connection (or session).

[22] Note that ad hoc networking can have particular meaning in certain RF networking discussions and contexts, but I am not referring here to any specific technique or networking technology.

[23] The range may vary depending on the technology being used and the hardware implementation, but this shall be discussed later.

[24] By directly, I mean that the RF propagates physically from one device to the other (and vice versa).

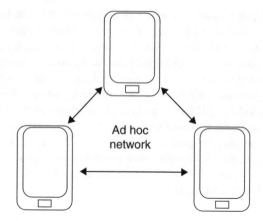

Figure 4.5 Ad hoc RF network.

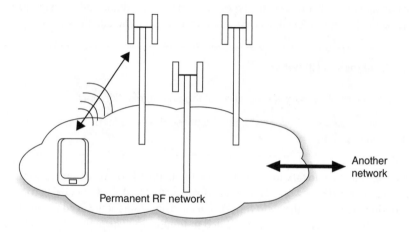

Figure 4.6 A permanent RF network.

I refer to the other type of RF network as *permanent*[25], requiring an installed infrastructure. The network is always available, always accessible and is ordinarily a conduit to other networks as indicated in Figure 4.6. In the early days of mobile telephony, the RF network was a conduit to the Public Switched Telephony Network (PSTN) system. As we shall see, with modern (second and third generation) cellular networks, the permanent RF network is still a conduit to fixed telephone networks, but is also a conduit to IP-based computer networks, especially the Internet and the World Wide Web.

It is tempting to think of the RF network as just a conduit and nothing more, as if its design objectives were to be a faithful representation of a robust wire-line connection (e.g. Ethernet CAT5 cable) but without there actually being a wire of course, as shown

---

[25] Although we recognise that RF network infrastructure, like GSM, can be temporarily deployed, such as at sports events or conferences. However, there is still the need for infrastructure, even in this case, to be permanent; if the infrastructure is removed, the devices cannot communicate.

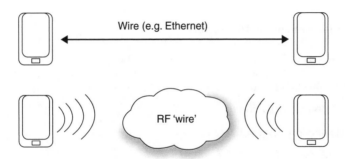

Figure 4.7 RF network as replacement for wires.

in Figure 4.7. However, as we shall see when we discuss the characteristics of the RF network, it has other powerful capabilities that unintentionally (and recently intentionally) have become quite useful to third parties, such as the ability to locate a mobile device (and presumably its user) or to deliver a text message from a non-mobile application.

In fact, it is quite deceptive to think of an RF network as being a wire-line replacement, the problem in perception probably coming from the use of the word radio frequency, which tends to elicit a mode of thinking based on the 'wireless' physics of connectivity and nothing more. It turns out that the cellular network has many components in its infrastructure and is a substantial intelligent network in its own right with powerful capabilities, some of which are listed below. The idea of the RF network as a powerhouse of assets is shown in Figure 4.8.

Some of these cellular network assets often go unnoticed, but are already important in mobile communications and will become more so for next generation services. Many of them relate to voice telephony, but voice has an important role to play in many applications, not just making calls.

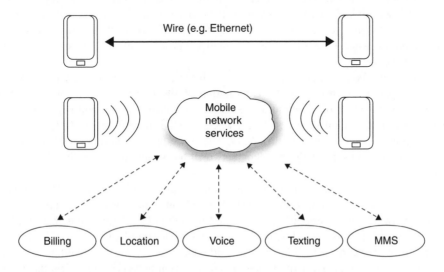

Figure 4.8 RF network as more than just wireless.

Other features of the RF network:

- Voice mail

- Text messaging

- Multimedia messaging

- Call handling (diversion, etc.)

- Call conferencing

- Billing

- Fraud management

- VPN connectivity

- IP address assignment

- Charging

- Authentication

- Location finding

- Customer care

It may seem strange to mention all the various cellular network features, especially items like customer care, which almost seem like incidental features of a cellular business rather than a feature of a cellular network. However, our minds should not become permanently focused on *applications*; rather, we need to think in terms of *services*. Taking the example of customer care, then it is perfectly feasible that we could develop a mobile application that took advantage of an existing customer care regime under the management of the cellular operator. If we think along these lines, we can envisage being able to submit customer care reports electronically so that customer care agents are able to understand in what state our application was when a user experienced difficulty (in the case of customer care manifested as technical help).

Let's not get bogged down at this stage with the operator-centric view of the world, which is very much driven by the exploitation of the infrastructure assets just listed, but let's instead think of all these assets as having intrinsic value and capabilities that may be useful to exploit in any mobile service, whether its offered by the mobile operator or through them by some other party. The powerbase or workhorse of our service may sit entirely outside of the operator network, but symbiotically utilise the operator assets.

The relevance to our present discussion is that we are examining what it takes to build mobile services, working our way through the possible attributes and features of the software services network (our interlocutor). We should hold the idea in our minds that all of the cellular network features could be made accessible to our mobile services, whatever they happen to be or wherever they happen to reside once we get down to building them. We will all have our own ideas about services, but what I want to do is make it clear that the RF network is rich with capabilities, which, once we gain programmatic access to them (i.e.

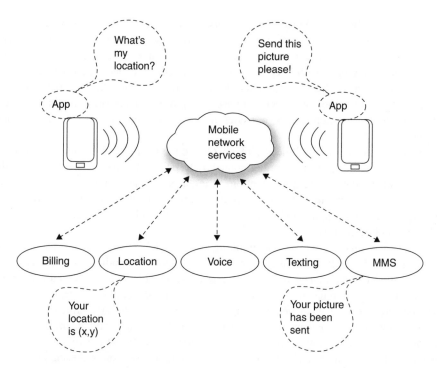

Figure 4.9 Mobile network services.

from our software applications), they potentially enable us to deliver very powerful mobile services, as alluded in Figure 4.9.

Let's briefly examine just some examples of exploiting this kind of access to RF network assets. Let's imagine that we want to use our standard voicemail system to announce that a new email message has arrived. Better still, perhaps we want to have the new email messages read out using text-to-voice technology, but integrated audibly into the menu of our standard voicemail service. This is one possible way of unifying message collection, but it needs a means of integrating into the voicemail service, which usually is provided by a voicemail service platform hosted by the RF network operator.

If the voicemail system in the RF network was open to access from our software application, then possibly we could implement the proposed email-reading system. As an extension to this service, we may like to consider the possibility of inserting a new option on the voicemail help system, like 'Press 5 for email announcements' or 'Press 6 for important messages', which would take us through to an MP3 'duke box' system on the Internet that we use to record messages from friends on our e-networking[26] bulletin board.

Historically, this level of integration into the RF network services has not been so commonly or easily available. The necessary service-access points in the RF network were not available for all (or many) of the various network assets that might be useful. Furthermore, the means of accessing these services was not conducive to popular programming and software paradigms, such as using Internet-centric models. On some networks, only the messaging centres (text messaging and multimedia messaging) have interfaces accessible

---

[26] *E-networking* is the term sometimes given to websites that facilitate business networking (i.e. meeting people).

to third-party applications. Mostly, the network resources are for use only by their owners or in the way the network owner (operator) makes them available to select customers. They are not available as configurable software services to other service providers. However, this is going to change, especially with 3G and something called Open Services Architecture (OSA) and its close technological friend, Parlay – topics that we shall discuss in Chapter 10 when we look at the RF network in depth. It is therefore something we need to bear in mind as an architectural consideration when deciding how to build our mobile services.

Not all mobile services will need direct access to the RF network resources. Some will simply make do with just the roaming connectivity power that the RF network provides. However, this is likely to be the exception. Straightforward access to a company Intranet might be such an example, although as we discovered in the previous chapter even such a seemingly innocuous application as corporate email had all kinds of exciting possibilities within the mobile context.

In summary, we have two ways of thinking about the RF network. One is as a wire-line replacement and simply a means of roaming connectivity to another resource. The other view is of the RF network as a mobility services provider, able to take care of a wider range of concerns than just connectivity. Which view we take is not simply a matter of preference. It will be determined by what we are trying to achieve, but will also be influenced by which resources (features) the owners of the RF networks make available and the means by which they are accessible. However, nothing short of total access to the network's resources is going to suffice. As we shall shortly argue in the concluding section of this chapter, mobile operators need to become adroit at converting their entire network into a services hosting environment for third parties or else face extinction.

4.4.4 IP Network

In the case of CD connectivity via the RF network, typically we are attempting to connect as a client to some networked resource beyond the boundaries of the RF network. This resource will be a computer of some description, be that a server, a desktop PC or some kind of embedded device – perhaps a dishwasher to turn it on, perhaps a remote sensing station to measure rainfall, perhaps an onboard trip computer in a car. The possibilities are vast.

Whatever the resource we wish to interact with, clearly it has to be connected to a network. There are myriad data networking solutions possible and available, but the prevailing solution is to base services and applications on the IP suite. Therefore, we shall quite rightly focus on the IP network as the next natural layer in our network model beyond the RF layer. To reach beyond the RF network, it needs to have an IP-cognisant interface. Thereafter, there is a whole world of networks to connect with, especially the Internet. In any case, this is already the natural assumption of the operators who own the RF networks. The prevailing standards for wide area RF solutions all assume interworking with an IP network, that the boundary between the RF world and the fixed one is IP based. Clearly, to ignore The Internet would be derangement in the extreme.

What this also leads to is the ability to assume certain approaches towards our service and software architectures, in particular the adoption of the Web-centric paradigm for software services and related models. At the heart of Web-based architectures is the golden protocol of the modern computing era, namely HTTP. This protocol had and still has a very simple design objective and consequently is able to provide an effective and highly scalable communications paradigm. This turns out to be useful for gathering support and momentum around web-based solutions, HTTP being relatively simple to implement. This is even truer today with so many products and tools that make it easy for software providers to adopt HTTP. The plethora of HTTP-aware solutions made available in recent times is dazzling.

HTTP is so popular that it is often used as a default communications protocol in many solutions, even ones where the objective is no longer to deliver web pages to browsers for visual consumption. This is increasingly so. For example, HTTP is at the heart of *Web Services*, which is an initiative to form a global IT consensus for how software services should talk to each other. Not surprisingly then, this model is what we see emerging as the preferred way for providing access to the mobility services in the RF network that we referred to earlier, such as the location-finding resources, text messaging centres and lots more.

HTTP is also gaining ground as an interconnection solution for M2M connectivity modes. It turns out to be relatively simple and cheap to embed HTTP communications into embedded computers, it even being possible to implement an entire web server on a single chip for a few euros. Other protocols may well make sense, but given the prevalence of HTTP support in software tools, components and technologies, it is a sensible proposition to adopt it wherever possible (provided it is appropriate, as it may not be for certain types of communication modalities, but this will emerge in Chapter 5 on IP-based protocols).

As with the RF network, it is disingenuous to view the IP layer as just another data pipe through to the target resource, which ultimately is probably going to be a database of some form or another in the classical CS paradigm, although as noted earlier, there are alternatives to CS that are gaining ground all the time, perhaps more so in the mobile arena than elsewhere. The IP layer has its own characteristics or capabilities, which we may well need to accommodate or utilise. For example, just as with the RF network, we need to consider security. There are ways of applying security models specifically to the HTTP protocol. Indeed, this has been an area of significant development ever since the Web became a widespread means of doing business and the need for secure financial transactions and secure authentication became apparent.

4.4.5 Content Network

The *content network* may seem a strange name for a network, but we should remember that the network model that we are developing here is an abstract high-level model that provides value in terms of thinking about how we build mobile services. For mobile services, we need such a model so that we can take a holistic approach to our consideration of how to utilise various technologies to deliver the desired services.

The purpose of this book is to sketch out frameworks that we can use to think about mobile services, as well as to design them. As we have already seen, the RF network has two prime capabilities. Not only can it support roaming connectivity, but also it has its own inherent resources and capabilities. In a more conventional approach to communications models, we would most likely omit these facts, which would become problematic. For example, we may postulate a service idea that includes voice or texting. Our network model, together with an understanding of the capabilities of each layer, will allow us to articulate that service idea into possible architectures; therefore, leading us in useful directions towards building the service.

Similarly, our consideration of the devices as forming a network in their own right also allows for a wider scope for mapping service ideas onto possible architectures. This also interplays with the networks that sit above and below the devices, such as the social patterns in the social network and the ability to form ad hoc networks in the RF layer. This view of devices as networks in their own right is valuable in aiding our thinking about how to build mobile services.

I am highlighting something that you may have already noticed, specifically, that the network model we have developed so far allows us to think of architectures that allow for interaction between layers as well as within layers, or any combination. This duality is important. This is also why I have elected to define the remaining layers in a similar vein, especially the content network, so that we maintain a consistency in our model; although, this particular layer has its own attributes that are worthy of note.

The content network is really the apparatus that lies beyond the IP network and which hosts the information that the client device wishes to interact with – for example, it is the user's email account sitting on an email server. It could also be the stock levels of soft drink cans in a vending machine. The types of content are quite diverse.

For any given mobile service, the content could have a single source, such as an inbox or a household security system, or could come from many sources, such as aggregated news, portal services, a plethora of networked domestic devices, or a hotel's availability service. We can well imagine that raising a question such as 'where's the nearest free hotel room' will require several databases of hotels to be queried. In such cases, it is easy to see how multiple sources imply the need for powerful networking capabilities, the means to gather the information from different sources. However, there are other issues relating to service that the content network has to take into account, such as the need to provide redundancy and other types of resilience that ensure that any given mobile service is always available or able to degrade gracefully in terms of the quantity, quality and timeliness of the information it can produce.

In our model the content network is not just the content, but it is the infrastructure required to host it and deliver it. Here we may typically think of servers, such as web servers, mail servers and database servers. These may require all manner of configurations depending on the nature of the servers and the implications of the service we are trying to deliver. Clearly there may well be a need for some servers to network with others, possibly across organisational boundaries, such as a hotel information service needing to consult directly with various hotelier IT systems. This is all part of what we are calling the content network.

The content network also includes any embedded devices that provide information, such as vending machine sensors, remote diagnostics systems, household security systems, postal delivery systems and so on. However, these types of systems, which are embedded into machines, we assign to a subdomain of the content network, called the *machine network* (which we explain in Section 4.4.6).

It is tempting to think of the content network as being passive. This comes from thinking in terms of the CS paradigm. We have already suggested that devices on the RF network will connect to the content network as clients, as indicated in our earlier introduction of the term 'Client Domain'. A client usually requests services and gets a response. This is true in many cases and is a workable paradigm for many mobile services and we will address this topic shortly; namely, how our network model supports our different interactivity modes (H2H and H2C, etc.) However, it is important to stress at this point that it is perfectly possible for the content network to initiate the interaction unprompted by the client, and this is a paradigm particularly unique to mobile and has lots of potential to enable services to work in harmony with users' needs, habits and circumstances.

The mobile content network has three modes:

1. *Pull* – the RF device initiates information requests from the content network
2. *Push* – the content network initiates information delivery to the RF device
3. *Peer* – the content network initiates information requests from the RF device

---

These communications modes may appear unconventional at first. *Pull* is the most straightforward and probably fits with our expectations from the content network. We ask for an email message and we get it, we ask for a web page and we get it, we ask for a stock price and we get it, very much a request and response cycle. These modes of interaction fit well with the CS architecture, which we shall discuss shortly.

However, *push* is perhaps not so obvious to figure out, although the concept is simple of course and easy to appreciate. Just to clarify that push is not the same as a polling technique. We may inadvertently think that some applications, like email, already use push technology, whereas they are actually based on pull technology. Email messages appear to be pushed to our mail client. However, this is not push. The email client establishes contact (polls) with the email server periodically to request (pull) the email messages to be sucked down for subsequent display. The RF network enables true push mechanisms wherein, without any initiation from the device or user, information can arrive at the device from some information source. If this information resides in the content network, or is generated by the content network, then clearly an interface is required to allow the content network to initiate push messages via the RF network.

We can consider the push mechanism(s) as a key enabling technology for all kinds of interesting mobile services, so we need to be aware of this aspect of our network model and enter it into our thought processes when we design our services.

There is yet another mode of networking to consider in the content layer. This is the *peer* mode. This is essentially like the pull mode in reverse. It is a mode where the RF device now becomes like a server, able to respond to requests for content that it is hosting itself. This could be useful in a number of services, but this approach is part of an emerging advance in software architectures called P2P computing[27], not to be confused with the H2H mode of interaction we have already discussed, which can be related, but not necessarily so because P2P computing can take place without human involvement in the information transactions.

An example of the peer mode of networking could be a software service in the content layer requesting contact information from a phone's address book, or searching a mobile device's picture memory for certain images.

---

[27] P2P computing is something more comprehensive than simply being able to assume either a client or a server role interchangeably, but we wish to focus on this interchange here and not the broader P2P issues, some of which we examine later.

4.4.6 Machine Network

The *machine network* is an oddity in terms of our networking model because there is no unified view for considering interaction with machines; it does not fit neatly into the model. It is shown as a subset of the content network because increasingly machines will become web-connected using tiny embedded web servers. Although the electronics beyond those web servers will be potentially sophisticated and certainly peculiar to each type of device involved, the web server will mask the inner nature of the machine. If we think of a home automation system and its sensory output, then requesting the sensor measurements from the system via a web server is no different to requesting a web page built using information from a database. Indeed, one way to think of connected machines in an abstract sense is as databases of real-time information.

Machine networking is becoming increasingly easier and can be done retrospectively at relatively low cost. Many machines have physical data ports, such as serial ports or similar. These can be easily converted to Ethernet ports using devices like terminal servers[28]. Once the device is on the Ethernet then it becomes easier to integrate into a local area network where a web server can reside and act as a web interface to the machine. Alternatively, the terminal server itself could contain an embedded web server. With solutions like this, many machines are potentially accessible via the Internet. Once that is possible, wireless access is achievable via the RF network's Internet gateway. Alternatively, an RF device could be attached to the machine itself, which is an especially attractive option for remote machines. However, with domestic appliances, these typically don't have any form of electronic interface other than the human one. For example, a central heating system has a temperature control, but not a serial port. Similarly, most alarm systems have access panels, but not a serial port. Networking domestic machines is a more challenging proposition that requires a deliberate attempt by manufacturers to incorporate networking options, and this is currently a slow development due to an uncertain business model for home automation.

4.5 MODES OF COMMUNICATION ACROSS THE NETWORK LAYERS

Having examined the network layers, we need to justify this model in terms of its support for our primary modes of interaction: H2H, H2C, H2M and M2M. We should understand how these interaction modes work across, and within, the layers we have identified.

Figure 4.4 shows the location of the primary actors in our network model. Let's now see how each of the layers supports these modes.

4.5.1 Human-to-Human Interaction (H2H)

H2H interaction can take place in one of several ways. Firstly, H2H can take place in the social network. We have already discussed this to remind ourselves of the potential impact of mobile services and devices as their influence extends into the social domain.

---

[28] A *terminal server* is a small device that has a serial port on one side (e.g. RS232) and an Ethernet port on the other side. Using software running on a PC, the serial port can be contacted via the terminal server via an Ethernet connection on the PC. As far as the PC is concerned, it is talking directly to a serial port (COM port). See, for example, http://www.lantronix.com

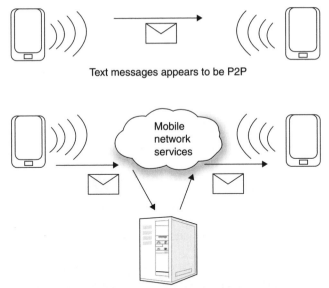

Figure 4.10 Despite appearances, text messaging is not a P2P application.

We mentioned verbally swapping phone numbers as an example, showing friends a picture message or video message, and so on.

Secondly, H2H interaction can take place using an ad hoc RF network, such as made possible using Bluetooth or WiFi technology. Interactive 'chatting' is possible using Bluetooth-enabled personal digital assistants (PDAs), perhaps to allow colluding attendees of a meeting to swap surreptitious notes during the dull moments of the meeting. Note that architecturally, this method of communication is P2P and will therefore require a specific software design approach. For example, just because we can form an ad hoc network using Bluetooth or WiFi-enabled[29] PDAs, doesn't mean we can automatically carry out any familiar communications task, such as conducting an IM dialogue. In fact, using the popular IM services, like MSN Messenger, it is not possible to use such a service in a P2P mode, as these services require an intermediary server, which in our model would sit in the content network or IP network. The requirement for an intermediary can go unnoticed. For example, text messaging, which at first seems to be device-to-device[30], is not a P2P method of collaboration. All text messages go via a store-and-forward messaging centre as shown in Figure 4.10.

So, H2H communications across an ad hoc network implies that all the necessary mechanisms to establish and maintain the communications have to exist on the devices and nowhere else in the network. This is potentially a complex problem and is the challenge of designing P2P protocols like JXTA[31] (see Chapter 6).

H2H communications is viable only within the SD in the device network, when we are specifically thinking about ad hoc networks. Clearly, there is no such concept as H2H communications within a PD, unless we are prepared to generalise to the case of self-to-self

---

[29] As we shall find out later, WiFi supports ad hoc networking as one of its modes of operation.
[30] Ignoring the obvious need for the RF infrastructure.
[31] JXTA is the open initiative by Sun to develop a set of workable P2P protocols. See http://www.jxta.org

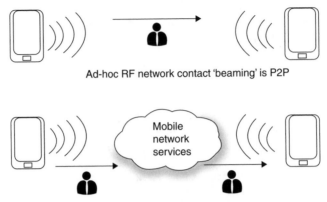

Figure 4.11 Contact 'beaming' is H2H interaction using P2P.

as an instance of H2H interaction, but that's overly pedantic. It is also preferable to retain an unambiguous definition of H2H; namely, that more than one person is involved. We shall tackle PD applications elsewhere.

It is tempting to think that the P2P computing paradigm only applies to ad hoc networks. This is not the case. The P2P paradigm can work with devices that are physically separated and unable to form an ad hoc network, thus implying the involvement of the permanent RF network and devices being connected in their CD mode. This is a perfectly viable option. As Figure 4.11 shows, two devices could interact in a P2P mode across a wide-area permanent RF network and there is no reason for both devices to be RF connected, one of them could be fixed[32]. Actually, there is no inextricable link between P2P computing and RF connectivity. Any computing device could work in a P2P networking mode, even across ordinary wire line networks.

The key to the P2P computing paradigm within the mobile context is probably the absence of a server. This tends to allow for a more impromptu approach to software collaboration, such as H2H interaction might entail, irrespective of whether the RF connection is present or not, ad hoc or not. However, it may be too misleading to suggest that there is never a need for a server, or some kind of intermediary computing platform. We might have noticed that in ad hoc networks there is automatically a degree of authentication as the device users (humans) can see each other (presumably) and there is also no need to find the mating device as physical proximity is both assumed and required. However, to mate with a peer device across a WAN, such as will often be the case in the permanent RF context, there is a need for mechanisms to locate the mate and quite possibly to authenticate it. Even with physically close devices, the need for a proxy function performed by a server may be useful. For example, in a conference-meeting hall with lots of attendees, the prospect of discovering all the possible devices to talk with may be technically cumbersome and could be assisted by a central server that is already cognisant of the location of devices in the neighbourhood.

We can think of the need to locate as being a type of *mobility management*, a feature already present in cellular networks, which are required to keep track of where a device is at

---

[32] Technically, both devices could be fixed as P2P is a general paradigm, not specific to the mobile context.

any one moment in time (whilst it is switched on). In a P2P session, mobility management is not the usual P2P lexicon for finding a mate to engage with, nor is the RF mobility management responsible for this function, although the similarities are interesting to ponder[33]. In P2P computing the concepts of service and device discovery are talked about, processes which have subtle differences from mobility management, but are essentially still providing a locating function (although it may not have anything to do with devices being mobile, just the information pathways to them being unknown). Our point of reflection is that it may be difficult to perform device discovery and any authentication process without an intervening server, so we should not assume that P2P computing is completely free of this need in general. Clearly, direct P2P communication is possible as this is the essence of device pairing in Bluetooth.

Other modes of H2H interaction can involve users who are on a fixed network, or can involve software services that provide the infrastructure for the H2H interaction, such as for IM and videoconferencing. In both cases we expect servers to be involved and we can imagine that, in effect, these servers are effectively the originators of content as far as our actors are concerned. Email is an example of a H2H interaction, although it is not real time and can also be used for H2C interaction, such as subscribing to email alerts that are generated by software services, such as a stock-market monitor. It can also be used for H2M interaction, such as a home security system sending out status alerts via email.

Any H2H software process that involves brokering, such as the store-and-forward capabilities of an email server, or the store and indexing capabilities of a bulletin board, can be viewed from the mobile user's perspective as if the other actor is sitting in the content network. There is no cognisance of what is happening beyond the content network. Perhaps an alternative view of the defining characteristic of the content network is that the sources of information are usually going to reside in databases. We don't really care how the data got into the database, whether a human initiated the input (directly or otherwise) or a machine, or some other computer in the content network, such as a syndicated news feed. By generalising the content network to represent an information-hosting capability – no matter the source, human or automaton – we are able to take a unified view of the architecture of mobile services and this makes our network model quite useful. We shall see later why this view makes perfect sense when we look at CS software architectures and their prevalence in software solutions. We can think of the client component as being our social–device network and the server as being our content network. Between the client and server are the RF and IP networks, notwithstanding that either of these networks can end up hosting services in their own right to make the CS process more powerful.

In summary, all modes of possible H2H interaction have been shown to be supportable by our network model and we now need to consider the H2C mode of interaction.

4.5.2 Human-to-Content Interaction (H2C)

As we traverse naturally from the device network to the content network, the path automatically suggests the H2C mode of interaction we need to take, but we need to examine its nuances.

---

[33] In fact, in a wide-area RF network, the issue of finding a P2P partner device on the RF network is related to the mobility management capabilities of the RF network, but this should become more apparent in later discussions.

When we want to grab content from its source, as we have already remarked, it is most likely sitting in a database somewhere in our content network. This would imply CD connectivity for the device and the use of a permanent RF network infrastructure. Most often, this will be the case.

However, the possibility exists for content to be sitting on another device and for it to be accessible in a P2P mode, so H2C interaction is not restricted to CS communications in the classical sense. Perhaps a user wants to access their friend's address book, having already gained permission to do so, (i.e. a level of trust exists between the owner of the content and the requestor). In such cases, there is no restriction on the type of RF connection; it could be ad hoc or permanent. I could be requesting access to a friend's address book and it could be on a device that is located nearby, or anywhere within the wide-area coverage of the RF network, or even somewhere on a fixed network. Possibly, it could be all three within one session of H2C communication, for a variety of reasons. For example, in accessing a friend's diary to book an appointment with them, the initial session could be established in P2P mode with a nearby device – assume that we are both sitting in the same coffee shop. However, after some communications back and forth, my diary application is notified that my friend's 'master diary' is sitting on a web-hosted calendar service, so my device changes mode and establishes a wide-area RF session to the content network. Of course, we could argue that the device should have gone straight to the central diary in any case, but why should that be the assumption? Also, we could argue that the friend's device would be in constant synchronisation with their central diary; thus eliminating the need for the first (P2P) step, but that's also an assumption that needs qualifying. Perhaps someone else had only a few minutes ago added an appointment to my friend's central diary and it had not propagated to his device, simply because his device is out of range, and thus, synchronisation is temporarily not possible and is suspended until coverage resumes. It could be that my friend's device is a WiFi/Bluetooth combination device and the coffee shop does not host a WiFi Hotspot, so synchronisation will not take place until the next Hotspot is entered (or by some other means). My device has a 3G connection and so it is possible to access the central diary (on the Internet) to get the most up-to-date appointments schedule for my friend.

This diary case is only one example of the possibilities of the device dynamically adjusting its communications mode to suit the desired task according to the current mode of interaction. To illustrate this further, let us for a moment consider the applicability of multimode behaviour to the previous mode of interaction, which was H2H. In the same coffee shop scenario, we may want to hold a three-way text-chat session with another colleague or friend who is sitting in an office somewhere else. In this situation, ad hoc Bluetooth or WiFi networking between my device and my friend's is combined with permanent networking over a 3G link, with my device acting as a gateway for the communications between my friend and the other party. An interesting aspect of this scenario is that this configuration is not compatible with standard (current) IM architectures because my friend's device is hidden from the chat server. Either my device could act as a proxy for a standard IM architecture, or, using P2P communications, it could act as a relay. In the latter case, my device could even deliver the service via a web interface and enable the entire session to be established irrespective of which IM protocols the other devices support, or possibly even irrespective of my friend's device to engage in P2P communication[34].

---

[34] For example, if my friend's device is able to surf the Web, then my device can 'become' the Web as far as his browser (device) is concerned and thereby allow for direct P2P communication to be established.

It is tempting to assume that the access of content requires a similar access mechanism whether it is sitting on a database in the content network or in a smaller database on a peer device. This could well be the case, but the essential difference is that P2P is generally a one-to-one association of user to content source, whereas a CS solution is likely to be a many-to-one association with possibly millions of users trying to gain access to an information source, possibly the same piece of information. Clearly, this dictates quite different considerations in terms of the information-delivery architecture. Scalability, availability, security and other such concerns will be more crucial in the CS scenario. We will get to this when we discuss the CS technologies later in Chapter 6.

4.5.3 Human-to-Machine Interaction (H2M)

The ability to interact with a machine has several possible modes:

- Interrogating a machine for its operational data (e.g. status)

- Receiving alerts from a machine based on operational status changes (state changes)

- Controlling the operational status of a machine via remote control

As we already noted when discussing the machine network, if the machines are networked using web-centric software paradigms (i.e. HTTP), then the interrogation of information is akin to pulling content down (pull) from a web server (content network). However, a machine itself may not be up to the job of providing a scalable or resilient web interface directly, so one consideration is the provision of web-based proxies in the content network that provide the direct interaction with the human (device), whether by CD or PD, as shown in Figure 4.12

The advantages of going via a front-end in the content network include:

- Enabling a more scalable architecture to be realised

- Potential to use powerful web-serving technologies (e.g. J2EE[35]) that are not easily integrated into a machine

- Easier to control security issues such as authentication and encryption

- Possible to integrate access to the machine into a wider service portfolio (e.g. Mobile Portal)

- Potentially lightweight protocols can be used for the machine-to-proxy interface, thus lowering the costs of implementation (including the link costs to get the information from the machine to the proxy)

In terms of linking the machine to the proxy, the opportunity exists to use wireless networking, and often this will be a more convenient option, sometimes the only option. For wireless-enabled machines, we have several possible modes:

- Remote connection via the content network using RF infrastructure. An example of this is an alarm that connects wirelessly back into a central control centre which itself sits in the content network.

---

[35] Java 2 Enterprise Edition (J2EE) is a set of technologies useful for implementing massively scalable and robust applications, particularly that support the CS mode of interaction.

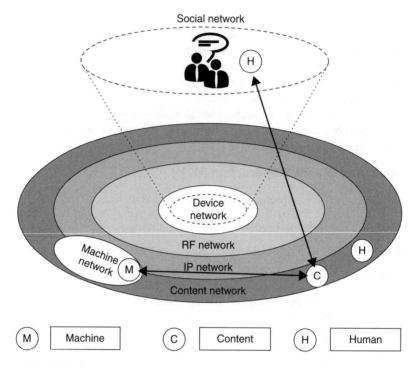

Figure 4.12 H2H via proxy.

- Remote connection via the RF network using RF infrastructure. An example of the ability to directly dial-up an RF modem embedded in a vending machine and then connect to its embedded computer. This is the more standard means of remotely accessing machines like vending machines. This mode can also include the use of text messaging to either send control messages to the machine's embedded computer or to receive alerts, or even to pull status messages using auto-reply text messaging.

- Local connection via the device network in ad hoc RF mode. An example of this is using Bluetooth to connect a PDA to a local printer in order to print out an email. There are plenty of other examples of this 'remote control' method of networking, such as control of the television, control of an automatic garage door, electronic fob entry to premises, and so on.

H2M interaction need not be limited to a CS mode. The local connection method just mentioned clearly implies a type of P2P connectivity. This could be widened to permanent RF networking, like the above option of remote connection via a modem, but using an IP-based networking solution. In other words, if the machine sits in the Internet then a mobile device could talk directly to the machine using an IP protocol, without any server necessarily being involved. In the classical P2P architecture, the communications pathways can move laterally to the next peer, if this is required. For example, a group of friends might each possess a hard-disk TV recoding device, similar to the TiVo[36]. If a programme were missed, then one of the friends could interrogate his own recorder to see if it managed to record the programme automatically. If it didn't, then the machine itself can contact a peer recorder in the group of friends to see if that device has the programme. This in turn could interrogate a peer, and so on. This is the P2P technique witnessed in such file-sharing applications as Gnutella and the infamous music-sharing service, Napster.

With H2M connections via the appropriate RF technology, the possibility also exists for real-time communication. For example, a night watchman for an accommodation block (e.g. apartments) could monitor a door-entry system that has a camera and two-way audio intercom. Ordinarily, the watchman might be stationed at the guard's desk where the desktop intercom terminal is situated. However, if the watchman needs to leave his desk post in order to patrol the building, then the intercom function could be enacted as a mobile application on his mobile device. The presence of a visitor, announced by a push button, could cause the mobile device to be alerted and for the remote monitor application to be initiated. This would receive a live video feed from the door camera and connect with the intercom via a two-way audio channel. Speaking into the mobile device would cause the audio to be emitted from the intercom speaker. What's more, if the guard decides to allow entry, then a command issued from the mobile device could trigger the electromagnetic latch on the door, thus allowing it to open.

This type of application is a powerful example of what H2M interaction can achieve within the mobile context. What makes this possible is a variety of technologies. The key, as in all H2M or M2M solutions, is in the design of the machine interface itself (i.e. the interface to the network). This does tend to necessitate a novel type of machine hitherto not seen in the markets. But the advent of ubiquitous mobile technology will surely promote these types of developments; the enmeshing of semi-intelligent and interactively responsive machines into 'the network'. The seamless transition from the RF network to the IP network seems a key to such services and applications. Matched with a seamless interface between embedded computers and the IP world, it is all the more likely that we will see H2M applications emerge with increasing frequency. A key contributor to the trend will be the possibility to directly plumb these remote machines into the RF network using embedded RF technology; thus, avoiding the grisly installation process and physical networking headaches associated with the plumbing of structured cabling and so forth.

In summary, the network model we have established seems to support the various modes of H2M interaction, although there may well be more that we haven't explored and some of these may emerge in further examples throughout the book.

4.5.4 Machine-to-Machine Interaction (M2M)

M2M communication is increasing viable as more and more goods leave manufacturing lines with computers embedded within them; computers able to engage with networks, should they be programmed to do so. The applications of embedded computing are constantly expanding and it is easy to overlook the fact that most of the computing devices in the world

---

[36] http://www.tivo.com

are hidden from view, buried away in some microwave oven, or hidden behind a panel in an elevator. Building control is an increasingly vital means of enabling sophisticated living and office quarters to function. Most buildings these days do not function without computers. Waterworks, elevators, fire systems, heating, windows, lighting and myriad other building utilities are pulsating with the hum of built-in computers amidst a fabric of sensors, actuators and controllers.

The M2M component is frequently seen as the ability to tie in all these embedded computers into a network so that they can connect to IT systems (i.e. other machines) housed locally or remotely and that have good reason to interact with the embedded machines.

So what is the role of wireless in M2M systems? There are two considerations for the use of wireless:

1. M2M systems offer a cost-effective and convenient means to bring static machines into the reach of an IT system. Essentially, this is thinking of wireless as a wired-line replacement, in order that a machine can be accessible to the nearest network infrastructure. The wireless link can be either short range, such as a WiFi connection, or long range, such as a GPRS or 3G modem – it depends on the physical location of the machine. Either way, the use of wireless to integrate static (immobile) machines is the more conventional solution that is being seen with increasing frequency, especially now the price of embedding RF (of one form or another) is decreasing.

2. M2M's other consideration for wireless is looking at how to exploit the mobility (or roaming) aspect of RF networks. There is no reason why we should not, with today's ubiquitous technology, enable any machine to become network-enabled, even a moving machine, such as a car engine management system, a forklift truck. Within this category, we can apply the idea of movement to previously static machines, which might lead to interesting application possibilities. For example, a medical diagnostic ultrasound machine that is portable could be networked to allow previous patient images to be retrieved from an imaging archive in the content network.

A prime example of an M2M application is the tracking of assets through a delivery service, something that has become fairly commonplace with most courier companies. An asset is tagged at the beginning of the delivery process and assigned a unique ID. Usually, a bar code is attached to the packaging and the ID is encoded in the bar code, or associated with it. At each stage of the delivery process, the appropriate courier agent scans the bar code. The ID is given to the customer who can enter it into a web-based application to track the whereabouts of the package. When the package moves through a transit warehouse, it is easy to scan the bar code and for the location of the package subsequently to be ascertained, such as 'at London airport'. In the past, the challenging part for the courier companies used to be tracking the final delivery to the destination address. However, using a wireless terminal (machine), the delivery agent is now able to get an acceptance signature and report the final delivery (and the signature) back to the courier's IT system (machine).

An alternative to bar code-based tracking is to use an RF device to constantly track the package by remaining itself with the package. This is made possible by the location-finding potential of the RF network, such that the RF device's whereabouts can be regularly interrogated. Clearly, it would not be cost-effective to put an RF device into each and every package in transit. For valuable assets, this might not be a problem and the customer could be charged extra for the service. Otherwise, the other approach is to ensure that the packages themselves are always travelling in vessels that have integrated RF-tracking technology. This could be the delivery van itself for example, which could have an RF device mounted somewhere on its body. Parcel carrier palettes with integrated RF-tracking devices are an alternative. Tracking of goods need not be restricted to delivery services. Any asset that needs tracking could potentially be tracked using an M2M solution. Perhaps equipment hire companies would like to add RF-tracking to their equipment, either directly, or to the equipment cases. Even people can be tracked using this approach and we talk about this when we look at location-based services in Chapter 13.

4.6 OPERATOR CHALLENGES

In the final section of this chapter we now turn to the challenges facing the mobile operators in order that we can take these into consideration when constructing an approach towards building mobile services, or towards building the underlying technologies that the services utilise.

Figure 4.13 shows us the 'old-world view' of the network operator, which stemmed from a very simple business model. The operator builds a cellular network and sells mobiles that work on the network. The owners of the mobiles get billed for using it in a very straightforward manner, such as charged according to time spent using the network (i.e. 'talk time').

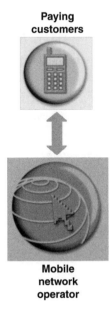

Figure 4.13 Operator 'old-world' view.

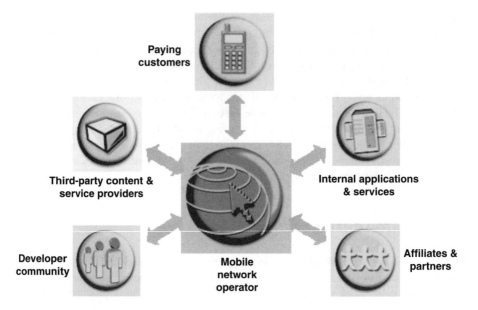

Figure 4.14 Operator 'new-world' view.

Figure 4.14 show us the 'new-world view' that the operator has stumbled into. Suddenly, there are a lot more things to cope with. In terms of the key challenges facing the operator, the following list is worth reflecting upon:

- *Give users something interesting to use and pay for*
  - Not necessarily what they 'want',– as they don't know what it is they want yet (creativity is required!)
  - Compelling enough to generate revenue
- *Balance service delivery and subscriber demand with capital expenditure on the service delivery platform*
- *Cope with uncertainty in the next generation*
  - Content and service types
  - Pricing models
  - Consumption patterns
- *Providing a rich 'open' end-to-end service package*
  - Building a third-party service platform that enables any application to be launched from the network, meeting both the operator's and the third party's needs
  - Embrace a technological approach that the developer community will welcome
  - Reach the 'event horizon' of a leading IT company, not a utility company

- *Entering into 'networked business' models with*

  - Affiliates
  - Third-party service providers
  - And even competitors

These challenges are reflected in the networked business scenario shown in Figure 4.14. Without the need to discuss the above points in detail, we can summarise the essence of 'the problem' facing operators by focusing on the first point and the last two points, keeping in mind that if any of what we are trying to do involves ubiquitous service coverage, then more than likely these problems directly impact on our ambitions to succeed in this industry.

The first point is really another way of saying that we don't yet have a killer app. Granted, we are not sure that the notion of a killer app exists, but nonetheless, whatever service or services are going to transform the mobile market, currently there is no indication as to what they might be. The reason that they don't yet exist relates strongly to the last two points in the list of challenges, which also point the way to possible solutions. One of the major challenges is technical in nature, whilst the other is commercial and perhaps the most challenging of all, if not pivotal in determining how this industry might move forward.

Given that we don't know what the killer app is, and we don't yet have the killer cocktail either, the sensible approach is to ensure that we at least maximise the chances of either emerging. Contrary to what some of us might think about developers being the vanguards of moving this industry forward, this responsibility clearly lies on the shoulders of the mobile industry (operators and equipment vendors) and NOT the developer community or its entrepreneurial commanders. This inevitable conclusion is reflected in the last two points in the list of challenges. The first point really says that the operator has to provide a 'platform' upon which it is extremely easy for competent developers to launch services. And it is not chicken-and-egg. The platform needs to be built and this book is really in essence an exploration of a set of platform technologies that can create a fertile environment within which creativity can thrive. This platform has to be both rich in features and open. We should qualify these two qualities before moving on to explore the final link, which is the commercial basis for gaining accessing to the platform.

We should think of the entire operator's network as a 'service delivery platform', which is my preferred usage of this term, no matter its various emergent definitions or common usage[37] in the industry (as discussed later in Chapter 6). The richness should be reflected in the ability to gain programmatic access to a diverse set of the network's capabilities, such as text messaging, picture messaging, micropayment, billing, location, call-processing, games vending, voice mail and just about any useful asset that the operator has in its network, which most definitely extends to include all devices running on the network. The openness should not be mistaken to mean 'open source', which it might well do. It really means *open access* to the platform. One aspect of openness is adherence to IT technologies that are widely used in the industry, avoiding as much as possible the temptation to introduce new incarnations or, worse still, entirely new approaches altogether; unless these are agreed upon by the entire mobile industry by due and open process. It is tempting to go as far as saying that this restriction should be a tacit creed for all operators. Part of the openness also

---

[37] The term *service delivery platform* doesn't really have an accepted common usage yet anyhow, but is being pushed by various companies as a category in which to market their wares.

means to use open standards as much as possible, such as the current trend towards using Java-based platform technologies.

The last point in the list of operator challenges is a critical one, perhaps the most essential. No matter how easy it might become to develop, test and deploy a mobile service on the operator's network, unless the means exists to establish a meaningful and productive commercial relationship without friction, then the technological advantage of a feature rich and open platform is likely to be of little consequence. Whichever part of the network platform that the service provider needs to access, there has to be an easy way of gaining commercial access to it that is economically viable for both parties.

On this point, I return to the vision that Dick Snyder (former Director of Marketing for Lucent Technologies) and I had many years ago when we first discussed the Zingo wireless portal that my company built for Lucent Technologies. At the time, our idea was rejected by operators and vendors alike who were still thinking in the 'old-world view' of the classic value-chain model that was regurgitated ad nausea in analyst reports of the time. The reason for rejection was that our 'new-world view' turned the classical value-chain model on its head, which in itself was nothing radical because we were only aping what we saw emerging on The Internet.

Whilst we may be tempted now to think that 'new economy' ideas sunk with the dot.com bust, the truth is probably far from it. It is my view that what Amazon has done with its marketplace affiliates scheme is what operators need to do with next generation networks. Just as Amazon gives access to its marketplace, which it heavily invested in, built, advertised and attracted the customers to in the first place, operators need to give access to their marketplace. Note the emphasis is on *marketplace*, not network. Giving access to the marketplace is completely different to giving access to the network, although both are clearly related. As a marketplace affiliate with Amazon, I am one step away from gaining access to a customer and from them gaining access to me, or my 'shop'. Amazon brings the customer onto the network, but when it comes time to purchase the goods, Amazon thrusts my wares and my shop to the fore. All this takes place at lightning speed and with dazzling efficiency and Amazon doesn't even care, nor know, what product I am selling, just that Amazon will receive their cut, whether I sell one product, or thousands. The model is amazingly simple, but not without considerable technical investment on behalf of Amazon and a real commitment to networked business. Despite all the amazing web technology used to build the Amazon platform, the most effective part of the implementation is the commercial process. If you have not already done it, I seriously suggest signing up to become an Amazon affiliate just to see how easy it is to become a business partner with Amazon. Try doing the same with any operator. Or contrariwise, imagine Amazon asks its affiliates to spend months gaining commercial access to what amounts to a single software component to allow me to put items in an Amazon user's shopping cart, but that's all, and I have to build the rest of the business infrastructure and relationship myself! Try being an affiliate on that basis. But this is what operators currently expect of all but the paid-up-by-the-millions strategic mega partners.

Until, and unless, this changes, the killer cocktail will not emerge. Granted, the platform and level of affiliation, even its nature, might be different for a network operator than for Amazon, but the essence is what has to be replicated and this should be a major focus of network operators, the main areas of creative thought, consultancy, analysis and investment: open networked business! This book can only describe what lies underneath such a business model (namely, the technologies used to build the operator platform).

4.7 THE WEB 2.0 CHALLENGE

As we are examining operator challenges, this is a good place to discuss Web 2.0. We discussed the wider significance of Web 2.0 in Chapter 3, so here we shall look at where the Web 2.0 paradigms fit in our current exploration of architectural concepts. Thus far we have looked at different service concepts based on modes of interaction (i.e., H2H, H2C, H2M and M2M) and we have mapped these to the different networks in the mobile ecosystem. In the previous discussion, we have taken a more commercial view of the operator network in terms of the actors likely to be interacting with the operator's piece of the ecosystem, such as users, developers, partners and so on.

Within the current context of exploring operator challenges, there are two questions that arise from considering the impact or relevance of Web 2.0. Firstly, does Web 2.0 have an impact on the modes of interaction? Secondly, where does Web 2.0 fit in relation to the operator challenges that we have just been discussing? It probably isn't that helpful to start discussing what Web 2.0 is. As I mentioned in Chapter 3, finding a precise definition is problematic. The following excerpt from Wikipedia's entry on Web 2.0 seems pertinent:

> *O'Reilly regards Web 2.0 as business embracing the web as a platform and utilising its strengths (global audiences, for example). O'Reilly considers that Eric Schmidt's abridged slogan - don't fight the Internet - encompasses the essence of Web 2.0 — building applications and services around the unique features of the Internet, as opposed to building applications and expecting the Internet to suit as a platform (effectively "fighting the Internet").*

In other words, one way to think about mobile services is this – if I'm going to go out and build a whole bunch of mobile services, why not use the Web as the platform to do so? To help answer this, I think it is worth elaborating on why the Web is a platform. What does this mean and how has it come about?

Today's Web is awash with the following:

- Lots of fast-moving ideas

- Lots of cheap and cost-effective resources

- Lots of people doing stuff (i.e. users, programmers, etc.)

- Lots and lots of low-cost software tools, APIs, code, coding projects

- Lots of easily connectable end-points: human (H) and machine (M)

- Lots of content

Any way you want to think about, it is hard to deny that this all adds up to a lot of opportunity. The core of a mobile operator's business has been, and still is, the ability to connect two people anywhere and anytime via a phone call or text message. The challenge is that almost any other service that goes beyond this can also be done, in some way, on the Web. The Web platform has all of the above attributes, none of which apply top operators. This is painfully obvious to developers and I can cite my own experiences here. When I've attended operator developer-forum conferences in the past, I'd be lucky to come away with any tools whatsoever to help me build an 'operator-platform' service. On the Web, I could never ever

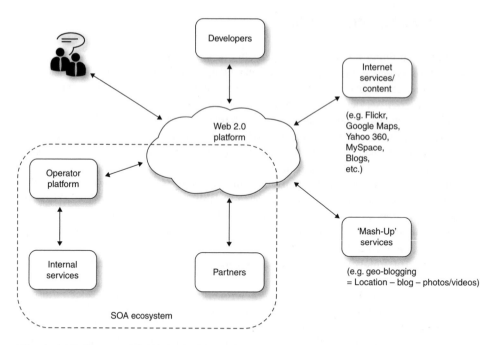

Figure 4.15 Operator Web 2.0 platform.

exhaust all of the tools and resources (including thousands of potential collaborators) to allow me to build a 'web platform' service.

So, going back to the two questions posed earlier. Firstly, Web 2.0 does have an impact on the modes of interactivity in the mobile ecosystem. It makes them much easier to implement. If I based my H2H or P2P service on internet and web platforms, I can build it cheaper, easier, faster. Secondly, Web 2.0 very much needs to figure in the consideration of operator challenges, most simply because Web 2.0 is in itself a significant operator challenge. The challenge is simply how to utilise the Web platform and still continue revenue growth. I believe that this is possibly *the question* that operators need to think about. So, for the purposes of concluding this chapter, which is introducing approaches and architectures to mobile services, we can now redraw our operator environment scenario to look like that in Figure 4.15.

5

IP-Centric Mobile Ecosystem and Web 2.0

5.1 INTRODUCTION

This chapter is about the Internet and the role it plays in our mobile services model. Having introduced our mobile ecosystem in Chapter 2, the Internet may seem a curious place to start. We could have addressed devices first, perhaps the wireless network, or maybe even the content network having stressed its importance in previous chapters. We are going to start with the Internet, particularly the Web, due to its particular significance in influencing the shape of next generation mobile services. At the end of Chapter 4, I suggested how Web 2.0 should be viewed as a platform within our mobile ecosystem. Its advantages and power are impossible to ignore. I don't think any operator denies this now, so I shan't spend any more time discussing its importance. Of course, how the incorporation of Web 2.0 into the mobile ecosystem leads to increased revenue is another question.

A lot of the power of the Web is embedded in its networking power (see sidebar 'The power of networks').

Sidebar: The power of networks

Charles Boyd[1] identified three maxims that affect how businesses will have to manage their activities for competing in the Internet era, but these truisms are generally applicable to all aspects of networked life and apply to mobility. Boyd mentions Moore's Law, Metcalfe's Law and Coase's exposition on business transactions.

---

[1] Charles Boyd is teacher of strategic management at Southwest Missouri University, USA. I found his website had many interesting articles on digital economy issues.

Moore's Law   Gordon Moore is founder of Intel Corporation. He elucidated an observation that in its revised form became known as Moore's Law[2]. In it, he states that every 18 months, processing power (of silicon microprocessors) doubles while cost holds constant. Moore's perceptive insight has proved to be generally true and remains so for the foreseeable future. Telecommunications bandwidth and computer memory and storage capacity are experiencing a similar fate because they are heavily reliant on silicon processing devices.

Mobile devices are no exception. It is not widely appreciated how much processing power is required in digital wireless modems. Some people may think that wireless networking is improving because of advances in the underlying mathematical theory of radio transmission. Generally, that is not true. It is more to do with the ability of modern chips to affordably implement what wireless theory has predicted for many years. At the time that GSM handsets were launched, they had more processing power in their tiny packaging that was available in the state-of-the-art PCs, which at the time saw the introduction of the newly designed and exciting Pentium chip (P60).

This makes it very affordable for individuals and businesses to be equipped with the electronic means to network easily. The price threshold for ubiquitous communications is within the reach of consumers, even very young consumers.

Metcalfe's Law   Robert Metcalfe designed the Ethernet protocol for computer networks. Metcalfe's Law states that the usefulness, or utility, of a network equals the square of the number of users. This principle is a combination of a straightforward mathematical fact and theory. The factual part is simply noticing that in a network with N nodes who are able to connect to each other, then the number of connections possible per user $N-1$, thus making $N(N-1)$ total connections:

$$\textbf{Total network connections} = N(N-1) = N^2 - N$$

The theoretical aspect is in assuming that some value can be attained to a user by utilising any of the possible connections open to them. It doesn't really matter what value in particular, but the benefit of the aphorism is in noting that the usefulness of the network increases in proportion to the square of the number of users.

The standards factor   We can see that mobile telephony is a combination of Moore's Law and Metcalfe's Law working to make mobile life possible. As already noted, it was Moore's Law that made mass consumption of mobile telephony possible. As we shall see later in Chapter 12, when we discuss the RF network, the ability to support large numbers of mobile users in a particular geographical area is only possible using digital communications techniques that require huge amounts of processing, both in the devices and the infrastructure (base stations). The realisation of Metcalfe's Law in mobile telephony is through the adoption of standards, such as GSM, so that we end up with a large network rather than lots of smaller ones; thus, the N is able to scale quickly and we end up seeing the benefits. In GSM, this has been driven by the decision of most operators in the world to adopt GSM, so Metcalfe's Law is given global momentum through the squaring of massive numbers.

---

[2] http://www.intel.com/technology/mooreslaw/index.htm

It almost does not need stating that the Internet has clearly benefited from Metcalfe's Law – it is the archetypal example given when discussing Metcalfe's Law. So clearly, if we have global wireless data standards for accessing the Internet, which we more or less do through 3G, then we have a very powerful network indeed. This is our justification for putting the 'mobile Internet' as our centrepiece of the mobile network model – i.e. the IP layer.

We can sometimes see the effect of the square law as it becomes obvious when we pass some point of inflexion where the rate of increase has a 'snow balling' effect. In GSM, this became obvious with the use of text messaging in many countries. Once the various operators within countries decided to allow internetworking of SMS, its usage boomed[3].

Transactions   As we have already discussed earlier in the book, the ability to network can be utilised in several ways, reminding ourselves here of the H2C, H2H, H2M and M2M modes of interaction. It may have been in our minds during the above discussion of networking maxims that networking is all (and only) about people connecting to other people. That's only part of it. We should not forget the other interaction modes identified in Chapter 4. Clearly, if there is value to be gained from interacting with a machine, then the addition of machines to the IP network also increases the value of the network. Some machines may already be networked, in which case we are combining the power of networks.

Connecting say M networked machines to N networked people, we end up with a combined value $N^2 + M^2 + 2NM$. One example that immediately springs to mind, and is a network that we often overlook (as we take it for granted so much), is the Visa Network (card payment), which is a network of machines. Combining this network with the Internet brings enormous pooled value.

Financial transactions are only one type of value attainable in a network, so we should not think exclusively about financial interaction, but nonetheless it is clearly an important consideration.

Ronald Coase won a Nobel Prize for his explanation of transaction costs in a 1937 article titled 'The Nature of the Firm' where he concluded that firms are created because the additional cost of organising them is cheaper than the transaction costs involved when individuals conduct business with each other using the market.

Moore's Law and Metcalfe's Law have made many types of transactions cheaper. This trend is sure to hasten because digitisation has cut transaction costs all the way to zero in some cases. Once a product or service is in place to be sold on the Internet, for example, there is zero cost for each additional transaction. The advent of marketplaces on the Internet, like Amazon and eBay, has brought amazing transactional power to the individual. This model is now appearing in the mobile operator landscape with similar marketplaces, such as the market for downloadable Java games. I-Mode in Japan has already harnessed the power of the marketplace by its revenue-sharing model, enabling thousands of producers to sell their wares into the I-Mode user base (network). We

---

[3] There were other factors lending a hand towards the success of text messaging, such as price factors and lower cost of access by the introduction of pre-paid mobile phone accounts, but the power of internetworking agreements soon became obvious.

should recognise that in both cases, the power of the network is the combined power of standardised devices and the billing network of the operator enabling charges to be levied and shared.

To understand this and its significance for mobile services, we should reflect a little on what networking is all about. The information presented in the sidebar 'The power of networks' helps us to understand the general principles of how access to networks has the potential to add value. However, the analysis can leave us detached from the purpose of networks. After all, a network is not very powerful if it has millions of users, but they simply plug in and do nothing thereafter. This would be like the electricity grid (which technically we could call a network, at least physically).

Clearly, we are interested in *communications networks*, so the power of the network is in its ability to support communication, or information exchange. Thereafter, the users can decide for what reason they want to communicate and what they want to say, but what determines the value of communication? We can explore some ideas in order to see if we can gain insight that will help us to understand better the power of mobile networking.

Firstly, let's take the example of business, or social, networking to illustrate some useful principles. We know that with business networking, the adage 'it's not what you know, it's who you know' tends to be true. We understand that value often lies in knowing someone who can do something for us, which may involve introducing us to someone else (i.e. more networking).

Knowing people is the essence of social and business networking, but knowing particular people is even more essential. The realisable value of our network is more to do with the actual connections that are useful to us. For example, if we are trying to find a job in a particular line of work, then the only connections in our network that matter are the ones to people who can connect us to openings in our desired field of work. This tends to suggest that the power of the network is not related to its total size, but by the number of relevant connections. We may also find that irrespective of what we are trying to achieve, certain people always figure more strongly in our network and we gravitate towards our connections with them.

How do people find out about jobs? Are relatives, close friends, acquaintances, employment agencies, or newspaper advertisements the most useful sources for finding out about job vacancies? Mark Granovetter, a sociology graduate student, tried to answer this question in his PhD research. He found that most often, people found out about job opportunities through acquaintances. The information people possess about job possibilities is affected by their placement within social networks, not by access to job advertisements.

Granovetter found that many of the paths between the job and the new employee were surprisingly indirect. Moreover, not only were the paths surprisingly long, the person who told them about the job was often someone they did not know very well. Granovetter named this as *the principle of the strength of weak ties*[4]. This tends to suggest that if we want to find a job, then we are better off if we are well placed in large social networks with people we don't know that well. This reinforces the idea of 'who you know', but demonstrates that our ability to participate in large networks in the first place should add value.

---

[4] Granovetter, M., 'The strength of weak ties'. *American Journal of Sociology*, **78** (May 1973): 1360–1380.

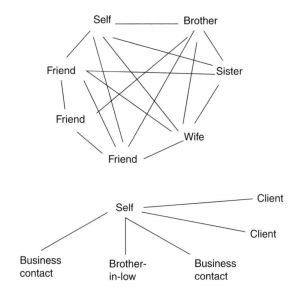

Figure 5.1 Weak ties and strong ties in social networks.

Granovetter was not able to explain conclusively the reason that weak ties had more value, but he offered an elucidation that has some merit and may be useful for our consideration of networks in mobile environments, as we shall see later.

Granovetter posited that the close ties, where we ordinarily expect more value due to their greater willingness to help (among other reasons), are also likely to be connected to each other. However, casual acquaintances are less likely to know each other. This is shown in Figure 5.1. The benefit of disassociated contacts is that they are likely to belong to their own networks and that information circulated in each network is more likely to be distinct within each network. In the close associates' network, information is more likely to be redundant as each contact has likely already heard it from someone else in the network. In other words, we will gain more information from loose associations as an aggregate than close associations.

The principle of the strength of weak ties suggests that it is going to be very useful for a user to be able to join networks easily and for conditions to be favourable for groups to form and to become manageably large. Clearly, any mechanism by which we can belong to a wider set of networks through weak ties, the better our chances, it seems, of gaining access to value. The Internet certainly offers this potential and ability. Currently, mobile networks do not offer this ability intrinsically. With today's mobile networks and mobile devices, there is no process or service to find others and network with them. Networking with someone almost always requires that you have actually met them, or know them, first.

One aspect of weak ties that seems important is being able to discover who knows what, or, who knows whom. *Discoverability*, as we shall see, is an important part of the networking process, and we will need to address this issue, especially within the wireless context where it possibly poses more challenges. We shall return to this topic later when we look at how users are identified in mobile digital space; this point is covered in Chapter 13 on location-based services (LBS), which not only discusses LBS, but also acts as a catalyst for fusing some of the ideas currently under review as well as other themes in the book.

We should not confine our understanding of personal networks to the SD or to H2H interaction (as discussed in Chapter 4). The adage 'it's not what you know, it's who you know' may no longer apply. In the networked world, it may well be possible to access other people's information without their *explicit* consent or involvement; something we shall discuss in Chapter 13. Using P2P architectures, it may also be possible to traverse personal networks that are associated by friend-of-a-friend connections: 'It's not who *you* know, it's who *they* know.' Irrespective of the means of locating the information, in a connected world H2C interactions are also important. Here, the importance of the content is in its value to the task in hand, which in a social networking context could be all kinds of information.

'Content is not king'[5] is a paper by Andrew Odlyzko in which he argues that:

- The entertainment industry is a small industry compared with other industries, notably the telecommunications industry

- People are more interested in *communication* than entertainment

- And, therefore that entertainment 'content' is *not the killer app* for the Internet

This observation would seem particularly relevant to the nascent mobile services industry. The main non-voice applications for mobile phones are probably text messaging and ring tones. Text messaging is a form of communication and most of the content is personalised. Text alerts from information sources only accounts for a small percentage of text-messaging volumes, which are measured in the billions for the UK market alone[6]. Ring tones are a form of entertainment, but probably more accurately thought of as items of fashion and self-expression, albeit of a highly limited form. The most popular use of premium-rate[7] messaging in the UK is dating. In other words, the mobile data market is already social-network centred, focused on the individual's need to communicate, even in terms of outward self-expression (e.g. ring tones, phone fascias, etc.). This notion has been verified by numerous studies[8].

If we take two mobile phone users today, how do we know anything about them beyond their phone number? We probably don't, so it is not even possible to instigate a discovery mechanism, as there is no information to support discovery, just phone numbers and, at best, physical addresses of users. However, what we do have is an extremely powerful resource indeed, which is the database of call records since the mobile networks started. This tells us who is talking to whom and can also be used to find groups with *low transitivity*[9]. This may or may not be useful as it is, but combined with other information and the power of networking capabilities deliverable via Web 2.0, we have the ingredients for exceptionally powerful networks that can provide the basis for all kinds of transactions.

So, when we think of a mobile network, for that network to be truly powerful it should be capable of supporting personalized networking. It should allow groups of users with similar

---

[5] http://www.firstmonday.dk/issues/issue6_2/odlyzko/
[6] http://www.text.it
[7] Premium-rate messaging is also called reverse-billed messaging, which is where the user pays to receive a text message, usually for a much higher price than sending one, such as 100 pence reverse-billed versus 10 pence to send (or less).
[8] See Katz, J. and Aakhus, M. (eds), *Perpetual Contact – Mobile Communication, Private Talk, Public Performance*. Cambridge University Press, UK (2002).
[9] These are groups with loose ties.

interests to cluster together and for their collective interests to be supported and magnified by the network (what Rheingold[10] called 'swarming'[11]).

To conclude this discussion of networking and its importance in the mobile ecosystem, I'd like to consider for a moment what a mobile service is actually. Historically it has been a means for two people to talk or text. What we are now saying is that mobile services are more to do with networking. Communication is a subset of networking, or follows in any case from networking. Hence, a possible definition for mobile services moving forward is as follows:

Possible mobile services definition:

The ability to add value to our lives through networking whilst freely moving anywhere we are likely to go in conducting our usual day-to-day business and social lives.

OR (shortened version):

The ability to network and transact at any time and from any place.

5.2 THE INTERNET AND WEB 2.0

The Internet is a universally accessible network of information, people and machines and it does represent a significant part of the future of mobile services.

We should understand the essence of the Web. We should also understand and appreciate its impact on computing. This is because the Web will undoubtedly figure strongly in facilitating interesting and significant mobile services. Firstly, the Web protocols are prevalent and easy to integrate with, so it is relatively simple to build mobile services using a web-centric model. However, probably more significant is the way in which the Web offers a huge amount of networking power. That is its appeal to mobile service developers, or should be. Many observers are coming to similar conclusions that the 'killer attribute' for mobile services is something to do with 'anytime networking', not the consumption of licensed content, which was a strong assumption at one point. Of course, content is important, but we have to face facts that most of it on the Internet is free. Moreover, we have seen a tremendous gravitation towards the user's own content, which really has always been at the heart of the Internet.

It seems certain that the 'walled garden' model will fail, should operators wish to pursue it. There is nothing inherently wrong or 'evil' about walled gardens and they are applicable at various points in the development of new services. However, protection of a service can only work whilst it is possible to add significant value and maintain a leading edge through protectionism. As we discussed in Chapter 4, given the incredible and accelerating power

---

[10] Rheingold, H., *Smart Mobs – The Next Social Revolution* (pp. 174–182). Perseus Publishing, Cambridge, MA (2002).

[11] I don't much like the current obsession to use animalistic terms to describe human behaviour. Human qualities are often overlooked in the pursuit of technological progress, where human involvement is usually confined to consumption. Not much is said about other aspects of human need. Given the record numbers of text messages sent in the UK on Mothers' Day and Valentine's Day, perhaps we should refer to the collective messaging phenomenon as 'huddling' rather than the distinctly bestial term 'swarming'.

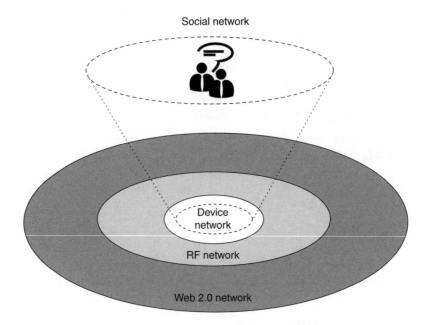

Figure 5.2 Internet at heart of mobile services universe.

of Web 2.0 as a platform, it seems increasingly unlikely that anyone could compete with it. Operators need platforms of various types. Some of these will remain closed and protected, even proprietary. There is also legacy to consider. Voice and text networks were mostly built around non-IP protocols and paradigms, so these will remain for some time to come.

However it takes place, we need to use the Web 2.0 platform as the foundation of the mobile ecosystem model, as shown in Figure 5.2. It may seem an obvious perspective or approach, but operators may be tempted to put their mobile network at the centre and work outwards. This is understandable and in some ways still realistic, given the investment in building RF-network infrastructure and capacity, but this way of thinking will not succeed in the end.

Today, we have truly powerful mobile networks that have already networked hundreds of millions of people globally, and with global roaming potential. Yet, despite the power of the RF network (we shall examine this in depth in Chapter 12) the most powerful network is the Internet, and so it has the greatest potential to add value to users. Adding the RF network brings additional value, no doubt, and much of this book is concerned with how to bring about the multiplication of value in combining the potential of all the networks shown in our mobile ecosystem model.

The emphasis we are placing in this part of the book is on the Internet, especially the Web and Web 2.0. It is a valuable centrepiece to our mobile services universe, which is why I have chosen to tackle this part of the mobile services network model first, namely the IP network layer, before looking at the other layers (such as RF and device layers).

In order to understand how we might build and utilise a web-centric mobile ecosystem, we first need to examine the underpinnings of the Web, which remains its incredibly powerful yet simple architecture. Whilst doing that in the following sections, we will look at what aspects of the Web provide challenges for wireless, but defer our detailed discussion on how to solve these challenges to later chapters, especially Chapter 7.

5.3 THE CHALLENGES OF LIBERATING DATA

In this section, I want to go back to basics, as it were. By this, I mean that I'd like to take you to a point where we can truly appreciate why the Web works, from a technical point of view. I strongly believe that once we understand a technology from the ground up, as if we invented it ourselves, we then gain a lot of benefits in figuring out how to use the technology. This is true of the Web. Most of us use it and we understand the concept behind it, but we don't really know the next layer down. I'm going to explain the fundamentals of the Web. Even if you understand the Web, it might still be worth a read. I have never found anyone on a course that I have given who hasn't benefited from this lesson, including the 'experts'.

With computing of any kind, we are concerned with processing and accessing information (or content). Today, thanks to the IT revolution in general, virtually all useful information resides in databases. That is the first observation that we need to keep in mind:

> **Observation No. 1** – most electronic information in the world sits in databases.

Databases are extremely efficient systems for storing collections of related information in a structured way that can be easily retrieved, organised and updated. The most widespread type of database is the relational database, where information is stored in tables. Each table will contain data that has a particular grouping, such as a table of customers, a table of products, and a table of prices and so on.

Relationships can be made between tables in the form of relational links, say regarding the way in which a customer from the customer table can be identified with products in a product table (products they may have purchased, for instance). Database servers allow any meaningful subsets of linked data to be rapidly updated and accessed, serving a high number of users making different requests in a number of different ways. A common database example is to think of an inventory of products, such as books or DVDs. But, there is no end to the collections of data that have found their way into databases.

Clearly, we need to be able to access remotely, across a data network, the information held in databases. Let's be clear that when we talk of accessing the information, our primary concern for the moment is to view it (visualise), although we should keep in mind that we may want to change it, or add to it.

When viewing information, we want to be able to see it in a format that makes sense. It should definitely be human-readable and preferably formatted in a way that is an aid to understanding and interaction. For example, we would like to be able to view data in tables where appropriate, with appropriate text rendering and so on.

We would also like to interact easily with the information if we need to alter our view of the current dataset, say from books on science, to books about a particular topic in science, by a particular author. In other words, we need navigability, a means to 'drill down' further into the information or look at other parts not currently visible to us.

In order to view the information and navigate its space, we obviously need the ability to pull down information from the databases to our device in the first place. A generic architecture for our system is shown in Figure 5.3, namely a *database server*. We need to consider the possible mechanisms required for implementing this system and how we might implement them.

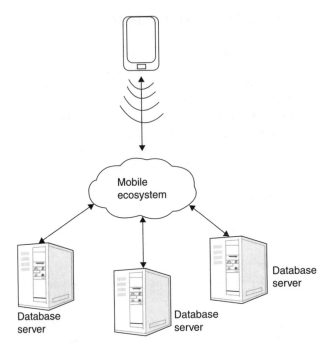

Figure 5.3 We want to view information stored in databases.

If we consider the primary problems that need to be overcome when viewing information that comes from databases, this discussion will enable us to understand not just how the Web works, but why it works and what the thinking was behind it[12]. I have identified four main challenges to overcome:

Challenges of visual interaction with databases

1. *Formatting* – how to format the information retrieved from the databases into a viewable format, and subsequently how to view it. (Note that the information in the database would most likely *not* have been stored with visualisation as a guiding factor in its structure (data types), but only with data representation in mind. For example, storing someone's name requires only a string of say 50 characters, typically. Usually, we do not need to consider other information such as what colour we wish to use to display the name in a particular application. Such display meta-information would typically come from the application that is responsible for generating the human interface.

---

[12] The Web model was not originally designed for accessing databases, but to access collections of pre-formatted pages of information (i.e. documents) that sat on a hard disk on some remote PC. However, assuming that these pages 'sit in a database', or will be populated with data from a database, is a useful generalisation and one that reflects a common occurrence in today's web systems and does not deny the original design aims of the Web.

2. *Visualisation* – how to add the visual formatting to the information requested from the databases, once it has been retrieved.

3. *Protocol* – how to tell the database what information we want to view.

4. *Delivery* – how to get the retrieved information back to the user so that it can be viewed according to the formatting.

In discussing each of these issues in general and their solutions, we shall also examine the special challenges for mobile access as we go along. A full discussion of the particular solutions for wireless is presented in Chapter 6.

We should first be aware that what we are about to discuss involves the origins and workings of the Web, *not* the Internet. The Internet has been around for a long time and its underlying principle is that any machine that can 'talk' IP, can talk to any other machine in the world that can talk IP. We should understand that IP has become the default and most widespread protocol set for computers to talk to each other; this is really the main thing to note about IP, and that there is a global data communications infrastructure that enables most PCs, irrespective of geography of locality, to access the Internet. In the computing cosmos, IP is just as English is in the international business world, the lingua franca of exchange. It is like the computing passport for global access. There is nothing inherently advanced about IP that makes it the obvious choice. Its prevalence and legacy make it an obvious candidate for network enabling any computer application today. A detailed understanding of IP is not required to comprehend this discussion (or the book), but an appreciation[13] would be useful.

Just like any language, what IP does is enable things to talk; it says how they should talk but not what they should talk about – the conversation. This requires higher layer protocols and this is what we are going to look at now for the purposes of agreeing on how we can use IP to talk universally and generically to databases and display their contents on our device screens, both in the wired world and in the mobile one.

Observation No. 2 – IP is now the universal language for computers and devices to talk to each other.

Now let's look at how to address the four challenges that we identified above for gaining universal visual access to databases.

5.3.1 Challenge 1: Making Database Information Human-readable

The information in the database does not contain (usually) anything about how the information should be presented to the user. It is just raw data. For example, customer records may describe customer name, address and so on, but nothing about where and how such information should appear on a screen, such as font, screen position, colour and the rest. In fact, as far as displaying information on the screen is concerned, the native database format would probably appear as completely unreadable and will definitely need transforming to a human-viewable format. This problem is shown in Figure 5.4.

---

[13] For a basic introduction see Levine, J.R. *et al.*, *The Internet for Dummies*. IDG Books, Foster City, CA (2002).

Database native format   Human readable format

Figure 5.4 Information stored in database generally needs transformation to be visualised.

Is there a generic means by which we can extract information from the databases in a format suitable for human viewing on computer screens, such that we don't need a different viewing program for each application that accesses the database, or for each type of data? We don't want to have to install an accounts manager viewing program, a customer management viewing program, an inventory control program, a TV-listings program, and so on, each time we have a different application that uses information in a database. This would be unwieldy to manage and most likely an expensive solution. In a mobile context, it would in any case be untenable because most mobile devices cannot support installation of programs in the field (after production and sale of the phone). We are looking for a universal and generic means to view information sets from databases.

Solution 1   We use something called a *browser* that can accept information from our databases along with some kind of accompanying layout information that specifies fonts, text positioning, columnar layout, tabular views, colours, italicisation, etc. The layout information is not really the data; it is called *metadata*, which means data about the data. What it tells us about the data is how to lay it out, how to display it. The layout metadata gets mixed up with the actual data in the same file and this gets sent up to the browser, as shown in Figure 5.5. Clearly, there must have been a process to extract information from the database and another process to add the markup.

We could standardise on the layout data in terms of its format, but the actual data comes straight from the database, be it the accounts, the inventory or the weather forecast.

The layout data is what we call *markup* as it 'marks up' the data, such as: 'take this piece of data and make it bold', 'take this block of data and put it in a table, width 500 pixels, 4 columns', 'take this piece of data and make it into a large heading font'. The markup needs to conform to a standard method to ensure that anyone using the appropriate browser (i.e. conforming to the markup standard) can access our database information, as the browsers will be expecting the layout data to be in a pre-agreed format, or language.

This layout metadata is formalised into what is called a *markup language* and in the case of the World Wide Web this language is known as *HyperText Markup Language* (HTML). Please see the sidebar entitled 'A quick look at HTML' to see an example of what HTML looks like. We will see why it is called *hypertext* ML in a minute and later in the book we shall describe its mobile variants, like WML[14], in more detail.

---

[14] WML stands for WAP Markup Language, but there are several alternatives for mobile browsers, as we shall discuss later in the book.

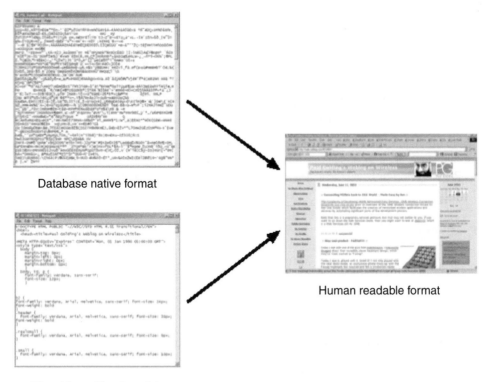

Database native format

Human readable format

Visual formatting template

Figure 5.5 Combining database information with visual formatting information.

A web server does the transformation process of adding the markup data to the actual data, as well as grabbing the data from the database. As mentioned earlier, clearly we need two things:

1. A programmatic way of controlling the database extraction

2. A means of adding the markup metadata to the extracted information

---

Sidebar: A quick look at HTML

HTML uses the concept of tags to annotate data. Not being the actual data, but rather data about the data, this markup is referred to as *metadata*. *Meta* is the Latin word for *indicating change*, and that's exactly the effect it has on the marked-up information, transforming it into a structure that is capable of being displayed by a browser.

To take an example, if we want to indicate that a new paragraph has began, so as to allow the browser to demarcate the paragraph with a suitable portion of white space, then we insert the tag <P> at the start of the paragraph and the tag </P> at the end. In general, the tags all take this form, using the angular braces '<' and '>' to delimit each tag, and the forward slash '/' to indicate a closing tag in a tag pair, most tags coming in pairs (<>) to show the beginning and ending of a particular marking.

Another example would be <B> and </B> to indicate bold text. Tags can also have parameters, such as <font color=white>Description</font> where these font tags indicate that the colour[15] of the word 'Description' is white.

So if we take a data set from our database, which could be something like this:

Deluxe Golf Trolley, 248, Burgundy

our program running on our web server, having extracted this data from the database, can add general data, such as text to make it clear that the 248 means 248 British Pounds (Sterling), and the markup tags. So we could end up with something like:

```
<P>
<B>Product:</B> Deluxe Golf Trolley<BR>
<B>Price:</B> <font color=white>248 British Pounds
(Sterling)</font><BR>
<B>Colour:</B> Burgundy
</P>
```

The tag we previously didn't consider is the <BR> tag which indicates a line break (i.e. a new line should start). Otherwise, we have already explained the other tags and the final display in a browser might look something like:

Product: Deluxe Golf Trolley
Price: 248 British Pounds (Sterling)
Colour: Burgundy

All of the available tags that we can use in HTML are already defined for us, we cannot change them. This is so information will appear the same in all browsers. The standard for HTML is maintained by the World Wide Web Consortium (W3C[16]).

So our web server has to provide these functions. Upon demand, the user requests information from the database via the web server, which accesses the database on behalf of the user and inserts the markup data and passes the combined information – called a *page* – to the user. The user's device receives the page in a browser, which knows how to interpret the markup in order to display the embedded information according to the markup. A page designer (i.e. a person who can design a suitable visual layout and design) determines how the information should be presented on the screen and uses appropriate markup chosen from within the available (fixed) HTML vocabulary. A programmer takes on the task of programming web pages that can access the database and then merge the retrieved data with

---

[15] Note that HTML tags use American (US) spelling conventions, so the use of the spelling 'color' in the tag is not a mistake here.

[16] After the invention of the Web by Berners-Lee, the World Wide Web Consortium (W3C) was created (October 1994) to lead the World Wide Web to its full potential by developing common protocols that promote its evolution and ensure its interoperability. See http://www.w3.org/Consortium/

the markup; which in itself is probably generated programmatically rather than stored as a template file, although this could be, and sometimes is, done.

Important Points to Note about the Browser Approach   Before we go on to look at the next challenge for gaining universal visual access to databases, there are two important points to note about the browser approach.

Point 1   You may have noticed that we have now introduced a new networked-computer element in our generic information delivery architecture that was not previously assumed or shown in Figure 5.3: namely a *web server*. So we now have what is known as a *three-tier architecture* as shown in Figure 5.6. This is the so-called client–server, or CS, architecture for software applications.

> **Observation No. 3:** The Web paradigm is technically based on a CS architecture.

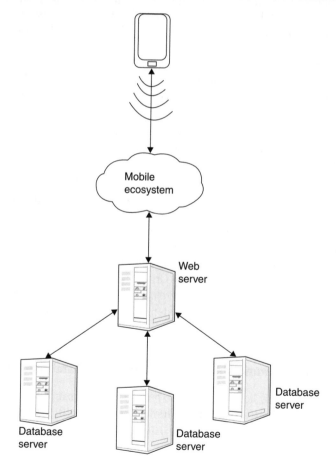

Figure 5.6 Web server in the middle.

In our three-tier architecture, both the web server and the database server are both sitting in the content network layer of the higher abstraction network model that we established and discussed earlier in the book.

Point 2 A browser is a generic display program capable of presenting any information that has been marked up using HTML. Therefore, it is a universal client not restricted to any one particular application or information set. This is its real power and is a reason for the popularity and significance of the Web. Because the browser can display any information (HTML formatted, of course), we are able to benefit from connecting with any web server where the owner has decided to use HTML, which they are free to do because HTML is an open standard and easy to learn, even for a novice. Previous computer paradigms had often assumed that a proprietary software program was needed to facilitate each and every application with its associated information set and that these programs were confidential and proprietary in terms of their construction and how information was represented internally.

The shift from using proprietary information processing software programs to a universal open information standard (HTML) and associated client (browser) constituted a simple yet significant shift in computing that heralded the explosion of the Web. This explosion had enough momentum so that its impact is now being felt in the mobile world, the consequences of which are explored in detail throughout this book. As I hope will become clear, the growth of web-enabled[17] mobile paradigms will be self-fuelling, as the power of the Web becomes magnified by mobility (ubiquity), which in turns magnifies the appeal of mobility: a virtuous upward spiral will emerge.

Wireless challenges

As we shall explore in more depth later when we look at WAP in detail (Chapter 7), the problem with HTML for markup is that it has been designed, and subsequently evolved, with large, colourful rich displays formats in mind. As is obvious, mobile phones do not have anywhere near the graphical display capabilities that come on the average desktop. Hence, we can imagine that HTML is probably too verbose and over-specified for delivering markup information to browsers on phones. It is also the case that phones do not have the same human interface characteristics as computers, such as a mouse. Therefore, the means to interact with the data on a mobile device may not be consistent with assumptions intrinsic to HTML.

5.3.2 Challenge 2: Adding Visual Formatting to the Database Information

In general, even if we knew how we wanted to present the information to the user, database server products do not have any means of formatting the data for viewing. There is no concept of a Graphical User Interface (GUI) that the database server can apply to the

---

[17] Web-enabled mobile paradigms are what we need and not mobile-enabled web paradigms.

information to make it viewable. This task is not the domain of the database server, which is optimised and designed just to serve raw data for consumption or formatting by another software program upstream. This suggests that conversion from database format to human format, as represented by the arrow in Figure 5.4, is not a feature of the database server, and consequently, an external server of some kind is required to fulfil this function.

Solution 2 It has already become clear in the previous discussion that we need a new server – the web server – to assist with extracting the information from the database and then adding the HTML markup before sending the requested information back to the browser. The way to think about the web server is as a store of page templates, representing the various views that the user may require of the target information. These templates only store the presentational information – the markup – but do not yet contain the actual information that the user requires from the database. There are placeholders at each point in a page template where the information from the database can be inserted.

The role of the web server then is to handle requests from the browser for information, which comes in the form of a request for a page. The server extracts the required content from the relevant database and merges this with the appropriate markup template to produce a complete web page for display in the browser. This process is shown in Figure 5.7.

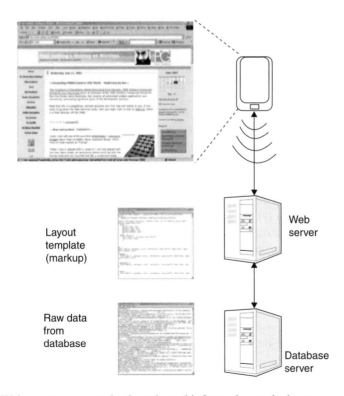

Layout template (markup)

Raw data from database

Web server

Database server

Figure 5.7 Web server presents database-bound information to the browser.

Wireless challenges

As we shall explore in more depth later in the book, when we look at WAP (Chapter 7), the challenge for wireless is to maintain the use of existing web servers already utilised by wired devices during the rising popularity of the Web. This is an obvious advantage. Having invested in web servers and learnt how to use them and program for them, it would be a pity to have to introduce new types of servers to cope with wireless devices. We would like to be able to benefit from the quite extensive investment that has already been made in web technology, infrastructure and – if we can – existing page design where we feel existing pages may in essence be useful to view whilst on the move (i.e. on our mobile phones). As we shall find out later, the lynchpin of the Web legacy is the HTTP protocol. We can think of software as becoming HTTP-centric, which is the essence of the networked computing revolution.

5.3.3 Challenge 3: The Need for a Protocol

We need a means to connect to the database via the web server, including the ability to send in queries to the web server, receive the resulting content and to locate the servers on the Internet in the first place. There might be thousands of databases (web servers) accessible to us, but how do we make sure that our device is talking to the right one. This is an *addressing* problem, just like calling someone on the phone. There are thousands (millions) of phones to call and we solve the addressing problem by giving each phone a unique number (telephone number). We need a similar addressing system for our web pages. Furthermore, just like phone directories, we need a means of looking up where on the Internet our server is in order to connect to it to issue the request. Once connected to the right server, we need an agreed language (protocol) in which to speak to the server to ask if for the information set we require to view. A web server may have thousands of page permutations, so it needs to know which one the browser requires from one request to another.

Solution 3   What we have is essentially a navigation and communication problem. How do we communicate from the browser to the web server and how do we navigate from one page to another. We have HTML page templates waiting to be programmatically filled out with information from the database, and then dished up to the browser, but how do we know which page is being requested by the user? This is the job for what has become known as the *Uniform Resource Locator* (URL), which is a web-specific addressing variant of the more generic *Uniform Resource Identifier* (URI[18]). It is a means of addressing the page on a particular server, so that the server knows to use a specific page template in responding to a request.

We would like to alleviate the need for the user to specify in 'database language' exactly what data is required from the database to fulfil the request. This would be a very

---

[18] If we are just talking about web addresses, like 'http://www.web.com', then we can use the term URI and URL interchangeable, as happens throughout the book. For a definition, see http://www.wikipedia.org/wiki/Uniform_Resource_Identifier

cumbersome approach and probably unworkable except by database gurus (who are a small minority, of course). Even then, this approach is not possible if the guru has no knowledge of the database structure, or how the database servers expect to be queried for information[19].

Databases are full of interlinked tables of the various information we are interested in viewing and manipulating. For example, for an online bookshop, perhaps there is a table of books linked to a table of authors linked to table of publishers linked to a table of prices, and so on. Let's say that the bookshop owner wants to view information about a particular customer and outstanding book orders against that customer. The information required to construct that view might be stored across several tables in the database. The way a database server works is that the information is requested by stating which tables we want, which columns in those tables and some qualifying condition to select the row or rows in these tables, such as 'customer number = 01728'. There is a special language to construct the request (query), called *Structured Query Language* (SQL), which is supported by most database server products on the market.

The average web user is not going to have a clue about SQL. If they had to specify the SQL each time they wanted to view a web page, this would be unworkable. Here's a sample of SQL that shows why:

```
SELECT    [Equipment Inventory].SerialNumber,
   Products.ProductName, Customers.CompanyName,
   [Equipment Inventory].DatePurchased
FROM       Products INNER JOIN
                   Customers INNER JOIN
                   [Equipment Inventory] ON
Customers.CustomerID = [Equipment Inventory].
   CustomerID ON Products.PartNo =
   [Equipment Inventory].PartNo WHERE
   ([Equipment Inventory].SerialNumber = '&num')
```

Despite its verbose appearance, this fragment of SQL is actually a very simple query against an uncomplicated database of hospital equipment, used to lookup when a particular accessory (with serial number '&num') was purchased by the hospital. Imagine if the user had to type this query string into the browser in order to obtain the information: a daring proposition, even for someone who knows SQL!

What we need is a means of labelling a page with an address that we can enter easily into the browser such that the above SQL runs implicitly and the results are merged with the markup and sent back to our browser. This is exactly what happens on the Web. We assign URLs to pages and we simply request the URL from the web server. For example, to attain the information requested in the above SQL sample, we might simply type something like 'MyHospitalInventory' into the browser, which sends a command to the server, something like 'GET MyHospitalInventory' and the corresponding page, myhospitalinventory, runs

---

[19] Unless the guru designed the database in the first place, they would not know how to structure queries to extract the required information.

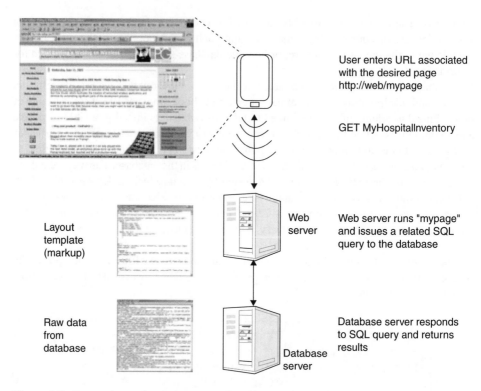

User enters URL associated
with the desired page
http://web/mypage

GET MyHospitalInventory

Layout
template
(markup)

Web
server

Web server runs "mypage"
and issues a related SQL
query to the database

Raw data
from
database

Database
server

Database server responds
to SQL query and returns
results

Figure 5.8 Converting web page requests into database queries.

on the server. Within the page template is the SQL needed to query the database and the markup to format the queried data. The query is run and the output formatted according to the template and sent back to the browser, as shown in Figure 5.8.

Let us say that the user now wants to change their current view of the data, perhaps to expand on a particular subset of the information for a more expansive view of the detail. There are probably countless ways that the user or users (as we have to contend with the entire user population for our application) may want to view and interact with the data. It would be extremely cumbersome and impractical to require each user to remember the entire set of different page names required to create these views. Even with simple page names like 'mypage1', 'mypage2', or meaningful names like 'mycustomers', 'myinventory', the task would be daunting and impractical. Thankfully, there is a way for the user to be alleviated of remembering the names of any pages beyond an initial entry page into the application (usually called the *home page*). This can be done using hypertext.

What the inventor of the Web, Tim Berners-Lee, came up with was the ability to embed the page names inside the web pages in such a way that the user only has to click on a visual link to invoke the browser to fetch, or *GET*, the next desired page from the web server. This saves having to type the name into the browser or from having to know or remember it in the first place. Furthermore, the method used by HTML to encode these links enables the page names to be masked behind clickable text or images that could say something meaningful like 'click here[20] for more information about this customer' as

---

[20] Actually, using an explicit linking phrase like 'click here' is considered by the hypermedia aficionados to be against good design principles.

opposed to 'get MyCustomerInDetail' or some other obscure label (obfuscation being what software engineers traditionally do best). In this way, we can continue to interact with the server without having to remember complex URLs or without having to know anything about the SQL behind the resultant queries that the web server runs against the database server.

It is worth stating that even if there are thousands of different pages, there is no need to link to all of them on every single page as a means to allow the user to navigate easily to any page in the entire navigational space. The natural flow of tasks in any application usually entails transitioning logically from one related state to another and the number of 'next state' options in any sequence is usually naturally limited to a few possibilities rather than the entire page set. Branching to another sequence in the application flow is usually done by returning to some fixed and known starting point (initial state), which invariably has come to be regarded as the function of the home page in websites. Web usability gurus talk about the desirability of designing a navigational model that allows a user to branch to any part of the application space in no less than three clicks on average. The sparseness of an application's navigational options from any one state to another is what makes hyperlinking an effective means to navigate through an application, even though the origins of hyperlinking are in branching off to separate information spaces, such as one might do using a footnote or external reference in an academic paper.

As just noted, although it works well within an application, hyperlinking is not limited to destinations on the same web server; links can refer to other web servers anywhere on the Internet. Therein lays one of the most powerful characteristics of the Web: the ability to refer to another completely independent resource without the permission of that resource (i.e. no authentication is required on the target server). This allows a massively connected information space to develop, which is how the World Wide Web grew so rapidly. Branching to any destination on the Web is intrinsically supported by both the URL format and the associated GET method that Berners-Lee came up with to fetch pages. This combination includes not only the ability to address a page on a server, but the web server itself on which the page resides, as specified using a host name, such as those we are familiar with today, like 'www.myserver.com'. The default behaviour of HTTP and a web server is not to authenticate a user, which means that it is easy to access any web resource unfettered by such concerns.

Given that the web server is programmatically generating the web pages, it can dynamically embed the URLs beneath the user-friendly hyperlinks according to the user's current position within the particular navigational flow they have charted through the database. For example, if the user is paging through a sequence of customer records, a link can be included on each page that says 'next' under which is the URL for the next page, corresponding to the next chunk of records. On the server, the web software is keeping track of where the user is in the customer record set and creating the appropriate 'next' URL to ensure that it will link to the appropriate database query required to fetch the next record set in the flow. Of course, the user is not cognisant of such details. Their view of the process is visual and symbolic.

Berners-Lee did not invent hypertext or hyperlinking, but what he did was to include a powerful means of incorporating the technique into the HTTP/HTML paradigm, which is why the 'H' in both cases stands for *hypertext*. Implied in the specification is a requirement for the browser to render hyperlinks on the user's screen in a way that is actionable. Initially, the familiar underlined blue font was the default indication of a hyperlink used in conjunction with an altered-state cursor (e.g. pointing finger) when hovered over a link. Images can now be used for links and there is no restriction on link text having to be underlined and blue, although the convention remains a useful one that aids usability by familiarity.

In order to understand the tie-up between HTTP, HTML and specifying database queries, we return to the example of the SQL query to lookup a customer order for books. There is one element of the mechanism that we did not yet address, which is how to encode into the page name which SQL query we want to run on the server. Although we can implement a mechanism to associate page names with SQL queries, we also need a means by which to feed in any required parameters that a particular SQL query needs to operate, such as the customer name or customer number. This is one of the original design goals of the Web that Berners-Lee was interested in, namely the ability to include *search parameters* in a page request. The method devised is to append the search parameters to the URL, so we end up with constructs like:

GET MyCustomerInDetail ? CustID=123

Here, the question mark denotes that the information following it are parameters, and then we have a parameter name ('CustID') followed by its particular value ('123'). In this way, we only need one page on the server that runs the corresponding SQL query, but we can feed it any parameters we like. As we shall see in Chapter 7, the actual GET command is embedded in the protocol requests that are sent from the browser, so this is something the user never sees. What they do see is the URL, which contains the server address for specifying which server is referenced, and then the page name and then the parameters, all prefixed with the protocol name (HTTP), so we end up with the familiar pattern:

http://www.myserver.com/MyCustomerInDetail?CustID=123

It is unlikely that the user would ever have to type this into their browser. They would see a user-friendly hyperlink, such as 'view customer'. Clicking on this link would invoke a GET request to the above URL.

5.3.4 Challenge 4: The Need for a Delivery Mechanism

We have already indicated that we need a browser to view the formatted information. The browser has to communicate with a web server to request the required information set. The web server then in turn requests the required data from the database server (using HTTP), programmatically adds the formatting markup (using HTML) and now it needs a way to send this information back up to the browser in response to its request. HTTP provides the necessary mechanism to return data. The protocol is a two-way process. A command, like GET, is always met with a response, which leads us to think of this protocol as a 'fetch-response' mechanism, as shown in Figure 5.9.

Solution 4 Web servers sit on the network constantly listening for requests. A web server is a software application that runs on a standard PC or workstation, particularly computers running the Windows and Unix-variant operating systems (and others, too[21]). Web server programs are multithreaded, which means they can listen to more than one request at the

---

[21] Originally web servers were based on Unix and still remain popular on variants such as Linux, notably the open-source server called Apache. Windows is also popular, but web servers have migrated to high-end business platforms like IBM's AS400.

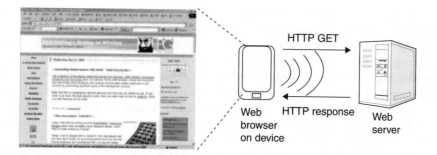

Figure 5.9 The get-response nature of web communications.

same time and handle more than one request at once. This is an essential part of their design, as they necessarily have to be capable of handling more than one user, otherwise they would be very limited for hosting scalable web applications.

Requests are made to the server and responses are sent back, as shown in Figure 5.9. The requests are actually called GET methods and HTTP specifies the exact structure of requests and responses, which is an Internet Engineering Task Force (IETF) protocol. In addition to the GET method, Tim also added another simple command called *POST* that is used to send text information along with the page request. This is a mechanism for uploading information to the server, such as could be fed into the database or used as parameters in the processing of a request. There is a mechanism for getting data from the user into the POST message, which is the familiar form on a web page. When the user hits the SUBMIT button, the contents of the form are sent to the server using POST.

We should note again that the HTTP protocol does not require any authentication between the browser and the server. In effect, the browser acts anonymously and the web server responds (see sidebar entitled 'Security (insecurity) of anonymous use'). All inbound requests that are valid (i.e. properly formatted requests for a page that exists on the server) will be responded to without asking for user credentials. Authenticating has to be added on as an extension to HTTP or at a different layer to the protocol, e.g. either within the IP packets themselves, or within the application, such as using the URL parameters to identify users, like 'GET mypage ? user=Tom' and then using forms to gather user credentials and subsequent server logic to filter out unauthorised users.

Sidebar: Security (insecurity) of anonymous use

We have mentioned in the main text that the HTTP protocol, unlike other IP protocols such as Post Office Protocol 3 (POP3) and File Transfer Protocol (FTP), does not require authentication. We should be clear about what this means and also whether or not it has any weaknesses. We should understand that HTTP can operate with a complete absence of any authentication mechanism, which is the so-called *anonymous user* mode of operation, and is very much the default for the majority of web servers and the most widely used mode for serving pages. There is no authentication at all in anonymous user mode; it is not that authentication is simply being bypassed with default usernames and passwords, just that there really is no authentication at all.

We already identified that by not placing any burden of authentication on either the client or the server, we make the Web more agile and this has helped contribute to its growth. HTTP is a very lightweight protocol that removes much of the paraphernalia that might otherwise get in the way of serving or viewing pages. The lack of authentication has also helped towards the culture of openness on the web where sharing of information is encouraged, made easy and very much the norm. The potential downside to this is that protection of data, where required, is somewhat of an afterthought. Whilst we should applaud the anonymity of the Web, we should recognise that, at times, some protection might still be required and useful.

Where authentication has proven to be required is in the establishment of online communities, many of which have naturally emerged from the grouping of interests as more and more users engage the Web. The other essential use of authentication is in membership of sites that offer additional services to subscribers, possibly for a fee, and so authentication becomes closely linked with the ability to make money. Generally, this is true for all financial transactions on the Web; authentication becomes important.

Some simple models for authentication have arisen and which are easily supported by HTTP. There are a variety of ways to implement authentication, but, in all cases, we don't get very far in our consideration of authentication before the wider topic of security becomes a dominating anxiety; not least, our natural concern for passing authentication information (usernames and passwords) securely. To do so, we have to consider the related issue of encryption. These are topics we discuss in more depth in the coming chapters of the book.

The Internet is extremely easy to connect with and this means that it is easy for unscrupulous people to plug in and get up to their antics, such as spying on authentication information in order to gain illegal access to net-based resources, some of which could be quite sensitive. An increase in surreptitious activity can be expected, and has been observed, the more people who join the Internet community.

Clearly, we need effective means of securing Internet, especially web-based financial transactions. Such mechanisms need to be extendable into the RF and device networks, or at least provide the means to interoperate successfully and without compromise.

A lot was said during the early WAP days about potential security issues. In fact, the marketplace was in confusion with some pundits saying that WAP was secure, and others saying it wasn't. Some commentators even mentioned that WAP security didn't matter because, in the case of GSM, the RF network offered enough security of its own (meaning the bit of the RF network that is the actual RF connection, the so-called *air interface*, which is encrypted in GSM and other digital cellular standards).

Later in the book, we shall address these issues and I hope to eliminate some of the confusion and myths about wireless security, an area that yet again seems in turmoil in the WiFi domain, this time also due to lack of understanding (and some poor system and service design, including lack of end-user education and support). Security, but its very nature, continues to be a source of anxiety and one that can impede progress in data communications if it is poorly understood (or even if well understood).

We shall look at the HTTP protocol in more depth later (see Chapter 7), as it is so vital to the infrastructure of modern mobile services, not only in the contexts we have been discussing so far, but in new and exciting areas like gaining access to resources within the RF network, such as location of the users.

There is one final aspect to the HTTP protocol that we should consider, namely its ability to support content format negotiation. Whenever a GET request is made to a web server, the browser is expected to issue information about the type of content that it is able to display. This enables the web server with the opportunity to dish up content that is formatted accordingly. In mobile services, this potential becomes increasingly important, as there is a greater diversity of device content-handling capabilities. This is why we have seen the mechanism for content negotiation become more sophisticated since the advent of HTTP, with working groups coming together solely to formulate standards initiatives in this area alone. It is such an important area that initiatives have grown out of both the Internet world and the mobile world, particularly the 3G Partnership Project (3GPP) and the WAP Forum.

5.4   DID WE NEED HTTP AND HTML?

People are often confused about why we needed, or need, a new protocol called HTTP to facilitate the information sharing process that we have just been describing. After all, Internet protocols for information exchange already existed, including ones for passing text back and forth between two computers, like the email protocols (POP3, SMTP) or file-transfer protocols (FTP) and so on (Newsnet, etc.). It is sensible to ask why these protocols were not adequate for the request-response processing of HTML files.

The answer is that HTTP and HTML are closely matched with the issues we have just been examining, whereas the other protocols are not. HTTP with HTML has the ability to:

- Facilitate file requests from servers using a simple addressing scheme that can be easily mapped to a programmatic request on the target server; for example, to get information from a pre-written HTML file on a disk (*static HTML*) or from a database (subsequently manipulated into a HTML file, called *dynamic HTML files*).

- Support anonymous information requests, which makes for an agile system.

- Refer users seamlessly from one resource to another, on the same server or even on another server anywhere on the IP network, all made easier by anonymous access (lack of authentication).

- Support content format negotiation.

- Provide a foundation for universal data access; a common interface for disparate information sets and applications.

Within HTTP there is the implied idea that the information being requested is usually HTML. This is why we have, for example, the intimate tie-up between submitting a HTML form and the POST command; a *form* being a generic means of attaining information from a user. There is also the assumption that hyperlinks can be used to jump to the next resource

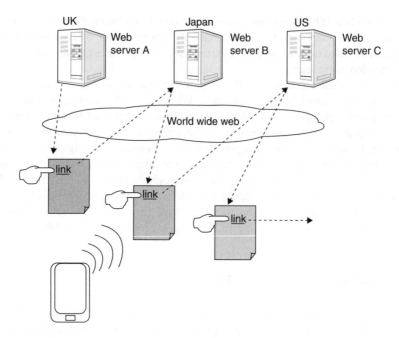

Figure 5.10 Hyperlinking with HTML/HTTP creates the World Wide Web (WWW).

request and this is supported in the protocol, including, importantly, the ability to encode in hyperlinks a means to address resources anywhere on the Internet, as shown in Figure 5.10.

However, it should be made clear that there is no other reason for the requested content to be limited to HTML. This will become significant later on from our discussion of Web Services in Section 5.7.

The Web model is very simple – it is a constant stream of GET or POST commands resulting in pages being constructed and returned from the server. Since its release, HTTP has remained a straightforward protocol that adequately supports this simple paradigm. Because the protocol is stateless[22], does not require authentication, and is text based, it is relatively easy to implement in software and this has contributed to its popularity. The simple paradigm also turns out to be generally useful for a wide range of applications beyond the originally intended human-navigated journey through an electronic document landscape.

Later on in the book, when we discuss CS software technologies, we shall see that web servers have evolved into very powerful distributed software platforms. These are able to work together in clusters (server farms) to cater for huge demand, complex failure modes and to provide many other powerful and complex infrastructural functions that we do not wish to discuss here, but which are important. We shall also see that the programmatic means of combining markup with database information are also a lot more powerful than perhaps indicated in the discussion so far, and for many good reasons that we shall also consider later.

---

[22] The meaning of *stateless* in this context shall become clearer in Chapter 7 where we look at HTTP in some depth.

More wireless challenges

On the surface, there appears to be no need to change the fundamental hyperlink paradigm when switching to wireless access for a handheld device, so we might wonder if we need a special version ('wireless web'), or will the standard one do. Of course, the driving factor towards wireless access is that there is already so much momentum behind the Web that it seems inevitable that we would want to access its benefits whilst on the move, or generally from anywhere. It seems evidently a good prospect to base the wireless information-access world, if there is a distinction from the wired one (something we shall consider in the next chapter), on a paradigm that has already proven to be effective for information sharing in the tethered desktop environment. However, if we proceed down this avenue of pursuing wireless access to the Web, then we need to think about the following potential problems:

a. Do the GET/POST commands running over IP involve any unnecessary overheads that would take up valuable bandwidth on a relatively slower wireless connection? We would not want to devote too much of our connection resource to merely issuing commands back and forth as opposed to sending the actual information we want to access. We shall find out in Chapter 7 that HTTP/IP turns out to be a relatively inefficient protocol from this point of view (i.e. raw bandwidth requirements).

b. Is there any means by which we can compress the data that's being sent back and forth to improve speed on a slow link?

c. What about the URLs? Are these going to prove to be too long to have to enter by hand in a small device (possibly with a tiny alphanumeric keypad) during the times that we do have to enter them by hand? We will examine this issue later in the book.

d. What if our communications pathway gets disrupted due to temporary loss of RF connectivity, which is common on some types of RF network? Will HTTP withstand such disruptions?

e. Is the visualisation model for HTML usable for small devices? (For some allusions to the answer to this question, see discussion about the WAP services model in Chapter 3.)

5.5 OVERCOMING WEB LIMITATIONS WITH WEB 2.0'S AJAX, WIDGETS AND OTHER GOODIES

The Web is a wonderful system for liberating data, but what are its limitations? The biggest challenges with HTTP/HTML are the delays in the get-response cycle and the fact that any updates to the information on a page requires a whole new get-response cycle and re-rendering of the page. As we have seen, the basis for the architecture of the Web is the idea of accessing and viewing information as pages. Clearly, this is a publishing-inspired idea, which isn't surprising given the publishing origins of the Web. However, with general purpose computing, interfaces are frequently not page based. For want of a better word, they are widget or component based. In other words, the interface is usually divided into areas of screen space, each with its own display and interactivity functions. Think of the iTunes interface as an example. There are multiple areas ('components') in the interface, each with a different function. There are the play controls, the library window, the playlists and device window, plus others. It is not at all 'page like' as we see in a web browser.

If we try to use a web page to deliver a more generic user interface akin to iTunes or other applications then it becomes difficult. HTML doesn't support the rich graphics, but aside from that the main challenge is the responsiveness of the interface when using the HTTP/HTML paradigm. Let's say I did create an iTunes type of service online and wanted to use the browser as the interface, not a dedicated program. Imagine the user can see his or her playlists in one area of the page and then clicks on one of them in order to view the contents of the playlist in a separate part of the page. As we have learnt, to get information for that playlist, we have to go back to the web server to ask it to get the data from the database and then package the data into the appropriate mark-up template to allow it to be displayed. Subsequently, this page is sent back in the HTTP response and then the browser will take time to reload this page to replace the old one. The user can now see his or her playlist. The main problem here is the delay in fetching the new data and the additional delay in re-rendering the page so that the user can see the playlist. The whole page is re-rendered even though we only want to update part of it, which is the part where we display the contents of the playlist. Then, if the page happens to be very large, there is the added delay in transporting the page data back to the browser and it will also take longer to render into the browser window.

A solution to the problem is to allow the browser to ask the server just for the new data, not the new page, and then to insert this data into the playlist area of the page without having to re-render the entire page. Better still, what if the browser can request all data for all of the playlists ahead of time and store them locally in the browser so that when a playlist is requested, the user doesn't experience the delay in fetching from the server – it has already been pre-fetched. This is essentially what AJAX is about. AJAX stands for Asynchronous JavaScript and XML. Why? The asynchronous bit means that the browser can fetch data from the server without waiting for the user to hit a button or link first. The XML bit is because that's how the data is packaged when it comes back from the server. What about the JavaScript? This is a computing programming language that the browser can execute. It is a language specifically designed to allow elements within a web page to be processed. It is general purpose enough to be used for a variety of programming tasks. For example, I could use it to implement a form in a web page that calculates tax depending on the entries from the user. In the AJAX model, JavaScript is used to run a program in the background that does the incremental fetching of data from the server, stores it in readiness for rendering and then finally updates the relevant part of the page by injecting the data into the portion of the HTML template that needs updating, such as the playlists area in our example.

Figure 5.11 shows a simplified example of the AJAX process. Using HTTP, the browser JavaScript engine *pre-fetches* the playlist data from the web server. The server returns the data in XML format. The engine then stores this data in its local memory cache in readiness to send it to the browser renderer to update the playlist area, as indicated in the diagram.

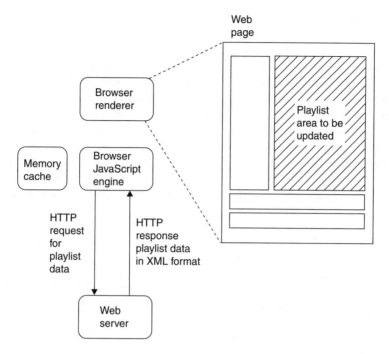

Figure 5.11 AJAX architecture.

Now, when the user clicks on one of the areas in the page to display the playlist in the playlist area, the data is already stored in the memory cache and is ready to go. The delay in fetching the data has been removed from the user-experience. Only the playlist part of the page is updated with the new data.

What then are the challenges for using AJAX for wireless? The increased response time in updating the interface has come at the expense of fetching data in advance. If the user had 100 playlists, then this would require the JavaScript to go fetch 100 playlist XML snippets from the server. This would be a naive implementation because of the significant overhead incurred with 100 HTTP cycles. A better implementation would put all the playlist data into one XML file and this would require only a single pre-fetch in the background. However, it might still be a large amount of data to transfer. The browser must have sufficient memory to store pre-fetched data, which is not a problem on a desktop PC, but could be very problematic with a memory-starved mobile device. Fetching the data also requires bandwidth, which is not a problem on a fat broadband pipe, but with many wireless connections that overhead might cause problems, particularly with additional costs of transferring the data, which might not get used in any case (if the user only looks at a few playlists and not all 100 of them).

The idea of pre-fetching data is ancient. It is a trick already used inside microprocessors to speed up execution. It has also been used in mobile for a long time, especially because when mobile data was first possible the links were very slow. Therefore, with some applications, most notably enterprise ones, it was worth pre-fetching the data. This was done in a variety of ways before AJAX was possible. As a programming method, which is what AJAX is, pre-fetching is a well-known technique and some wireless programmers use it in embedded applications that are loaded onto the phone (i.e., not browser-based applications). In particular, the idea of pre-fetching has become popular with various implementations of *On-Device Portals* (ODP). These are applications that run on the device and present various digital content, such as ringtones and videos, to the users for them to purchase. ODPs are used instead of browsers for accessing the operator's licensed content. This is because the user-experience can be much better with certain embedded approaches, as we shall discuss in Chapter 11. Pre-fetching is used to populate the ODP with the latest and most popular content so that users can experience instant gratification when trying or buying.

Recently there have been attempts to fuse the ODP and browser approaches, essentially by implementing an ODP using web-programming standards (e.g. HTML). An increasingly interesting idea within this model is the use of *widgets*. These are tiny programs, usually with a very narrow and single function, such as presenting the local weather. These can be viewed instantly. There are various ways to do this. On the Mac OS X, widgets occupy a virtual screen space that can be switched to instantly. Again, data is usually pre-fetched to the widgets, which are running in the 'background' and therefore always updating themselves with the latest data. On mobiles, it is still early days for widgets, but there are various attempts to implement them. The advantage of a widget is its persistence and instant availability, however that is managed on a mobile. A similar problem still remains with mobile in that widgets will eat up bandwidth, which will have an ongoing battery drain and a possible incremental data cost. However, as mobiles continue to become more efficient in terms of power consumption and data connections, the use of these techniques becomes more feasible.

5.6 SIDESTEPPING THE WEB WITH P2P INTERACTION

Thus far, in our discussion, we have assumed that the databases we are trying to access are not local to the devices, but are hosted on database servers in the content network, accessible via the IP network, using intervening web servers somewhere on the Internet. In our high-level network model developed in Chapter 3, these network entities (servers) are situated in the content layer of the network (see Figure 5.12). The HTTP connection has been operating over the IP layer, and is going to become our dominant focus when we scrutinise this layer in more depth throughout the book, especially in Chapter 7.

As Figure 5.12 shows, P2P interaction does not appear to directly involve the content network. In all probability, the information sets that we want to access in P2P mode are on the devices themselves, and not on database servers in the content network. We can still think of the information as sitting in databases, but these are device-bound and will be unlike the powerful database servers in the content network. It is tempting to think of the P2P mode as a bit like folding up the content network into the devices themselves, but this denies the entirely different characteristics of a P2P architecture compared to CS interaction.

With mobile P2P applications, the content itself will almost certainly be restricted to certain types and forms. Users will share files, such as ringtones (see sidebar 'Ringtone files'), MP3 files, pictures and video clips. File sharing is the original P2P application, as popularised by Gnutella for general file swapping and Napster for MP3 music files, until the music industry shut it down. There are specific types of information, such as address books and diaries, which are likely to figure strongly in P2P applications. We shall examine this in more detail when we look at location-based services later in the book.

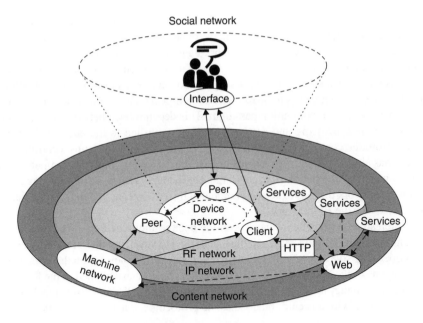

Figure 5.12 The web server in the mobile services network ecosystem.

Sidebar: Ringtone files

Ringtones have proven to be very popular with mobile phone users – part of the personalisation of phones that has become an important element of mobile service design consideration.

One of the reasons that ringtones have been popular is that it has been relatively easy to get them. This is because ringtones can be described in very small files, small enough to allow them to be sent via a text message. Nokia, who incorporate its support in their phones from very early on in the GSM market development, invented the most popular ringtone format. The format is the Ringing Tones Text Transfer Language (RTTTL) specification. An example of using RTTTL to describe musical notes is shown in the following caption.

The caption shows the musical notes both on a musical stave, and the textual description of these notes according to the RTTTL format.

It is easy to imagine the swapping of ringtones in a P2P network, especially if a group of friends are in both a single peer group and in adjoining peer groups. A tone file could be searched for on immediate peers, and if not found there, then on the adjoining peers, and so on.

For phones capable of sounding more than one note at a time (polyphonic) then the Scalable Polyphony MIDI (SP-MIDI – where 'MIDI' stands for Musical Instrument Digital Interface) content format is a similar type of scheme for textually describing musical sequences.

The other aspect of P2P computing that makes it different from the CS approach is that the reference to other resources achieved by hyperlinking in the Web model is not so straightforward within a decentralised networking model. Let's say that we want to access a certain ringtone file we have heard about, called 'Groovy Toon'. Our device would look for 'Groovy Toon' on peers that it knows about (i.e. how to connect to). If it does not find the file on those peers, then it asks to be referred to their peers, and so on. This peer hopping is a crucial part of the P2P computing paradigm. This decentralised networking arrangement is one where the processing and service provision is pushed to the edge of the network and away from the centre, as shown in Figure 5.13. For obvious reasons, centralisation of resources has always offered efficiencies in any kind of system (not just computing ones), especially when the number of users gets large. Deliberately removing this advantage places strain on the network that has to be accommodated by other means and may have deleterious consequences for the RF network.

It is interesting to consider at what level the peer communications take place. Systems like Gnutella use the IP network to establish peering. However, another possibility exists for devices operating solely on the mobile network, especially if they are on the same network (i.e. run by one of the operators). A P2P connection could be established in the RF network layer without leaving the RF network infrastructure. With P2P protocol suites like JXTA, there is no reason why this can't be done as the protocols are network-agnostic.

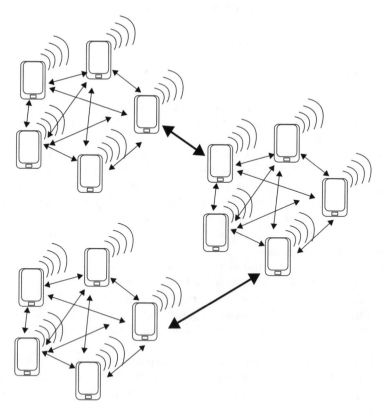

Figure 5.13 P2P networking.

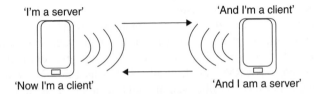

'I'm a server'   'And I'm a client'

'Now I'm a client'   'And I am a server'

Figure 5.14 Thinking of P2P as client–server back-to-back.

Efficient P2P communications within the RF network is something the operator could make available to the developer community by offering device-programming interfaces that support such a mode. The relative pros and cons would be interesting to understand.

Purely as an aid to understanding P2P communications, it may be useful to imagine that much of the time P2P communications is a one-to-one bi-directional CS arrangement, as shown in Figure 5.14; although, we should keep in mind that otherwise these two paradigms are very different. The differences shall become obvious when we look at how to build mobile application servers for CS usage and later when we look at the architecture of devices. We should also keep in mind that we are not confining our view of P2P to the publishing model that we described earlier when developing our understanding of the Web. Whilst it is perfectly feasible for one device to browse content on another within a P2P context, the P2P paradigm has other objectives. Such objectives might include file swapping and other sharing paradigms to do with social interaction, whether for pleasure or business, such as variants of instant messaging and whiteboard sharing, where the lack of intervening central servers might facilitate an ad hoc approach to networking that makes it easier for new services to be adopted by interested user groups.

5.7 GOING BEYOND PUBLISHING WITH WEB SERVICES

There has been significant momentum concerning the adoption of the Web paradigm for all kinds of applications. It is so easy for anyone to publish information, particularly if the resulting web pages do not require database queries and can be constructed as static pages using a web-publishing tool, such as CityDesk 2.0[23] from Fog Creek Software or the popular Microsoft FrontPage[24], which is part of Microsoft Office.

As of 14 September 2003, when I first wrote this chapter, a glance at the homepage of Google[25], the most popular search engine on the Web, showed the following line:

Searching 3,307,998,701 web pages

That was a lot of pages back then, Google no longer indicate how many pages, but it is obviously a much bigger number.

---

[23] http://www.fogcreek.com/citydesk/
[24] http://www.microsoft.com/frontpage/
[25] http://www.google.com

The popularity of the Web has resulted in widely available infrastructure, including web servers, routers and firewalls configured (and optimised) for web traffic. The other infrastructural momentum is that which has arisen around the number of software solutions with embedded web access or HTTP awareness. This ranges from high-end enterprise systems, such as popular CRM[26] products like SAP, which now has a web interface, to software development tools and libraries, of which there are a plethora offering built-in web support. It is easy for a relatively novice software engineer to construct a program that is able to network with the Web. The motivation is to consume the information into the innards of an application, not to send it to a browser, although that is where some of it may end up depending on the design and purpose of the system. The use of the Web to exchange information between programs, as opposed to a browser and a server, is called *Web Services*.

The prevalence of the infrastructure also means it has an increasingly low-cost base. It is very cheap to 'web-enable' a product, and not just software products. It is increasingly common for machines to be networked using web infrastructure. In fact, an entire web server is now available as a chip. In machine-to-machine applications, the trend is towards using a web model to connect machines, rather than proprietary models that have been in use for quite a few years.

What we have been considering in the preceding discussion about the origins of the Web is a view of the Web as a tool for human-centric document publishing or information sharing. The 'request–response' protocol paradigm of HTTP has been used to request chunks of HTML, the intent being to eventually render the output visually in a browser.

What is becoming apparent is that the Web can provide a useful backbone for any applications to swap information. For example, an inventory-checking system in one business could easily use the Web to talk to a stock-ordering system in another business, thus automatically ordering new stock.

For these embedded applications, HTML is no longer an ideal means of annotating data for this type of application interchange. The application requesting the information from another is more interested in the structure and meaning of the raw information, rather than how to display it (it may not be displayed at all). The motivation is to consume the information into the innards of an application, not send it to a browser, although that is where some of it may end up depending on the design and purpose of the system.

In the case of stock inventory, rather than markup information to structure its visual display, it is preferable to embed markup to indicate the business-specific structure of the data, so that our application can identify interesting primitives within the information, such as item names, part numbers, and current prices and so on.

The way of marking up the structure of data for application interchange is to define our own markup language, or tag set, specific to the particular application that's being implemented. This is possible using something called *Extensible Markup Language*, known by its acronym XML.

Within certain limits defined by the basic grammatical assumptions that XML imposes (such as the use of angular braces '< >' and so on), we can have any tags we like, and it is

---

[26] CRM stands for Customer Relationship Management – check out Chapter 3.

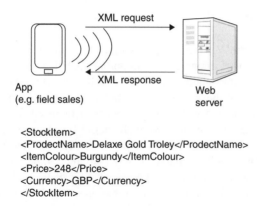

```
<StockItem>
<ProductName>Delaxe Gold Troley</ProductName>
<ItemColour>Burgundy</ItemColour>
<Price>248</Price>
<Currency>GBP</Currency>
</StockItem>
```

Figure 5.15 Applications talking across the web using XML.

up to our applications to make meaning of them. Therefore, for the example of inventory again, we could propose the use of custom tags like:

```
<StockItem>
        <ProductName>Deluxe Gold Trolley</ProductName>
        <ItemColour>Burgundy</ItemColour>
        <Price>248</Price>
        <Currency>GBP</Currency>
</StockItem>
```

This kind of data interchange using XML could be between any applications that are able to connect with the Web. Figure 5.15 shows a mobile device running a field sales application that is gathering updated stock information automatically (i.e. without user intervention). Any device connected to the Web can engage in Web Services dialogue (see also Chapter 7), it does not have to be wireless. When we look at the RF network later in Chapter 12, we shall see examples of where Web Services are very important for mobile service delivery, even though the Web Services aspects do not involve wireless, but take place between servers in the content network and application gateways in the RF network.

We shall look at XML later in the book, but we needed to introduce it now in order to understand how the Web has evolved since its inception. It has grown from being a purely human-centric Web to one that is also application-centric. Instead of pulling down HTML files from a server, Web Services request and post XML files. We may have realised from the above example, that XML is a plain-text format and has no special dependency on a particular operating system or programming language, which means it is another example of a universal model that means even more computers can connect to the application-centric Web, thereby increasing the value of the Web even further.

> A Web Service is any service that is available over the Web using XML to exchange messages, and not tied to any particular operating system or computer language.

5.8 SEMANTIC WEB

We have looked at how the Web is a major networking force today, not only for human-centric uses where visualisation of information is important, but also for application-centric uses where basic exchange of information enables physically separate applications to collaborate.

Most of the content on the Web today has human consumption in mind and therefore the semantic meanings of the web pages are what their visual form conveys implicitly. For example, a web page displaying a photo album only means something to a person who knows what the pictures contain and to whom the content has relevance and this information about the photo album is only available visually. An appointment in a web-based events diary only has meaning to a person who knows what the event is about and to whom it is relevant. In other words, the meaning of information on the Web is usually conveyed by whatever is visibly evident on the page. Furthermore, these meanings are only accessible to human consumers, not to applications running on machines, because only humans, not machines, can see the pages.

It would be useful if applications could understand the meaning of information on the Web. For example, we might want to search the Web for news information relating to a particular field of interest. Probably, whenever we look at a news item on a page, we know that it is news. We can tell by certain formatting and context, even if the word 'news' is not itself displayed on the page. However, we need a way to enable an application to determine that a particular web page contains a news story. We can do this using the *Semantic Web*.

> **Semantic (*adjective*):** of or relating to meaning in language

The Semantic Web is an attempt to add lots of metadata to the information contained in web pages so that applications can recognise the meaning of the information described (i.e. comprehend it). Physically, the Semantic Web is not a different web from the one we have so far introduced and discussed, it is an alternative context with semantic qualities accessible to applications, not just humans. To understand this concept better, it is probably best if we introduce some of the ways that the Semantics Web uses to convey semantics.

Firstly, it should come as no surprise that the manner of describing semantics is through XML. The beauty of XML is that it is both human and machine-readable. It is relatively easy to write a software program to recognise XML tags, find them in an XML message (called *parsing*) and then extract the information contained in the tagged fields. In fact, the use of XML has become so prevalent in general, that there are lots of software tools and libraries to make XML processing easy and cost effective.

Secondly, we need a framework to formalise an approach towards describing semantic information. This has emerged recently in the form of the *Resource Description Framework* (RDF). Whilst XML allows us to add any structure we like to information, it doesn't actually tell us what the structure means. This is where the RDF comes in.

In just the same way that we describe things in human language, using constructs such as sentences that have a subject, a verb and an object, RDF provides a similar ability that is accessible to machines. For example, we could use RDF to make an assertion about a particular web page by giving it properties, such as 'this page contains news', and say things like 'this person is the author of this news item', where 'this person' is someone's name

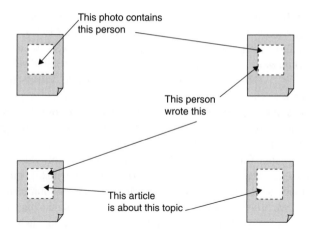

Figure 5.16 Adding meaning to content create a Semantic Web.

found in another web page. We can better understand these semantics by examining the concept pictorially, as shown in Figure 5.16.

In Figure 5.16, the writing on the arrows represents the *annotations* of the underlying web-based information. These annotations enable use to understand the meaning of content in the pages. Without these annotations, we would not understand the meaning of the information implied by the symbols, nor would we understand the relationships between them that enable us to gain more insight into the meanings. This is exactly the same problem that computer applications face, unless we state the meanings. In the Semantic Web, we use RDF to make these annotations (to add semantics).

The need for machine-digestible semantic annotations first arose because it became increasingly difficult to find information on the Web due to it growing so huge. From Figure 5.16 we can see that the semantic labelling can take advantage of the interconnectedness of the Web. An RDF description can say things about one resource on the Web that links its meaning to another resource. As the figure shows, we can say that a certain page contains a photo of someone whose biography is contained in another page. When we get to that page, we can use semantic labelling to give meaning to the content, such as indicating that it is biographical in nature. Within the biography page, XML can identify biographical data, like first name, second name, date of birth, educational history and so on.

```
<Biography>
        <FirstName>Joe</FirstName>
        <SecondName>Bloggs</SecondName>
        <BirthDay>24</BirthDay>
        <BirthMonth>8</BirthMonth>
        <BirthYear>58</BirthYear>
        <Education>
        . . .
        </Education>
</Biography >
```

We shall examine this topic in more depth later in the book. However, here we can briefly mention some uses of RDF already in common usage, or beginning to emerge in the Semantic Web. Perhaps the most well known goes by the acronym of RSS, which stand for either *Really Simply Syndication* or *RDF Site Syndication*; the reasoning behind the different names is that there have been two major, and sadly separate, efforts to define an RSS semantic framework.

Putting quibbles about acronyms aside, what semantic purpose does RSS fulfil? RSS is an attempt to allow any changes in website information to be summarised in order that applications can monitor a website for changes of interest. We might think that we do not need a particular semantic framework to do this, as we could possibly watch web pages to see if their publication dates change; this information (i.e. publishing date) being a standard part of the HTTP protocol, as we shall see later in Chapter 7. However, monitoring date stamps is not useful enough. We probably want to know what information on a page has changed. For example, in the case of news items, there could be several additional news items on a page, so we would rather find out about these in particular and not be satisfied just to know that the page has been updated. We may want to know what has changed and who has changed it? We could then look out for particular news items on a particular topic by a particular author; or rather, we can use a software program to do the watching for us. RSS makes this possible.

Perhaps I have fallen into the trap of using news as an example of using RSS, in that being such a common usage of RSS, it is tempting to think that monitoring news is its only purpose. Actually, RSS is an umbrella semantic framework for tracking any website changes. For example, if we think about the possibility of posting picture messages to the Web according to where the pictures were taken, we could use RSS to add semantics for monitoring what is happening in a particular geographical location where pictures may often be posted.

If we think deeply about posting pictures from mobile phones, there is a whole range of semantic information that might be useful and for which we would need a framework. We might need to know who's in the pictures, what the pictures are of, and where they were taken (i.e. ordinate information). We could use this information to gather pictures of certain people in particular places. We could also try to find people who were in the same place, but their pictures taken by different people. The variations are probably many and a discussion of this type of application can be read in Section 13.8, where we introduce other uses of RDF within a mobile services setting, placing emphasis on location and multimedia messaging.

The Semantic Web is an important development of the Web phenomenon and it adds incredible extra dimensions to its usefulness, many of which will become especially important in mobile services. What the Semantic Web does is to enhance the networking power (Metcalfe's Law) we already discussed earlier in this chapter. It should enable us to think differently about a topic that we mentioned earlier, which is the discoverability of information in a network and the impact this has on its power. Discoverability is greatly enhanced by the addition of semantic information interweaved with our core data.

5.9 XML GLUE

Earlier in this chapter, we looked briefly at XML. It is fast becoming a universal way of describing data that passes from one application to another. This technique is not bound to any particular protocol or network, nor is it bound to any particular network architecture, whether CS, P2P or any other. From an applications perspective, as long as applications can produce and consume XML, then we don't really care how the XML passes between collaborating agents, nor where they sit in the network.

Realising the universality and network-independence of XML, it is tempting to suggest the notion of a virtual XML 'highway'. This is just an academic concept, but as we progress through the book, the idea that all software applications exchange XML messages becomes a powerful archetype by which to think about software collaboration. To implement such a highway globally would require an addressing and routing scheme. Of course, one already exists with IP, and with the advent of IPv6, we now have a big enough address space to accommodate a unique address for every device on the planet. However, we have already alluded that IP is not always the best networking option, particularly for P2P transactions within the device network (more on this later).

It is likely that all wireless applications will eventually swap information using XML vocabularies. Currently, this is not the case. For example, we have already seen that the RTTTL tone format is not XML (nor is SP-MIDI). In fact, there are plenty of non-XML message vocabularies currently in use. XML is an open standard upon which many other open software standards now rely, and for all manner of internal and external information descriptions, not just message flows. We shall be looking at many of these descriptions throughout the book, but for now, it is worth highlighting that XML is likely to form the main backbone for passing data through our mobile network, as shown in our network diagram in Figure 5.17.

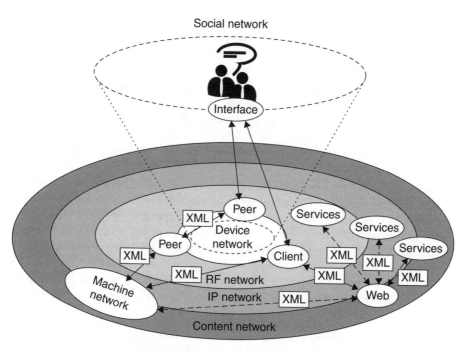

Figure 5.17 XML in the mobile ecosystem.

5.10 REAL-TIME SERVICES

Although the Web has been the dominant force behind the explosive growth of the Internet, HTTP is not the only Internet Protocol (IP) out there. For some types of application, especially those more sensitive to pushing lots of information across the network in a time-sensitive manner (i.e. real time), other protocols are important. In this section, we look at some key protocols and paradigms for real-time interaction using the Internet.

5.10.1 Multimedia Streaming

Thus far, our view of the Web has suggested a fixation with a page-request model as the dominant application model, whether the page be constructed from HTML or, as we have just discussed, a 'page' of XML[27]. Additionally, it seems that we are primarily interested in pulling down chunks of information one page at a time, doing something with them (such as reading or further processing by our application) and then requesting some more.

Depending on the availability of resources within the entire Web infrastructure at any one point in time, we might expect variable performance in the infrastructure responding to page requests, most likely on a user-by-user basis. Perhaps a server gets busy, or perhaps a communications pathway is heavily congested, or perhaps a backend database server is busy doing a computationally intensive task, such as indexing a database. During such times, page responses might be slow. This will lead to variable performance. For example, we might get a page returned within two seconds on one try and then within twenty seconds on another.

Wireless challenges

In a mobile services context, users are generally less tolerant of delays and so this is a matter for further consideration, suggesting that mechanisms that can ameliorate the variation in delay ('jitter') would be useful to implement in mobile networks.

With applications where the performance of the system manifested in time is not critical to the user's successful usage of it, we can think of these as being non-real-time applications. However, there is a class of applications called real-time applications. These are where the delivery of data within certain limits is essential for acceptable performance. These are 'time critical' applications, such as listening to audio files streaming across the Internet. Later in Chapter 12, when we discuss the RF network, we shall explain the concept of voice digitisation, as this topic underpins the design of digital cellular networks for real-time speech transmission. For now, all we need to know is that it is possible to take digitally

---

[27] In fact, as we shall see in the following chapters, HTML in its newest form, called *XHTML*, is now an XML-based language, which previously it wasn't, despite appearances (i.e. stark similarities).

recorded voice and slice it into small audio files, which when played successively can be heard as one contiguous audio presentation faithful to the original recording. We would not notice the interruptions between the small audio files, as they are seamlessly stitched back together using audio processing techniques (also known as Digital Signal Processing, or DSP).

We can do this sliced transmission in two ways. We can wait for all the audio files to arrive at the receiver before being reassembled into an entire audio track. As long as we do not have a particular time limit for hearing the audio, this mode will work and we can think of it as being non-real-time mode; when it's ready, it's ready, and then we can listen to it.

The other method for transmitting an audio track is to receive the mini audio files and try playing them as we receive them, one at a time, stitching them back together 'on-the-fly', as we go along. Here, we can appreciate that the performance of the network is more crucial, especially its delay characteristics. For example, if we receive the audio in two-second[28] chunks of sound at a time, by playing these back in succession (without buffering[29]) means that from the time we start playback of a chunk, we have two seconds to receive the next file and stitch it on, before the user needs to hear it. Otherwise, if it takes more than two seconds to receive and cue up the next chunk, we will have to endure a period of interruption in the audio. For example, if the network becomes loaded during playback and we have to wait eight seconds to receive the next chunk, then we have a six-second gap to fill, which for most audio applications is probably intolerable. Were this to happen regularly during playback then the whole experience would become insufferable.

Using the most primitive forms of IP communication (i.e. the basic packet-data mechanism), there is no guarantee that we shall receive the audio files in order, especially due to routing differences[30]. The Internet is a network of networks, and so the ways of traversing those networks from A to B are many. If audio file N and audio file $N + 1$ take different routes, it is possible that $N + 1$ could arrive before N, especially if N has a particularly long route dogged by resource problems such as congestion; this possibility is shown in Figure 5.18. Therefore, in addition to ensuring the timeliness of files arriving (i.e. mitigating excessive and variable delay), we also need to ensure that we have a method for re-ordering audio chunks (or any other real-time media chunks) should they get out of order.

Fortunately, IP-based protocols have been designed to allow real-time streaming of data for media applications, exactly to overcome the problems we have just been exploring. Protocols such as Real Time Streaming Protocol[31] (RTSP) and Real-time Transport Protocol[32] (RTP) are available for such applications.

The difference between these two protocols is that RTP is the actual means of enabling the streaming process to take place vis-à-vis the chunking, sequencing and buffering process just alluded to in the foregoing discussion of real-time media data flows.

---

[28] In actuality, for various performance reasons, the chunks will be in the order of milliseconds, not seconds, but for discussion purposes it seems easier to imagine audio breaks and the playback process quantified in seconds.
[29] *Buffering* is a data communications term meaning to store in a queue.
[30] The Internet is a vast labyrinth of physical network segments tied together with devices called *routers*. Routers do not necessarily send all the traffic from one source via the same route.
[31] See http://www.rtsp.org or consult the specification [RFC 2326] at http://www.faqs.org/rfcs/rfc2326.html
[32] Consult the specification [RFC 1889] at http://www.faqs.org/rfcs/rfc1889.html

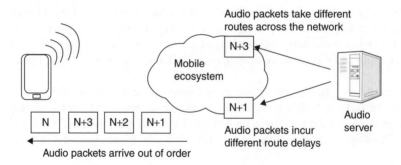

Figure 5.18 Streaming of audio files across the Net.

To quote from the RTP specification:

> *RTP provides end-to-end network transport functions suitable for applications trans-mitting real-time data, such as audio, video or simulation data, over multicast or unicast network services. RTP does not address resource reservation and does not guarantee quality-of-service for real-time services.*

Multicast means to send out the media stream to more than one receiver at the same time, whereas *unicast* is to a sole receiver. The last sentence from the extract tells us that although the RTP provides a means to implement the mechanics of 'chunked real-time transmission' (streaming) across the Internet, it does not provide any mechanisms for guaranteeing performance. That may seem a strange thing to testify given that our earlier discussion of the 'real-time problem' seemed to suggest that performance guarantees ('Quality of Service' or QoS) were exactly what we were looking for, such as might be required to avoid interruptions to the continuous playback of the media file at the receiver. However, the problem is with the essence of the Internet itself. At its core, the foundational proto-cols did not include QoS techniques. Thereafter, the Internet infrastructure (e.g. routes and switches) has grown without these mechanisms in place and this inheritance remains, a problem only recently addressed by the newer version of IP, called IP version 6 ('IPv6'), but this is not yet widespread. Wherever a Qos mechanism is available, this would need to be supportable by the RF network, too (see the sidebar entitled 'Wireless challenges for real-time streaming').

What RTP does provide is a means to enable streaming of information where the stream-ing process is cognisant of the underlying information sources, be that audio or video. For example, it can cope with chunking according to the time base of the media, including synchronising disparate sources that may be encoded using different compression tech-niques, either due to different source natures or as compelled by more stringent bandwidth availability for some sources.

RTSP is not a streaming protocol per se, despite its name. It does not facilitate the real-time transport of media files. It is better to think of it as a 'remote control' solution. If we think of the media source server on the Internet as being like a CD player, then at the receiving (client, or 'media player') end, we need a means to control the CD player, such as 'play', 'pause', 'rewind', 'select track 2' and so on. This is the function of the RTSP, to allow the exchange of these types of commands between the media player and the media server.

Sidebar: Wireless challenges for real-time streaming

In our mobile services network, as shown in its IP-centric form in Figure 5.2, information coming via the IP network still has to traverse the RF network, and potentially the device network, before arriving at applications on the device, such as a digital audio player for the real-time application we have just been considering.

Therefore, if we are able to utilise IP protocols, such as RTSP[33], which accommodate real-time services (e.g. media streaming), we need to make sure that they are sustainable across our RF and device networks, such that these networks do not become the weak links. This indicates that the designers of the RF networks for next generation mobile services have to ensure that IP protocols can be supported in general and real-time ones in particular, without degradation. However, certain RF-related issues may present greater challenges, such as greater and more unpredictable contention for resources in an RF network, thereby adding greater burden on our network not to disrupt real-time services.

Some of these issues relate to a concept called Quality of Service (QoS), which is about providing network resources such that certain performance guarantees can be made on a user-by-user basis.

Later in the book, when we look at the RF network (Chapter 12), we shall examine the concept of content-transmission techniques that can adapt to the prevailing network conditions. Indeed, this is a particular challenge for designing mobile applications that remains with us despite massive improvements in RF network technologies.

5.10.2   Session Initiation Protocol (SIP)

Session Initiation Protocol, or SIP, is a protocol new to mobile. It supports the necessary signalling between two IP-connected entities to enable them to establish a media-exchange connection, such as a streaming connection just discussed in the previous section. SIP is sometimes called 'rendezvous technology', as it allows network entities to 'meet' for the purposes of exchanging data, but doesn't get involved in the actual exchange of the data itself, as shown in Figure 5.19. SIP is a packet-based signalling protocol that can facilitate packet-based (by virtue of being IP) telephony, whenever the negotiated media connection is used for voice. The use of SIP in this fashion gives rise to a form of internet telephony, more generically referred to as Voice over IP (VoIP).

The SIP protocol is at the heart of the IP Multimedia Subsystem (IMS), which represents the next generation in mobile networking technology, as adopted by the 3GPP[34]. As their names suggest, IMS and SIP are about more than just voice. What the 3GPP had in mind originally was multimedia, although now it is widely recognised that SIP can be used to

---

[33] Real-Time Streaming Protocol – http://www.rtsp.org/
[34] Third Generation Partnership Project – www.3gpp.org

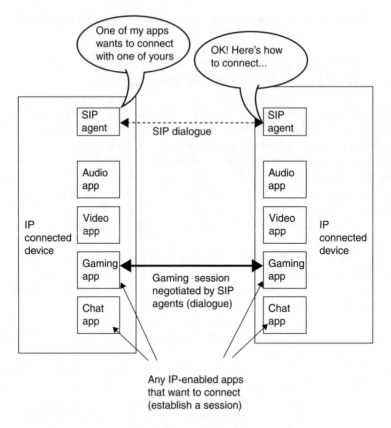

Figure 5.19 SIP agents negotiate IP-based service connections.

establish all manner of connection types and is likely to become the backbone of mobile communications in next generation networks. SIP enables any media connection to be established between two IP-connected entities. The SIP entities negotiate the connection (i.e. which sockets to use and which IP protocol) and negotiate the media type, voice or video, or both. Other sessions are possible, like gaming and IM (see Figure 5.19). We shall describe SIP in more detail in Chapter 14.

6

Client–Server Platforms for Mobile Services

In this chapter, we will look at how to deliver mobile services using the client–server (CS) architecture. CS architecture is the predominant approach for many mobile applications due its widespread usage in the Web 2.0 platform, which is now seen as the basis for building wireless services. We have already discussed that this is not necessarily the case for all services and applications, but that the significance of the Web demands our attention. The next most significant platform is probably the IMS architecture; we shall look at this in Chapter 14.

The underlying technology should be highly flexible and, as much as possible, allow the service creators to invest most of their available resources into only creating those parts of the service that are distinct and add value, rather than creating general infrastructural components. In our following discussion, we suggest that the entire mobile network (the device, RF, IP and content networks) is really only a chain of software services. In essence, a good deal of the required services are intimately associated with assets that the operators own, therefore the responsibility to make these accessible as a platform lies on their shoulders and no one else's. This is still a major challenge. We shall look at possible approaches to this problem after our consideration of the underlying software technology. There are two approaches that have emerged. One is from the telecoms world itself, which is the concept of using a SDP to manage controlled access to operator resources. The other approach comes from the Web 2.0 evolution, which is the mash-up of services. These approaches are similar in many respects, but we shall explore how they might work together. As we discussed in Chapter 3, it is difficult to ignore the Web 2.0 platform.

By adopting the CS approach to software services, the operators are able to take advantage of all that has already been done by software experts towards tuning this architectural approach to software implementation. As we shall see, considerable achievements have already been made by these experts. The CS approach is well established, especially since the advent of the Web, so there isn't any need to convince anyone of its effectiveness. How it gets used in wider ecosystems, like Web 2.0 is an ongoing question and we can only address part of it in this book. Later on, we will also consider P2P and other application paradigms relating to how software services materialise in the device network, such as J2ME and IMS.

6.1   THE GREATER CHALLENGES

Figure 6.1 reminds us of the basic structure of our mobile network within which the CS architecture is utilised to deliver mobile services. We can see that the device communicates over the RF and IP network layers to the server in the content network. Figure 6.2 is an alternative view of this model taken by slicing through the networks. We can use this alternative view to highlight several observations about the CS approach.

In most mobile services, we have a device at the top of the chain of events, as shown in Figure 6.2; we shall look at devices in depth later in Chapter 10. The next component in our chain is a set of protocols – a common language between the device and the server, represented symbolically by the arrow in the top half of the figure. The arrow represents two mechanisms. Firstly, the method of communication between the two ends, which for the current discussion will focus on HTTP and its wireless variant WSP from the WAP family of protocols. Secondly, we need a way of formatting the information passed by the protocol. As Chapter 5 discussed in some depth, the format of the data stored on the server or in a companion database server is not suitable for direct display on a device; it will be neither understandable by the device nor in a presentable format. As the figure shows, attempting

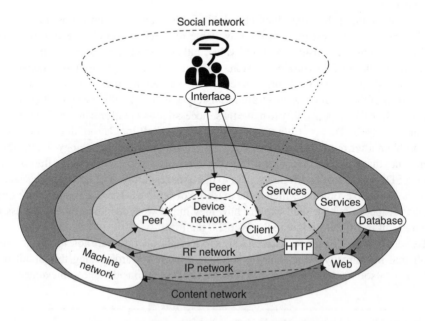

Figure 6.1 Mobile network topology.

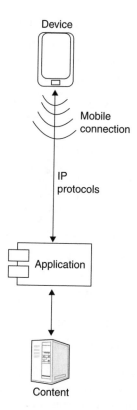

Figure 6.2 CS architecture.

to view the information in its native state will result in incomprehensible gibberish being displayed (the file snapshot on the far right). The alternative is to agree on a common formatting language, which in our case will focus on something like XHTML (Extensible Hypertext Markup Language), which is now the standard for presenting information in browsers on next generation mobile devices.

At the bottom of the diagram, we have the content, which we represent as being hosted by servers other than the ones we want to run our applications on (which is most often the case, as shall become clear). The data-hosting servers are more than likely hefty commercial database products, such as Oracle or the Microsoft SQL Server or other industrial-strength scalable database platforms.

What we then need is the application itself, which is the 'brains' of our system. All other things being equal, which invariably they will be[1], the application is where we have the greatest scope to differentiate. It is up to us what we program and how we present the application to the user – there is no need to settle for off-the-shelf solutions.

The final component in our mobile service is a bit pipe to carry the information from the application to the device and vice versa, this being assisted by the common protocols, like HTTP and the common data format, like XHTML.

---

[1] More often than not, all the mobile operators will have similar infrastructure equipment from similar vendors and they will all offer the same handsets or mobile devices. Applications are the biggest point of divergence.

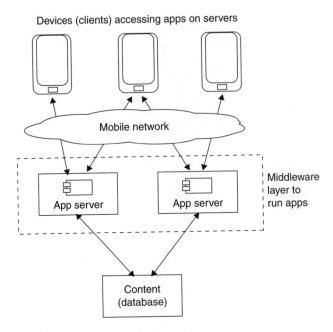

Figure 6.3 There's usually more than one client and more than one server.

Figure 6.3 reminds us that we are not just dealing with one client, at least not if we want to offer a useful service with lots of users and resultant revenue; targeting services at mobile operator customers, we have *millions* of potential users. Furthermore, in a typical hosting environment, we will need to host more than one application. Hence, we can appreciate that we have a formidable design task ahead of us if we are going to reliably support a large user base potentially accessing many applications. Scalability is a key requirement.

The challenges of handling more than one client and more than one server are further complicated in mobile delivery platforms by the frequent need for different parts of the system to collaborate by sharing common information and services. For example, an application to provide a guide for entertainment in a city may need to link to a mapping application on a separate server. If the user requires turn-by-turn, navigational directions, this may require reference to a route calculator running on another server, or to traffic information from yet another server. Perhaps a user decides to visit a cinema first and wants to view the movie trailers on their mobile device. This may involve linking to another server to access video clips. Somewhere in all this flow of tasks, we may have to charge the user for something, perhaps the guide information, the map, the directions, the video clip or all of them. We may also have to authenticate the user across *all* of the applications so that we really know the user is who they say they are, especially if we are going to access any personal information stored about the user in a subscriber database (most likely on yet another server). From this example, we can see that a mobile service might require a high degree of interaction and collaboration between different systems, as shown in Figure 6.4, which shows disparate systems exchanging messages with each other. As we shall see in Chapter 13, when we consider LBSs in some depth, messaging is a process where disparate software entities can communicate with each other without the need to wait for the message recipient (other application) to be ready at that precise moment to process the message.

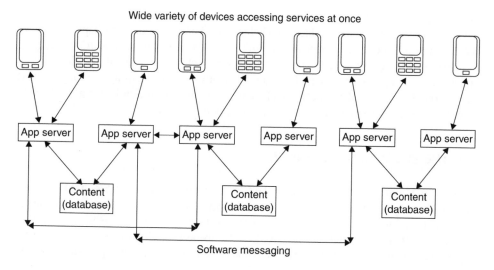

Figure 6.4 In a complex system, there are many applications networked together.

This is called *asynchronous* messaging and has become increasingly important in software architectures used in mobile systems.

Figure 6.4 also illustrates an important point about the devices accessing our applications. There are likely to be many different devices with varying capabilities. Not only are there myriad device models at any one time being sold in the market, but there is a continuous legacy of devices in circulation with fixed functionalities that cannot be upgraded in the field, thus further complicating matters. This is a challenge quite unique to mobile CS solutions.

The sophisticated software platform coping with diverse devices in large numbers is complex enough, but as Figure 6.5 reminds us, we still have a wireless data network to consider; which in itself is potentially complex and perhaps not as transparent as we would

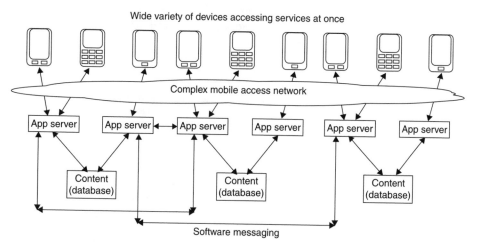

Figure 6.5 Complex system made more complex by the RF network.

like or might expect. For WAN services, such as 3G cellular networks, this system is not only complex in terms of its interfaces and behaviours, but, as we noted in Chapter 5 (and we will examine in depth later), it has its own internal resources that our applications may have to interface with. These include embedded services such as a voicemail, location-finding and text messaging. As we note in the figure, for many operators setting out along the next generation path, many of the network elements and software technologies are relatively new. This has its own challenges, but plenty of opportunities, too!

6.2 THE SPECIFIC CHALLENGES

If we return to our earlier sliced view of the CS architecture, Figure 6.6 highlights some of the key challenges. At the user interface end, namely the device, the greatest challenge is for our applications to be *useable*. Of course, this is heavily influenced by the applications that run on them, but device design should not impede usability. In fact, it should offer opportunities to enhance usability. We shall discuss this topic in more depth when we look at devices in Chapter 10.

At the very other end of the chain, the content itself should be *relevant*. This is perhaps obvious, but easily forgotten. Relevancy is not just about having the right content types and genres available, it is about making sure that the user's view of the content is relevant to

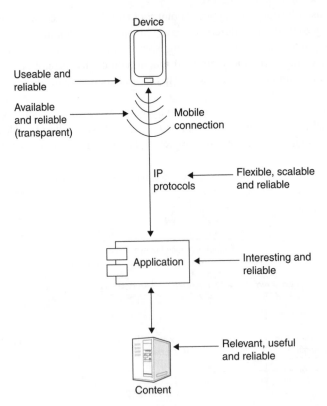

Figure 6.6 Some challenges for mobile services.

them. For example, when viewing train times from a particular train station, clearly the user wants to view times that are relevant to that particular time and place. It would be no use giving information that is slightly out of date, or for the wrong train station. With LBSs, ideally the user would not have to input which train station they are standing or sitting in.

For the physical link between the device and the backend, the priority is that the link should be *available*. As much as possible, the various limitations of the RF link should be *transparent* to the user. For example, if availability is limited to a narrower pipe than usual, due to high network congestion say, then the user should still be able to use the services and benefit from them. The services should *adapt* to the available pipe, not merely expect the user to tolerate poorer performance than usual. Users may not have such a forgiving tolerance of the variability of mobile services, especially if they are relatively expensive to use, either in monetary times or user effort.

Finally, in considering the application layer itself, which is running on our backend servers and responsible for dishing up the relevant content and responding to users' requests, then the most apt adjective that comes to mind when describing the key challenge is to make the applications *interesting* to the users. What this tells us is that the primary concern of the applications developers should be in putting as much of their effort as possible into making interesting things happen, or allowing interesting things to happen. Creativity is important.

There are many usage scenarios to consider and there will be lots of interaction with other systems. We will have to consider all the different presentation formats according to the vast array of devices that may want to access the content. We will have to think about how we are going to charge for the content; not just pricing, but the mechanisms to handle financial transactions. Perhaps the marketing team comes up with a scheme for banded billing, service bundles, add-on upgrades, pay-as-you go plans and many other wonderful pricing schemes dreamt up in a huge spreadsheet. We probably don't want to have to keep updating our charging scheme once we have programmed the application, so we need to provide a pricing interface that enables our spreadsheets to prime the charging 'engine' directly; it could then be updated as often as the marketing people wanted. As quickly as new schemes are dreamt up, they are in operation the very same day. This is all beginning to sound a bit daunting and we might be tempted to put the book down and go do something else. Before doing that, let us first see where this is headed.

Before we move on to look at how to build our applications platform, we should briefly reflect on its physical environment. Figure 6.7 gives us a very simplified view of the CS architecture within its physical domain.

At the top of the figure, we have the device network, which has a variety of devices types. The devices can talk to each other in their own private (or shared) network and this may at times complicate matters in the CS approach, but let us ignore such complications for now. The rounded rectangle around our device network represents the sea of RF connectivity in which the device network floats, which could be via a wide-area network (e.g. UMTS) or local-area network (e.g. WiFi). The RF network adds its own complexities too, but we shall discuss them later in the book in Chapter 12.

The RF network interfaces with an IP network, which is the main infrastructure we use to reach our content network (backend). In this book, we are not concerned so much with the physical nature of IP infrastructure, like routers, caches and the like, although we will look at some architectural issues, like how to cluster servers to enable a scalable mobile application deployment. In this book, our main concern with the IP network is in understanding the IP-related protocols that we need to take advantage of the IP infrastructure in a wireless

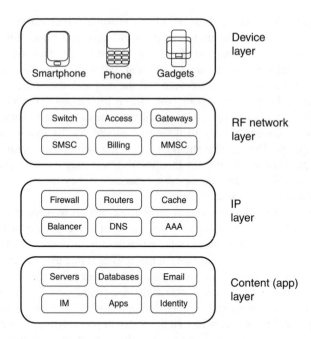

Figure 6.7 Example of the physical environment.

networked environment. This is principally HTTP and its wireless-optimised cousins, like Wireless Session Protocol (WSP) from the WAP family of protocols and standards.

Once we get through the IP network, we are now in the content network. This is largely a farm of interconnected servers running our custom wireless applications, web servers, databases and the like. We should also note how the content network could stretch out its tendrils into many other collaborative networks, including enterprise networks sitting behind firewalls somewhere else on the Internet (or via a private-leased line).

6.3 SERVICE DELIVERY PLATFORMS

Let us list just a few of the challenges (there are a lot more) common to building any mobile application:

- Handling different devices types

- Coping with legacy devices

- Handling a huge user base with different account schemes, possibly with each scheme customisable by the user themselves

- Dealing with disparate data feeds including a mixture of content types, like pictures, ringtones, audio files and video clips

- Transforming data feeds to formats appropriate to our service needs

- Formatting user data to be appropriate for display on their device and according to their circumstances

- Keeping track of users usage patterns and preferences

- Providing current location information to enhance any mobile application that needs location enabling

- Providing a unified and openly accessible means for an application to charge the user

- Providing a means for pricing information and schemes to be kept current and to be automatically reflected in prices sent to users and added to their bills

- Providing payment mechanisms for users to engage in various commercial transactions

- Enabling users to be added to the system (and removed when necessary)

- Providing discovery mechanisms so that mobile services can be discovered and subscribed to by users

This list of potentially common requirements leads us generally to the idea of building a mobile SDP that has the capabilities to provide a lot of these infrastructural assets without having to explicitly include them (program them) into each application. Furthermore, and perhaps most crucially, we can try to keep our developer community free to innovate and provide useful and interesting applications by providing a lot of common house-keeping functions for them.

It seems clear that SDPs as identifiable products are already emerging in the software marketplace from vendors providing these components to mobile network operators. Indeed, this trend seems well underway and the concept of needing an SDP is solidifying in the marketplace. In this book, we shall not only develop the theme of the SDP in its own right, but we shall also explore the theme of technologies suitable for building SDPs, much of which shall centre on the J2EE technology suite.

We shall address these issues later in the book, but first we need to spend more time surveying the ingredients of a mobile service and application. Having done that it will become clearer as to what some of the roles of an SDP might be.

A useful approach when examining mobile services is to move away from a technological view of the infrastructure towards a software-services-based view of the infrastructure, as illustrated in Figure 6.8. As the figure shows, we can propose four main service areas:

1. *Creation services* – this is the availability of software services that enable us to create new services. For example, the ability to manage an appointment book in the user's diary. Whether this is a web-based diary that sits in a networked database, or be it a diary application that runs directly on the user's device, the implementation is not that important so long as it is possible to call upon services somewhere 'in the network' to manage the diary. This enables us to create services based around diary events without having to program a diary application in the first instance. Combine this with other creation services, like a picture-messaging service, and we can glue the two together to enable a wedding anniversary reminder service by allowing pictures to be sent from one user to another on a particular anniversary occasion. A delivery florist service might be interested in such a capability. Each of these component services are potentially complex and time-consuming to develop. However, when made available as services, our developer community is free to concentrate on interesting ways to combine these software services to create exciting mobile services. All of the housekeeping functions like handling different device types would be provided automatically. What we need then are services that allow us to easily create new mobile services from components that already exist in our network.

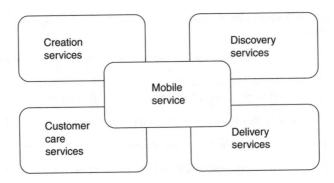

Figure 6.8 Two views of the mobile services environment.

2. *Discovery services* – after creating our service, which is then hosted by the SDP, we need a way for our users to discover its existence. This is perhaps the next stage in the product life cycle. To an extent, we are probably talking about a marketplace for services – a concept that makes most sense within a portal framework – which is a place where users frequently go and they can access all manner of content and services, new and old. This is the most common discovery metaphor used by mobile network operators. In terms of a service for discovery, there is a need for mechanisms by which the application developer could make their service known to users. It would again automatically alleviate the developer of developing certain housekeeping components. For example, a means for customers to register with the new service would be available as part of a common portal framework. If we wanted users to be able to personalise their services requirements within a range of possible options on offer, a common portal framework could manage this personalisation. The same framework could be utilised to manage the selective offering of different service elements according to device types and to manage differential pricing and charging for the various service permutations. If the new service is tailored towards a particular demographic group or lifestyle category, then the framework could manage this process in terms of ensuring that only the target group are especially able to discover the new service. Portal SDPs are already common, such as the Integrated Mobile Marketplace (IMM) from July Systems[2].

3. *Delivery services* – having created our service and made it discoverable to the target user community, our users need to be able to access the service itself. Delivery can take many forms, from enabling password access to the service, to enabling content to be downloaded to the device, or both. However, another aspect of delivery is the ability of the service platform to allow inherent scalability. For example, if the new service turns out to be popular, then the platform should be able to apply sufficient resource to allow the service to scale. This should be intrinsic in the platform. There also needs to be mechanisms to facilitate any basis for a commercial agreement; for example, asking the end-user to accept terms and conditions, specific licensing agreements and payment agreements should all be part of a common services framework.

4. *Customer care services* – once the service is made available to the user, we will need offer them support and charge them according to our pricing strategy and charging

[2] http://www.julysystems.com/

model. This would vary in terms of complexity, from a simple one-time access fee to the service, to a sophisticated event-based fee structure. For example, we could charge for downloading a game and then for accessing new levels and for high scores to be registered. Additionally, we could charge for multiplayer access and other variations, depending on what suits the gaming concept in hand. We don't want the developer community to be bothered with developing charging mechanisms, so this is provided as part of the common framework. The framework needs to provide the freedom to determine how the payments are made, such as various revenue-share options via mobile operator charging vehicles (like reverse-billed text messaging) and more straightforward credit-card payment gateways, depending on what makes most sense.

In re-thinking our view of the mobile services network to be a collection of software services that our application can access, we are reminded that a mobile service is ultimately a network of collaborating software programs acting in concert across the entire mobile ecosystem. In reality, the architecture of the entire network can be very complex indeed. The operator portion is shown idealised in Figure 6.9. The idea behind this architecture is that any of the software services (e.g. MMS, streaming, content filtering) within the ecosystem can be utilised to create user applications. These services are made accessible via APIs that are most likely web-services based. They are glued together to form new services within the SDP layer shown near the top of the diagram. There is a layer beneath this that exposes legacy APIs to the SDP as web-services APIs. The SDP also exposes the service components to external software services, such as Web 2.0 services, via a secure and controllable set of web-services APIs. This kind of Service-Orientated Architecture (SOA) is becoming increasingly necessary to achieve low-cost, maintainable and agile services environment. Note that the services layer shown near the middle of the diagram is logically placed within what we have called the RF network inside our mobile ecosystem. The Access/IP networks at the bottom of the diagram is just the RF portion of the network.

6.4 SOFTWARE SERVICES TECHNOLOGIES

If we are going to start building service delivery platforms for mobile service environments, then we need the software building blocks with which to construct such platforms. It's easy to talk about ideas like SDP's common software services, like access to charging mechanisms and so on, but how is this going to take place? What will be the actual implementation steps required to both realise and access such a service?

To understand what we might need, let's return to our CS architecture for a moment and imagine that we are building such a system from scratch on which to assemble our mobile applications. We wish to understand the issues we may have to solve in order to build and deliver a CS system suitable for mobile services.

Figure 6.10 shows the classic CS architecture[3]. Let's try to think of all the design issues we would have to solve in order to make such a system work. You can try writing these down yourself on a piece of paper, but let me have a go first in order to convey the gist of the exercise.

---

[3] Why do we simply jump straight in with this architecture? The CS architecture is just a fancy name to say that we don't want to install applications either on every single device, or on one server per device. Both of these are inefficient for many applications, especially those that imply a degree of centralised activity that is common to all users.

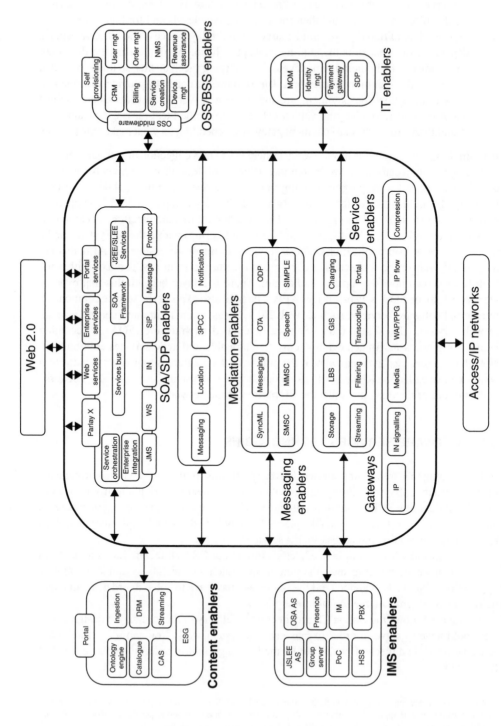

Figure 6.9 Possible architecture for operator network.

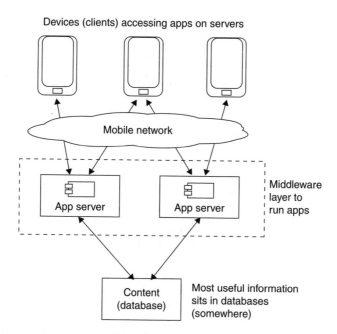

Figure 6.10 CS architecture at its simplest.

6.4.1 Example CS Design Issues

Let's take the server on the left of the diagram (Figure 6.10), we can see that it has two clients attempting to access the software running on it. For arguments sake, let's say that the application allows a user to see how much money they have been spending on their mobile phone bills (voice calls). We can imagine that the billing information is stored in the database at the bottom of the diagram.

Thinking carefully about the design issues, and remembering that we have to program all of the above system from scratch (i.e. the server bit, not the database or the client software); some interesting challenges become apparent. These are challenges that many of us may not have pondered on before, unless we have stood back and examined the fundamentals of CS computing. For example, let's just think about the challenge of handling more than one client. We could probably write a program to present a table of billing items to the user (for the moment ignoring how we access the database and present the information, which are some of the other challenges). Let's say that this program receives an incoming message[4] from the device that says 'Show bill'. A poor implementation of the program would probably only support the processing of one 'Show bill' message at a time.

Figure 6.11 shows what happens if our application can only process one message at a time from a client. While it's busy processing one message, other requests are unable to be handled and so other users get *blocked*. Thinking about this in the crudest software terms, we can imagine that we have a processing loop in the software that sits idle waiting for inbound messages, as shown by the loop (A) against the piece of pseudo code in Figure 6.12.

---

[4] By message, we mean something generic at this stage, not tied to any protocol.

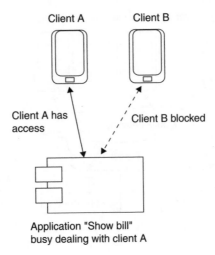

Figure 6.11 Application processing only one request at a time.

As soon as it gets a message, the loop begins execution (path B). Thereafter, the application proceeds with execution of the code within the loop (still B) and so can no longer listen for new messages until it finishes (C) and can get back to the beginning of the loop where the listening process is reactivated (A again). In other words, during the processing part (path B) the code is busy. Consequently, during that time, other requests are blocked.

This seems like we have gone back to computer science school for beginners to learn about loops, but this simple example is rather powerful in illustrating our current enquiry into the CS architecture. Clearly, we have a problem with the implementation as shown, because it appears to imply that we can only handle one user at a time. A variety of solutions exists to solve this type of problem, but our challenge is not to suggest the solutions at this stage. We are merely attempting to identify key challenges, which have to be solved one way or another, so this is just one example of the design challenges posed by the CS architecture. A more important point is that we would have to address this problem if we were going to build such a multi-user system from scratch.

If you are feeling ambitious, then please go ahead and try to brainstorm other design issues. If you are not in the brainstorming mood, then continue reading whilst we uncover more of them.

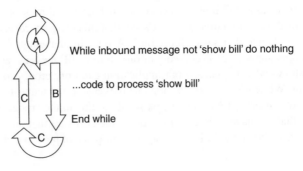

Figure 6.12 Loop code.

If we carry on with the above example then the next challenge we face is how to get the message to our application in the first place. For now, let's assume that we can build a Transmission Control Protocol over Internet Protocol (TCP/IP) link between the device and the server, so we don't have to worry about the really low level data-networking details. However, we still have to think about how we trigger our processing loop. The first challenge is to agree upon a format for the 'Show bill' message. Let's not complicate matters, so why not have just the string 'show bill' all in lower case letters in clear text. That should work, but the next challenge is how to grab the message from the TCP/IP connection. For example, we could use TCP/IP to send the string in a datagram across the connection from the device[5], but how do we know the other end is ready to receive the message. For example, if we try sending the message whilst the application is busy processing the loop (B) for another user, then the application may miss the message. If that happens, then how do we know that we missed the message and that we should initiate a retry? In other words, we need a protocol to send our message, a protocol that inherently solves all these issues. We are not going to identify a suitable protocol (though HTTP will work fine), as we are not trying to solve problems here, only highlight problems for the purposes of scoping the task of designing a CS architecture.

Let's say we found a way to solve the blocking problem and that we could service many 'Show bill' requests at once. There will come a point where we reach the resource limit of the computer running our application. At that point, the ability to handle 'Show bill' requests will saturate[6] and some users will experience degradation or even denial of service. The manifestation will likely be a delay in service as our network protocol will most likely engage in a series of retries on the user's behalf to within a certain time limit. In a wireless environment, that is particularly irksome, as the tolerance to delays is less, as usability guru Nielsen tells us:

> How quickly?
> 0.1 seconds: immediate
> 1 second: uninterrupted flow
> 10 seconds: limit of attention span
> > 10 seconds: coffee break, do something else, . . .
>
> Nielsen, *Usability Engineering*[7],
> Chapter 5

We can see that if we block our users for a period longer than 10 seconds, then we have probably lost their attention, which means we have probably lost their attention.

Previously, we had identified a blocking problem at the application level, now we have a problem at the server level. Just as we might envisage that we need more than one copy of the application attending to our user population, similarly, it would seem to make sense to deploy more than one server running our application. In fact, we probably want to install

---

[5] Note that this implies that our device and RF network can both support TCP/IP, but we shall come to this topic in Chapter 7.

[6] The way a computer runs many loops 'at once' is to run them in succession, but very quickly such that each instance of the loop appears to be running adequately fast. Eventually, if too many loops are placed in a queue, the process will slow down.

[7] Nielsen, J., *Usability Engineering*. AP Professional, Cambridge, MA (1993).

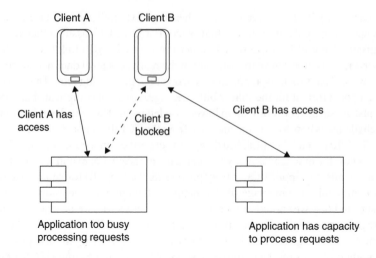

Figure 6.13 Being shifted from one server to another, the application loses track.

a whole raft of servers if this is going to be a highly popular application with potentially millions of users. That may seem fanciful, but in a well-visited mobile portal hosted by a mobile network operator these numbers are realistic.

Using a *cluster* of servers instead of one server seems like a straightforward proposition. However, it has its own challenges. We have to think about how we assign users to servers. This clearly implies some kind of switching function to route traffic to the appropriate server in order to achieve what we call *load balancing*. This is so that one server does not end up saturated with requests whilst another is hardly breaking a sweat.

However, what if there is more than one message from a user? Perhaps we can envisage the 'Show bill' message followed by 'Show item', where the second message is used to get information that is more detailed on a particular billing item in the bill. As Figure 6.13 shows, one of our users manages to be served by server A when the 'Show bill' message is sent. After a short while looking at the results on the user interface, the users wants to see more detail, so the 'Show item' message gets sent for one of the items[8]. As the diagram shows, the server A is now too busy to process the request, so somehow we manage to reroute our message to server B using a clustering process that is yet unknown (we will get to this later). However, this is where we now run into a problem. The application on server B was not the original application that serviced our 'Show bill' request, and so it has no record of servicing such a request. Therefore, it cannot fulfil the 'Show item' request, as it does not know which bill is being referred to. Another possible problem is that server A may have opened the billing table for our user and has locked it (in the database server), so server B would not be able to gain access anyway.

Of course, this problem has many solutions, but we are deliberately playing up naive implementation ideas for the purposes of highlighting design problems particular to the CS architecture.

We could continue to scrutinise the CS architecture and uncover many such problems. To save us the time of doing so, here is a list of challenges and some hints in brackets as to the

---

[8] This is just a high-level thought exercise, so we ignore how the actual item is identified.

solutions. Don't worry if you don't understand the buzz words or acronyms in the brackets – these are just here to whet your appetite and maybe to begin the process of connecting with solutions that you may have heard about, but don't yet know what they are.

CS challenges:

- How do we handle lots of clients trying to access the same application? (*multithreading, resource pooling, load-balancing*)

- How to implement more than one copy of the application across several servers and maintain contiguous processing and a framework for distributing load? (*low-level distributed software mechanisms, JMS, RMI, load-balancing, clustering*)

- How do we optimise server access to our backend databases so that this link does not become the bottleneck? (*resource pooling, connection pooling, load-balancing*)

- How do we operate a system where redundant database servers are being used to replicate data across different physical sites?

- How to handle different types of client? (especially wireless)

- What protocol to use for the CS communications? (*HTTP, RMI, Java Messaging Service*)

- How do we implement security at all levels of our system, such as authenticating users, setting access permissions on applications and preventing one application illegally accessing data from another?

- What protocol do we use to implement the server–database connection and how do we optimally write software to handle it, again without producing bottlenecks? (*JDBC*)

- What if a server fails – how do we swap over to another one without affecting service availability? (*fail-over*)

- What happens if something goes wrong with our wireless connection and we are unable to complete a critical sequence of tasks – how can we ensure we do not end up in an erroneous state? (*transaction monitoring*)

What we may notice about the above design challenges is that they are generic in nature, not specific to any particular type of service or application, mobile or otherwise. A pharmaceutical or financial application in a wired environment will have similar issues to contend with in a CS configuration. Therefore, we would really like not to devote our attention and resources to solving these problems, as they are *not adding any particular value to our mobile cause*[9].

---

[9] We are specifically talking about these low-level software process tasks, not about service delivery components common to mobile applications, such as a charging mechanism. The charging mechanism is a value-added software feature and this would be built using the low-level software process principles we have just been discussing.

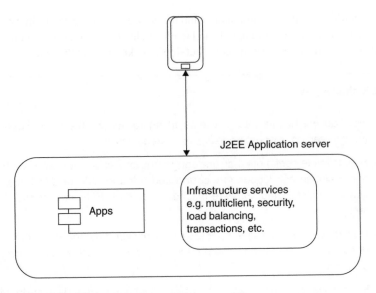

Figure 6.14 J2EE application server concept.

Fortunately, these problems are very well known and prevalent in many areas of the software industry. Therefore, solutions exist to tackle them for us, most notably in recent times the J2EE platform[10], which is probably the most popular solution in use in the mobile solutions community.

6.5 INTRODUCING J2EE – THE 'DIRTY STUFF' DONE FOR US!

Ideally, we want to be free to develop our mobile application software without having to develop software to take care of all the infrastructure issues we outlined in the last section. The ideal solution would be to use a software platform that already provides these 'software infrastructure' services for us. This is exactly what a *J2EE* (Java 2 Enterprise Edition) application server does, as shown in Figure 6.14.

The J2EE platform comprises of four software themes:

1. Java language support across a wide variety of underlying operating systems

2. Infrastructure services provision for enterprise applications

3. A programming model to enable infrastructure services to work

4. Set of software services to provide powerful interfacing capabilities to enterprise information service tier (e.g. databases, mail servers, etc.)

Our interest for the current discussion is especially with items two, three and four. For now, we are not interested in the Java programming paradigm. This is discussed later when we look at using Java to program applications directly on the devices (Java 2 Micro Edition – J2ME). For now, we will just take the underlying J2EE mechanisms for granted (e.g. Java

---

[10] http://java.sun.com/j2ee/

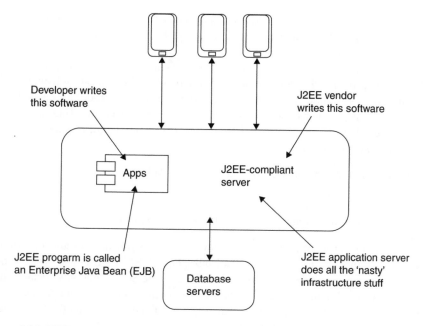

Figure 6.15 J2EE programs are called Enterprise Java Beans (EJBs).

support) and not be so concerned with how the programmer writes code. We are more interested in how J2EE enables us to build a mobile services platform, so the infrastructure services concept is more relevant to the current discussion.

The essence of the J2EE approach is that developers are left to concentrate on writing software that is business-specific, whilst leaving all the 'dirty stuff', such as security management and distributed processing, left to be pre-programmed 'out of the box' infrastructure services. To do this, we have to impose some rules on the software development process. Firstly, we have to program using the Java language. Secondly, we have to package our custom applications as components called *Enterprise Java Beans*, or EJBs. Don't get worried about the jargon. An EJB is just a program and for the most part we can think of EJB as just another word for 'program'. The good news is that as long as we stick to the packaging rules (known officially as the *EJB Developer Contract*), then we can simply plug our EJB into any J2EE-compliant server (*application server*) and the server will take care of all the infrastructure stuff, which for the purposes of this discussion we shall call *middleware*[11].

As Figure 6.15 shows, we can envisage the EJB program as though it were a pluggable component in a J2EE system. The EJB plugs into the server and everything should work together in a harmonious fashion. We are free to focus on our EJB programming. The infrastructure services, like handling clustering and multi-user access, are all programmed by the system vendor who provides the *J2EE application server*. The beauty of this approach is that the system vendor is an expert at infrastructure services programming and this is all

---

[11] The term *middleware* tends to get used rather liberally to describe a host of ideas. In some cases, an entire solution programmed to sit between our user device and some enterprise system (e.g. mail server or database) can be called middleware.

they do, so we expect them to be good at it. That means we get solid and robust middleware to support our application, leaving us free to focus on the business-specific functions of our system, which we call *business logic*.

Because the J2EE specification is an 'open' specification[12], many vendors can enter the marketplace to provide J2EE application servers, and indeed this is what has happened. This means that wireless solution providers do not have to rely on one vendor for the underlying software platform on which to build mobile services delivery platforms. Thus, it is possible to avoid vendor lock-in, which is better for business for a number of reasons[13]. There are many J2EE vendors, such as:

- BEA Incorporated, who provide *Web Logic Server* (WLS)

- IBM, who provide *Websphere*

- Apache Jakarta (Open Source project), who provide *TomCat*

It is up to the reader to confirm which solution is the best for their needs depending on a variety of factors. In this book, we shall talk about J2EE in general terms, but occasionally allude to specific features from BEA's Web Logic Server, not only because this is a highly popular solution already amongst mobile solution providers, but because I have most familiarity with this product more than any other. Otherwise, no particular product is recommended or endorsed by this book.

To clarify a potential misunderstanding, we shouldn't jump to the erroneous conclusion that we lump all of our entire application code together into one EJB. That's not the case at all. Typically, we would divide our application design naturally into components that have functions grouped according to a common purpose. The Java programming language supports object-oriented design methodologies, so typically, we would refer to these components as objects and they would usually reflect functional entities in the problem space we are working with. For example, returning to our earlier design problem of displaying a user's mobile phone bill on their device, previously retrieved from a database server, then we might expect an EJB call 'Bill'. This software object would respond to our messages like 'Show bill', 'Show item' and so on. If the application also allowed our user to configure their preferences for viewing billing information, we might expect an EJB called 'Preferences'. This might respond to messages like 'Show preferences', 'Set preferences' and so on. We can see how a design methodology emerges around the idea of identifying objects and the messages they respond to (and generate), but we do not discuss object-orientated design in this book.

6.6 WHY ALL THE FUSS ABOUT J2EE?

J2EE is certainly gaining a lot of momentum in the mobile services world, so we might well ask 'why all the fuss?' Firstly, we should reflect upon the circumstances we find ourselves in with next generation wireless services. As already ruminated, the plethora of next generation

---

[12] 'Open' means that it is publicly available for anyone to read and thereby implement a J2EE solution (application server).
[13] If we rely upon one vendor and they give us poor service, we are stuck. Similarly, they may tie us into a very punitive vendor relationship that becomes costly over time due to monopoly of supply.

mobile service opportunities and emerging technologies is already quite mind-boggling; we might even say perplexing (which is perhaps why you're reading this book). Therefore, as much effort as possible needs to be expended in creating new services that meet all of the various criteria we have been suggesting throughout this book (e.g. usability, scalability and so on). This means we do not want to spend effort on building anything that is not going to add direct value to the effort, such as optimised load-balancing algorithms and interfaces to databases, mail server interfaces or XML content feeds.

Recently, several problems have hit mobile operators at the same time. The ARPU has been trailing off as the voice markets begin to saturate and consequently pricing becomes more competitive in attempts to lure customers from one service provider to another (*churn*). At the same time, non-voice services have become possible thanks to mobile data-networking technologies, new device technologies and a whole host of software technologies to chose from. This is a problem to the extent that operators are not sure what to do with the new technologies. With all this going on, there is pressure to produce new services and bring them to market quickly and cost-effectively.

In this environment of change and uncertainty, it makes good sense to invest in software platforms that provide a good deal of the software infrastructure services that we need to build reliable and scalable service delivery platforms. The existence of the J2EE open standard has already proven attractive to many mobile solution providers working in the mobile service markets. J2EE has therefore apparently emerged as the de facto standard for the mobile industry as the preferred technology with which to deliver powerful mobile services. J2EE is seemingly being adopted right across the range of systems required to support a mobile services business, not just the customer-facing services, but internal business and operations support services, too.

J2EE has rapidly become a widespread solution for applications delivery throughout the software industry, so there is a growing network of support, including programmers, tools and other related solutions. An increasing number of independent software vendors supplying mobile solutions are building them using J2EE, so the technology itself is already emerging as a preferred platform.

6.6.1 The Challenges of Integration

In the case of a mobile network operator, the general ethos of service provision seems biased towards applications that include a good deal of ability to integrate with other applications and services, wherever that makes sense and at whatever level. That sounds vague, but it should become clearer as the extent of the software services horizon within the mobile context becomes more apparent as we progress through the book. To repeat an earlier example of possible integration, users would expect to be able to connect with a viewing application for cinema trailers (and movie reviews, ticket booking, etc.) from a general 'what's on?' application. In fact, probably the user doesn't know, or need to know, that these are two separate applications, but the integration may require a common context to be maintained. We may also require a diary application to store calendar events. After booking a cinema ticket, the user may want to enter the time and date into their personal mobile diary. This will be yet another application. Again, the user should not have to leave (i.e. logoff) the 'what's on?' application in order to enter (e.g. logon) the 'diary' application and then manually enter the event details. This would be a wholly unsatisfactory user experience.

More than likely, all of these applications will appear via a common mobile portal, this too being an application in its own right. All of the common services, such as the personal calendar, will be available via suitable APIs, and this will become clearer in the following chapters of the book.

There are many challenges to integrating applications in the manner we have just suggested.

- We need a framework for integration. We need to decide how one application talks to another.

- If we want to apply certain constraints on applications, like support for a certain device set in a particular way, then we may need a framework to enable this to be updated without having to update each application.

- We need a means to provide unhindered access from one application to another, such that common applications, like a diary, do not get loaded to the point of breaking.

- We need a way for disparate software vendors to build applications to their internal specifications, but which still work within the integration schema in the hosting environment.

These challenges are each significant and collectively even more so. Figure 6.16 gives a feel for the levels of interconnectedness for just a few applications in the possible services portfolio. We should reflect upon the implications of what the diagram suggests, keeping in mind that the number of applications shown is small and that perhaps a mature mobile portfolio will stretch eventually to hundreds of applications. The aggregated access to common 'housekeeping' applications, like the diary or device authentication, will cause the load on these applications to be very concentrated. All the issues we have been discussing in this chapter will certainly come into play, not only for this application, but also for all

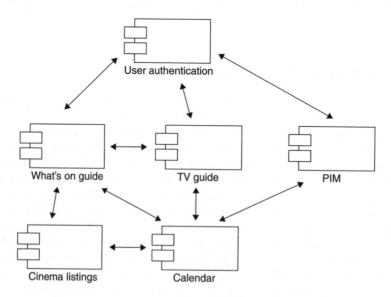

Figure 6.16 Interconnectedness of applications in mobile portfolio.

of them. The entire application portfolio may have to handles millions of users 365 days of the year, 24 hours a day, sometimes experiencing intensive peak loads.

The case has been made, at least conceptually, for requiring low-level software infrastructure services to assist with providing a robust applications environment. The need for a standard approach for integration becomes acutely apparent if we again review the diagram, this time to notice that each application is potentially coming from a different vendor. Were each vendor to adopt a unique approach towards interfacing with the service delivery platform, we can only begin to imagine the chaos that will ensue and the size of the integration task ahead. It would probably not be practical to cope with this problem in a cost-effective manner, if at all.

The adoption of J2EE as a common platform environment has made this type of integration effort much more manageable and resource-efficient. The problem is actually at two levels. As Figure 6.17 shows, it is not likely that all of our applications will be physically co-hosted in the same place on the same set of servers. By necessity, some applications will be hosted elsewhere and will need to connect back into the applications framework, most likely over the Internet. However, there is also an added element to consider and cater for. We may have painted an idyllic picture wherein we deploy wonderful J2EE technology all over the place like magic dust. This is not the case. There will be a great number of

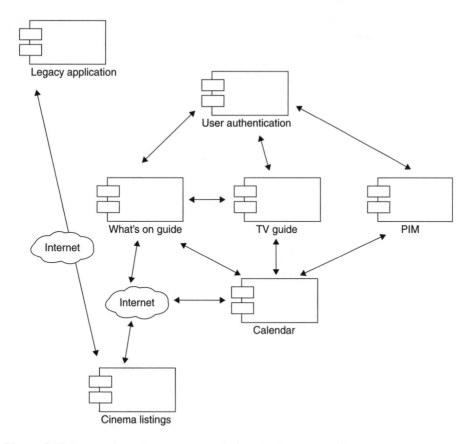

Figure 6.17 Interconnectedness compounded by the Internet and legacy access.

legacy systems to accommodate. By legacy, we mean anything that already exists and is not running on the J2EE platform.

A mobile network has many resources already deployed in the network that will need exposing to the mobile services platform. Some of these legacy functions might perhaps be wrapped up as housekeeping applications, such as an existing user directory service, or some may be applications in their own right, like multimedia messaging. Therefore, we need to be able to cope with interfacing with legacy systems, not just in terms of primary connectivity, but making sure that we can maintain security and scalability performance consistent with the rest of the platform.

We already discussed how the RF network in the case of a WAN, like a cellular network, already has many network assets that the operator has invested in and can facilitate interesting services. It is essential that we can tap into these resources from our service delivery platform, and so support for interfacing with legacy systems becomes a powerful requirement in such a system, as shown in Figure 6.18.

The case for J2EE is even more compelling if we are able to support effective access to legacy systems. It turns out that this is exactly one of its strengths. What we need to implement our environment shown in Figure 6.18 is predominantly two things:

1. A set of software mechanisms to enable the interfacing

2. A set of agreed protocols to pass meaningful messages between the collaborating software entities

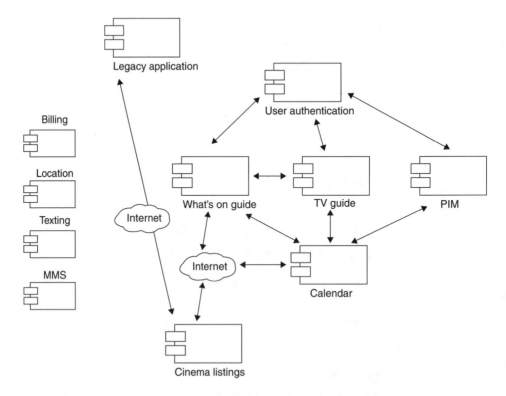

Figure 6.18 Many legacy systems in the RF network need to be accessed.

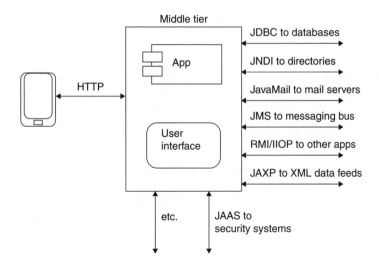

Figure 6.19 Rich set of interfaces available from J2EE services.

Our second requirement will be discussed later in the book (see Chapter 12 on RF network) when we look deeper into service delivery platforms, and in particular at the OSA aspects of UMTS 3G networks and the use of something called *Parlay X*. This is an initiative to make the mobile network resources available to external software entities, enabling these entities to access the mobile network and build services on top, as if the entire mobile network itself now becomes a platform for running external custom applications, like a giant 'mobile network operating system'. This is clearly not an easy task. Just as with low-level software we need a set of mechanisms and interfaces to enable infrastructure services, similarly, we need suitable software mechanisms to enable OSA to take place. These have been defined already by a technology initiative (forum) called Parlay[14], which, fortunately, has defined a particular instantiation of the interface that utilises a loosely coupled Web Services interface.

Leaving aside the OSA/Parlay aspects for now, we still need to address the first issue highlighted above, which is providing the low-level software mechanisms to enable interconnectivity with disparate systems. Life would be hard if we had only one interface, called 'J2EE Speak' say. All of our subsidiary systems, support systems, housekeeping applications and custom applications would have to be programmed to use 'J2EE Speak'. This is highly undesirable. However, this is where J2EE excels due to its rich set of system interfaces that come as standard as part of the J2EE infrastructure services, some of which are shown in Figure 6.19.

Some of the interfaces supported by J2EE include:

- *JDBC (Java Database Connector)* – this provides the programmer with a set of standard message formats and mechanisms to allow compliant database server products (of a wide variety) to be accessed. Due to the prevalence of database applications, from the beginning, J2EE has supported an effective database access model that has been present in Java for some time and builds upon the principles of the widely successful Open Database

---

[14] http://www.parlay.org

Connector (ODBC) industry standard for accessing databases. From Sun's Java website, we read:

> *JDBC technology is an API that lets you access virtually any tabular data source from the Java programming language. It provides cross-DBMS connectivity to a wide range of SQL databases, and now, with the new JDBC API, it also provides access to other tabular data sources, such as spreadsheets or flat files.*

JDBC is a highly developed database-access solution and includes the ability to create, modify or delete all manner of data entities stored within a database server. The handling of data can include entire database manipulation in addition to low-level record access and manipulation within databases. Database servers from popular vendors such as Oracle and Microsoft are accessible via JDBC, which itself is an open specification, thus allowing any database server vendor to provide J2EE[15] access (drivers) to their database product. The JDBC API itself has a low-level form that enables these drivers to be programmed.

- *JNDI (Java Naming and Directory Interface)* – this provides the programmer with a set of standard software message formats, interfaces and mechanisms to allow access to external entities that store information in directory structures, such as we might need to store the details of all subscribers in a mobile network, as found in GSM and UMTS networks in the Home Location Register (HLR). The HLR is a central repository of key mobile subscriber information, such as equipment, SIM and STD numbers as well as other data like encryption keys for user authentication. Often, vendors of HRL products provide directory protocol support, such as Lightweight Directory Access Protocol (LDAP). A product such as Lucent Technologies Flexent Distributed HLR has such support, as mentioned in its product information:

  > *Use interfaces like Lightweight Directory Access Protocol (LDAP) to allow managed, two-way relationships between operators and third party vendors.*

- *JavaMail (Java eMail interface)* – self-explanatory, this interface provides software features to enable connection with any email system that supports the common Internet standards for mail transport (e.g. SMTP, POP3, IMAP4, etc). Clearly a mobile user will more than likely have an email account, either their own account hosted by their chosen supplier, or an operator account, or both (or many other possible options). The availability of a software service to handle the mechanics of email communication is clearly desirable within a mobile context. Moreover, protocols like SMTP are also usable for communicating with non-email systems, such as submitting push messages to a WAP Push Proxy Gateway (PPG).

- *JMS (Java Messaging Service)* – this provides the programmer with a means to send asynchronous software messages to other software systems, like a kind of application-to-application text messaging system (or email). Software messaging is a powerful means of enabling disparate software processes and systems to communicate, especially where

---

[15] Other Java platforms, like the desktop standard edition (J2SE) also provider JDBC support, but not the Micro Edition (J2ME), as we shall see later in the book.

the synchronous information-pull mechanism of something like HTTP is not suitable for either submitting requests or fetching results. In fact, fetching information is probably not applicable to the software messaging paradigm. We shall discuss JMS in more depth in Chapter 16, whilst looking at location-based services in Chapter 13. In this context, the nature of software being interrupted by location-update events is very well matched to the software messaging technique and provides a useful context within which to explain JMS. As we shall see, messaging is a key consideration in the design of the J2EE platform.

- *JavaIDL (Java Interactive Data Language), RMI/IIOP (Remote Method Invocation/Internet Inter-ORB Protocol)* – this is a low-level software mechanism for enabling one application to use the services of another as if they were both on the same machine (location transparency). It is a major underpinning technology in the J2EE platform as it enables applications to be distributed across a cluster of servers, for various reasons. This technique will be discussed later in the book.

- *JAXP (Java API for XML Processing)* – as the name clearly suggests, this chunk of infrastructure software provides the programmer with powerful services for processing XML messages from other systems. JAXP enables applications to parse and transform XML documents independent of a particular XML processing implementation. Mobile application and tools developers can rapidly and easily XML-enable their Java applications using JAXP. In a mobile service delivery platform, a good deal of information exchange is required to configure the applications within the common environment. Using JAXP, the information exchange is implemented easily using XML.

- *JAAS (Java Authentication and Authorisation Service)* – this is a software service that enables our mobile applications to authenticate and impose access controls upon users. It implements a Java version of the standard Pluggable Authentication Module (PAM) framework, which enables new authentication technologies to be deployed without the need to change our applications to take advantage of them. For example, a message 'Authorise User' will still be supported, but underlying it will be a new (pluggable) method for carrying out the authentication process. The JAAS is flexible enough to the allow integration of any legacy method of authentication that an operator or service provider might already have in place.

- *HTTP (HyperText Transfer Protocol)* – already discussed in the book, the HTTP protocol stack is a powerful connection paradigm and the standard way by which applications will connect with client devices. The entire protocol handling mechanism is available as a built-in software service on a J2EE platform. Many of its features will be discussed elsewhere in the book.

The existence of these powerful interfaces in the J2EE platform make it an ideal base upon which to build powerful applications that can integrate with other systems, which is, and will be, the nature of any scalable and useful mobile services. The intrinsic capabilities of the J2EE platform to support these interfaces in a highly scalable and robust manner, whilst providing all manner of key underpinning software building blocks makes it an ideal choice upon which to build mobile services. This shall become much more evident, if it isn't already. We mentioned at the outset of the book that we aim to become 'smart integrators', using existing and proven technologies to build value-added services with great speed, moving rapidly with the tide of emerging next generation technologies and not against it. J2EE is a smart set of technologies that provides a worthy candidate to be a major part of the puzzle in building effective mobile services.

6.7   HANDLING SIP WITH JAVA

In Chapter 5, we introduced the importance of SIP. We shall explore the protocol in greater depth in Chapter 14. In many ways, it is similar to HTTP. There are two endpoints where one initiates a request to the other, which duly responds. The key difference is that, unlike HTTP, the responses don't include any of the data that the two ends want to exchange. The actual exchange is an entirely separate IP connection, which could potentially be any protocol, though most usually a protocol used to sustain an ongoing session of some sort, meaning some type of real-time exchange or 'conversation', such as audio, video or IM.

Given the similarity to HTTP, we would expect Java to offer support for SIP, which is does. However, there are two APIs for supporting SIP. The first is a model called the *SIP Servlet*, which is very similar to the Web Servlets, and the second is called *JAIN SIP*.

The JAIN initiative is a community of industry experts, coordinated by Sun Microsystems, developing a set of open, standard Java APIs to handle call signalling. For the purposes of this discussion, we don't need to get into the details of the differences between JAIN SIP and SIP Servlet. It is more useful to think of the SIP Servlet as being a type of program in its own right that comes alive in response to a SIP message coming into an application server. The servlet can also initiate SIP dialogues. On the other hand, JAIN SIP can be thought of as a component that gets used by other programs (officially called *containers*) to allow them to communicate with other systems via SIP. These containers include EJBs, so our powerful J2EE application server can do everything we've discussed so far and mix this up with SIP programming if we want to. JAIN SIP could also be used in a Java 2 Standard Edition (J2SE) desktop environment, such as inside an Internet telephony client.

In a telecoms environment, it is important that our applications can handle high numbers of transactions with low latency. A Service Logic Execution Environment (SLEE) is a well-known concept in the telecommunications industry. A SLEE is a high-throughput, low-latency event processing application environment. JAIN SLEE is the Java standard for SLEE. A SLEE is designed from the ground up to focus its resources on the processing of asynchronous events, which includes a SIP message (inbound or outbound). Of course, a web server is also processing events, but these are restricted only to HTTP get-response cycles. We shouldn't assume that such servers aren't scalable; they are, but the whole architecture has grown up around the HTTP protocol and various bits of support logic to make this work well. If we took a web server and then tried to replace HTTP with SIP, we would rapidly encounter all kinds of limitations in the architecture and would end up with a system that is non-optimal for SIP dialogues. Performance would soon become an issue. The SLEE is not only capable of handling SIP dialogues in an efficient manner, but any kind of real-time and high-throughput event. The architecture is centred around event processing. It is easy to develop a mental block about an abstract concept like events, so you might like to think of events as inbound and outbound messages, but typically that require a fast response of some kind or another. Messages could be SIP, in which case the SLEE container would use the JAIN SIP API to handle them. However, the messages could be something else, like billing messages from a billing platform, or network support messages from a performance-monitoring platform or whatever else we might find in a typical telecoms environment.

In Chapter 14, we shall introduce the IMS architecture, which is built around a triad of server types called Service Functions (SF), all of which need to handle high volumes of SIP messages in an efficient manner. We could easily use JSLEE to implement these SFs. Many vendors, such as jNetx and Open Cloud, have done exactly that. It might be tempting to assume that we can throw out J2EE and go straight to SLEE. However, these are complementary technologies. It really depends on what type of application and service we are trying to build. Typically, a J2EE platform is generic in nature and can be used to implement a range of services, including those that require SIP in some way. SLEE is optimised for telecoms applications and many SLEE implementations will be bundled with a range of other telecoms features, such as billing protocol support. Some of these features might come with expensive licences (they are often licenced anyway by the SLEE provider from third parties) and be unnecessary for many types of application.

As the trend towards convergence continues, we shall find ourselves needing a range of architectures and technologies. For example, we might have an existing application that is mostly web-based, which could be extended into the SIP domain although is still web-centric. Perhaps the SIP is used only to register the user with a SIP-based service, but the actual SIP service is handled elsewhere, probably using a SLEE platform. There would be

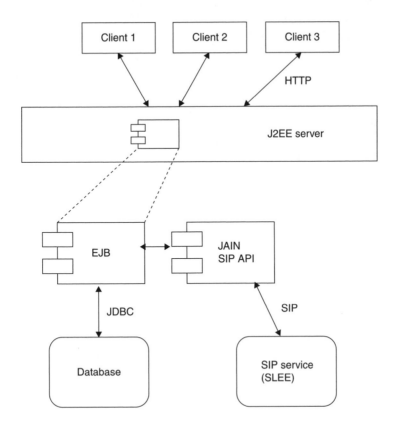

Figure 6.20 Handling SIP with J2EE.

no need to adopt SLEE in this case. The existing web-centric service might be J2EE based and heavily use EJBs, in which case we can use JAIN SIP in one of the existing EJBs to add the SIP logic to the workflow, as shown in Figure 6.20. Alternatively, we might want to allow our service to receive incidental SIP messages from a SIP platform and the SIP Servlet would be a good way to add this logic to our service.

7

HTTP, WAP, AJAX, P2P
and IM Protocols

7.1 THE RISE OF THE WEB

The browsing paradigm made possible by HTTP and HTML has become a key mechanism
for liberating digital information. A virtuous circle arose out of the universality of the
browser, once its installation reached critical mass. More browsers in circulation heightened
the appeal of offering services that could be accessed via the browser. Many applications
and services have since become 'web-enabled'. It is almost impossible to think of a desktop
application that doesn't access the Internet in some way, even if only to check for software
updates. Moreover, tools and software supporting HTTP have become so widespread that
HTTP is a common protocol for many types of software systems, not just browser-based
services.

As we have learnt in Chapter 5, HTTP can be used to gain access to any data on remote
servers, including snippets of XML. If we combine these two capabilities then we have
an effective mechanism for network-enabling mobile devices. As we shall discover in
Chapter 11, software is now widespread to make HTTP easily accessible to programs
running on mobile devices. This makes HTTP a useful backbone for mobile services. There
are only two other protocols that are important: SIP and RTP. We shall discuss SIP and its
uses fully in Chapter 14. We introduced the basic operation of RTP in Chapter 5. We shall not
cover it in any more depth in this book because, despite its importance in enabling real-time
media services on mobiles, it is not a generic protocol like HTTP. Moreover, increasingly we
shall find that programmers can access high-level software libraries to access media-player
functions on mobile devices; thus, further removing the need to understand RTP and its
sister control protocol RTSP.

Next Generation Wireless Applications, Second Edition   Paul Golding
© 2008 Paul Golding

Many machine-to-machine solutions have adopted HTTP as the means to exchange information, clearly demonstrating that this mode of exchange is no longer concerned solely with retrieval of visual information. In an M2M configuration, we don't have a human at the other end wanting to view information; we have a machine wanting to consume and process it. This is a trend that we also see emerging with the growing popularity of Web Services, a generalised means for any IT systems to exchange information using HTTP and the Web infrastructure.

There are yet other uses for HTTP. OTA synchronisation of contact and diary information using the new Synchronisation Markup Language (SyncML) can be done using HTTP. Querying a mobile location center to get the coordinates of a mobile phone is done using HTTP. In fact, the list of uses for HTTP is growing daily and has extended way beyond its original scope (hyperlinked document retrieval). These days, even a lowly soft drinks vending machine might be connected via HTTP.

This universal adoption of HTTP for many applications has taken place because software implementation folk are now very used to the paradigm and there is plenty of know-how, tools, software and infrastructure available to implement and support it. Indeed, the availability of web servers, web server code, HTTP code and a whole host of related utilities often means that there are very real cost and efficiency benefits of using HTTP. This only adds to the growing momentum behind the protocol.

HTTP and HTML (now XHTML) have become very widespread in general computing, and now also in mobile computing. It is therefore valuable to scrutinise its workings a little deeper than we did in the previous chapters. As we shall learn, one problem with HTTP is that it was designed and has since evolved[1] very much with the existing Internet, data networks and abundant computing resources as the assumed ingredients at its disposal. This assumption does not hold for wireless networks where we are still restricted by limited User Interface (UI) capabilities and device-computing resources.

In Chapter 6, whilst discussing HTTP and HTML at a very high level, we identified several challenges for the wireless environment. These were mainly to do with the limitations on connectivity (slower links, possibly intermittent in nature) and limitations, or variations, in user interface capabilities (screen sizes, keypad constraints, etc.). In Chapter 10 we shall learn about device architectures and technologies in more detail, and in Chapter 12 we shall learn how the RF network functions. Then we shall better understand the nature of these limitations, although many of the superficial differences between the mobile computing world and the fixed one should already be obvious.

To address the limitations of the browsing model for wireless services, a collaborative industry effort arose that eventually spawned a new protocol family under the umbrella title Wireless Access Protocol, or WAP. In this chapter we shall chart a course through the browser paradigm, explaining in some detail how HTTP and HTML work whilst explaining how we get to the WAP optimisations and possible alternatives.

7.2 HOW HTTP AND HTML WORKS

This browser paradigm is really one of 'link–fetch–response'. This string of words nicely reflects the concept behind the paradigm. The idea is that we first *link* to a resource, usually

---

[1] It has evolved very little as its original design concept and specification has not needed much in the way of revision. It is also extensible in other ways outside of its core protocol.

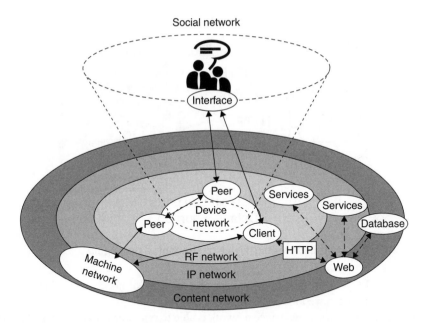

Figure 7.1 Device talking to server via the mobile network or networks.

via the user actually clicking on a hyperlink displayed in the browser, or by entering an address that uniquely points to the linked resource. Linking to the resource causes the browser to initiate a *fetch* of the linked-to resource, which may be a web page, or some other content type, like a descriptor for a ringtone or a game (later on, we shall look at the download mechanism for media files and games in more detail). The fetch request to a server hopefully causes the server to generate a *response*.

HTTP was very much designed with this cycle in mind, and by assuming that an application called a *browser* was on the requesting end doing the fetching, so the details of the mechanism have been designed to facilitate a page browsing metaphor. However, we should caution that the requestor might not be a browser, and hence the general term is *user agent*, which is the terminology used in the HTTP specification[2].

What we assume for HTTP is that the underlying transport mechanism used to shuttle fetch and response messages is the IP. This provides us with the essential means to connect one device with another or one software program with another running on physically separate devices. We do not describe IP here and the reader is referred elsewhere[1] to understand the details. All we need know for this discussion is that IP provides the means to shift packets of data between two programs. In Figure 7.1, we are reminded of our network topology and that our device layer interacts across a multitude of interfaces to get to the content layer. In this chapter, the IP layer is the focus of our attention. We are not concerned with how the RF network functions other than for now we assume that it is able to transport IP packets. In Chapter 12 we shall examine how the RF network supports IP, but again, for the moment we won't concern ourselves with this issue. However, the IP connectivity offered by an RF network has unique characteristics that force us to consider better ways to implement the

[2] HTTP 1.1 found at http://www.w3.org/Protocols/rfc2068/rfc2068

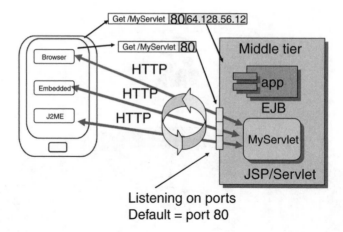

Figure 7.2 HTTP gets used in all the device programming paradigms.

higher protocol layers, such as those under consideration in this chapter. This shall soon become apparent when we look at WAP.

In the figure, we can see the device network is able to run three primary types of application: browser-based, embedded or Java[3]. We examine each of these programming paradigms in depth in Chapter 11, but what we need to know for the current discussion is that each of the paradigms will most often interact with the content network using HTTP, or a similar protocol like WSP from the WAP set of protocols.

In Figure 7.2, we see how all the main application paradigms can utilise HTTP to talk with the backend server. The figure also shows how the backend has its own internal architecture, here reflecting the J2EE method of delegating HTTP handling to programs called servlets and JSPs (Java Server Pages), which we describe in detail in Chapter 8. In the figure, there is a hint at how the HTTP protocol works. We can see messages, called *GET* requests, being sent from the user agent on the device to the server. This is the name of the HTTP method that initiates requests for resources from the responding end of the HTTP dialogue, which is typically a web server but is officially known as an *origin server* in the vernacular of the HTTP specification.

The figure also shows that the GET requests are routed via the IP network, with the IP address being appended to the requests, to enable routing to the appropriate server. The physical server that hosts our origin server has an IP address, but also has some important additional addressing information called a *port number*. The default for HTTP is port number 80.

From now on, we shall not concern ourselves much with the IP addressing and port numbers. We will assume that the user agent has the means available to exchange HTTP messages with the target origin server. What this usually means, as we shall see when we look at devices in Chapter 10, is that the device will have another application running on it, besides the browser, called an *IP stack*. The user agent actually makes its requests to send and receive messages to the IP stack, which in turn packages up the messages using the underlying IP protocols and routes these data packets to the RF modem included on the device.

---

[3] More specifically we mean Java MIDlets, but we shall get to that topic in Chapters 10 and 11.

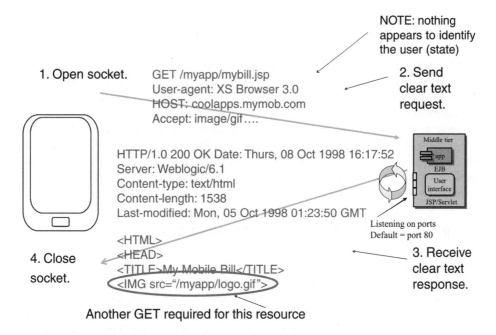

Figure 7.3 Basic HTTP request–response cycle.

Our default, or idle condition, in the mobile network is to assume that our device and content network have an available means to communicate and that messaging pathways are open and ready to transport messages at the whim of the user agent. We also assume that our device will already have a valid IP address assigned to it, as will the physical target server (we shall look at how IP addresses are assigned to devices when we look at the RF network in Chapter 12). Our server is sitting, actively listening out for messages on its port (port 80). It is ready to respond as and when these messages come in. Let's now look in more detail at how this process works.

In Figure 7.3, we can see the very basic anatomy of a HTTP request–response cycle. Let's go through it step by step. An IP connection is established between the user agent and the origin server in readiness to pass messages – this is called *opening a socket*[4].

To make a page request, what amounts to a small text file is sent to the origin server from the user agent. Its contents contain a list of text lines, each one containing what we call a header field. We can see that the opening header starts with the keyword 'GET', thus, denoting that this message is a GET method. Next to the keyword is a parameter. This is a path name to a resource on the origin server. The resource, some kind of executable file, is required to generate content that the server subsequently sends back to the user agent. In this example, we are showing a file name with a .jsp extension, which indicates that the resource is a type of Java program. In cases like this, the program must execute on the origin

---

[4] Two software processes communicate via TCP sockets, which is originally a name given to the association of a Unix communications process with a TCP/IP link. It is really the idea of an system interface provided by the underlying platform (e.g. operating system) to applications, enabling them to bind to a TCP/IP stream to a given port number and IP address. These bindings are uniquely identified so that several can be supported at once. The binding can be considered a socket.

server and programmatically produce the content. In other words, the program itself is not the content; we don't want to fetch the JSP file, we want to fetch its outputs after it runs on the server.

After receiving the GET request, our origin server either executes the file if it is an executable program (e.g. JSP or servlet), or fetches its contents directly (e.g. from a HTML file), and then attempts to send the output back as a text-file message to our user agent. The response message begins with a header, just like the request message did, but this time with the resource output appended, everything sent in clear (i.e. human readable) text. If this is the end of the session, then the two endpoints can dispense with the active IP connection by closing the socket. For this discussion, it is not important to know what a socket is, nor what they mean in low-level software terms. Such concepts are of interest to programmers who implement IP stacks and to the programmers who subsequently access these programs via an API.

7.3 IMPORTANT DETAIL IS IN THE HTTP HEADERS

Before going on to analyse the basics of the HTTP mechanism in the mobile context, let's make a few observations about the process just described.

In the request cycle (see Figure 7.3), we can see a header called *User-agent*, which is a string that informs the origin server what type of user agent is making the request. For example, it could indicate that the user agent is a particular browser vendor running on a particular mobile device. This information is potentially useful to the origin server, particularly in the mobile context. This is because there are many different types of device, from limited display mass-manufactured handsets to very feature-rich larger format smartphone devices. Even within the same device class, there can be vast differences in display capabilities between different generations of device, as shown in Figure 7.4. Clearly, we would not want to send a response with detailed colour graphics to a device that is unable to display them. Thus, we can use the user agent header to assist our origin server with deciding what content to dish up according to device characteristics.

In theory, this process is also aided by the *accept* header that we can also see in the example (Figure 7.3). This is a list of content types that the user agent can accept. For

Figure 7.4 Differences in browser capabilities for mass-market phones. (Reproduced by permission of Nokia.)

example, if the device can display GIF (graphic interchange format) images, then it says so in the accept header. If it can't, then this is omitted and the origin server can avoid sending such content. This is not the same as content *adaptation* that we might perform using the user agent header. In the case of adapting the content dynamically, we may adjust the entire output to suit the user agent interface. This is what we might call *capability negotiation*. In the case of the acceptance of content, we may still have the same content being generated by our application logic (i.e. the code within our JSP), but the origin server blindly does not bother sending any content that the user agent has said it can't accept. So, in the case of embedded images that can't be accepted, the server simply doesn't bother with sending them, whereas ideally what we need is for the page to be formatted differently in this case.

The original request may lead to a response that contains a web page with embedded images. Images referred to in a web page file (HTML file) are not stored in the file itself, they are referenced using image tags that point to the images stored somewhere else on the web server (which could be on a different server to the origin server). Therefore, in order to display the entire page properly, including the images, the user agent issues a set of repeated requests to go get the remaining content, one image at a time, request by request, as shown in Figure 7.5. In other words, each page element results in a separate HTTP request. The results from these requests are visually aggregated by the browser.

Each embedded item is going to result in another request. Over a wireless connection this is not desirable because we don't want to waste precious bandwidth on the overhead of making several requests just to view a single page. On the other hand, this staggered fetch could be an advantage because we can load the HTML into the browser and start displaying the page before we have received all the content. In other words, we achieve a kind of

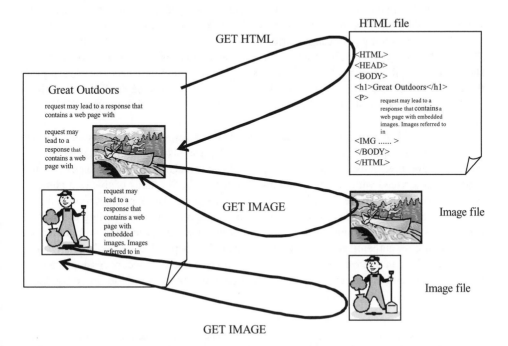

Figure 7.5 Browser page made from several files means several requests.

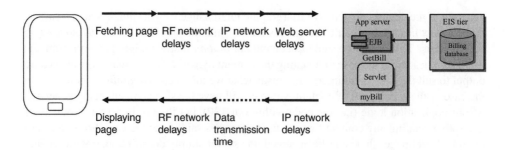

Figure 7.6 Example of delays in a network.

multitasking – whilst the reader starts reading, we load the images in the background. We shall look at this issue in more depth as we proceed with the discussion, since there are other factors affecting page-loading and viewing time (latency).

The overhead of making several fetches versus one fetch is not necessarily trivial. Just by looking at the HTTP headers, perhaps we are tempted to think that the overhead is only a few lines of text; like three GET strings instead of one in our example, and ditto for each HTTP header. This is true in terms of the actual amount of data fetched, but the network latency is the problem, not necessarily the amount of data that needs to be transported.

In any network there are delays, as shown in Figure 7.6. These are due to a variety of reasons, but are often due to data buffering, or network congestion, or waiting for resource allocation to occur, or possibly by virtue of the system design (such as unavoidably long intervals of interleaving[5] on an RF network). When we add up all these delays, they can become the dominant factor in the network latency compared to the actual time spent sending the data (see Figure 7.6), which is why making unnecessary round trips in any type of network is not good design, particularly in a network that is prone to high latency (like many RF networks). In browser-based communications using RF networks, we really should aim to minimise the number of network requests to a minimum necessary to carry out the task in hand. There are a number of design strategies that can be deployed, as well as overhauls of the protocols to facilitate a more reliable method, which is what WAP is all about, as we shall shortly see.

A key part of the header is the opening field of the response. This states the HTTP protocol version supported by the server, followed by a status code. In the example given, the string '200 OK' indicates the status code. This particular code (200) means that the request was serviced without any problems. The format of the status is always a numeric code followed by a readable English summary of its meaning. We are perhaps more familiar with some of the other codes as we often see them cropping up when surfing the Web; a code like '404 File Not Found' is one that most of us are painfully familiar with. Later on, when we discuss authentication and security issues (in Chapter 9), we shall see that '401 Unauthorised' is an interesting status code that is used to initiate a secure session whereby the user agent (and in turn the user) must undergo authentication in order to verify that a request for content is legitimate.

---

[5] *Interleaving* is the process of spreading out a chunk of data over many blocks of data, only using a small portion of each block to send the data. This is usually done to minimise the chances of a network-error burst taking out an entire chunk of data.

In the HTTP response shown in Figure 6.3, we can see the HTTP header contains a header *Last-modified*. This enables the user agent to know how fresh the file is. This is useful for caching of information. This is a process where any resource we have previously fetched from a URL is stored in local memory on the device. This process is illustrated in Figure 7.7. Step 1 is the ordinary GET process to retrieve all the content from the URL in the headers. In Step 2, the browser stores the content into an area of nonvolatile memory on the device called a cache. If later on the user requests the same resource, as in Step 3, then the browser is able to detect that this content is already sitting in its cache. However, prior to loading the content, the browser has to check that the content on the server has not been modified since it was last retrieved from the server, which is Step 4. This step does not involve performing a fully blown retrieval of the content, as that would be self-defeating. In this step, aspects of the HTTP protocol are used to conditionally fetch the requested resource. The user agent adds an *If-modified-since* header, applying the timestamp from the *Last-modified* header being the last time the resource was actually fetched from the server. If this is not more recent than the timestamp for the same content in the cache, then the serve issues a 304 (Not Modified) status code and the browser loads the content from the cache instead, as shown in Step 5.

This caching strategy saves unnecessary fetching of content, which not only saves the user power (battery consumption) and money (airtime consumption[6]), but also makes page loading much quicker because the browser is able to pull the objects from device memory very quickly indeed. There are other ways to implement tagging of content for the purposes of validating its freshness in the cache, but these are beyond the scope of this book.

We do not wish to ponder too much on the various HTTP headers and their usages, as we are more concerned with the specifics of using this protocol over a wireless link, or in the possibility of alternative protocols that might be more efficient. To determine what we mean by efficiency, we defer this discussion until later in the chapter when we look at Wireless-profiled HTTP from the WAP family of protocols. Here we will also look at some of the other headers and how they are especially relevant to wireless communications, even though this was not the original intention of the protocol designers. We shall also examine how the most recent version of HTTP, which is 1.1, is better suited to wireless communications than its predecessor (1.0), although most web servers and web clients these days use HTTP 1.1.

7.4 THE CHALLENGES OF USING HTTP OVER A WIRELESS LINK

Having discovered how HTTP enables chunks of information to be retrieved from an origin server, we might well ask what impact a wireless link has on the process. Firstly, let's look at what the characteristics of an RF connection might be with respect to establishing HTTP-like communications over the link. During a communications session between a mobile device and its communications companion (e.g. server), the data to be passed is sporadic; sometimes there is something to be sent, sometimes there isn't. But when there is data to be sent, such as a picture file, then the two ends have to synchronise to ensure that the communication is successful.

---

[6] More accurately, we mean money spent on transferring data, which is most likely billed by data amount, not transmission time (though the two are related, of course).

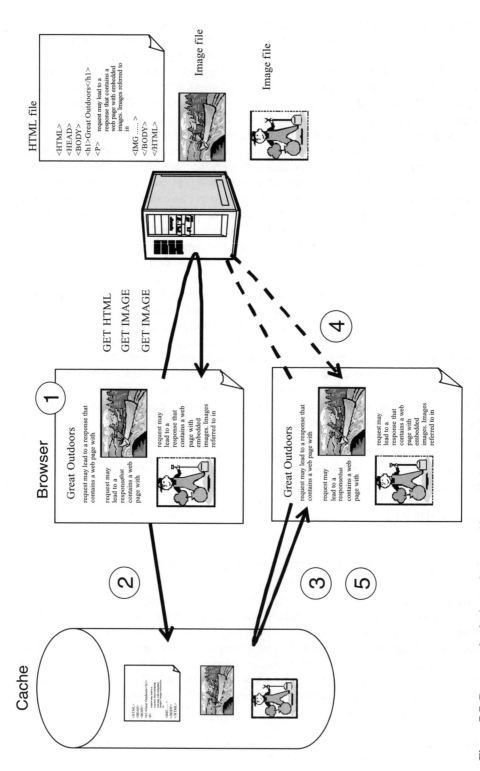

Figure 7.7 Browser checks its cache to see if content is already loaded.

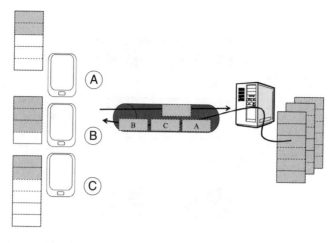

Figure 7.8 Links can be shared using a packet approach.

It is generally the case that, even in relatively reliable wired transmissions, it is not possible to transmit data without experiencing some errors in the transmission, which manifest themselves as small parts of corrupted data. Also, there are times when the link is not available at all and these times vary in length from very transitory to lengthy outages. These two effects combined are a problem, but one that can be ameliorated by sending data in chunks (packets) and not in one go. This is convenient anyway because of the sporadic nature of most data communications, as just mentioned. By sending data in packets we can make sure that the link is not tied up between two parties whilst others are also waiting to use the link. When a packet has finished transmission, the link is relinquished for someone else to have a go. In this way, many devices can share the same data connection, as shown in Figure 7.8 by devices A, B and C all receiving packets over the same link. This segmentation of data into packets is taking place at a level below our HTTP communications plane using a technique called TCP/IP; Transmission Control Protocol over Internet Protocol, which we first mentioned in Chapter 6. We can imagine that the three files shown to the right of the server in Figure 7.8 are each requested by a single GET-method request from the corresponding device. It is the underlying TCP/IP software that is doing the segmentation for us and this is quite transparent to our user agent and origin server software. As far as our user agent is concerned, it issues a GET request to the server and expects an atomic response. In other words, the user agent receives the entire response in one go, unaware of the low-level packetisation of the response.

The benefit of sending data in packets is that we are able to implement some kind of flow control to take care of intermittent link performance. We can ask for packets to be re-sent if too many errors were experienced. We can also increase packet size to accommodate conditions when the chances of error are less, or decrease if we think that resending is likely in order to avoid wasting time sending the same data over and over again, even though only some of it might be corrupted. This point is best illustrated by considering the limitations due to RF variability and showing how packet-based transmission can be used to overcome them. What we find with RF communications is that due to the nature of the RF link, the transmission characteristics for data are very variable, even though powerful signal processing techniques are used in attempts to maintain the best performance possible on the

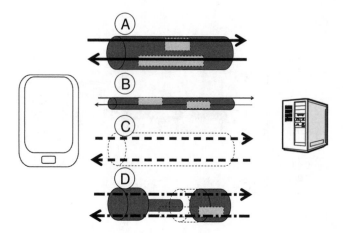

Figure 7.9 Variable packet size and delays due to variations in RF link performance.

air interface[7]. The variation in the link and the effects this has on the data communications are shown in Figure 7.9.

As indicated in the figure, sometimes we can achieve very good link performance, as shown in A by the thickness of the data pipe. The link is reliable and available for a good period of time and the link quality is good enough to send data at high data rates. In this situation we can send data in large packets because we are able to sustain good communications for longer periods of time and with a lower chance of error.

From the condition of channel B, we see that the link is very slow, as indicated by the narrowness of the data pipe. This may be a temporary condition or possibly even a permanent feature of some RF links, like GSM data calls. In this situation, we tend to send data in small packets so that we do not have to resend too much data if a packet gets corrupted, thereby attempting to maintain a reasonable throughput on the link.

In situation C, we can see that the channel is simply not available. This delay in availability may be variable. This is not a problem per se for packet-based communications as we generally have a hand-shaking procedure that checks that the recipient is ready to receive before we attempt transmission. This flow control means that data does not get lost; it just gets queued up at the transmitter. However, what we do not want is for the transmitter to switch off because the link is down; the link absence may be for only a short while. Therefore, part of the flow control mechanism is to keep retrying the handshake. Clearly, this interval should not be too long; otherwise, we may end up not transmitting even though the link has become available again, thereby inadvertently adding to the latency problem.

Finally, condition D shown in the figure reminds us that we may well experience any combination of the above scenarios during a particular packet transmission attempt. The link may slow down or even disappear for an interval whilst we are transmitting. This may occur within the context of trying to send a file or within the context of sending a packet.

---

[7] It is tempting to think that error-correction techniques used in wireless communications overcome entirely the hostile conditions of the RF channel. This is not the case. There are always bit errors on an RF link. Error rates vary according to the link conditions and how much error correction is used. However, high error correction means less data gets through due to the overhead of inserting redundant bits into the data stream that enable the correction process to function

In the former case, we can appreciate that the ability to dynamically adjust packet size and things like retry attempts (and intervals) may be useful. In the latter case, the ability to cope with link variations during a packet transmission should not cause an undue failure in the communications progress. Our protocol should be able to cope with both these situations.

We have seen that RF communications suffer from variable link performance. However, similar problems, but with different causes, already existed in ordinary data communications networks; hence, why IP-based communications already has the necessary flow control procedures in place. Therefore, for a moment we may be tempted to think that we don't need to do anything different for RF channels as opposed to wired channels (link Ethernet over twisted-pair cables). This is not the case, primarily for two reasons.

Firstly, some RF links can become so hostile that the standard IP flow control algorithms are no longer suitable for coping efficiently with the punitive conditions. As an example, if the link becomes very prone to errors, then the optimal packet size may be smaller than the IP protocols have allowed for, so we end up with packets that are too big and data is repeatedly and needlessly re-sent. Another example is in the segmentation of large files into packets. If the process is such that the corruption of a packet requires the entire file to be re-sent, then the process is going to become extremely cumbersome for a relatively slow RF link. Secondly, mobile RF links are invariably a lot slower than their wired cousins, so it is essential to make best use of the link. If the IP mechanisms require a lot of data to be sent just to manage the flow control, then the process is eating up valuable bandwidth that would be better spent sending the actual data, like the picture or game being downloaded. All of these issues have been thoroughly scrutinised by the WAP forum, which is now under the auspices of the Open Mobile Alliance (http://www.openmobilealliance.org). Subsequently, a set of protocol optimisations has been designed to accommodate packet-based transmissions over mobile RF links. Due to its prevalence, HTTP has been used as the basis for the WAP protocols.

7.5 WAP DATA TRANSMISSION PROTOCOLS

In this section, we shall look at the WAP protocols that enable us to implement a link–fetch–response communications paradigm, like HTTP, but in a manner that works well on a mobile RF link. Before we look at the details of the protocols, let's first talk a little about protocol stacks and how data communications are implemented in software terms. This enables us to better understand the concept of service layers within a communications protocol.

7.5.1 Protocol Stack Paradigm

In our discussion of HTTP, we saw that the protocol consisted of a set of communications primitives based upon request methods, such as GET. However, we have just learnt that the underlying mechanism of TCP/IP is a bit more complex. Perhaps, and most probably, a single GET request may not make it in one attempt to our origin server. It may need to be divided up into segments and sent one piece at a time, allowing for retries and handshaking, as shown in Figure 7.10.

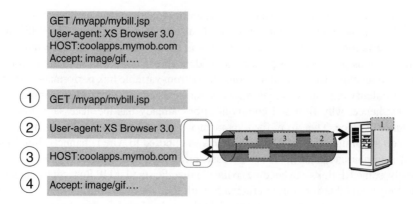

Figure 7.10 Segmenting a HTTP GET request into packets.

The exact size of the packets may not be as we have shown them in the figure, nor is it likely that the packets will be delineated neatly along the boundaries of HTTP headers as shown; this is just to illustrate the principle. The key point is that the user agent is not responsible for this segmentation process, or the underlying flow control. The user agent is a software application that only knows about HTTP methods and how to send them, like a GET request. So we might ask to whom has the request been issued. Whilst logically it has been issued to the origin server, physically the user agent has actually sent the GET request to another piece of software running on the device and it is this process that handles the communications across the link on behalf of the user agent, as shown in Figure 7.11.

The TCP protocol uses an even finer grain underlying protocol, which is the IP itself. The IP level is really about how the packets get constructed, whereas the TCP level is more about how they get used to establish a reliable communications pathway using flow control. Therefore, there could well be a software process to handle TCP, which itself utilises a separate software service to carry out the low-level IP packet processing. We can see that

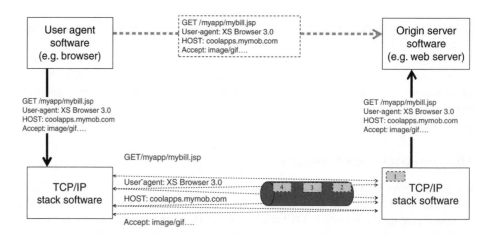

Figure 7.11 User agent talks HTTP to another software process, but thinks it's talking to the origin server.

this arrangement is a layer of responsibilities and functions within the communications process, with one functional layer stacked on top of another, which is why the set of software gets referred to as a *protocol stack*.

In Figure 7.11, we can see this stack, although we are only showing the division between the HTTP part and the TCP/IP part, grouping the TCP and IP processing together (which is common in any case). As we shall see later in Chapter 10 when we look at devices, the TCP/IP stack software is probably part of the device operating system and can be accessed by any user application running on the device. In this way, our user agent only needs to implement the HTTP protocol handler in software and use an appropriate interface (API) to pass the HTTP requests to the TCP/IP stack software. As far as the user agent is concerned, logically it is talking directly to the origin server, as shown by the dotted pathway across the top of the diagram.

7.5.2 The WAP Stack

Having examined the idea of a layered protocol model with various functional layers taken care of by components in our software stack, we can now look at the optimisations to this process proposed by WAP. If, as we claimed earlier, the mechanics of the TCP/IP process does not work well for mobile links, then we might propose removing the TCP/IP stack and replacing it with something else. This is certainly one approach that has been advocated, as we shall see. However, there is a design limitation that we should think about first, which is related to the existence and prevalence of HTTP.

The advent of mobile services is something fairly new and so we have the luxury of developing and implementing new protocols in devices without fear of remaining compatible with old ones, simply because there aren't any to be compatible with. However, at the other end of the connection, we would rather stick to the use of *existing* web infrastructure and not have to deploy something new to cope with modified protocol stacks. Certainly, if we take the J2EE platform, as discussed in the previous chapter, then we want to be able to tap into all its benefits without deviating from the standards upon which it is built, especially HTTP. This was exactly the design limitation that the WAP inventors placed upon their protocol design; that the WAP-compatible devices should still be able to communicate with standard (i.e. HTTP-compatible) web servers so that existing applications developers could use existing and familiar design approaches, infrastructure and tools. This, no doubt, was a wise decision.

However, immediately we see that we might have a problem. On the one hand, we need to do away with HTTP running over TCP/IP as it is too cumbersome for hostile RF connections, but on the other hand, we want to stick with HTTP running over TCP/IP because that's how web servers work. These requirements seem incompatible, but there is a solution to this problem, which is to use a gateway or proxy. As the name suggests, this device maintains a standard HTTP session with a web server by proxy, on behalf of a device that does not then need to communicate using HTTP over TCP/IP, as shown in Figure 7.12.

What we end up with is the user agent able to communicate with the web server in the most optimal fashion possible. Thanks to the software stack method, the user agent software still thinks it is talking to the origin server, which due to the proxy can still be an ordinary web server. Web application designers can continue to use their well-honed skills and the available infrastructure and web-related products to deliver services to our mobile user. The origin server could well be a J2EE application server as discussed in the previous chapter, so we maintain all of the benefits we identified with going with the J2EE approach.

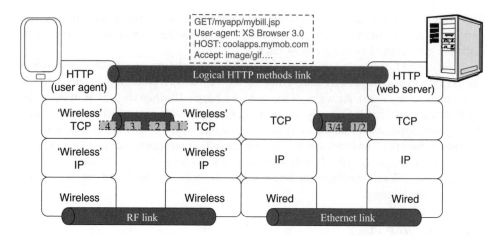

Figure 7.12 WAP Proxy 'talks' TCP/IP by proxy for the wireless stack.

The diagram shows how the 'wireless' TCP and IP layers are maintained between the device and the proxy and then converted to the standard TCP/IP protocols. This would tend to indicate that these protocols have to be fairly similar in design concept in order for the conversion process to be possible in an efficient manner. However, clearly there will be differences, which is why the diagram deliberately illustrates a different packet flow on either side of the proxy. This is what we would expect. The flow on the left side is optimised for a wireless connection, whereas the flow on the right-hand side is optimised for a wired connection.

WAP has gone through a series of design evolutions: WAP 1 and WAP 2[8]. In WAP 1, the main design criteria were link optimisations within a context of the very slow mobile RF services available at the time. For example, with early WAP devices on cellular networks, the only connection available was to establish a circuit-switched data call. On GSM devices this was theoretically a maximum data rate of 9600 bits per second, which often meant no more than 2–3 Kbps in practice. For a network like GSM, the data bearer itself could support an IP connection, so at the low levels, this could be utilised for data communications. However, many alternative RF data systems at the time, particularly in the United States, were unable to support IP communications (e.g. the two-way paging networks).

Sidebar: Long Thin Networks (LTNs)[9]

Cellular networks are characterised by high Bit Error Rates (BER), relatively long delays and variable bandwidth and delays. TCP performance in such environments degrades on account of the following reasons:

- Packet losses on account of corruption are treated as congestion losses and lead to reduction of the congestion window[10] and slow recovery.

---

[8] There are minor version numbers, like WAP 1.3, WAP 1.3 and so on, but for simplicity I prefer to stick to the major version numbers to highlight that we are dealing with two quite different sets of design recommendations.
[9] Extracted from http://www.faqs.org/rfcs/rfc2757.html
[10] Windowing is discussed in the main text when we look at Wireless-profiled TCP (W-TCP).

- TCP window sizes tend to stay small for long periods of time in high BER environments.

- The use of exponential back-off retransmission mechanisms increases the retransmission timeout resulting in long periods of silence or connection loss.

- Independent timers in the link and transport layer may trigger redundant retransmissions.

- Periods of disconnection because of handoffs or the absence of coverage.

Research in optimising TCP has resulted in a number of mechanisms to improve performance. Some of these mechanisms are documented in Standards Track Request for Comments (RFCs) and have been accepted by the Internet community as useful and technically stable. The Internet Engineering Task Force (IETF) PILC[11] group has recommended the use of some of these mechanisms for TCP implementations in LTNs [RFC 2757[12]].

This led to a WAP 1 protocol stack that was highly optimised for wireless communications and that did not assume anything about the underlying network, other than it could support datagram-based (i.e. packet, but not necessarily IP) communications. It was the responsibility of the WAP implementer for a particular wireless solution to implement the necessary adaptation software for the WAP protocol stack to work over the underlying data bearer. Because there is no universally supported datagram protocol for all wireless solutions, there was a need in the WAP specifications to define a datagram protocol from scratch, which was denoted Wireless Datagram Protocol, or WDP.

The *IP-based User Datagram Protocol* (UDP) is adopted as the WDP protocol definition for any wireless bearer network where IP is available as a routing protocol, such as GSM, CDMA and UMTS. UDP provides port-based addressing and IP provides the segmentation and reassembly in a connectionless datagram service. It does not make sense to use WDP over IP whenever UDP suffice, and is already supported. Therefore, in all cases where the IP protocol is available over a bearer service, the WDP datagram service offered for that bearer will be UDP. UDP is fully specified in [RFC 768[13]] while the IP networking layer is defined in [RFC 791[14]] and [RFC 2460[15]].

The bearers defined in this specification that adopt UDP as the WDP protocol definition are

- GSM Circuit-Switched Data
- GSM GPRS
- ANSI-136 R-Data
- ANSI-136 Circuit-Switched Data

- GPRS-136
- CDPD
- CDMA Circuit-Switched Data
- CDMA Packet Data

---

[11] Performance Implications of Link Characteristics (PILC). See http://www.ietf.org/html.charters/OLD/pilc-charter.html
[12] http://www.faqs.org/rfcs/rfc2757.html
[13] http://www.faqs.org/rfcs/rfc768.html
[14] http://www.faqs.org/rfcs/rfc791.html
[15] http://www.faqs.org/rfcs/rfc2460.html

- PDC Circuit-Switched Data
- PDC Packet Data
- iDEN Circuit-Switched Data
- iDEN Packet Data
- PHS Circuit-Switched Data

- TETRA Packet Data
- DECT Packet/Circuit-switched Services
- UMTS Circuit-Switched Data
- UMTS Packet Data

For the current discussion, our main concern is the higher layers, in particular how the HTTP/browser paradigm is achieved in the wireless environment.

WAP 1 specifies an alternative to TCP, called *Wireless Transport Protocol* (WTP). The good thing about WTP is that it was designed from scratch with the fetch–response paradigm in mind for browser-like applications, so it is appropriately structured. However, it is still intended as a generic transport protocol for wireless applications, not just for browsing, so browser-specific ideas are left out and are instead incorporated into a higher-layer protocol called the *Wireless Session Protocol* (WSP).

As the capabilities of wireless carriers have evolved, particularly with the advent of 3G solutions like UMTS, the need for highly optimised protocols begins to lessen. The extent to which they are no longer needed is debatable, but it is worth reminding ourselves that even with 3G, the reality of realisable data rates is still far below what has been prevalent for years within cabled networks in the Internet arena, which is why there is still a need to think about optimising the protocol stack for mobile applications. However, there is always an appeal to utilise standard protocols from the IP family as these are so well supported and understood that the benefits of using them are desirable. Therefore, with WAP 2, the evolution of WAP has been towards using the standard HTTP/TCP approach but with some minor (and often optional) modifications, but notably ones that will not take the implementations outside of what the existing specifications can support. The analogy that comes to mind is the use of higher octane unleaded fuels in cars. These fuels allow for an optimised performance where a car can take advantage of it, but they will still work in standard cars. This approach has led to *Wireless-profiled TCP* (W-TCP) and *Wireless-profiled HTTP* (W-HTTP).

We shall begin with the WAP 2 protocols first as these closely resemble what we have already been discussing thus far with HTTP and so enable us to build upon these ideas whilst uncovering some of the design challenges that will subsequently enable us to understand the design ethos of the WAP 1 protocols for completeness (and given that WAP 1 is currently a widely used protocol).

7.5.3 Wireless-Profiled TCP

TCP is about adding error handling and congestion handling on top of the crude underlying packet transport provided by IP. With TCP, we gain the ability to get feedback from the receiver about how well the data is being received, so that we can resend any lost information. Also, in a packet-based network where lots of users attempt to share the same link, hoping that there are always enough gaps in the transmission of others to send our own packets, there is the danger of congestion. If congestion occurs, then we need the ability to detect it and to control packet transmission in order to ease the congestion. This is another capability of TCP.

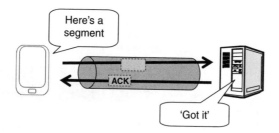

Figure 7.13 TCP send and acknowledgement (ACK).

With TCP, we take the data we want to send and we divide it into segments. A key mechanism for TCP is *windowing*. This is an important concept to grasp in order to understand the wireless profile optimisations, or the need for them. Whenever a segment is sent from the TCP sender, the sender wants to know whether or not it was received correctly so that, if required, a segment can be re-sent. This error control is a key function of the TCP protocol layer. To implement this feedback, the TCP receiver is required to acknowledge (using an ACK message) that segments have been received correctly, as shown in Figure 7.13.

The process of send and acknowledge constitutes a feedback loop. An important feature of the loop is its response time. There will be a certain delay in the segment getting across the network to the receiver and another delay in the ACK getting back. In wireless networks, this delay can be particularly long for a variety of reasons, mostly to do with how the RF resources get shared between users. This is a problem in itself for wireless networks, as discussed in the sidebar entitled 'Long Thin Networks (LTNs)'[16].

Regardless of the delay length, it would be a very inefficient process to have to wait for an ACK before the next segment gets sent. This is because whilst waiting for the ACK, the transmitter will not be transmitting anything and so the available bandwidth is not being used. This is particularly problematic for wireless networks where the bandwidth is a very limited resource. It would be unwise on a slow link to spend half the time just waiting around for acknowledgements instead of sending data. For the user, it would mean that loading browser pages takes longer than is theoretically possible and this would detract from the user's enjoyment of a WAP service.

To solve the problem of waiting for the ACK, the transmitter simply carries on transmitting without waiting for a response, segment after segment, up to a controllable amount of segments called a *window*. This is the windowing process, as shown in Figure 7.14. Basically, the TCP sender assumes that there will not be a problem with the segments that get sent, so it just carries on transmitting to make best use of the bandwidth whilst waiting for an ACK. What the receiver does under this arrangement is to wait until all the segments in the window have been received before sending an ACK, and the ACK is now used cumulatively to acknowledge all the segments rather than one at a time. This process gets pipelined so that the sender slides the window along and sends the next window. It checks for the ACK from the previous window and keeps going as long as the ACK is received before a timer expires, which gets set for each window.

---

[16] Note that some of the terminology used in the sidebar might not be accessible until you complete the reading of the current section.

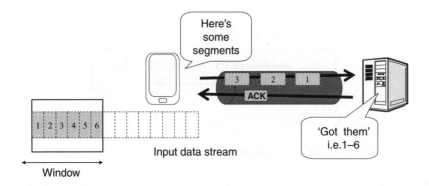

Figure 7.14 TCP segments get sent several at a time in a window.

There are several problems with this approach in a wireless network. For a LTN, the delay in receiving the first ACK may be much longer than in ordinary wired networks. Of course, since most TCP implementations assume a wired network, the sizing of the *initial window* is too small for the wireless case. In physical terms, this means that the transmitter will stop transmitting at the end of the window and wait for the cumulative ACK. It will still have to wait because the wireless connection may involve considerable delays. The windowing process was invented to avoid waiting, so this situation is undesirable. It can be overcome by making the window longer and so this is one of the mandatory changes to TCP implementation for WAP 2, turning TCP into W-TCP.

What should be apparent is that the cumulative ACK implies that if something goes wrong in the window transmission (and reception), then the entire window of segments will get re-sent as part of the error-avoidance strategy. The problem with this approach is that not all the segments may have been corrupted, but they will all get re-sent, so this results in redundant data transmission. The increased window length suggested by WAP 2 now compounds this problem; we shall have to re-send a lot more data if things go wrong! Hence, for WAP 2, the TCP profile utilises *Selective ACK* (SACK) [RFC 2018[17]] where each segment is acknowledged explicitly. Therefore, only the corrupted segments need be re-sent; a process known as *selective retransmission*. We shall return to this topic when we look at WTP from WAP 1.

Clearly, the size of the window is an issue that affects link performance. In an ordinary cabled network, it may be necessary to adjust the window if the network gets congested. In a congested network, there will simply be no point in transmitting lots of segments in the hope that they will get through. The amount sent should be backed off. In TCP, there is actually no mechanism for discerning if a packet is lost due to corruption or congestion. This can be taken advantage of in the wireless case to reduce the initial window size, should it prove too big for sustained link performance. In other words, we set the initial window to be large and we rely on the congestion control to back it off if it proves too large.

On the other hand, congestion control can be a problem for a wireless link if it gets used to correct for packet loss rather than genuine congestion. If we decide to back-off our transmission rate, either by adjusting the window or by deliberately imposing a waiting time before transmitting, then we need to attempt to recover from this situation as soon

---

[17] http://www.faqs.org/rfcs/rfc2018.html

as conditions improve. TCP has its own algorithms for recovery. The problem is that the conditions under which congestion occur versus packet loss in a RF network are different, so recovery strategies should be adjusted accordingly. Packet loss can be very transitory in an RF network compared with congestion scenarios. If several packet losses occur during a RF degradation episode, the TCP link can drastically curtail performance under the assumption that severe congestion is present on the link. This is made worse by the slow recovery attempts that are appropriate for mitigating congestion. In other words, an interruption to the RF link may cause undue performance losses as an artefact of TCP rather than actual link performance. This is clearly undesirable.

The performance of TCP under wireless conditions has been studied on many occasions (again, refer to sidebar entitled 'Long Thin Networks (LTNs)'). The results from several of these studies have been incorporated into the specification for W-TCP. There are many possible optimisations and we have only discussed two of them; large initial window and selective ACK. These are the only two that are mandatory in the W-TCP specification. There are plenty of other optimisations that are recommended and their affect on performance will depend on a variety of factors that are beyond the scope of this book.

7.5.4 Wireless-Profiled HTTP (W-HTTP)

In discussing W-HTTP, we need to resume our explanation of HTTP in order to understand some of the benefits that HTTP 1.1 offers over its predecessor 1.0, especially as W-HTTP is modelled upon HTTP 1.1.

Pipelining of Requests on a Single TCP Link We have briefly examined TCP in the preceding discussion. What we did not examine is the process of establishing a TCP connection, other than creating a socket, which is the software binding to the TCP/IP protocol stack according to the desired destination address for the packets and so forth. To establish a TCP connection between two end points, there is some initial handshaking that goes on to make sure both ends are ready to exchange data. Part of the start-up process in TCP is called a *slow start* wherein the performance parameters that affect TCP communications are set to pessimistic values in case the network is congested from the start, not wanting to exacerbate the problem with a new transmission. The effect of the slow start algorithm is exactly as its names suggests – the TCP link performance is initially suboptimal.

When we looked at fetching a web page, we said that there are often objects (called *inline objects*), such as images, that are stored as separate files on the web server and need to be fetched via separate HTTP requests; each object requiring a separate HTTP GET request. This creates quite an overhead in response time due to the slow start nature of TCP. With HTTP 1.0, as we saw in Figure 7.3, after a successful response from the origin server, the connection gets closed by the server. Therefore, to fetch a referenced object, like an inline image, the client has to open a new TCP connection with the server. This takes time, not only in the TCP handshaking, which is a lot of control packets being sent back and forth, but in the overhead of the slow start every time a new TCP connection is established. As we shall see later, WTP does not have this start-up problem.

To improve performance with HTTP 1.0, the solution is for the user agent to open concurrent TCP connections and request the inline objects via independent HTTP requests. However, this causes congestion and processing overheads. To improve matters, HTTP 1.1

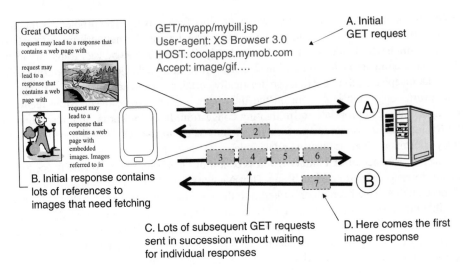

Figure 7.15 Pipelining of HTTP requests over a second TCP connection.

has the ability to keep the same TCP connection open and to pipeline the HTTP requests, which means that many GET requests can be sent in a row without waiting for the responses, as shown in Figure 7.15. This makes maximum use of the open TCP connection. This aspect of HTTP 1.1 is carried forward into W-HTTP. In fact, all of the core features in HTTP 1.1 are specified as required by W-HTTP.

The pipelining used in W-HTTP is particularly beneficial for wireless devices as it requires lesser processing overheads than opening many concurrent sockets, as would be necessary with HTTP 1.0 to achieve a better user experience.

The improvement of HTTP 1.1 over 1.0 is shown in example measurements taken from an experiment to investigate HTTP 1.1 optimisations[18]. These are shown in Table 7.1. The first set of columns list the measurements taken from retrieving a sample file for the first time in which there are plenty of inline image references. We can see that the performance of HTTP 1.1 is clearly better than 1.0 when pipelining is used. Note that these results are for HTTP 1.0 using multiple concurrent connections (up to 6), whereas all the HTTP 1.1 results are for a single TCP connection. The second set of columns show the results for a subsequent fetch of the same page, but this time verifying whether each resource is newer than the cached one in the user agent's cache. Here we can see even better performance of HTTP 1.1, which is because other features of 1.1 can be utilised to improve performance (for example, using *byte ranging*, which is not discussed here but gets discussed later when we look at segmentation and reassembly (SAR) in downloading Java applications to mobile devices).

In terms of optimising HTTP 1.1 performance, we still may want to use more than one TCP connection, but we can avoid the many connections that HTTP 1.0 implementations found necessary to spawn. It is useful to open an initial TCP connection to grab the XHTML content from the origin server, as shown by connection A in Figure 7.15. As discussed in Chapter 6, in a J2EE environment we generate the XHTML content dynamically using software (e.g. enterprise java beans). The process of writing the dynamic content onto the TCP

---

[18] http://www.w3.org/Protocols/HTTP/Performance

Table 7.1   Example performance measurements comparing HTTP 1.0 with HTTP 1.1.

	First-time retrieval				Cache validation			
	Packets	Bytes	Seconds	Overhead (%)	Packets	Bytes	Seconds	Overhead (%)
HTTP 1.0	559.6	248 655.2	4.09	8.3	370.0	61 887	2.64	19.3
HTTP 1.1	309.4	191 436.0	6.14	6.1	104.2	14 255	4.43	22.6
HTTP 1.1 pipelined	221.4	191 180.6	2.23	4.4	29.8	15 352	0.86	7.2
HTTP 1.1 pipelined with compression	182.0	159 170.0	2.11	4.4	29.0	15 088	0.83	7.2

link back to the device takes time, but we can still receive some of the content before the entire XHTML file arrives. By opening another TCP link (connection B) we can begin to pipeline requests for any of the inline content that the browser already sees referenced in the initial portion of the XHTML. In this way, we can achieve an even higher degree of parallelism and thereby improve the overall response time. The net result is a better user experience for the end-users, which is an important consideration in the design of any mobile service.

In terms of pipelining of requests, or even multiple connections via HTTP 1.0, it is important to realise that the browser is normally capable of displaying the initial XHTML content prior to receiving the inline images (or other embedded content), so the user gets to see the partial page displayed. This improves the apparent response time as far as the user is concerned and so is a desirable implementation feature.

With WAP 2, the specifications dictate that caching of fetched content should be supported and be compatible with caching methods for HTTP 1.1, which we have already discussed. Additionally, the WAP gateway, if used, should be capable of caching data itself and there are some extra headers that WAP agents can use to ensure cache maintenance on the gateway can be controlled.

Using Compression with W-HTTP (HTTP 1.1)   In terms of improving user experience, there are yet more things we can do to improve the underlying performance, such as the use of compression; an optional feature in HTTP 1.1 that should be implemented in W-HTTP user agents wherever possible. The advantage of compression is that the content in the responses is compacted. The HTTP headers themselves do not get compressed, but the XHTML content does. Even where link-level compression is available, such as V42bis, by using something like *zlib* compression[19] on the HTTP payloads, performance can be improved and without relying upon the wireless infrastructure to implement link-level compression (which cannot be relied upon[20]).

[19] http://www.gzip.org/zlib/
[20] Just because a particular technique is referred to in the technical specifications for a cellular network does not mean that it will be implemented on all networks, nor on all devices.

With HTTP 1.1, there is a HTTP header called *Accept-encoding: deflate*. If this header is included in the initial GET request, then the origin server knows that it can compress the response payload. In other words, the user agent is indicating that it is able to accept compressed content as it has the capability to uncompress it. Compression is not required for image fetches as the image file formats themselves, such as PNG[21] and JPEG[22], support internal compression by default.

Compression in itself can reduce the size of the XHTML content, and on a slow wireless link this will improve response times. But there are some subtleties with this approach that are interesting to note. Firstly, compression requires more processing time prior to displaying the XHTML, as we need to run the algorithms to uncompress the content. This is an added delay not only for displaying the initial XHTML content, but also in our user agent being able to extract inline references and initiate the request pipeline to get the rest of the content (although on pages for very small device displays, we expect the number of inline images to be small). On a very slow device, this may prove problematic and the gains of compression can be lost, but there is no hard and fast rule. As with all of these protocol optimisations, the actual implementation affects overall performance.

There is another reason that compression is more important than it might seem at first, which is to do with the low-level operation of TCP and how it interacts with the pipelining process. The details of the problem are beyond the scope of this book, but we can briefly mention the principles. It is more efficient in the pipelining process if we can buffer up several requests and send them at once. This reduces the amount of control packets that need to be sent. The buffering process will incur a delay if we have to wait for the buffer to fill before sending out (flushing) its contents. The problem with this is that we may have an ACK sitting in the user agent buffer that the server needs before it can send out more responses. In which case, we may incur an extra delay in waiting for a timer to expire that indicates that the buffer should now be flushed (in the absence of enough content to fill it and cause a flush automatically). By compressing the XHTML, the receiver gets more of the page in one go and consequently (on average) more inline image references. These references result in more GET requests, which will fill up the outgoing buffer and cause it to flush sooner. It has been shown by experiment[23] that this can improve performance.

Using Cascading Style Sheets with XHTML-MP   We shall look at the technology of WAP markup languages in Chapter 8, but for now we should mention something about the use of *Cascading Style Sheets* (CSS) to improve link performance. CSS allows the browser to understand how it should layout text and what formatting to apply. By supporting CSS, it is possible for rich layouts to be achieved without resorting to the use of image files to achieve similar results. CSS are textual descriptions of layout criteria and, as such, they can represent certain visual effects in a more compact fashion than using image alternatives.

For example, the image shown in Figure 7.16 with the caption 'solutions' requires 682 bytes to be stored as a GIF image.

[21] Portable Network Graphics – see http://www.libpng.org/pub/png/
[22] Joint Picture Experts Group – see http://www.jpeg.org/
[23] http://www.w3.org/Protocols/HTTP/Performance

solutions

Figure 7.16 Typical web page image.

Using XHTML+CSS, the same content can be represented with the following phrase:

```
P.banner {
 color: white;
 background: #FCO;
 font: bold oblique 20px sans-serif;
 padding: 0.2em 10em 0.2em 1em;
}
```

```
<p CLASS=banner> solutions
```

In WAP 2, the markup language recommended is called XHTML-MP (XHTML-Mobile Profile), which is a derivative of XHTML-Basic, which, as we shall see, is a specification for using XHTML for display-limited devices. However, XHTML-Basic does not include the use of CSS. The WAP forum has specified support for a limited version of CSS in XHTML-MP, so we should be able to take advantage of its features by enabling us to make visually interesting pages that load faster. The style information itself can be stored separately in a file called a *style sheet*, which has to be fetched using a GET request. This is apparently an added overhead. In the pipelining scheme, the overhead will be reduced. However, the real advantage in performance comes by using a single style sheet across many pages in our application. This is because the style sheet will only be fetched and loaded once, so there is no need to keep fetching it for each subsequent page that references it. The use of CSS is a potentially powerful performance-enhancing feature for WAP 2.

7.6 WIRELESS PROTOCOLS – WTP AND WSP

7.6.1 Introduction

Having discussed the wireless profiles for TCP and HTTP, we have covered the main direction for WAP support in the future. However, the WAP 1 protocols are already in wide circulation in millions of devices and will continue to be used for some time to come as they are extremely efficient, especially on slower WAN links, like GPRS. Many newer devices will support both WAP 1 and WAP 2 stacks. These are not backwards compatible, so they have to be implemented as separate stacks on the device – WAP 2 will not handle a connection with a WAP 1 proxy (and vice versa). This does not mean that we need two browsers! The trend is for a single browser implementation that can handle all the various markup languages that have evolved within (and alongside) WAP. The browser will use either a WAP 1 or a WAP 2 stack to access the corresponding stack on the other end of the link, but it cannot use a WAP 1 stack to access a WAP 2 source (and vice versa).

WTP and WSP are not compatible in any way with TCP and HTTP – they are separate protocols, although similar in many respects. The design ethos for both these protocols is to facilitate the fetch–response paradigm in the most optimal manner. This is not like TCP, which is a generic protocol that has been designed to support many varied higher layer applications, not just web browsing (HTTP). TCP can support protocols like FTP for file transfer, telnet for remote terminal access, Simple Mail Transfer Protocol (SMTP) for email and so on. WTP has been designed to support WSP, so in some respects it already mimics the fetch–response structure of WSP (and HTTP). WTP is built on a messaging principle rather than the streaming[24] principle that TCP/IP supports.

Both WTP and WSP have both been designed from scratch to cope with LTNs, so they have overcome some of the limitations of TCP and HTTP, as we shall now explore.

7.6.2 Wireless Transport Protocol (WTP)

WTP is a transaction-based protocol. It has been designed with the assumption that the wireless device and its server will engage in an ongoing conversation that can be broken down into distinct transactions, as shown symbolically in Figure 7.17.

The transactional nature of WTP makes it ideally suited to fetch–response applications, like browsing. WTP supports a variety of features that make the protocol ideally suited to the greater challenges of the potentially hostile RF network.

The protocol actually supports several modes of transaction, two of them primarily concerned with one-way data flow (push) and the other concerned with requesting data in the fetch–request manner (pull). We shall primarily focus on examining the characteristics of WTP via the latter transactional mode, as this is the one used to support WSP, the WAP equivalent of HTTP.

WTP does not require the three-way handshaking that TCP uses to establish a link. WTP simply sends an invoke message at the start of a communications cycle. This invoke message is always used to initiate a transaction on the link, whether for the first transaction or the hundredth, so the inefficiencies of the TCP slow start are done away with in WTP.

WTP allows for a much stronger verification regime for ensuring that messages have been both received *and processed* by the application that is using WTP, which might be a WSP stack program for example. This is unlike TCP, which only allows for verification of reception, but not that the higher layer application using the protocol has successfully consumed the message. This can be compared with text messaging from one mobile to another. With the Short Message Service (SMS) in GSM, the sender can ask for a message delivery receipt to indicate that their original message to the recipient has actually been delivered to the device. This is like a delivery acknowledgement, and is paralleled in WTP by the delivery of an ACK packet (Packet Data Unit, or PDU). However, WTP goes a step

---

[24] Here we do not mean real-time streaming that multimedia applications require, we mean that the structure of the protocol is agnostic of the content, viewing the input as a stream of bytes to be sent in a reliable manner. Whereas WTP views the content as being transactional, or conversational (message based), like HTTP, and groups protocol primitives around the completion of transactions.

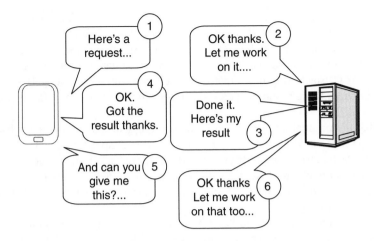

Figure 7.17 WTP supports dialogue at the application level.

further. It has a further level of acknowledgement that allows the higher layer application to say that it has consumed the message. The nearest parallel is to imagine a mechanism in text messaging whereby the sender could be notified that the recipient has actually opened and read the message, just like some recipients will, by their own volition, send an OK message back.

Figure 7.18 shows us the WTP transaction sequence in some detail (though some of the finer details are ignored here for brevity). Step 1 is the user application, which could be another layer in the stack (e.g. WSP), invoking a transaction. The transaction is initiated over the link (Step 2) and passed to the user application at the other end (Step 3). Note that we do not wait for an acknowledgement (ACK) before sending data, we have included it

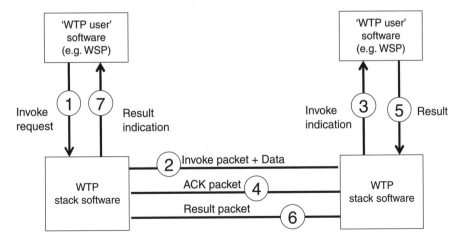

Figure 7.18 WTP provides strong verification of message transaction.

with the invoke payload going across the wire to save time. The WTP stack software will send an ACK (Step 4) to acknowledge successful receipt of the data. Then there is another level of acknowledgement, which is referred to as a *Result* within the WTP nomenclature. In Step 5, the application that has consumed the received data is sending a result request back to the WTP stack to indicate that all is well with the data. This gets sent back to the WTP provider (Step 6) and emerges as a *Result Indication* (Step 7) to the originating application.

Note that the original invoke was to initiate a complete transaction and subsequently more data packets could have been sent by the WTP user application as part of a single transaction. The receiver is aware from the start that more packets are expected and so will begin a process of tracking where it is in the progression of the transaction. All packets are marked with a *Transaction ID* (TID) to commonly associate them with a single transaction wherever that may be so.

There are various details of the WTP protocol that make it ideally suited to wireless transmission channels. In the above example, we could dispense with explicit ACK packets wherever we expect a result packet to be sent as this implies that the data was received, thereby making explicit ACK redundant. Result and ACK packets also allow data to be appended to their tail ends, data that is not part of the transaction. This is referred to as 'out of band'[25], as it is not really part of the main dialogue. This allows other processes at either end of the link to swap information on the back of the WTP transaction. For example, this could be used to send RF measurement information, or some other link-quality indicator. This might be useful for the higher layer application to adjust its behaviour accordingly. Perhaps during episodes of poor link performance, a browser could automatically switch to accessing a sparser set of web pages, such as a text-only version of a site.

7.6.3 Concatenation and Segmentation

The transaction capabilities of WTP are quite advanced and support two methods of dynamic adjustment to the underlying RF bearer capabilities. Firstly, WTP supports a concatenation technique that allows several WTP packets to be aggregated before being sent, so as to fit them into a single (larger) RF bearer transmission unit. This will allow maximal efficiency to be achieved on certain types of bearer, especially those that support very slow datagram services (e.g. some paging networks).

Secondly, WTP supports a Segmentation and Reassembly (SAR) technique that allows very large transactions to be distributed over successive WTP packets. The WTP stack itself can reassemble the segments before passing them to the upper layer (WTP user). There is often some confusion that arises when dealing with SAR techniques as this can be done at different layers in the entire protocol stack. As we have learnt, WTP packets flow over a WDP link, which is the WAP-defined, low-level datagram protocol. Wherever possible, WDP defaults to UDP, which is its equivalent using IP to structure the packet flow. IP has SAR capabilities. Later in the book, when we look at downloading Java Archive (JAR[26])

---

[25] Strictly speaking, this is not out of band communication, but it is similar. The point is that data can be sent riding on the back of an existing transmission, so no specific connection is required to be set up to carry the supplementary data, which is also quite independent of the main transaction.

[26] A JAR file is how a Java application gets packaged up for distribution. Here we are specifically referring to packaged J2ME applications that are designed for small devices and can be loaded OTA.

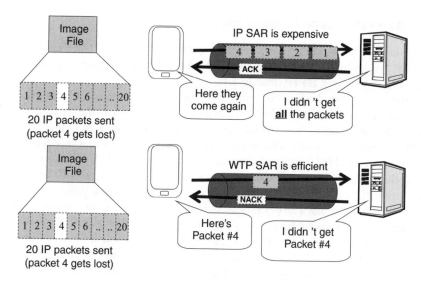

Figure 7.19 Selective resending of WTP SAR is more efficient.

files to mobile devices, we shall see that HTTP 1.1 also supports SAR using its *Range Requests* HTTP headers to ask for only segments of a resource rather than the whole thing.

Given that all these levels can support SAR, which one should we use? This is application-dependent to a degree, but a key feature of WTP SAR is its selective retransmission capability, which is something we have seen before with the selective ACK (SACK) option in HTTP 1.1 (which is made mandatory for W-HTTP).

7.6.4 Segmentation and Reassembly in Action

Let's say we have a 30 KB picture file that we want to download to our device to use as wallpaper on the display. Using WTP SAR, we can divide the file into 20 units of 1.5 KB and send them as a single transaction, notifying the receiving WTP stack that a segmented transaction is taking place. It is perfectly valid to have a multipacket transaction that is not a segmented higher layer message, so we should not think that transactions bigger than one packet are automatically using SAR.

The WTP provider on the receiving end will buffer up the segments in readiness to reassemble them. With basic SAR capabilities of WTP, only 256 segments can be accommodated, so this places a limit on the SAR process[27]. In terms of implementation, the WTP stack must have enough memory allocated to allow for successful buffering.

If some of the 20 packets in our image file get lost during the transmission, then the WTP stack issues a *Negative ACK* (NACK) to specify which packets have not been received properly. The WTP provider at the transmitting end can then retransmit the lost packets, but not those that were already successfully sent. Contrary to this, with IP segmentation, any loss of the IP packets cannot be selectively recovered. All the packets are required to be re-sent. This contrast in behaviour and performance is illustrated in Figure 7.19.

---

[27] This limit can be overcome using *Extended SAR*, but this is an optional feature of the WTP protocol.

7.6.5 Wireless Session Protocol (WSP)

WSP is very similar to HTTP in concept and implementation. This is what we expect given that WAP devices are designed to fetch content from standard web servers, which, we as know well, only support HTTP (mainly 1.1).

Similar to HTTP methods, WSP supports all the HTTP 1.1 headers, except that there is a requirement to ensure that headers are listed in a certain order in the request. This is to allow for the use of *code pages*, which the grouping of different headers into collections. With WSP, commonly used (well-known) headers are replaced with a shorthand form to condense the overall HTTP request for transmission across the RF link. For example, instead of using the string 'GET' in the header, the hexadecimal code '40' is used. Instead of the header string 'Last Modified', the hexadecimal code '1D' is used instead. With HTTP requests, it is common that a lot of the headers remain the same from one request to another, especially if the requests are within the same domain (e.g. a WAP portal). By grouping the unchanging part of the request into code pages, these can be cached and substituted for a short-form version that indicates a particular page is being used. When received by the device or the proxy, the code page is looked up from the cache and its values can be extracted and used.

Most of the HTTP header region is assigned binary token equivalents to achieve the shorthand form as a means of compression. This applies in both directions. For example, the status codes of responses are also encoded. Some of them are listed here, together with their hexadecimal tokens ('Ox' stands for hexadecimal):

```
"200 OK, Success" = 0x20
"201 Created" = 0x21
"202 Accepted" = 0x22
"203 Non-Authoritative Information" = 0x23
"204 No Content" = 0x24
"205 Reset Content" = 0x25
"206 Partial Content" = 0x26
"300 Multiple Choices" = 0x30
"301 Moved Permanently" = 0x31
"302 Moved temporarily" = 0x32
"303 See Other" = 0x33
"304 Not modified" = 0x34
"305 Use Proxy" = 0x35
"307 Temporary Redirect" = 0x37
"400 Bad Request - server could not understand request"
    = 0x40
"401 Unauthorized" = 0x41
"402 Payment required" = 0x42
"403 Forbidden -- operation is understood but refused"
    = 0x43
"404 Not Found" = 0x44
```

Clearly, the use of tokenised headers is not directly compatible with HTTP 1.1, even though semantically WSP and HTTP headers are identical, except for some WAP-specific

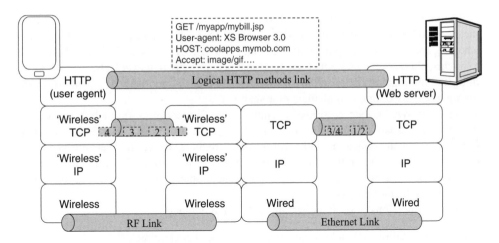

Figure 7.20 WSP requires a proxy.

extensions. Therefore, unlike with W-HTTP, we have to rely upon a WAP proxy to convert our WSP sessions to HTTP sessions, as shown in Figure 7.20.

WSP relies upon the reliable transport mechanism of WTP, namely the strongly verified transaction capabilities discussed earlier. As noted in our discussion of WTP, it is possible not only to acknowledge messages, but also to verify that the consumer has successfully attained a meaningful result from the transaction, as shown earlier in Figure 7.18. This method of transaction is used by WSP in its *connection-mode*. An alternative mode – *connectionless-mode* – provides less reliable transport, but can be used without the need to establish session credentials and so is more efficient for applications that require one-off interactions on an infrequent basis. This is useful for applications that require data to be pushed to a client, which we shall discuss push separately in the following section.

Other aspects of WSP that are notable are its ability to support asynchronous requests, similar to the pipelining of HTTP, so it would be possible to achieve overall efficiencies of pipelining end-to-end. WSP supports this queuing technique and does not require the responses to be in order (i.e. in step, or synchronous, with the requests).

WSP also implements multipart responses where the content can consist of several objects grouped together into the body of a single response. This uses a technique called *multipart MIME*, which we shall discuss later when we look at the Multimedia Messaging Service (MMS). It should be noted that multipart transfers are used specifically for media types that are by their nature multipart and can be handled properly by the user application sitting on top of WSP. We should not think that this is a technique for grouping together several requests from an origin server into one 'concatenated' response.

7.6.6 WAP Push

So far, we have been looking at methods for a client communicating with a server on a transactional basis where the application on the device (e.g. the browser) is requesting content. The content could be varied, from pages to display in a browser, to ringtones and other downloadable objects that could be subsequently utilised on the device.

The underpinning paradigm is the link–fetch–response, where the link aspect is some means of addressing a remotely stored resource that we subsequently attempt to fetch from a hosting entity that can respond with the content. Protocols like TCP/HTTP can be utilised to do this, although they have inefficiencies that become apparent over LTNs. These can be mitigated by utilising optimisations within the scope of the protocol specifications, albeit via some largely optional implementation features.

However, WAP goes even further in enabling the delivery of useful mobile services by offering an alternative communications paradigm called *WAP Push*. This is the ability to send content to an application without it making a request. As we shall see, this paradigm is also useful for going in the other direction to enable users to 'push' unsolicited content to other users. The advent of devices such as camera phones makes the utilisation of this paradigm all the more likely.

Some applications of push messaging to the client include:

- Email notification – 'You have new mail from Fred – click to read'

- Location-sensitive advertising – 'SuperMart's 10%-off everything – click for map'

- News alerts – 'Budget headlines now available – click to read'

- Travel alerts – 'Paddington 5.20 cancelled – click for timetable'

Here we are showing the use of push as a means to display alerts to the user, but this is not the only use of WAP Push. The push mechanism is a protocol feature; it is part of the WAP stack, it is not an end-user application. However, using push to send displayable alerts is an application. It is probably the most likely use of WAP Push and is definitely worthy of consideration in a wide variety of contexts, especially as WSP enables content to be pushed that can include web pages. This is why our examples above are actionable; they all contain links that can be clicked as a response to the alert. This would cause a browsing session to be initiated, whether by HTTP or WSP. This is one of the powerful features of the WAP Push mechanism.

The push mechanism is inherent in WSP/WTP. A WAP-compliant device should have a WDP port dedicated to the reception of WAP Push messages and always be ready to receive WAP Push messages. The messages can be identified by the sender as destined for a particular device application using an application ID field in the message. What is great about WAP Push over WSP/WTP is that it can take place over any bearer that supports WDP, which includes SMS. Recall that with RF bearers that support IP, WDP defaults to UDP in those cases and then WTP runs over UDP. However, the way that most RF networks work is that *an active IP session*[28] has to be established before IP communications can take place. Hence, if a device has no active IP connection on its RF network interface, then we are not able to utilise a UDP connection to support push messaging. However, we can run WDP over SMS, and devices that support SMS are always ready to receive text messages; there is no need to establish a messaging context (session) first, other than the device being within RF coverage of course.

WAP over SMS enables us to send WAP Push messages without an IP session. A special type of WAP Push message can be sent to ask the device to establish an IP session for

---

[28] We shall look at what an active IP session means later, but we can imagine it is like a mobile device needing to plug into the IP network before it can talk over IP, even though there are no plugs in a wireless network of course!

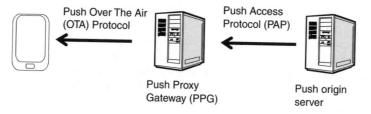

Figure 7.21 WAP Push architecture.

subsequent push communications where more substantial amounts of content could be pushed. We are able to detect that an IP session has been established by asking the device to register with a PPG, which is the network element that is responsible for sending push messages to the device, as shown in the WAP Push architecture diagram in Figure 7.21.

As Figure 7.21 shows, the PPG is the network element that sends the push messages to the device. Messages are submitted to the PPG using HTTP 1.1 and a method called POST (see Figure 7.22). With the POST method, content can be attached to a HTTP request. This is different from the GET method, where requests consist of header information only (HTTP headers), but do not have a body. With GET methods, we are only expecting to see a body in the HTTP responses from the origin server, the body being used to carry the requested content from the resource identified in the requesting URI header field (part of the GET method). The actual message contents submitted to the PPG using POST must conform to the Push Access Protocol (PAP).

The PPG reminds us of the WAP gateway we saw earlier, and in some ways it has similar functionality. However, it has the added capability of being able to store messages to forward them later to a device that is unreachable (i.e. out of RF coverage). Because we are using WSP, even in connectionless push mode, we are still able to receive an acknowledgement from the device that the push message has been delivered to it (but not to the higher layer application). This enables a retry mechanism to be implemented, which makes the whole push process reliable.

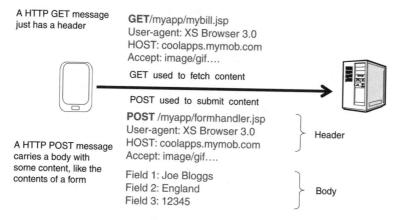

Figure 7.22 HTTP POST method used to submit content.

If we know that the objective of a particular WAP Push message is to visually notify the user of an actionable item (e.g. link to another server), then we can send a particular message type called a *Service Indication*. This contains an embedded URI that points to a resource on an origin server. The objective of the push message is to ask the user to visit this resource, usually via a browser. The user can elect to click on the link and this will automatically initiate an IP session and subsequent WSP or HTTP (or W-HTTP) session to the referenced resource.

For WAP 1 compatible devices, WSP works fine for asynchronously pushing messages to mobile devices. However, with WAP 2, both the standard or wireless-profile forms of HTTP and TCP do not support push mechanisms. Therefore we have to implement a push channel using an alternative approach. The way this is done is by turning HTTP on its head and asking the mobile device to act like a HTTP origin server so that it can accept POST method requests. If this is done, we are then able to send it content over a TCP/IP link.

It is fairly obvious that this would require an active IP session between the device and the network. Therefore, in the absence of an IP session we can still revert to the WAP Push over SMS. The use of more than one protocol to achieve a task is perfectly plausible. It is just a matter of implementing both protocol stacks in software on the device. It is perfectly acceptable to use WAP Push over WTP/SMS to initiate a subsequent browsing session with an origin server via HTTP/TCP (or W-HTTP/TCP).

7.7 AJAX

Before going on to talk about other protocols used in mobile services, it is useful to look at another technique for improving browser-application performance, called *AJAX*. What AJAX does is enable part of a webpage to be updated without going back to the web server to fetch the entire updated page. Instead, a mechanism is used to fetch only the changed data and then to insert this into the page already displayed in the browser, as shown in Figure 7.23. This technique is useful for reducing latency, which is something we are sensitive to with mobile RF connections. The performance improvement is achieved in several ways. Firstly, if we only have to fetch a small amount of data versus a whole page, then we won't have to wait so long to fetch the data. Secondly, as we don't have to re-render a whole page in the browser, we won't have that delay to worry about either. This is great for the user because they continue to see the UI displayed in the browser, which gives a feeling of continuity and responsiveness to the application. Thirdly, the technique used for selective updating of the page using JavaScript isn't limited to waiting for the user to make a request. Data can be pre-fetched ahead of time. For example, if the user is browsing some kind of list (such as pop videos) we can fetch the next page of listings before the user scrolls down to view them. Furthermore, the JavaScript program that is controlling the fetching and updating can store data locally before displaying it, which can further speed-up the responsiveness of the application. For example, say the listing is 1000 items long. The first page load needs to be quick, so we preload 30 listings with the initial page load. However, we only display 10 of these in the limited space available in the part of the page dedicated to displaying the listings. As the user scrolls down, we can load from the other 20 listings into the UI. Meanwhile, and even before the user started scrolling, we can pre-fetch another 30 listings so we are ready to display these as soon as the user scrolls.

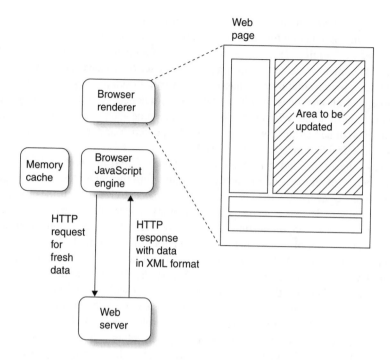

Figure 7.23 Using AJAX to fetch web page data.

If used wisely, AJAX can improve the user experience in a mobile environment. However, there are a few problems. Firstly, most mobile browsers don't currently support AJAX, so the technique can't be relied upon for general mobile web application design. Secondly, a naïve AJAX implementation can actually cause us problems, especially if we are paying for the data connection by the byte. Some AJAX implementations pre-fetch huge amounts of data and store it 'off screen' in readiness for updating the webpage UI. This could mean that we load a lot of data that we don't ever end up viewing; thus, bulking our web pages with lots of redundant data that we may well have paid to fetch and then never seen.

7.8 PEER-TO-PEER (P2P)

P2P is a new and exciting area of computing and due to its infancy, clearly established protocols have yet to emerge that are as firmly established as HTTP is for the Web. P2P is not that widespread anyhow, so there has not really been a chance for a set of protocols to emerge, but one of the efforts that is attracting a lot of interest is the JXTA project, which we shall discuss shortly.

7.8.1 Defining P2P

We have already introduced P2P in the early chapters of this book, so it seems a bit late to be defining it now, but it is an interesting topic nonetheless. We can dispense with some of the more meaty academic discussions that might surround this topic and concentrate on P2P in the mobile context, which is after all where our interest should lie; although, we should always have in mind that mobile-to-fixed communication can still take place in P2P mode. The philosophical origins of P2P don't quite fit the mobile context anyhow. There is a thought process that taking into account the increasing prevalence of broadband connectivity and the equally widespread availability of massively powerful desktop computing concludes that reliance on centralised processing might be outdated, or unnecessary in some cases. Not in all cases, but certainly some. With mobile networks, the same abundance of resources does not exist, although under certain conditions it could approach certain wire-line configurations, such as a powerful laptop networked via WiFi.

The key concept that underpins our consideration of P2P is that mobile devices can talk to each other, without the need to establish any link with the content network, or possibly without having to involve the IP network either. The most obvious case is the direct association of two or more devices in a personal area network sustained by Bluetooth connectivity. This seems to be the more interesting case to apply research and development effort in the short term. The reason for this is that it seems obvious that dual-mode devices, or even tri-mode will become commonplace. Dual-mode[29] being Bluetooth and 3G, tri-mode being the addition of WiFi to these two modes. It therefore seems obvious that increasingly devices will talk to each other directly, or will be able to network without the need to go through a server (although the absence of a server is not a necessary and sufficient definition of P2P[30]).

The important part of the P2P concept in the wireless space is that devices may want to connect with each other without going through a mobile cellular network; the reason being that the other modes of RF connection would provide potentially greater speed and be at lower cost, possibly no cost at all. This does not mean that we assume an either/or networking situation, where we use either the WAN or a LAN. Oddly, both could be used in a P2P application. For example, a connection between two devices could be established using a WAN and the switch over to a local connection in order to swap a file. The obvious example is music files[31]. What's interesting about this type of application is that the ability to determine the location of a device makes it all the more possible to find a neighbouring device to swap files with, assuming that this isn't already obvious by virtue of direct association with a nearby colleague or friend.

7.8.2 Some P2P Concepts

It is a common misconception to think that P2P networking is based on a single protocol. This is not necessarily the case. It really depends on a number of factors, but in the first instance it should be borne in mind that P2P networking can work in several different modes:

- *Point-to-point* – This is a direct communications pathway between two devices, although the devices could be co-located or separated in terms of a direct RF connection. A variety of protocols could be envisaged for point-to-point, but also there is no need to be limited to one, even within the context of a single communications session. For example, in a P2P session, devices may need to talk to each other in order to exchange control information.

---

[29] It is recognised that there is a trend in the PDA market towards WiFi with Bluetooth dual-mode devices, but here we are talking about the wide area connection being one of the required modes.
[30] For a more in-depth discussion of what is P2P, refer to: Brookshier, D. et al., *JXTA: Java P2P Programming*. Sams Publishing, Indianapolis, IN (2002).
[31] Note that despite the failure of Napster, music file swapping is still a possibility. *Digital Rights Management* schemes could make file swapping possible within a controllable framework.

This could be done using a protocol separate from the one used to exchange session information, such as music files. In the latter case, it could be that it makes more sense to stream information from one device to another, thus requiring a suitable streaming protocol, whereas control information would be exchanged using some other means.

- *Neighbour-linking* – Within a P2P network, devices may need to talk to each other to establish an interlinked mesh so that information and people can be found via the mesh. We shall discuss this idea at the end of the book when we look at *spatial messaging*, which is a type of application built on the location-finding capabilities of the RF network.

- *Broadcast* – This is the ability for one device to send the same information simultaneously to other devices in a defined group. It need not be restricted to a one-way process. All devices within a defined group could broadcast to every other device in the group. An obvious example of this is multiplayer gaming where the moves of a player need to be made known to all other players in real time.

- *Multipoint* – This is similar to broadcast, except that the means to get the information to each device group is via hopping from one device to another. This should not be confused with the neighbour-linking protocol(s) just mentioned, which are to do with discovering other devices ('out-of-session'), not communicating with them 'in-session'.

All of these modes of networking are required to implement a P2P system. In the mobile context there are also implications relating to the RF network and, in particular, the locality of the peer devices. For co-located devices that are able to communicate physically with each other directly (such as via Bluetooth), there are different considerations from those that are indirectly networked, realising that either mode can still support P2P communications.

Direct P2P connectivity exhibits unique challenges compared to indirect P2P, most notably the challenge of potentially having to support the entire communications session without any intermediary server functions whatsoever. For example, this would imply that the P2P clients would have to be able to find means to authenticate each other. This may not be a problem, because, as already noted, authentication in one sense is implied if the communicating parties are visually within contact and visual identification is possible. However, we should not ignore the cases where visual identification is not possible simply because the two parties hitherto had never met.

There are many issues, like authentication, that need to be considered when designing a P2P-based application in the mobile context. No one set of protocols has emerged yet to offer a comprehensive and satisfactory conclusion, and there is nothing stopping the adventurous amongst us from designing their own P2P protocols, which need not be IP-based. Indeed, one of the contenders for P2P protocols is JXTA, which is transport-agnostic in terms of the protocol set.

7.8.3 JXTA

Project JXTA started as a research project incubated at Sun Microsystems, Inc. under the guidance of Bill Joy and Mike Clary, to address the growing interest in peer-to-peer architectures as viable solutions to certain types of computing problems. JXTA is a set of open and generic peer-to-peer protocols that allow *any* connected device (mobile or fixed) on the network to *communicate and collaborate.*

Despite is origins at Sun and its use of the letter 'J' in the project name[32], the JXTA protocols and possible implementations are intended to be computer language agnostic, not solely or especially Java, though one naturally expects a bias towards Java within the open developer community attracted to this project.

JXTA proposes a set of protocols that address the common needs of a P2P solution, in the same way that TCP/IP proposes a set of protocols for hardware networking. However, JXTA technology connects peer nodes with each other, not hardware nodes. TCP/IP is platform-independent by virtue of being a set of protocols. So is JXTA. Moreover, JXTA technology is transport autonomous and can exploit TCP/IP as well as any other transport standards, which technically could include the native data protocols for UMTS within a 3G environment.

JXTA offers the following protocols:

- Peer Discovery Protocol

- Peer Resolver Protocol

- Peer Information Protocol

- Rendezvous Protocol

- Pipe Binding Protocol

- Endpoint Routing Protocol

These protocols offer various solutions to the P2P networking problem, and it is not intended to describe them in any detail here, only to introduce them and to identify some interesting aspects pertaining to mobility.

Before explaining the basis of the protocols, we need to introduce some other concepts from the JXTA P2P technology suite.

- *Advertisements* – An advertisement is an XML document that names, describes, and publishes the existence of a resource on the P2P network, such as a peer (a P2P participant), a peer group, a pipe (see further on in this list), or a service. We can see that advertisements enable peers to find out what exists on the P2P network, but no mention has yet been made of how to find these adverts (i.e. a discovery mechanism).

- *Peers* – A peer is any entity that can communicate within the P2P network using the JXTA protocols. This is analogous to the Internet, where an internet node is any entity that can talk using the IP protocols (some, or all of them). As such, a peer can be any device, including embedded devices like domestic appliances (e.g. burglar alarm).

- *Messages* – Messages are designed to be usable on top of asynchronous, unreliable, and unidirectional transport. Therefore, a message is designed as a datagram, containing an envelope and a stack of protocol headers with bodies. A message destination is given in the form of a URI, on any networking transport capable of sending and receiving datagram-style messages, which could include wireless or wire line, and any type of RF network. Messages can include credentials used to authenticate the sender and/or the receiver, but no set way of doing this is specified, but needs to be negotiated (like with SSL – see Chapter 9).

---

[32] In fact JXTA is a play on words, influenced by the word 'juxtapose' (to place side by side), hence it is not an acronym at all, more a sort of onomatopoeia.

- *Peer Groups* – Typically, a peer group is a collection of cooperating peers providing a common set of services. The specification does not dictate when, where, or why to create a peer group, or the type of the group or the membership of the group. It does not even define how to create a group. Very little is defined with respect to peer groups except how to discover peer groups using the *Peer Discovery Protocol*.

- *Pipes* – Pipes are virtual communication channels for sending and receiving messages, and are asynchronous. They are also unidirectional, so there are input pipes and output pipes. A pipe's endpoint can be bound to one or more peers. A pipe is usually dynamically bound to a peer at runtime via the *Pipe Binding Protocol*.

Retuning to the protocols for a brief explanation, they do the following:

- *Peer Discovery Protocol* – This protocol enables a peer to find advertisements on other peers, and can be used to find a peer, peer group or advertisements. This protocol is the default discovery protocol for all peer groups and so has to be included in the implementation of any peer software. Peer discovery can be done with or without specifying a name for either the peer to be located or the group to which peers belong. When no name is specified, all advertisements are returned, but clearly this is a costly option within the mobile context.

- *Peer Resolver Protocol* – This protocol enables a peer to send and receive generic queries to find or search for peers, peer groups, pipes and other information. Typically, only those peers that have access to data repositories and offer advanced search capabilities implement this protocol.

- *Peer Information Protocol* – This protocol allows a peer to learn about other peers' capabilities and status. For example, one can send a *ping* message to see if a peer is alive. One can also query a peer's properties where each property has a name and a value string.

- *Rendezvous Protocol* – This protocol allows a peer to propagate a message within the scope of a peer group.

- *Pipe Binding Protocol* – This protocol allows a peer to bind a pipe advertisement to a pipe endpoint, thus indicating where messages actually go over the pipe. In some sense, a pipe can be viewed as an abstract, named message queue that supports a number of abstract operations such as create, open, close, delete, send and receive.

- *Endpoint Routing Protocol* – This protocol allows a peer to ask a peer router for available routes for sending a message to a destination peer. Often, two communicating peers may not be directly connected to each other, as discussed earlier when we talked about the general concept of neighbour-linking. Any peer can decide to become a peer router by implementing the Endpoint Routing Protocol.

The use of JXTA for wireless is being looked at by some members of the JXTA community and has led some researchers to find interesting ways to combine JXTA P2P concepts with the emerging and evolving wireless technologies[33], such as the mobile version of Java, called Java 2 Mobile Edition (J2ME), which is a software technology we discuss in Chapter 11.

---

[33] Arora, A., Haywood, C. and Pabla, K. *JXTA for J2ME – Extending the Reach of Wireless with JXTA Technology.* Sun Microsystems, Inc. (March 2002) – https://jxta.dev.java.net/

The main challenge with JXTA in a mobile environment is ensuring that the protocols enable the most efficient use of resources within the resource-limited RF network. There are other concerns relating to the robustness of JXTA within a temperamental mobile environment. Neighbour-linking needs to be resilient to any of the neighbouring peers finding themselves in a hostile RF situation, including complete signal loss. The rate at which the network can heal and pipes be reformed will be critical to system performance. There seems a need to research the efficiencies gained by mapping P2P protocols onto the RF network in a way that is cognisant of network characteristics.

7.9   INSTANT MESSAGING (IM) PROTOCOLS

We shall briefly discuss these protocols because they are used in mobile services, but not central to a wide range of mobile applications; particularly, as they were all designed initially with only one application in mind, which is Instant Messaging, or IM. To go into the details of how these protocols work would be too detailed for this book, so instead we introduce the protocols, their backgrounds, uses and relationships. This is a useful grounding for those not familiar with the protocols for mobile IM, which, at times, can appear very confusing to the uninitiated. At the time of writing, there are three industry-wide *open* standards for implementing IM services. These are

1. SIP/SIMPLE

2. XMPP

3. IMPS

An overview of these standards and their applicability to the mobile environment is now given in the following subsections.

7.9.1   SIP/SIMPLE

SIMPLE stands for *SIP for Instant Messaging and Presence Leveraging Extensions*. This standard is the activity of an IETF working group ('SIMPLE WG'). Their charter is available online and they summarise it thus:

> *This working group focuses on the application of the Session Initiation Protocol (SIP, RFC 3261) to the suite of services collectively known as instant messaging and presence (IMP). The IETF has committed to producing an interoperable standard for these services compliant to the requirements for IM outlined in RFC 2779 (including the security and privacy requirements there) and in the Common Presence and Instant Messaging (CPIM) specification, developed within the IMPP working group.*

The Instant Messaging and Presence Protocol (IMPP) Working Group (WG) mentioned here is no longer active because its work is complete. The IMPP WG produced a number of specifications outlining generic and abstract operation and requirements for IM systems, including RFC 2778 and RFC 2779, as well as the Common Profile for Instant Messaging

(CPIM) specification[34], which is concerned with end-to-end messaging formats for the purpose of interoperability between IM systems. Due to a lack of consensus, the IMPP WG never produced any protocols, instead deferring to other working groups that at some time in the future might develop protocols to meet the requirements defined in RFC 2779. Thus far, two[35] IETF working groups have formed to define protocols: XMPP (see the next section) and SIP/SIMPLE.

Given the rising importance and popularity of SIP for real-time, session-based communications over IP, such as IMS (see Chapter 14), it was an obvious choice for extending into the IM domain – hence the formation of SIMPLE. SIMPLE is still an active working group in the IETF, with the bulk of specifications still in draft form. It is possible to implement a working IM system now that is based on SIMPLE, but whilst the specifications are still in draft form, there is a possibility that vendors would need to use proprietary interpretations, especially in order to realise a working solution that is feature-rich. (This has already proven the case with notable implementations, like Microsoft's Live Communications Server (LCS)[36].)

The interest in SIMPLE for mobile IM systems also arises from 3GPP architectural requirements for a presence service, defined by TS 23.141[37], which states:

> *The Presence Server shall support SIP-based communications for publishing presence information. The Presence Server is a SIP Application Server as defined by 3GPP [TS 23.228[38]].*

The SIP Application Server mentioned above is an entity within an *IP Multimedia Subsystem* (IMS) environment. IMS is an integral part of the evolution of the 3GPP core network and is fully described in Chapter 14. Therefore, at least as far as 3GPP is concerned, the future of IM within a mobile network is going to be a SIP one. Moreover, 3GPP liberally references specifications from the SIMPLE working group, so the future for IM according to 3GPP is SIMPLE. Alignment with SIP/SIMPLE should come as no surprise given the strategic importance placed on SIP (i.e. in IMS), not only in the mobile world, but now in the fixed world, too (i.e. European Telecommunications Standards Institute's (ETSI) TISPAN activity, which stands for Telecoms & Internet converged Services & Protocols for Advanced Networks).

In addition to its support in the 3GPP, SIMPLE is attracting increasing support from other industry forces, including Microsoft and IBM. In particular, Microsoft continues to base its IM and related real-time collaborative technologies on SIMPLE, not in the consumer (e.g. MSN Messenger), but in the lucrative enterprise markets (e.g. LCS). Like others, including 3GPP, Microsoft has recognised the versatility of SIP as a protocol suitable for a wide range of collaborative real-time services. The Open Mobile Alliance (OMA) has also recognised this. The OMA has a charter of defining services for the sake of commonality across the Mobile Network Operator (MNO) networks. The alliance has

---

[34] Peterson, J., "Common Profile for Instant Messaging (CPIM)", RFC 3860, August 2004.
[35] The two standards are competing, but the 'democratic' nature of the IETF allows for the submission and acceptance of more than one approach to a particular problem. Moreover, when XMPP was submitted to the IETF, SIMPLE was not yet formed. Besides, there is a very credible argument that the two standards can exist in the same market because of the legitimate existence of different interest groups (e.g. the existing 'SIP Community' has distinct interests that make SIMPLE an obvious candidate for IM solutions).
[36] Although it is not possible to confirm that Microsoft's extensions were a result of incomplete SIMPLE specs.
[37] 3GPP TS 23.141 'UMTS; Presence service; Architecture and functional description; Stage 2'.
[38] 3GPP TS 23.228 'IP Multimedia Subsystem (IMS); Stage 2'.

broad mobile industry support within both the MNO and the vendor communities. OMA has an active interest in both IM services and presence services. These are addressed through separate working groups in the OMA:

- Presence and Availability Working Group ('PAG')

- Messaging Working Group ('MWG') – Instant Messaging Sub-group ('MWG IM')

The existence of a separate group for presence (i.e. PAG) is because presence is seen as an enabler for a plethora of future mobile services, not just IM, so it is important to ensure that wider requirements are catered for.

The initial activities of these groups (PAG and MWG) to define an IM system were focused on the adoption of the work of the *Wireless Village* initiative, which is now fully merged into the OMA. The technical initiatives of the Wireless Village were re-branded as the *Instant Messaging and Presence System* (IMPS), which is explained later.

Due to the influence of the 3GPP and its interest in SIP/SIMPLE, as noted above, the OMA has been influenced to take steps towards defining an IM system based on SIP/SIMPLE. This work is under the OMA Work Item (WI) 74 'SIP/SIMPLE Based IM Service Definitions', as defined in OMA-WID_0074[39], which summarises its goals as the following five stages:

- *Stage 1* – Consolidate and coordinate instant messaging service requirements, based on SIP/SIMPLE. This activity should be conducted cooperatively with the Requirements WG, and thus avoid creating any duplicate efforts within OMA.

- *Stage 2* – Define a set of generalised instant messaging capabilities and functions, embracing through reference architectures from other groups, e.g. IETF, 3GPP and 3GPP2. Create an OMA instant messaging architecture based on SIP/SIMPLE, which is agnostic to the underlying network.

- *Stage 3* – Create OMA specifications building upon the generalised instant messaging capabilities and functionalities.

- *Stage 4* – Map these specifications onto systems defined by other groups, e.g. IETF, 3GPP and 3GPP2. The OMA MWG goal is to embrace and extend existing functionality.

- *Stage 5* – Create test cases for IOP based on these OMA specifications.

Hence, we can see that the future of IM systems for mobile deployments is heavily reliant on the work of the IETF SIMPLE working group, which feeds the OMA SIMPLE standardisation process.

7.9.2 XMPP

The Extensible Messaging and Presence Protocol (XMPP) is the IETF's other formalisation of an IM system based on IMPP. The core of XMPP is the base XML streaming protocols for instant messaging and presence developed within the Jabber[40] community starting in 1999.

---

[39] OMA-WID_0074-SIMPLE_IM-V1_0-20031202-A 'SIP/SIMPLE Based IM Service Definitions'.
[40] http://www.jabber.org

The Jabber project was intended by its founder (Jeremie Miller) to devise an open standard for IM systems, promoting interoperability (in contrast to the highly segmented world of commercial IM deployments). Working protocols from the group were defined in 1999, which was before the work of the IETF's IMPP began (in 2000). In 2002, the Jabber open-source community, under the auspices of the *Jabber Software Foundation* (JSF), submitted the base Jabber protocols to the IETF. The first submission was made in February 2002 as an informational Internet-Draft[41].

Then the JSF submitted a proposal to form an IETF WG devoted to formalisation of the Jabber protocols, under the neutral name of Extensible Messaging and Presence Protocol (XMPP), which was approved by the IESG in October 2002.

In November 2002, updated Internet-Drafts were submitted for 'XMPP Core' and 'XMPP IM', two central definitions of the XMPP protocol. Upon publication of the accepted XMPP RFCs, the work of the XMPP WG was concluded. However, the XMPP WG mailing list[42] remains active, and development of further XMPP extensions is pursued through Jabber Enhancement Proposals (JEP)[43] produced by the JSF, which is an open forum. It should be noted that XMPP implementations almost always include key JEP extensions in addition to the XMPP RFCs. Given the historical roots of XMPP within the Jabber community, the standard continues to evolve mostly within that community and using the JEP standardisation process. Therefore, the XMPP protocol set is based on a core set of IETF documents, which are extended by JEP documents. Together these form the 'XMPP' standard.

The XMPP protocols, unlike many other IM protocols, emerged from an *open-source software* community. This means that protocols *and* software became available through the contributing community, not just protocols. Therefore, as is to be expected, there has been a lot of support for XMPP in a variety of products and services as it is easy to implement due to existing code. However, it should also be noted that XMPP is a protocol that has been devised from the ground up to support IM functions, unlike SIMPLE, which is based on an existing non-IM protocol (SIP).

7.9.3 IMPS

OMA IMPS v1.x Specification work originated in June 2002 out of the Wireless Village initiative and merged into the OMA as a technical activity in 1 October 2002. After restructuring decisions by OMA Technical Plenary in September 2003, the maintenance of these specifications moved into the Messaging WG. Subsequently, the work has been moved into the Messaging WG IM sub-group (MWG IM) and into the Presence and Availability Workgroup (PA WG). This further dividing of responsibility has come from the growing recognition that the separation of the IM service from the presence service is not only technically feasible (as has always been the view of RFC 2779), but that such a separation makes sense from a service perspective, too (as also reflected in 3GPP TS 23.141). Presence is seen as a key *service enabler* for a variety of services, not just IM, but also VoIP, Push-To-Talk over Cellular (PoC), gaming and so on. Presentity-based communications is set to become a prevalent user-interface modality in next generation mobile (and wireline) communications (see Chapter 14).

---

[41] http://www.jabber.org/ietf/attic/draft-miller-jabber-00.html
[42] http://mail.jabber.org/mailman/listinfo/xmppwg
[43] http://www.jabber.org/jeps/

As mentioned above, the OMA's long-term vision for IM is reflected in work items in both the MWG IM[44] and PA WG[45] to develop solutions based on SIMPLE. However, in the meantime, IMPS is already at version 1.2, with 1.3 being drafted, and with vendors already delivering compatible solutions based on earlier (1.1) and current releases, whilst promising support for future releases (1.3).

7.9.4 IM Interoperability

Given the emergence of competing standards and commercial deployments for IM systems, interoperability has been a hot topic of discussion and activity within the standards communities for many years. Of additional interest for product vendors is interoperability with non-standards, especially those with large user communities, like AOL, MSN and ICQ.

Interoperability between different service domains using the same standards is not a problem as all of the standards listed in the previous subsections have addressed the issue of server–server connectivity across different service domains. Although, it is worth noting that interdomain operation of a common standard might still require a degree of common housekeeping and good administration 'etiquette' to ensure interoperability. In all the IM standards mentioned, there is no such thing as a common implementation, although IMPS comes closest to trying to define one (not surprisingly, given the keen interest in interoperability that its participants would clearly favour). Therefore, when operating across different domains, implementation issues might become pertinent to ensuring mutually successful operation. Interoperability between different IM types across different service domains is a more exacting challenge. However, there are some trends already in motion that are worthy of our attention for this analysis.

Both the XMPP community and the IMPS community are concerned with interworking with SIMPLE. In the case of IMPS, this is an explicit recognition of the OMA's long-term objectives to move to a SIMPLE-based IM solution. Therefore, the OMA has introduced work items specifically to ensure interoperability between IMPS and SIMPLE, including the drafting of a set of interworking requirements[46]. However, the motivation of this work is largely to ensure contiguous IM service operation as an operator migrates from IMPS to SIMPLE, so we should be clear that the context of the IWF is apparently a single trusted domain. For example, solving issues of user authentication and validation between IMPS and SIMPLE are far easier where a common database of users is available for both systems. This is the approach already taken by some vendors.

In the case of XMPP, one might argue that in the spirit of the Jabber community's open-standards interests, an attempt to ensure a workable mapping of XMPP to SIMPLE is a natural progression, despite being 'competing' standards. Moreover, as both of these standards are under the IETF 'umbrella', then one would expect interworking initiatives to be concomitant with the objectives of the IETF (although that is not always a valid assumption). In particular, both the XMPP and SIMPLE WGs based their submissions on the work of the IMPP, especially RFC 2778, 2779 and the Common Profile for IM (CPIM).

[44] http://www.openmobilealliance.org/tech/wg_committees/mwg.html#im
[45] http://www.openmobilealliance.org/tech/wg_committees/pag.html
[46] OMA-RD-IMPS-SIP-SIMPLE_IWF-V1_0_0-20050211-D   'IMPS-SIP/SIMPLE   Interworking   Function Requirements'.

As stated above, the whole point of the CPIM was to provide a baseline for end-to-end messaging that would allow for interoperability, as stated in RFC 3860:

> *This specification defines semantics and data formats for common services of instant messaging to facilitate the creation of gateways between instant messaging services: a common profile for instant messaging (CPIM).*

From this, we see that the main objective of CPIM is to allow gateways to be implemented between disparate IM implementations. In the first instance, this is achieved by ensuring that each IM standard defines its protocols in a way that will map to the common CPIM. Hence, that is the approach clearly outlined by SIMPLE and XMPP, where the standards can both be shown to demonstrate a mapping of their protocols onto the CPIM.

However, the XMPP community has gone as far as drafting an IETF Internet-Draft (memo)[47] to clarify the interoperability requirements between XMPP and SIMPLE. In particular, as stated in the memo:

> *The approach taken in this document is to directly map semantics from one protocol to another (i.e., from SIP/SIMPLE to XMPP and vice-versa), mainly for use by gateways between systems that implement one or the other of these protocols.*

Thus, the memo gives us a mapping of protocol semantics, which will greatly assist our analysis, which is largely semantically based (not implementation based).

It is worth noting that XMPP is, at its heart, a generic XML streaming protocol. As such, it is feasible to support XMPP messaging streams via SIP dialogues and such a mechanism was suggested as an early contender for SIMPLE. However, this does not mean that an XMPP is automatically compatible with a SIMPLE IM solution and care should be taken to avoid such mistaken assumptions.

Given the wide interest – and visible standards efforts – in ensuring interoperability between existing protocols and the emerging SIMPLE standard, the most likely scenario for IM interoperability in the future is the emergence of SIMPLE as the main hub of interdomain IM communications.

7.9.5 Protocol Acceptance (Support)

From the preceding discussion about the extant open standards for IM systems, we can now focus on the support for these protocols in terms of understanding wider interoperability scenarios in the real world.

IMPS IMPS already receives wide support in the mobile industry by virtue of its origins from that community, now nurtured within the OMA. Many handsets on the market include IMPS clients and there are many IMPS server products, from both traditional mobile infrastructure suppliers (e.g. Nokia, Siemens) and a host of third-party suppliers (e.g. Colibria).

---

[47] Saint-Andre, P., Houri, A. and Hildebrand, J.,
'Basic Messaging and Presence Interoperability between the Extensible Messaging and Presence Protocol (XMPP) and Session Initiation Protocol
(SIP) for Instant Messaging and Presence Leveraging Extensions (SIMPLE)' draft-saintandre-xmpp-simple-05.
See http://www.xmpp.org/internet-drafts/attic/draft-saintandre-xmpp-simple-00.html

There are also numerous IMP-client providers (e.g. Ecrio), including those aimed at embedded reference designs, such as TTPCom (with their AJAR suite – which is now owned by Motorola).

Of particular interest are the announcements by some vendors of IMPS server products who voice future product support for SIMPLE. Clearly, as the OMA SIMPLE initiative is still underway, no vendor can reasonably claim to support it, but the SIMPLE specifications themselves have been in draft sometime, some of them released, and it is possible to build early release SIMPLE IM systems, particularly where the feature set is restricted to a 'basic' set of messaging and presence semantics.

The clear theme of vendors attesting to SIMPLE support, either now or in the future, is integration into the 3GPP IMS environment. Again, the deployment of IMS within the operator community is practically non-existent, mostly based on trials to date. Therefore, such claims are also to be tempered with an appropriate degree of caution.

What matters is that the development direction is clear. The IMPS standard is here today and has wide support. Equipment is already in the field that is IMPS-capable, including numerous handsets. However, the future seems biased towards SIMPLE and this is reflected in IM-vendor product positioning where, increasingly, the trend is towards IMPS products with SIMPLE 'upgrades' or evolution paths (e.g. Colibria) and eventually with complete SIMPLE support end-to-end.

XMPP   Because of its open-source origins, XMPP receives wide support amongst the Internet community, especially as the original and popular open-source Jabberd server remains in active circulation and development.

Many high-profile companies such as IBM, HP, SUN and Oracle have already adopted Jabber/XMPP technologies and this trend has led to an interest in XMPP within enterprise markets; especially as XMPP has included robust security and encryption support from the start, based on existing IETF protocols like Transport Layer Secure (TLS). Additionally, XMPP has included federated identity support, which allows trusted domains to connect securely so that enterprises can safely link with branch, affiliate, supplier and customer networks. This is deemed essential for enterprise operation and will be an added layer of complexity for SIP solutions. Currently, federated identity is a non-consideration for IMPS solutions (i.e. it has not been considered and is potentially difficult to achieve).

There are also various companies supporting XMPP in the enterprise markets who have demonstrated interworking with SIMPLE for server–server support with SIMPLE networks, whilst also maintaining federated identity support.

The most notable recent supporter of XMPP is Google, who use XMPP (with a number of extensions[48]) to support their Google Talk service. Google have mentioned plans to federate with various other internet-hosted communications communities, such as EarthLink's Vling, SIPphone's Gizmo Project and, perhaps, the large XMPP-based services such as those offered by BellSouth, Wanadoo.fr, Orange, Portugal Telecom, SAPO, and so on.

SIP/SIMPLE   SIP/SIMPLE has gained wide support throughout the industry, mobile and fixed. Despite (or because of) its wide support, including 3GPP and OMA, the problem

---

[48] XMPP is, by design, an extensible set of protocols, so Google's extensions should not be seen as indicative of a deficiency in XMPP, unlike some proprietary extensions to SIMPLE that have arisen because of SIMPLE's immaturity.

with SIMPLE is the slow maturation towards a workable and final set of specifications. One issue in particular has been that SIP itself is not ideally suited to IM modes of message exchange. Apart from SIP itself being a comparatively verbose protocol (problematic for wireless transport), it (SIP, not SIMPLE) isn't ideally suited to IM communications, as now discussed.

SIP-based IM can be split into two forms. The first is called *pager mode* messaging, which uses the SIP MESSAGE (RFC 3428) method to send a simple, short SMS-style message. However, messages sent in this manner are not associated in a single SIP dialogue and it is therefore difficult to correlate multiple messages and is only useful for individual exchanges (e.g. SMS or SMS-style communications). Pager mode messaging has been defined for a long time (RFC 3428) and is widely used in SIP deployments today.

The alternative messaging style is called *session mode* messaging, which is used for multiple-message exchanges between two users with an explicit IM-session association at the SIP dialogue level[49]. Once the SIMPLE WG understood that a session-aware IM extension was required for SIP, they were faced with a choice. The existence of XMPP, which is a session-aware IM protocol, meant that SIMPLE WG could have adopted it as a very reasonable solution to the problem. The WG did consider this approach at one point (2003). However, for reasons probably relating to commercial interests, the WG proposed a session-aware protocol, called Messaging Relay Session Protocol (MRSP). This is still under construction within the IETF and its delay has been considered by some as a critical issue for the success of a SIP-related IMS solution. A recent increase in the activity in this area has resulted in a solution that is nearly ready for submission (to IETF) for standardisation.

Because of the addition of MRSP and increasing interest in SIMPLE from SIP vendors, SIMPLE is emerging as probably the standard of choice for future IM systems, especially within the telecoms community. However, XMPP continues to attract interest and there has even been a recent announcement that BT will be adopting it for their 21st Century Network (21CN) project. As already mentioned, XMPP is an extremely simple and attractive IM solution and perhaps Google's adoption, along with BT, makes it a more significant technology than it might otherwise be.

---

[49] You can think of this as meaning that two parties 'call each other' and then have an IM session and then 'hang up', rather than just keep interrupting each other with 'text messages'. Once the 'line' is open, exchanging many messages has a lower overhead than lots of individually addressed 'texts'.

8

J2EE Presentation Layer

In Chapter 6 we looked at how we can build scalable and robust mobile applications using a software platform technology called J2EE. The essence of J2EE was the division of software into infrastructure services and business-specific services, as shown in Figure 8.1. The infrastructure services are inherent in the J2EE platform. They enable us to produce mobile applications that are multi-user, scalable, secure, fully redundant, database-connected, and so on, without having to include all these functions in the software we write.

In Chapter 7 we looked at protocols to support the link between the devices and the platform. Our main focus was on link–fetch–response paradigms supported by IP protocols like HTTP/TCP and wireless equivalents from the WAP family, like WSP/WTP.

In this chapter, we shall examine in more detail how J2EE supports these protocols and how we might structure our business software to take advantage of the power of J2EE whilst simultaneously supporting protocols like HTTP.

We shall also look in more detail at how the browser processes the content fetched from the server. We have mentioned the browser application as a key component in many mobile services, but we have not yet looked at how it works in any detail, or how it ties back to the J2EE programming paradigm. There are also some challenges that are very specific to the mobile environment, such as how to cope with many different devices types. This is a challenge for our application designers and one that needs to be addressed. What if we design an application to access an email server; how are we going to make sure that it works equally well or satisfactorily well, on all devices? When we look at markup languages and their relationship with device capabilities, we shall examine such thorny problems.

8.1 SEPARATING PRESENTATION FROM BUSINESS LOGIC

When discussing J2EE, we introduced the concept of encapsulating our business-specific software into software units called Enterprise Java Beans (EJBs). We said that if put all of

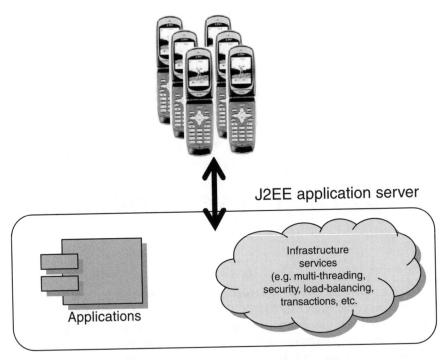

Figure 8.1 Division of software into infrastructure and business services. (Reproduced by permission of IXI Mobile.)

our business functions into EJBs and then hosted these on a J2EE application server, we could rely on the server doing all the challenging and complicated middleware stuff; such as enabling the bean to be accessed by many clients at once without collapsing under the strain. In Chapter 9, we shall look in detail at some of the underpinning technologies of J2EE in order to see how the middleware works.

Most of the time, when developing an application that runs on a J2EE application server, the entire set of software functions we end up needing will fall neatly into two functional areas. The first area is concerned with implementing the UI and the other is the 'brains' of the application, as shown in Figure 8.2.

We generally refer to the software concerned with the user interface as the *presentation logic*, or *presentation layer*. The rest we refer to as the *business logic*, or *business layer*. This relatively neat division of functionality is exploited by the J2EE architecture, to the benefit of the programmer, by providing distinct J2EE tools for each case. For example, in the case of the presentation logic, the powerful built-in HTTP capabilities of the J2EE platform can be offered to the programmer in a model that makes programming for HTTP connectivity easier. By separating out presentation logic from business logic, it also means that we can maintain these portions of software independently of each other, which is good for managing the software development process. It is especially useful for mobile services where the evolution of device types tends to be quite rapid, so we might need to be more agile with evolving the UI part of the service.

J2EE has three types of software program that we can use to create an entire J2EE application: *servlets, JSPs* and *EJBs*. We already know something about EJBs, so let's

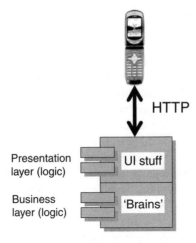

Figure 8.2 Presentation and business layers. (Reproduced by permission of IXI Mobile.)

look at these other entities to see what they are, their essential differences and how to use them.

8.1.1 Servlets and JSPs – 'HTTP Programs'

As shown in Figure 8.3, there is a simple distinction between EJBs and the UI entities (servlets and JSPs). The EJBs are always associated with accessing what is referred to as the *Enterprise Information Services* (EIS) tier. The EIS tier is where all the heavy database servers reside; directories, legacy systems, and mail servers and so on. We can access a good deal of the EIS tier via the powerful J2EE APIs, like JDBC, but it is essential that we do this

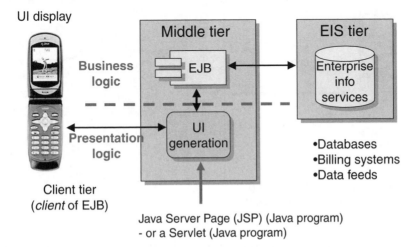

Figure 8.3 J2EE supports business and presentation tiers via different program types (EJB, servlet or JSP). (Reproduced by permission of IXI Mobile.)

from EJB code if we want to build a system that truly benefits from the J2EE-infrastructure services approach.

The servlets and JSPs act independently of the EJBs, but call upon the services of the EJB to carry out useful work that requires resources in the EIS tier. The servlet and JSP program types are specifically designed to facilitate browser-based applications accessing the J2EE platform via HTTP. In particular, it is assumed that any software written to run inside a servlet or JSP is going to produce a markup text stream that will be sent over an HTTP link back to the browser that initiated the program's execution in the first place when it sent an HTTP request to the J2EE server. This tells us that one of the roles of a J2EE application server is a web server (or *origin server* in the HTTP vernacular). In other words, servlets and JSPs are programs that sit on the J2EE server ready to run whenever a HTTP request is sent to them by the application server's web engine, as shown in Figure 8.4.

If we look at the sequence of events in Figure 8.4, we can better understand the role of servlets and JSPs and how they function in a J2EE environment. We will also be able to appreciate the intimate connection these program types have with the HTTP protocol.

Step 1 is the user initiating a request via a link to one of the servlets on the server. This happens when the user enters a web address in their browser (via the address-entry box) or clicks on a link in a page that has this request resource address beneath it, causing the browser to make an HTTP GET request to the relevant address.

Step 2 is the user agent responding to the user's request by generating a HTTP request using the GET method. We can see in the header for the request that the resource required on the server has a path name of '/myapp/prog1'. The URL for the request has the domain name 'coolapps.mymob.com' and the user agent uses this information to route the HTTP request to the corresponding server. When the request reaches the server via the RF and IP network layers, it is routed by the server's operating system to the J2EE server application.

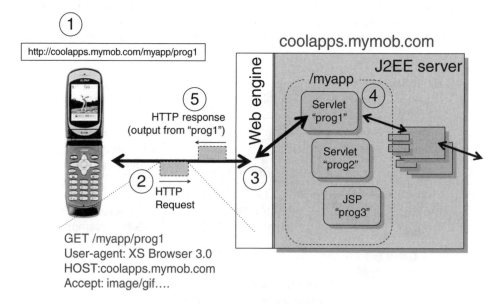

Figure 8.4 Servlets and JSPs are 'HTTP programs'. (Reproduced by permission of IXI Mobile.)

The operating system, or its TCP/IP stack component, knows to do this because the J2EE server has registered itself with the appropriate port (port 80) to say that it expects all traffic on that port to be forwarded to it. Forward communications to port 80 are processed by the web engine component of the J2EE engine. The web engine extracts the pathname from the request – '/myapp/prog1' – and 'looks' in the web folder '/myapp' to see if there is a program registered there with the name 'prog1'. Upon finding that there is indeed a servlet with this name at that path location, the web engine proceeds to schedule the servlet for execution on the J2EE platform, as in Step 3.

All the HTTP header information is made available to the servlet and it is the job of the web engine to pass this information to the servlet (or JSP). This is done by making the information available via a Java API, which in this case happens to be called *servlet*. This is a Java program that has methods that can be accessed by the servlet when it executes. One of these methods can be used to extract HTTP headers from the request that spawned the servlet's execution. If there were any parameters in the HTTP header that the servlet needs, then it has full access to them via the API.

In Step 4, the servlet is now executing on the server and has full access to all the J2EE APIs, including the ability to call EJBs to carry out business logic tasks. For example, this servlet could be responsible for creating a web page to list ringtones on a particular theme. The list and details of the ringtones are assumed to be in a database. An EJB can use the JDBC API to fetch tone names from the database. The returned descriptors might be in XML, so the EJB also extracts the appropriate XML fields, such as the title of the tones (e.g. from a <Title> tag pair).

Upon completing the bulk of its execution, including any calls to EJBs (which could be on the same server or another one), the servlet has now generated a page of XHTML (or other appropriate browser language) and is ready to stream it back to the device via HTTP. The servlet does this by writing the page to the HTTP connection using another method within the *servlet* API. This is Step 5. The HTTP response output from the servlet is now received by the user agent on the device and it gets displayed in the browser interface. The user can now see a list of ringtones. These may have been encoded in the page as links, probably linking to a related servlet or JSP (e.g. 'prog2') that can reveal the detailed information about a tone, like its cost and how to order it; hence the cycle of HTTP requests and responses continues.

8.1.2 Comparing Servlets with JSPs

Servlets and JSPs are optimised and designed for the HTTP (browsing) paradigm. Many of their features were originally conceived to meet the challenges of HTTP programming. Hence, wherever possible, we should aim to use these entities for building web-user interfaces for our web-based mobile applications. Technically, there is actually nothing preventing an EJB from servicing HTTP connections, but they are not ideally suited for this task. When programming for web-based interaction, there are some common challenges that programmers face. In keeping with the J2EE infrastructure ethos, some of these challenges have already been addressed by powerful facilities within the J2EE platform, which are part of the JSP and servlet programming environment.

The essential difference between a JSP and a servlet is shown in Figure 8.5, where we also make a rudimentary comparison with EJBs. We can see from the figure that what JSPs

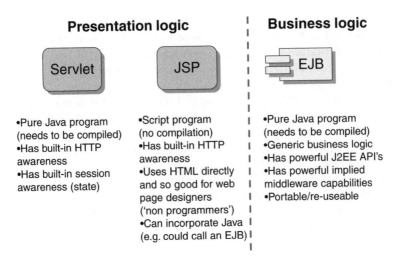

Figure 8.5 J2EE program types.

and servlets have in common is 'HTTP awareness'. As we saw in the last section, these program types can be executed in direct response to an inbound HTTP request from a user agent. More than this, these programs are designed so that there outputs are fed back to the requesting user agent automatically. The programmer does not have to make any special effort to track inbound requests and match outgoing responses, no matter how many users are concurrently accessing the program. J2EE automatically supports HTTP concurrency, making sure that the responses get back to the requesting agent.

The essential difference between a JSP and a servlet is that the JSP is constructed like a script and can include the direct use of XHTML (or HTML, etc.) in its construction, whereas a servlet is a pure Java program. The JSP is written, like a servlet, as a text file. The JSP subsequently gets stored on the server as a text file. When the J2EE server runs a JSP, it will simply attempt to open the file and output all the JSP's contents as the body of the response to the inbound request, as shown in Figure 8.6.

As we can see in Figure 8.6, the user agent processes a link (Step 1) with the appropriate address (URL) to eventually create an inbound HTTP request (Step 2) that gets routed (Step 3) to the corresponding JSP on the server. This JSP is run as a script on the server (Step 4). What happens is that the JSP file is opened and then each line is stepped through, line by line[1], until the end of the file. Whatever the J2EE web engine finds on each line is written straight back out to the HTTP body of a corresponding response to the inbound request, as seen in Step 5. The entire response, its header and its overall construction are all taken care of by the J2EE server – the JSP programmer does not have to program any of these infrastructure tasks. You can imagine the operation of a JSP as if our device's user agent simply picks up the contents of the text file and pulled them back up to the requesting device.

The exception to this simple flow is if the web engine encounters special scripting inserts in the JSP, of which there are two types. The first type (called *scriplets*) are demarcated with special beginning and ending markers: '<%' to begin an insert and '%>' to end an insert.

---

[1] Just like in the way that any kind of script, even a film script, is followed one line at a time from the beginning to the end; hence why a JSP is called a scripted program. It has a linear flow from top to bottom.

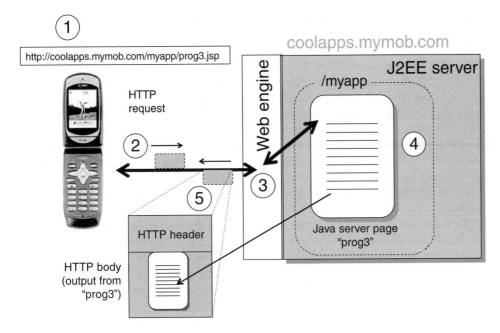

Figure 8.6 Java server page execution. (Reproduced by permission of IXI Mobile.)

What lies between these markers is Java code and it will get executed by the J2EE web engine. The output *from the execution of the code* is what gets sent out to the HTTP body for that part of the script, not the code itself. In Java, as with all programming languages, there is the concept of a program writing its output as a text stream that the programming runtime environment handles, such as routing the output to the screen. In the case of a JSP, the J2EE environment takes the output stream and inserts it into the HTTP output flow at the appropriate places.

Scriptlets have access to particular Java APIs that allow the code in the scriptlet to gain programmatic access to the request and response HTTP headers. Also, for inbound POST requests, the scriptlets can gain access to the fields of any HTML form that is posted in the request. The values from the form are already extracted from the input stream by the J2EE web engine and are made available programmatically to scriptlets, to make the programmer's life easier and the overall processing of pages more efficient.

The other type of special insert in a JSP is something called a *JSP tag*. Essentially, these are directives to the web engine to substitute information wherever one of these tags is found. We shall return to explain these in more detail later when we look at how to develop JSPs that can adapt their output to suit the display characteristics of various mobile devices.

A servlet is similar to a JSP. It is invoked in a similar way, as per Steps 1–3 for the JSP. However, at the execution stage, the entire servlet is run as a compiled Java program (or *class file*, which we explain in Chapter 10). The file itself is not returned in the body of the HTTP response. The HTTP body is constructed by the program itself, which uses whatever programming logic it chooses, including any number of references to other Java programs, including EJBs, to get the job done.

Essential Differences between Servlets and JSPs   Clearly, a servlet can only be constructed by a Java programmer, and probably a fairly experienced one at that. However, a JSP can be constructed by someone with a quite different skill set, someone who knows about markup languages and how they get displayed in browsers, noting that a JSP script is primarily constructed using markup languages. It is likely that some JSP programmers will be able to learn enough about JSP to have a go at handling the different inserts, like scriptlets. It is often the case that scriptlets are passed around from programmer to programmer like tiny off-the-shelf utilities, almost like components (but at a much smaller and less sophisticated level than EJB components). Just to confuse us a little, it is possible to reuse more substantial components to make JSP programming more powerful, which are called *beans*. These are not the same as EJBs! We can think of a bean in a JSP as being like a portion of scripted code, but without the need to actually include the code in the JSP, just a reference to it (i.e. the bean).

To understand beans a little better, we could imagine a routine in software that checks the HTTP user-agent header and looks the string up in a lookup table somewhere (could be a flat file for example, or a database). We could use the output from the routine as a condition on which to decide how to format the rest of the JSP file, so that an appropriately formatted page is sent back in the HTTP response. For example, if the device is not capable of handling colour images, then we shall send monochrome ones. (This is the technique of adaptation that we alluded to earlier and that we shall investigate in more depth later in this chapter.)

Perhaps another programmer has already figured out how to program this function. Having spent the time take to program it, they can encapsulate the code as a bean. Instead of the JSP programmer having to include all the code from the bean (cut and paste of the code itself), they can simply include a special JSP tag that points to the bean that resides elsewhere on the system. This is a very useful approach as it means that JSP programmers can insert ready-made components into scripts. It also means that the component can be maintained independently of the JSP (and vice versa).

8.2   MARKUP LANGUAGES FOR MOBILE DEVICES

So far, we have addressed the process of a mobile device fetching content from a J2EE server using a HTTP link. This model applies to many mobile service scenarios. As we discussed earlier, when we looked at WAP, the WAP protocols will also work with this model as they have been designed to work with HTTP very closely. As far as our J2EE environment is concerned, it will receive HTTP requests from WAP devices just like any other type of request. Hence, we do not need to resort to specific design approaches in these cases. We can still use JSP and servlet programs for our presentation logic.

What is unique to the mobile context is the nature of the information that the JSPs and servlets will produce. Clearly the information is targeted at devices that have various display characteristics, so we expect the need to adapt our designs to match each device accordingly. Ordinarily, web pages for the desktop world are very information-rich and can have very complex layouts. We have seen the complexity of browser display capabilities evolve from something akin to a word-processor document to something capable of supporting arcade-game levels of complexity in both display and interaction, all within the web browser.

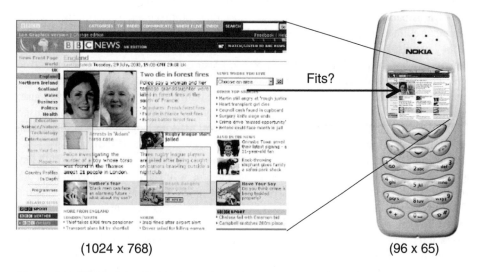

(1024 x 768)   (96 x 65)

Figure 8.7 Which fits which? (Reproduced by permission of Nokia and the BBC.)

Contrast this with a legacy WAP device, like the Nokia 3410, with its monochrome display and resolution of 96 × 65 pixels. Figure 8.7 shows us what happens if we try to fit a standard web page into the screen of a 3410, which is an extreme example, but still indicative of the general problem. Clearly, even if this were possible, the resolution is insufficient to display the page and it is simply too small to view. Going the other way, if we were to display the information at its real size and scroll around to see the page, then we can see how impractical that would be. The page image is shown in the figure with some sample screen locations overlaid on top, each sized to the 3410 resolution. Asking the user to scroll horizontally and vertically to see the page would be too much to ask for.

The other limitation is the ability to interact with the web page using a pointing device (e.g. mouse) and a keyboard. On devices like the 3410, there is no pointing device to speak of and the keypad is a poor substitute for a QWERTY keyboard.

The people who brought us the WAP protocols that we looked at earlier in the book have proposed various techniques to adapt the web browser display technologies to mobile devices. In this section, we shall look at markup languages aimed at mobile devices, notably those proposed by WAP 1 and WAP 2 specifications. Mobile-specific content was the vision for WAP, but using markup languages familiar to existing web programmers, which we shall look at in the subsequent section.

8.2.1 The HTML Foundation

As discussed earlier in the book, the Web originated with the combined emergence of HTTP and HTML. These two technical concepts were thoroughly intertwined from the start. HTML is a language that enables text and graphical information to be formatted for viewing in the browser. It is a means to control how the information gets displayed. For example, using HTML we can display text in columns, formatting some of it as headers, some of it as paragraphs and enlivening it with italics, boldface and underlining. We can insert pictures into the text and control whereabouts the pictures get displayed, and even what size they get displayed at. We are so used to viewing web pages that it hardly seems worth mentioning what they look like, but the point to grasp is that the formatting of any given page is controlled by describing to the browser how we want the page to look. Therefore, in a typical HTTP request, two types of information are returned in the body of the response: content for the user to look at, plus instructions to the browser on how to present it to the user. Often, we do not distinguish between the content and markup language when discussing web solutions. We may refer to both of them as 'the HTML', or 'the HTML file'.

As most of us are probably aware, HTML is a language that consists of tags. With any paginated information we can always treat the page as a linear list of content and work through it from top to bottom. Similarly, a browser opens an HTML file (or accepts the HTML stream via HTTP) and interprets it from top to bottom, or beginning to end. As the browser flows through the page, we can think of turning formatting instructions on and off to control the layout, such as 'start boldface' and then a while later 'end boldface'. Those of us familiar with old generation word processors are probably familiar with this concept, as these control codes were actually visible in the word processor file before the advent of WYSIWYG[2] solutions where the formatting itself became visible, not the codes.

With HTML, we similarly insert control codes into the page flow, but they are referred to as *tags*. In the case of boldface text, we use the tag <b> to turn on the boldface and </b> to turn it off.

```
HTML:
This is <b>bold text</b>
Displays as:
This is bold text
```

We can use a variety of tags to affect the text, including sizing, italics and so on. The <p></p> tags mark paragraphs, <i></i> mark italics, <u></u> mark underlining, <font size="4"></font> controls font size.

```
HTML:
<p>This is <b>bold text</b>, getting
   <font size="4">bigger</font>, getting
<font size="5">bigger</font>.</p>
<p>Next paragraph, say something <u><i>profound</i>
   </u>.</p>
Displays as:
This is bold text, getting bigger, getting bigger.
Next paragraph, say something profound.
```

---

[2] WYSIWYG = What You See Is What You Get

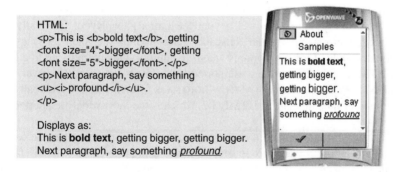

Figure 8.8 Text formatting in a mobile browser. (Reproduced by permission of Openwave.)

Our JSP or servlet programmer has to know HTML in order to insert the appropriate tags into the output stream of their program in order to achieve the desired layout in the browser on the requesting device. The challenge then is how to support the same concept for mobile devices. So far, the tags to format text do not seem to present any problems for a mobile device. We could easily support this type of formatting in a small display device, as shown in Figure 8.8, although we can see that already we are filling the entire screen area with just a few words.

In addition to the font formatting tags, there are a range of tags in a HTML file that are used to indicate other information that may be of use, such as the title of the page. A page title may not be necessary, but search engines can use them to index a search, so judicial usage of titles may be a benefit. They also show up in browser title bars, so they are useful references for the user. In the example shown in Figure 8.8, the title is 'About Samples', which is indicated in the HTML as <title>About Samples</title> and gets displayed in this case at the top of the screen. We can see that this information may be useful to give context to the user, so that they know what the page is referring to. For example, if this were a TV listings application, and we were viewing information about programs on BBC1, then we might think to use 'BBC1' as a title. In this way, the user will see a visual indication that the times being viewed belong to BBC1 and not another channel. This is probably a good example of how designing pages for mobile devices has different considerations than with desktop browsers. Something that seems relatively insignificant design detail in a desktop situation can have greater significance in the mobile environment and seriously affect application usability.

So far, so good; our HTML model seems to be working in the mobile environment. However, we have only touched upon a very small subset of the HTML tag set. Originally, HTML was a very sparse language and the options for formatting were limited. This was in keeping with the original precept of the Web, which was as an information oracle for scientific information. Anyone who has studied science knows that scientific papers are columns of text with the occasional diagram[3] inserted. As profound as the contents might be, the layout is not that stunning, not like the page of a youth culture magazine for example, or a car brochure.

---

[3] In fact, HTML 1.0 did not support images, but this arrived in HTML 2.0.

Not surprisingly, the evolution of HTML since version 1.0 has progressed towards supporting a wider variety of purposes, well beyond the journal paper layout. Originally, desktop browser manufacturers, in particular Netscape, started to introduce their own tags in order to meet the demands of more ambitious webpage authors. This initially led to a fragmentation of HTML, with different tags supported by different browsers. However, in 1994, the World Wide Web Consortium (www.w3.org) was formed and this eventually led to a major overhaul of the language to HTML 3.2, which is the most widely supported version of the language.

With HTML 3.2 (and its successor 4.0), the richness of formatting has increased considerably. For example, we are able to support tables using the <table></table> tag pair to indicate the table itself, and the <tr></tr> to start and end a row, with <td></td> to indicate a column cell within the row.

HTML:

```
<table border="1">
   <tr>
      <td width="50%">This is the upper left</td>
      <td width="50%">This is the upper right</td>
   </tr>
   <tr>
      <td width="50%">This is the lower left</td>
      <td width="50%">This is the lower right</td>
   </tr>
```
Displays as:
```
</table>
```

This is the upper left	This is the upper right
This is the lower left	This is the lower right

Tables like this can also be supported in modern mobile browsers, as shown in Figure 8.9. However, tables can be used to support complex layout arrangements rather than be used to display tables in their own right. This is where things may start to get tricky for a small display.

8.2.2 The Mobile Evolution (WML)

Originally, when WAP was first being proposed, there were no colour phones on the market. Most phones were monochrome and had limited display capabilities, mostly character based and with a correspondingly poor pixel resolution and greyscale depth. Because of these obvious limitations, much of the complex formatting possible with HTML 3.2 (as it was at the time), would simply not have worked on mobile devices. With hindsight, what may

Figure 8.9 Tables in a mobile browser. (Reproduced by permission of Openwave.)

have made most sense was to have gone back to an older version, like HTML 2.0 and tried to support this. In this way, web programmers would have still been able to use a familiar language and there wouldn't have been any concerns about formatting compatibilities, as the options were mostly limited to text formatting and insertion of images.

However, there was the need for some optimisations and this led to a new language being developed called *Wireless Markup Language* (WML), which actually ended up being different to HTML[4]. Despite its departure from HTML, many of the features of WML made a lot of sense and are worth exploring because the principles are still valid and because some of the features have ended up in what we have today (which is no longer WML, but we shall get to that in a moment).

The main ergonomic challenges for markup language applications on mobile devices are:

- No, or limited, pointing device

- Small display

- Tiny alphanumeric keypad

In addressing these issues, the WAP architecture proposes a standard interface model, which is the presence of a two-button interface to which *soft keys* can be defined in the WML code, as shown in Figure 8.10.

In addition to the soft keys, the user can select options from the keypad. In both cases – soft keys or keypad – the user is actually selecting a standard anchor tag in the WML that is used to link to another page. We can see the links displayed along with the numeric shortcuts (access keys) in Figure 8.11. In the figure we can see link 3 – 'Another Link' is highlighted. The user could scroll down the list of links using the arrow keys, or could select and activate the link in one step simply by hitting the '3' key.

The other optimisation that WML incorporates that makes it quite different to HTML, is the concept of decks and cards. Perhaps it has not been obvious until point out that the HTTP/HTML paradigm involves single-page requests per HTTP cycle. In other words, when the browser requests the resource from a single URI, it is expecting to display the returned content in its entirety in one go. If the user subsequently clicks on a link on the

---

[4] We are avoiding a discussion of the historical development of WML and its merits versus other approaches as this is largely irrelevant now.

Figure 8.10 Two-button interface. (Reproduced by permission of Openwave.)

fetched page, another HTTP cycle will be initiated to fetch the linked page. Clearly, we could foresee an arrangement of pre-fetching of pages to speed-up user response, but there is no such model in HTML, even though there is no reason why HTTP couldn't support it. On the other hand, with WML, this provision is allowed for.

To distinguish between requested pages and pages that are fetched in the same cycle, the WML model is based on *decks and cards*. A deck is like a page, but contains several cards (or only one) and these are the basic unit of content for rendering in the browser interface window. The reason for this mechanism is the small display size of some devices. It may

Figure 8.11 Access keys to make link selection quicker via numeric keys. (Reproduced by permission of Openwave.)

be so small that the reader can digest the contents very quickly and proceed to the next linked resource. If this is stored on the origin server, then another WSP/HTTP fetch cycle is required and this has an obvious overhead. If the display buffer to render content is very small, then the user will be requesting new pages from the server with a high frequency and the delays will become a dominant factor in the overall experience.

The way to think of decks and cards is that cards are like HTML pages and several of them can be fetched at once. Once fetched, only one card is displayable at a time, but the user can link from one page to the other. Because a card is already stored in device memory, it is relatively quick to go from one card to another.

Figure 8.12 shows us what is going on with decks and cards. As per usual, the browser makes a request (Step 1/2) for a particular resource on the J2EE server to run (Step 3); in this case it is for a JSP. The JSP is run by the web engine and it outputs code that is WML and forms a deck of cards. The cards are part of one contiguous file (deck). Because the WML file (deck) is one file, it gets passed in one go via the HTTP response (Step 5) and via a WAP proxy that we do not show here. Eventually, the cards end up at the device and the user can switch between them without the need for any further HTTP requests, as indicated in the figure by the cascading arrows between cards on the device.

WAP Binary XML (WBXML)   The potentially slow link speed of the RF network poses additional challenges. In the previous chapter we examined how WAP protocols can be used to optimise communications over an RF link, but we did not discuss issues related to the actual information being passed over the link. This does make a difference. In the case of HTML, a large proportion of the body of a HTTP response can be the HTML tags as opposed to the actual content.

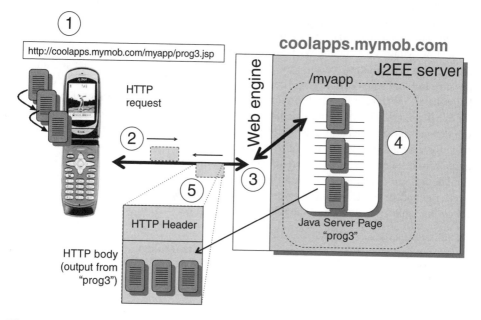

Figure 8.12 Fetching WML deck and cards from a JSP. (Reproduced by permission of IXI Mobile.)

If we consider the table example will showed earlier, then we can see how much of the data is simply metadata rather than the actual data itself. If we separate out the two types of data, we can see the difference more clearly:

Metadata (tags)	Content

`<table border="1">`
`<tr><td width="50%"></td>`
`<td width="50%"></td></tr>`
`<tr><td width="50%"></td>`
`<td width="50%"></td></tr></table>`
	This is the upper leftThis is the upper rightThis is the lower left-This is the lower right

128 characters (inc. spaces)	90 characters (inc. spaces)

In this short sample of HTML code, 59% of the information is the metadata. When transmitted over an RF link, we spend over half the time just transmitting metadata that the user never gets to see (and doesn't want or need to see) – it is simply baggage as far as the user is concerned and its inclusion results in a worsening of the overall user experience in terms of a tardiness in the responsiveness of the application.

With WML, the actual WML stream that gets transmitted instead of a HTML stream is binary encoded to achieve compression of the metadata. This is done in a manner similar to the compression of HTTP headers in WSP, whereby a hexadecimal code (token) is substituted for a well-known string in the WML.

The tokenising of the data steam is a relatively straightforward process. Fortunately, we do not have to ask our JSP or servlet programmer to implement this. Tokenising takes places at the WAP proxy. Our JSP or servlet outputs the WML in the body of an HTTP response stream back to the proxy. The proxy performs the tokenising and places the stream back out on the link, using WSP, and the WAP browser is required to de-tokenise before it can display the file. Strictly speaking, because the WML compression is a tokenised data stream, it is possible to extract the WML tags and content without the need to have the entire file available beforehand. This means that the browser display could be written to in parallel with receiving the WML stream. This may improve display response as far as the user is concerned, though this depends on the exact implementation of the browser on the device. One concern in doing this might be that the file turns out to be incorrect WML syntax. This might prevent the page from displaying depending on how forgiving the browser is of bad syntax. However, this should be circumvented by the strict validation of the file at the time of tokenisation.

The encoded WML output is referred to as *WAP Binary XML*, or WBXML, and the following table gives an idea of the tokens that would get used in our table example.

Tag	Hexadecimal token
`"<table>"`	1F
`"<tr>"`	1E
`"<td>"`	1D
`"<p>"`	20

In WML, the table tags are slightly different to HTML in terms of the allowable parameters. For example, we cannot specify the border parameter, as we did in our HTML example, which indicates to show visibly the borders of the table. The correct syntax for a similar table in WML would be:

```
<table columns="2">
  <tr>
    <td>This is the upper left</td>
    <td>This is the upper right</td>
  </tr>
  <tr>
    <td>This is the lower left</td>
    <td>This is the lower right</td>
  </tr>
</table>
```

	Metadata (tags)	Content
WML	`<table columns="2">` `<tr><td></td><td></td></tr>` `<tr><td></td><td></td></tr>` `</table>` [81 characters = 81 bytes[5]]	This is the upper leftThis is the upper rightThis is the lower leftThis is the lower right [90 characters = 90 bytes]
WBXML	1F 53 03 02 00 1E 1D 01 1D 01 01 1E 1D 01 1D 01 01 01 [18 bytes]	03 This is the upper left 00 03 This is the upper right 00 03 This is the lower left 00 03 This is the lower right 00 [98 bytes]

With the WBXML coding scheme, we have reduced our metadata overhead to only 16% of the overall response stream across the RF journey of the HTTP response. This is a significant saving, albeit for a solitary example.

Comparing WML with HTML and cHTML   The main issue with WML has been its digression from HTML, as WML is not compatible with HTML, although there is a degree of similarity between the tag sets in each language. The optimisations for WML are useful, especially the deck construct and the binary encoding, although the tokenised encoding is something that could be implemented for HTML if there were ever a need to do so.

The programmable soft keys is actually a feature developed specifically as part of the WAP concept, hence WML contains specific tags for gaining control over these keys. This can make applications altogether more useable, as one-touch operation of key commands

---

[5] Assuming UTF-8 coding, which means 1 byte per English character.

is possible. With HTML, or its ilk, there is no model for soft key interaction. The paradigm assumes the existence of a pointing device. In the absence of such a device, the user has to scroll down lists of links and select the one they want to visit.

The main concern with WML has always been its deviation from HTML, as web programmers are forced to learn a new language. This has been considered an obstacle to the take-up of mobile application programming using the browser model. How much this is true is difficult to say, but a comparison has often been made with the relatively greater success of a competing solution called *iMode*, which is originally a Japanese 'standard' developed by NTT DoCoMo.

The iMode system uses a language called *cHTML*, where the 'c' stands for 'compact'. It is literally a cut-down version of HTML, removing many of the tags, especially ones that would be unlikely to be useful in mobile browsers. The iMode service is more than just mobile web pages, including screen savers, email and so on. It is more akin to a portal service, but DoCoMo first of all invented the method of delivery. Japan has a long established record of doing her own thing, so the adoption (creation) of iMode in place of WAP is not at all surprising and is not necessarily a rejection of WAP on technical merits.

The cHTML is designed to meet the requirements of small information appliances with limited memory and processing power. It is designed based on the following four principles:

1. *Completely based on the current HTML W3C recommendations* – cHTML is defined as a subset of HTML 2.0, HTML 3.2 and HTML 4.0 specifications. This means that cHTML inherits the flexibility and portability from the standard HTML.

2. *Lite specification* – cHTML has to be implemented with small memory and low-power CPU. Frames and tables that require large memory are excluded from cHTML.

3. *Can be viewed on a small monochrome display* – cHTML assumes a small display space of black and white. However, it does not assume a fixed display space, but it is flexible for the display screen size. cHTML also assumes single character font.

4. *Can be easily operated by the users* – cHTML is defined so that all the basic operations can be done by a combination of four buttons; *Cursor forward, Cursor backward, Select*, and *Back/Stop*(Return to the previous page). The functions that require two-dimensional focus pointing like 'image map' and 'table' are excluded from cHTML.

The major features that are excluded from cHTML are as follows:

- *JPEG image* – an image format that utilises a high degree of processing to display (due to the handling of the compression method, which implies a mathematical process to display the image rather than simply reading the pixel values from a file).

- *Table* – the handling of tables requires potentially a lot of memory due to the nested nature of the tables and the implication this has on software data structures.

- *Image maps* – an image map is the ability to turn certain parts of an image into clickable links. This potentially requires complex data structures to maintain the clickable areas and also a lot of processing power to dynamically detect 'mouse'[6] cursor position on the image and compare against the image map.

---

[6] Most mobile devices do not have a mouse, and many lack any kind of pointing device at all, especially one that will work effectively in two dimensions.

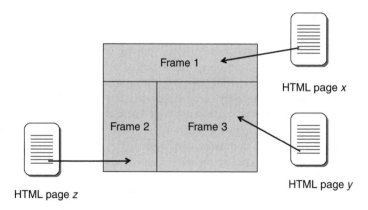

Figure 8.13 Frames in the browser used to display more than one HTML page.

- *Multiple character fonts and styles* – again, this will take up too much memory and the advantages of different fonts for a small device are arguable.

- *Background colour and image* – in early mobile devices, colour support was irrelevant anyhow.

- *Frames* – with desktop browsers it is possible to display several HTML pages at the same time by dividing the screen into frames, as shown in Figure 8.13. Again, maintenance of this construct in the mobile browser would require potentially excessive memory and processor resources.

- *Style sheet* – a style sheet is a means of defining styles, like font colour, font type, but in a block of code that is separate from the main HTML. This makes the maintenance of look-and-feel easier, as it is separated out from the content. For example, we would still use a header tag in our file, like <h1></h1> to signify the browser to display oversized text as a means to demarcate a section heading. However, we may decide that all such headings should be in Garamond font and burgundy in colour, perhaps to match a corporate-identity colour scheme. We can control such appearances using a style sheet. For early mobile devices, the scope for style was limited, so style sheets were not supported by cHTML. We shall return to style sheets when we look at XHTML-MP.

Despite repeated claims to the contrary, the use of cHTML is probably not the main reason for the apparent success of iMode in Japan. Other factors figure heavily, including the revenue share that developers can gain from providing their content via iMode. Social factors in Japan are quite different from other countries[7] and have meant an altogether different reliance on mobile devices than seen in most other regions. However, cHTML became a source of inspiration for how to choose a suitable markup language for the WAP 2 era, as we shall soon discuss.

---

[7] For example, the Japanese spend a lot of their time out of their homes, because often homes are very small. This 'out of home' living tends to engender a different social dynamic, as noted by Rheingold in *Smart Mobs*.

8.3 FULL CIRCLE – WML 'BECOMES' XHTML

As we discovered in the previous chapter, when looking at mobile protocols, WAP progressed from WAP 1 to WAP 2. In doing so, it moved a step closer to the fully blown desktop browser model, especially with respect to the protocols and the wireless profiles of HTTP and TCP, which are completely compatible with standard HTTP and TCP.

It would be folly to suggest that we dispense with wireless optimisations altogether. This is especially true when considering which markup language to use. Clearly, mobile devices have severe display constraints compared with desktop devices. This would suggest that utilising an identical approach is not going to work and this is a topic we critically examined in Chapter 2. Certainly, the idea of viewing standard web pages on mobile devices is anathema to giving the mobile user a compelling experience. All our notions of usability are contravened by asking a user to wade around a page that is up to twenty times too big for the display.

From WAP 1 to WAP 2, the era has seen the advent of many more devices capable of displaying colour and a higher density of pixels. Screen sizes have also increased on many devices. At the same time, as we have already noted when looking at protocols, the ability to connect a wireless device to the Internet has become easier. As we shall see when we look at RF network protocols, the mobile networks have evolved from circuit-switch technology to packet-switched.

The main advantage of packet switching is not actually the greater network speeds that are theoretically achievable[8]. With packet-switched connections, there is no need to initiate and maintain a data call, like we used to do[9] with modems at home when accessing our favourite ISP. In effect, the modern devices are already connected and ready to transfer data, like sending a GET request to a server and receiving the HTTP response. The 'line' only becomes active for that period of time, during which we pay for the data transferred.

For the end user, accessing data sources, like a WAP site, is a convenient process, as the line is always connected and so delays with connection set-up are avoided. Moreover, the process is affordable as there is no need to hog an expensive resource (a switched circuit).

Given the improvements in devices, we could posit technical arguments for why switching back to HTML from WML might be a good idea. However, the reality is probably that the likeliness of users being prepared to use mobile services is increasing all the time, due to the improvements we have just been discussing. Therefore, if we want to attract more widespread development of mobile websites, then it is probably a good idea to make this as easy as possible for existing web designers. This necessitates the adoption of a language that is compatible with existing approaches, such as HTML.

As we shall now examine, HTML itself has evolved during the WAP 1 to WAP 2 time frame. Any attempt to adopt HTML for mobile applications would need to be synchronised with the evolution of HTML itself. However, just as we did with the wireless profiling of HTTP and TCP, any performance improvements we can find that are within the confines of HTML technology would probably be worth pursuing, especially if they advance usability.

---

[8] Packet switching in itself does not improve data transfer rates. The higher speed referred to is a result of other improvements in implementation.
[9] Some of us still use telephone-line modems rather than broadband connections.

8.3.1 XHTML is Modular

HTML is a language for formatting the display of information. We have seen that the language itself comprises of tags. Tags are used to indicate formatting states within a stream of data that is going to be displayed in a browser, like <bold> to turn boldface on and then </bold> to turn it off some time later in the information stream.

Earlier in the book, we briefly looked at a generic way of annotating data using tags, which was called XML, or Extensible Markup Language. The use of XML has mushroomed of late. Many tools and APIs have emerged to make processing of XML easier for us. It is relatively simple now to adopt XML as the means of describing any information set that we care to exchange between two devices, no matter their application. Because of the increased momentum behind XML, and its obvious similarity to HTML, a decision was taken by the World Wide Web Consortium to make sure that HTML was fully XML-compliant.

XML uses tags to describe data. The tags are mostly undefined, which is the whole point and why it is called 'extensible'. However, the rules for defining the tags are very much fixed. Sticking to the rules is like obeying the principles of grammar and punctuation, or syntax. Defining our own tags within the rules is like formulating our own vocabulary and doing this in conformance with XML syntax constitutes making a well-formed XML document. To achieve conformance with XML, HTML has become XHTML, but the features of the language (i.e. the vocabulary itself) remains unchanged.

By moving to XML, there is an opportunity to modularise the language. This is a slightly strange concept at first, but makes perfect sense; as we shall see in a moment when we reflect upon why bother with standards at all.

With any XML document, we need a means to determine what the allowable tags actually are. This is done by writing a special document called a *Document Type Definition* (DTD), or a Schema, which is an alternative approach. For example, in the DTD for XHTML 1.0 is where we will find that the <bold> and </bold> tags are actually allowable. DTDs also tell us about tag relationships, or nesting. If we take an example XML vocabulary to describe anything to do with films (movies), we could imagine a file snippet as shown in Figure 8.14. We can see the tags in pairs and information contained within them, noting that some tag pairs (e.g. <song></song>) contain yet more tag pairs, whereas some tag pairs (e.g. <title></title>) contain the annotated information itself.

Figure 8.14 Example of using XML to describe films.

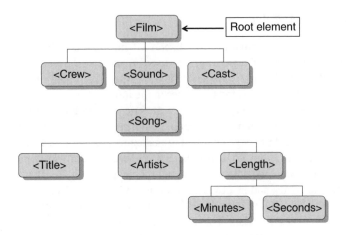

Figure 8.15 Tree structure resulting from nested tags in XML.

This nesting of XML tags leads to a tree-like structure in the file, as shown in Figure 8.15, which is a useful data structure to accommodate as it reflects the implicit organisation of many data sets.

Within an XML document, there might be the need to refer to several DTDs, each one taking care of a particular subset of the vocabulary that is allowable within our file. Taking our fictitious Film-XML (FML) as an example, the topic of films is clearly quite broad. We could envisage being able to describe so many detailed topics related to a film. For example, we could include information about where a film was shot, each location, the details of the props used, and the extras in the film taken from the location and so on.

Were we to include every conceivable facet of film making into our XML definition, it might become a bit unwieldy, especially if we wanted to extend it at a later date. An alternative approach is to use a DTD for each distinct topic subgroup within the overall information space concerning films. This is modularisation.

Modularisation allows for compatibility to be maintained between systems whilst allowing for extensibility within an application space. For example, various software solutions could be envisaged that relate to the film-making industry. If we want to standardise on the information exchange between these applications, then we could define and adopt our FML, similar to the example we have just been looking at. However, a particular group of applications might only be concerned with the audio aspects of film making. These applications would want to exchange information about sound tracks, so they could use the openly published DTD for sound tracks and thereby achieve a guaranteed level of compatibility by sticking to the DTD and the resultant XML vocabulary and structure.

However, perhaps a particular part of the industry recognises a need to go further within the realms of describing sound tracks, to include information about the recording equipment and sound formats. Instead of adopting a new standard altogether, the interested parties could agree upon an extension to the existing Film-XML, simply by adding a new module to include the means to describe the data they are interested in exchanging. They could publish the DTD for this extension back to the public commons and for the sake of clarity name their hybrid with a different name, something like FML-A, with the 'A' standing for 'Audio'. It is also possible that within FML-A, the agreeing parties could have jettisoned some of the existing modules within FML that are not especially relevant to the applications catering for audio. This is the benefit of the modular approach – it allows both subsetting and supersetting of the base (host) language.

8.3.2 XHTML Basic

Having just looked at the modular approach adopted by XHTML, we can now move on to look at a particular incarnation of XHTML, getting back to our mobile network and away from the world of films, interesting though they might be (perhaps we should make a film based on this book ☺).

XHTML Basic is a version of XHTML aimed at small information appliances, particularly mobile devices (though not exclusively). To quote from the W3C document defining XHTML Basic[10]:

> *HTML 4 [foundation of XHTML] is a powerful language for authoring Web content, but its design does not take into consideration issues pertinent to small devices, including the implementation cost (in power, memory, etc.) of the full feature set. Consumer devices with limited resources cannot generally afford to implement the full feature set of HTML 4. Requiring a full-fledged computer for access to the World Wide Web excludes a large portion of the population from consumer device access of online information and services.*

In our foregoing discussions we have already mentioned at least three approaches to markup languages for mobile devices: HTML, cHTML and WML. We could be selective about how we use HTML, but because there are many ways to subset HTML, there is a danger of ending up with many subsets; as many as possible are defined by the various organisations and companies with competing or disparate interests in this matter.

Without a common base set of features, developing applications for a wide range of web clients is difficult. This is because we might design a user interface assuming we can use tables, only to find that some of our users do not have table support in their browsers. Clearly, such a fragmented approach will ultimately work against all of us: users and developers alike.

The impetus for XHTML Basic is to provide an XHTML document type that can be shared across communities (e.g. desktop, TV and mobile phones), and that is substantial enough to be used for content authoring that works well enough within the limitations we may find applicable to our application. However, by choosing XHTML as a basis for XHTML Basic, we automatically inherent the modular approach, thus new community-wide (i.e. shared common 'standards') document types can be defined by extending XHTML Basic in such a way that XHTML Basic documents are in the set of valid documents of the new document type. Thus an XHTML Basic document can be presented on the maximum number of browser clients.

In our earlier discussion about cHTML, we listed some of the absent features dropped from HTML and briefly discussed possible reasons for the omission. Certain types of markup were not predisposed to implementation in a mobile browser running in a resource-constrained environment. Most of the limitations are related to memory and processor

---

[10] http://www.w3.org/TR/xhtml-basic/

constraints. We treat the OTA constraints separately by applying compression techniques afterwards, just as we did with WBXML and its tokenised-compression method. In other words, the need to overcome device limitations is uppermost, whereas the need to overcome file size problems is treated separately and is not a major consideration for the language definition.

From our consideration of resource constraints and their impact on the design of cHTML, we can appreciate the similar need to limit XHTML for mobile devices. Many of the design decisions behind XHTML Basic mimic the considerations for cHTML, not surprisingly, since cHTML was submitted for consideration by the W3C[11].

Let's briefly examine the official W3C comments on the rationale for XHTML Basic before going on to look at how the WAP community used this as a foundation for XHTML-MP (Mobile Profile), for good reason, as we shall see.

The design rationale for XHTML Basic covers the following points:

- *Style sheets* – these are recommended, but limited in scope. We shall defer our discussion of these until we reach XHTML-MP, where a more comprehensive treatment is required (and relevant).

- *Form handling* – basic form handling (*Basic Form* XHTML Module) is supported with some limitations.

- *Presentation* – many simple Web clients cannot display fonts other than monospace. Bi-directional text, bold-faced font, and other text extension elements are not supported. It is recommended that style sheets be used to create a presentation that is appropriate for the device.

- *Tables* – basic table handling is supported (*Basic Tables* XHTML Module). Note that in the Basic Tables Module, nesting of tables is illegal.

- *Frames* – these are not supported.

- *Scripts and events* – something we have not discussed so far in relation to markup languages is the concept of scripts and events. We can see how XHTML tags can be used to drive the display of content. However, perhaps we want to do something more complicated. In desktop browsers we can insert scripts into our pages that enable us to programmatically manipulate the markup content once it is on the device. We can think of this as being like being able to run the JSP or servlet on the device, simply to allow some degree of processing of the content (though clearly this analogy does not extend to accessing the J2EE platform and its APIs, etc.). The scripts can programmatically gain access to any of the content in the XHTML page. For example, we could move an image around the screen.

Perhaps the most classic example of widely used scripting is the now famous 'rollover' effect, as shown in Figure 8.16. This is where the image under the cursor changes when we place the cursor above it, a technique typically used to give a degree of positive feedback that we have selected an option. The figure shows the mouse cursor over the image (Step 1) causes the *onmouseover* event to be fired in the associated container for this image element, which in this case is an anchor (<A>) tag (hyperlink). Inside the opening <A> tag, we

---

[11] http://www.w3.org/TR/1998/NOTE-compactHTML-19980209

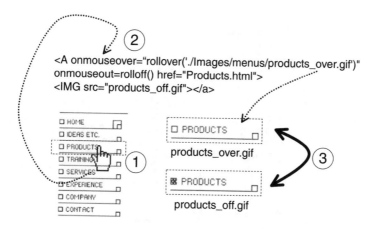

```
<A onmouseover="rollover('./Images/menus/products_over.gif')"
onmouseout=rolloff() href="Products.html">
<IMG src="products_off.gif"></a>
```

Figure 8.16 Rollover effect is achieved by events and scripting.

have specified the *onmouseover* attribute, thus letting the browser know that if this event occurs, then to call our scripting function *rollover*, as indicated in Step 2. This function is not shown, but is written using another language called JavaScript and is listed at the top of the HTML page where the browser knows to find it and to run it when this event fires. Consequently, the original image gets swapped for another image (Step 3), which is how the rollover effect gets accomplished.

The event model and scripting capability that we have just mentioned implies that our web browser is a lot more complicated than perhaps envisaged. It does a lot more than simply render HTML to the screen. Figure 8.17 gives us a better idea of what a browser might consist of.

We have five main elements[12]:

1. *Browser display* – the software that handles the actual user interface and can render XHTML (or others) into a meaningful graphical representation on the screen, most likely using the graphics APIs of the device (which we will discuss later when we look at devices in Chapter 10).

2. *HTTP/WAP stack* – if not already present as an accessible API on the device, then the browser needs its own HTTP/WAP stack to enable interaction with the origin server. This part of the browser may well also include the cache to cache pages, images and other fetched objects.

3. *Event handler* – having defined the possible events that the browser could respond to or generate (e.g. like *onmouseover, onmouseoff, onpageload*, etc.), we need a software process to detect events and then pipe these to the scripting engine.

4. *Scripting engine* – we need an API that fully exposes the markup document to access via a scripting language, which itself needs an execution environment. The scripting engine provides this facility and typically we refer to tags or events from the script using a *Document Object Model* (DOM). We can think of this as modelling the entire

---

[12] Note that these are somewhat arbitrary for the sake of illustrating the detailed operation of a modern browser. Actual browser implementations may follow alternative architectures, although the functions would be similar.

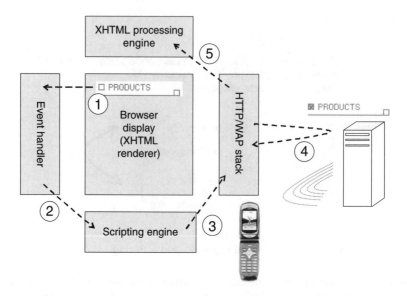

Figure 8.17 Browser bit more complicated than we thought. (Reproduced by permission of IXI Mobile.)

document as a class and then exposing its structure via properties and methods that our script can call.

5. *XHTML (HTML/WML) processing engine* – this component actually parses and validates the inbound markup document and converts it to a form to pass to the browser display module.

In Figure 8.17, the sequence of a typical rollover effect is shown. In Step 1, the image within an anchor tag (<A> tag) is pointed to by the pointing device. This causes the event handler to detect the event and to pipe this (notify) to the scripting engine (Step 2). The scripting engine receives the event and executes the appropriately tagged code segment that is bound to the event within the markup (e.g. via an event attribute in the <A> tag), as shown in Step 3. In this example, we imagine that the executing script accesses the pointed-to image in the document and replaces it with an alternative image (the 'rollover image') from a URL specified within the script. This causes the scripting engine to fetch the image from the URL (Step 4). The final step is to process the content for display (Step 5).

We have just described the complexity of a browser that supports scripting and events. Because of the implications in terms of device resources, in XHTML Basic, there is no scripting support, so the browser architecture is less complex.

The problem with scripting and events is not just the heavy resource implications, but with defining a suitable event model and scripting API for the document (i.e. defining the DOM). Events and API possibilities in certain devices may not make sense in others. For example, a mobile device with voice-calling capabilities would benefit from the ability to link an embedded number to the device address book. We would need an event to indicate that we have clicked on content tagged as a phone number[13]. However, this would imply a

---

[13] This feature is supported in WML and on some WAP 1 devices.

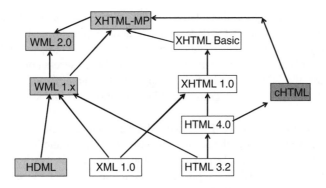

Figure 8.18 Relationship of different mark up languages.

particular scripting API that is not necessary on devices that do not support voice, such as dedicated email devices for example.

However, richer mobile devices that support full XHTML will generally include scripting. Wherever scripting is supported, then the object model for it will most likely be simplified, as there are still plenty of things that a mobile device can't do compared with a desktop environment; something that we shall consider when we look at devices later in the book.

8.3.3 XHTML-MP (Mobile Profile) – The Final Frontier

Before discussing the essential elements of XHTML-MP, let's recap the relationship and evolution of markup languages thus far, something that we can show graphically, as shown in Figure 8.18.

As we have discussed on several occasions, the advent of the browser paradigm began with HTML, which became widespread and feature-rich at version 3.2. This version migrated to version 4.0, which was converted to XHTML 1.0, using XML and modularisation to develop a host language that could be used in its own right or as a basis (or container) for other languages. The HTML standards community (W3C) took it upon themselves to define a subset of XHTML, called XHTML Basic, aimed at resource-limited devices.

As the figure shows, in parallel to the HTML progression, the WAP community were busy with their own activities, initially inspired by several vendors' technologies, most significantly *Handheld Device Markup Language* (HDML) from Unwired Planet. To promote a more vigorous standardisation effort, eventually the WAP forum coalesced efforts with W3C. At the same time, NTT DoCoMo had already submitted i-Mode's cHTML as a technical note for the W3C to put into consideration for lightweight applications.

What we have ended up with is XHTML-MP (Mobile Profile). This is a specification owned by the WAP community, which is now part of the OMA. Whilst the specification is standards based in terms of its dependence on XHTML Basic, the specification is not a W3C specification. This in itself does not pose a problem, especially as the whole point of XHTML Basic was to provide a host language that other communities could extend to suit their needs, which is what the OMA has done.

Of course, XHTML-MP is an open specification, not an edict. As such, there is no need for anyone to conform. However, whilst the aim of the language is to support browser applications on very resource-limited devices, the aim of the specification is to engender sufficiently widespread support that a large population of devices can display XHTML-MP pages. This will make it easier for developers to target applications at mobile users. The language is easy to use, due to its reliance on the prevalent XHTML (HTML) foundation, and using it pays off as a large number of users will be able to access content authored with it. In this regard, the 3GPP has adopted it as part of the standards for UMTS, the global 3G standard. It is already a recognised standard for most 2G and 2.5G standards, such as GSM.

Before considering the details of XHTML-MP, we should discuss WML 2.0. This specification uses XHTML but with the unique semantics of WML 1 and is used for *backwards compatibility only*. In other words, new applications based on WAP technology should not be using WML any longer. WAP 2 applications should only be created using XHTML-MP. We therefore do not discuss WML 2.0, although we will look at browser-specific extensions to XHTML-MP that are motivated by some of the features of WML that did not make it into the XHTML-MP specification. There are also similar features from cHTML that need noting.

8.3.4 Using XHTML-MP

XHTML should be very familiar to web designers and authors. Many of the familiar design features are present. From XHTML Basic, XHTML-MP inherits the following abilities to format content:

- Structure
- Text
- Hypertext
- List
- Basic forms
- Basic tables
- Images
- Meta information

Additionally, it supports:

- Subset of the fully blown XHTML Forms module
- Subset of the XHTML presentation module
- Style sheets and style attributes

Except for structure, it is easiest to explain these by example. The structure support is the ability to divide the XHTML content into components, consisting of a title block, header block and the body itself. Most of the content that gets displayed in the browser interface is contained in the body block and comes from the rest of the XHTML elements we have listed above. The title information can be displayed in the browser and the header block can contain important meta information. This is a way of specifying HTTP headers that should be contained in the response when the web engine writes out the XHTML stream in a HTTP response. For example, we can use a meta tag to explicitly set a validity period for the purposes of caching. This is particularly useful for information that changes regularly, like news pages.

With the basic text display capabilities of XHTML-MP, previously shown in Figure 8.8 and repeated on the left of Figure 8.19, we could have a basic and navigable WAP site, such as the one shown in Figure 8.19, which is a WAP version of a weblog.

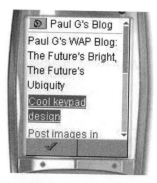

Figure 8.19 Basic XHTML site. (Reproduced by permission of Openwave.)

In Figure 8.19, we demonstrate the basic features of text formatting (e.g. plain, **bold**, *italics*, large fonts, etc.). We can also see the use of the all-important hyperlink, such as the highlighted 'Cool keypad design' link shown, which would take us to the next page in the weblog.

The greatest feature of XHTML-MP is its richer formatting and graphics support than seen with earlier WAP browsers. For example, in Figure 8.20 we can take an image (b) and insert it into the page. Moreover, the image types supported can be GIF, JPEG and PNG, which are all standard formats that web authors are familiar with using. We can even take an image (a) and use it for the background of the page.

Both these ideas are amalgamated in Figure 8.21.

A major feature of XHTML-MP is that we can add the formatting that affects style by defining styles in what is called a style sheet, or a Cascading Style Sheet (CSS). The real beauty of the style sheet is that it is a separate file from the XHTML page. Initially, this is a problem because that means that our browser has to initiate a separate HTTP fetch cycle to go grab the style sheet. What's more, it has to do this before we can start to see how the initial page should look in the browser, so it is difficult to preload the page, or carry out these tasks in parallel (using pipelining of requests that we discussed earlier).

We use the style sheet to tell the browser about the look-and-feel of any pages that reference it, as shown in Figure 8.22. We can define style information for more or less every

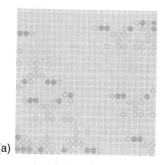

(a)

(b)

Figure 8.20 JPEG images supportable by XHTML-MP.

Figure 8.21 Adding images to XHTML-MP pages. (Reproduced by permission of Open-wave.)

display element that the browser understands. For example, to make the title of the page shown, which is 'Paul G's WAP Blog', we use the following tag in the XHTML-MP:

```
<p>Paul G's WAP Blog:</p>
```

This tells the browser that this is a paragraph. This is not quite the same as the grammatical idea of a paragraph in English. It is really a block element that contains text, inline images and related content. The XHTML-MP rules (from the associated DTD) determine what tags can be nested within the paragraph tag. As we mentioned inline images, then this must be

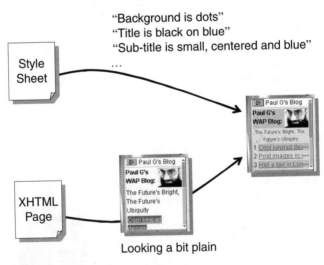

Figure 8.22 Separate style sheet tells browser about look-and-feel of pages. (Reproduced by permission of Openwave.)

one of the permissible nested tags, which is how we insert the image seen in the top right of the screen shot. To do this, we use the image (<i>) tag[14], as below:

```
<p>
<img src="pgtiny.jpg" align="right" />
Paul G's WAP Blog:
</p>
```

The <img> tag is using a couple of attributes. The first is the source attribute, which tells the browser where to find the image. Here we have the value, which is actually a URL – 'pgtiny.jpg'. Because there is no domain name or other path information in the URL, then the browser assumes it to be in the same path from where the XHTML-MP page was fetched, so the browser initiates another fetch to go and get the image. Don't forget that the browser can issue this fetch as soon as it finds this tag when processing the input HTTP stream containing this XHTML-MP content. In other words, it can use pipelining to speed up fetching the image before waiting for the entire XHTML-MP page to be loaded (streamed in). The other attribute is the *align* attribute, which controls alignment on the page with respect to the browser window, the options being to align to the left, centre (*center*)[15] or right.

To apply a look-and-feel style to our pages, we can define different styles in the style sheet and then reference them in our pages. For example, let's say that for page titles, such as our 'Paul G's WAP Blog:', we want to display this in bold font on a blue background. We can define a style called pagetitle (the name is up to us within the syntactical rules of CSS). This would look something like (the line numbers would not appear in the CSS file):

```
1       .pagetitle {
2               color: black;
3               background-color: #99CCFF
4               font-weight: bold
5       }
```

This information would appear somewhere in our CSS file. Line 1 is the name of the style. The name begins with a period ('.') to indicate that this is a custom style. Line 2 indicates that the font colour is black[16]. Line 3 specifies the background colour for the paragraph and Line 4 indicates that we want the text to be boldface. Line 1 and Line 5 contain the opening and closing parentheses (curly brackets) to demarcate the style.

---

[14] Notice how the tags are in lower case. This is not a strict requirement of XML syntax rules, but case-sensitivity is. Hence, once defined as lower case in the DTD, then we have to stick to lower case. Otherwise, in theory, our browser should reject the XHTML-MP file as it is not *valid* XML.

[15] XHTML, like many similar specifications, is produced in American English, so common words like 'centre' are spelt with American spelling (e.g. center).

[16] There is a slight inconsistency in the syntax here as we do not have to write font-color, just color.

To use this style in our page, we use the attribute *class*, which can be used in most tags to define the applicable style, so we end up with our paragraph tag looking like this:

```
<p class="pagetitle">
<img src="pgtiny.jpg" align="right" />
Paul G's WAP Blog:
</p>
```

So now when the browser displays this paragraph block, it knows to use the style class called "pagetitle", which it finds in the associated CSS, which is linked to the XHTML-MP page using the link tag:

```
<link href="style.css" rel="stylesheet"
   type="text/css" />
```

The key attribute here is the href, which contains the URL to the style sheet, which the browser again assumes as relative to the current page and fetches it.

The benefit of this approach is that every page in our browser-based application can have a common look-and-feel by all referencing the same CSS; each file simply links to the same URL where the CSS is to be found, as shown in Figure 8.23.

With a common style sheet, whenever we want to update or change the look and feel of the application, we only have to change one file. This makes our application easier to maintain.

The style sheet can be stored in the browser's cache. This means that once loaded, we don't have to fetch it again. This speeds up the application. Not only is the style information available locally on the device, so we don't have to fetch it, but the XHTML-MP pages themselves are also sparser because on average they contain less information. On a relatively slow RF network connection, this can make a noticeable difference.

A key feature of WAP, or any browser application, is the ability to gather input from the user. This is done using forms, recalling from earlier discussions that the content of the form gets sent in the HTTP stream using the POST method. XHTML-MP supports forms, as shown in Figure 8.24. This shows a page layout to gather input from the user who wishes to submit a question (see form field 'Question') to be answered by a consultant within a

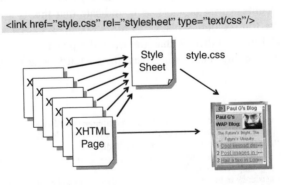

Figure 8.23 Achieving same look-and-feel by linking to the same style sheet.

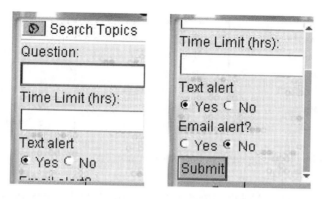

Figure 8.24 XHTML-MP supports forms.

certain time limit (see form field 'Time Limit (hrs)'). In addition to the text boxes, forms can support all the usual form widgets, such as radio buttons (the 'yes/no' options, as shown) and select boxes and all the rest.

8.3.5 Browser-specific Enhancements to XHTML-MP

As with all technologies, competing companies always attempt to bring their own interpretation of what's needed in the marketplace, despite the existence of standards. This is also true of WAP and XHTML-MP.

Perhaps the world's leading supplier of mobile browsers is Openwave, which is no surprise, given the origins of this company; it was originally Unwired Planet, the inventors of HDML, which was the inspiration and technical basis for WML.

If we look at the Openwave Browser (Universal Edition), we can examine its support for XHTML-MP and become aware of some of its interesting features.

The extensions to XHTML-MP supported by Openwave Mobile Browser fall into the following categories:

- Legacy HTML elements (tags) and attributes (for their counterpart XHTML tags)

- Useful elements and attributes from XHTML 1.1 (i.e. fully blown XHTML)

- Useful elements and attributes that are taken from WML

- Openwave's proprietary extensions (only two of them)

When looking at some of these extensions, it is difficult to understand the rationale behind including them. Possibly, as with so many technical 'enhancements' to products, particular application scenarios arose that could not be adequately addressed with the basic technology set. Hence, someone decides to graft an extra capability onto the side of an existing standard just to accommodate these exceptions.

One of the most elegant extensions can also be found in iMode's extension to cHTML, which is the ability to reference local source images. These are images that are already stored on the device in its physical memory during product assembly. These images are actually icons, some of which are used to convey emotions, just like the emoticons used so prevalently now in text messages, emails and IM. Because these images are built into the device, there is no need to fetch them from a web server using HTTP, so they load very

#	Icon	Name	#	Icon	Name	#	Icon	Name
1		exclamation1	2		exclamation2	3		question1
4		question2	5		lefttri1	6		righttri1
7		lefttri2	8		righttri2	9		littlesquare1
9		littlesquare1	10		littlesquare2	11		isymbol
12		wineglass	13		speaker	14		dollarsign
15		moon1	16		bolt	17		medsquare1

Figure 8.25 Some of the Openwave browser icons.

quickly. The Openwave Mobile Browser includes hundreds of icons that can be used in XHTML-MP documents. In fact, the WAP forum itself specified a set of WAP *Pictograms* that are also supported by the Openwave Browser, although Openwave has a larger set of proprietary icons. Similarly, iMode supports a custom set of icons called *emoji*.

WAP browser icons can be used wherever an image can be used within an XHTML-MP document, including as inline monikers for lists. A sample of the icons built in to the Openwave Browser is shown in Figure 8.25.

Using these built-in graphics enables the web-page author to add a touch of colour or graphical reinforcement to information, which when used diligently, could possibly enhance usability; something that remains a key consideration for mobile applications. We have used some icons to redo some of our sample weblog application, as shown in Figure 8.26.

Icons can also be used inline as bullet points in lists, as shown in Figure 8.27, where we have taken the 'righttri2' and the 'bolt' icons as given in Figure 8.25. We insert these into the bullet lists via the style sheet of course, thus being able to change all the bullets in one easy step across all pages using the list element in XHTML-MP.

```
UL
{
        list-style-image: localsrc(''bolt'')
}
</code>
```

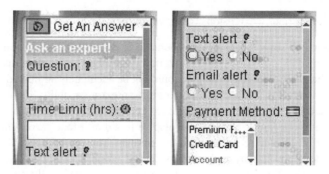

Figure 8.26 Using icons on mobile web pages.

This style is different from the custom-defined style we used earlier, where we made up our own style name – "pagetitle". Here we are specifying the style that is to be used for an existing XHTML element, the *unstructured list* element, demarcated by the <UL> tag.

We can see how we build this list by looking at the following XHTML-MP, which illustrates several other features.

```
<ul>
<li class="blue">
<p class="nowrap"><a href="story1.html">Cool keypad
   design</a></p></li>
<li>
<p class="nowrap"><a href="story2.html">Post images
   in space</a></p></li>
<li class="blue">
<p class="nowrap"><a href="story2.html">Hail a taxi in
London</a></p></li>
<li>
<p class="nowrap"><a href="story4.html">Free trip to
   Mars</a></p></li>
<li class="blue">
<p class="nowrap"><a href="story5.html">Japanese
   Cool</a></p></li>
</ul>
```

Notice how we begin and end the entire list using the <ul></ul> tag pair, which is how our browser picked up the reference in the CSS for the UL style[17].

To place an item in the list, we use the <li> tag element, some of which we have added the class attribute to, using our own custom style – "blue". This is placed in alternate list items in order to give the striped effect seen in Figure 8.27, which makes it easier to read the items on a small display. The other style worthy of note is in the embedded text within the list items, which use a paragraph <p> tag to allow a style to be applied that specifies no wrapping of text onto the next line. This is why the list items appear to run off the side of the screen. However, if we look carefully at the left and right image in Figure 8.27, we can see how placing the focus on a list item causes it to scroll sideward, like one of those LED signs in a bank. This is called a *marquee effect*, which is supported by the Openwave Browser, though not all browsers.

As mentioned, iMode also supports browser icons, called *emoji*. In Japanese, *ji* means character. Thus, *kanji* are characters originally borrowed from the Han Chinese repertoire, *gaiji* are 'foreign characters'. *Emoji* are characters invented by NTT DoCoMo for people to use in text messages on their mobiles. The most obvious example is the well-known 'smiley face', often encoded in ASCII as :) and called an *emoticon*. Thus, 'emotion' + *ji* gives *emoji,* some of which are shown in Figure 8.28.

---

[17] Note that CSS syntax is not case sensitive, which is why we end up with UL in capital letters, which still works even though the element in XHTML must be lowercase.

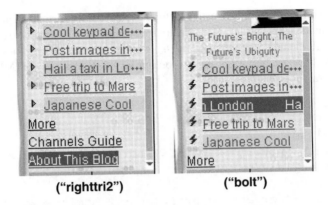

("rightri2")   ("bolt")

Figure 8.27 Using browser icons for bullet points.

Since DoCoMo uses standard Web infrastructure, including basic HTML and HTTP, the question arises concerning how these icons are encoded. They use Unicode's 'Private Use Area', a built-in range of character codes that's there for people who want to use their own non-standardised characters.

To my relief, as Figure 8.29 shows, at least emoji includes some characters related to mobile telephony, whereas WAP and Openwave icons do not (unless I have missed them)!

8.3.6 Guidelines for Mobile Webpage Authoring

Having now looked at the evolution of mobile web-browser markup languages, we can briefly examine some of the issues facing us when designing mobile applications using the browser paradigm.

We have already noted on many occasions throughout the book, that usability is a key consideration for mobile design, more so than with the desktop scenario where there is a greater threshold for tolerating poor design.

The golden rule is to make life easy for the user, which generally means several things:

- Do not deviate from accepted user interface norms.

- Minimise user scrolling and 'deciphering' of content to find what they are looking for.

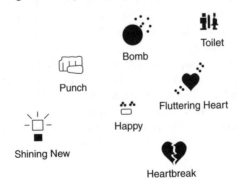

Figure 8.28 Some emoji characters.

Figure 8.29 Some emoji icons for mobile telephony.

- Think 'task orientated', not 'style orientated': i.e. let the user easily achieve what they set out to do, not simply be impressed by the visuals (looks).

- Make the page set as sparse as possible to achieve the task in hand, be prepared to cut things right back to the bone.

- Avoid the use of jargon wherever possible. Don't start using words like 'central' instead of 'home'. Only use jargon if it is a well-understood part of the user's vernacular.

Looking at the specifics of design, here are some things to think about in the sections that follow.

Beware of the Gateway   Don't forget that a WAP gateway may still be present in your system, even if you didn't put it there. This is especially true if you are writing applications to work with, in, or alongside a mobile operator's portal. Some gateways are very strict in checking for valid XML, so if you're sloppy with the XHTML-MP coding, then pages that work fine in emulators may not work when published live. The best approach is to use a validating editor or a validation tool, which should be a standard design practice. Some gateways also behave strangely and may start doing things like transforming content, especially images. Be prepared for this and the impact it may have on your design. Hopefully, you will use sparse and easily discernable images anyway (see below).

Think Task List, Not Brochure   In designing page layouts, think about the vertical flow and making it as efficient as possible for the user to complete the task presented for each page(s). Just like with newspaper writing, keep all the important information at the very top of the news item, or at least a summary of the rest of the article. Use similar principles in your design, not just for news articles, but for every content type. Think about partitioning the content so that not too many important ideas or chunks of information are present on any one page. Try to direct the user towards being decisive. For example, don't give them the option to see information that you could easily have included anyway on the current page. It is often tempting to sculpture the design around what is easy to code (especially in the JSP/servlet sense) rather than what is easy for the user to accommodate. Asking a user to jump to another page to see the prices of items is not a very good idea if you could have easily included the prices on the current page, just with a little more coding effort at the backend.

Beware of App Killers[18]   Unfortunately, some browsers will not support a feature that you may have relied upon in your design, such as background fills or background image, so if you used a white for font colour, you may end up not getting what you expected, as shown in Figure 8.30. Image (a) looks fine; it is exactly as we intended it. In image (b) we

---

[18] Not 'killer apps' of course.

(a) Looks good   (b) Looks worse   (c) O dear!

Figure 8.30 Lack of browser features wiped out the title.

see that style sheet support is limited and we are unable to display the background fill on the title text, resulting in poor readability against the light background. Image (c) has really done it for us; there is no background support and the background image has also dropped out, leaving us with an invisible title – oh dear!

We could test our design in every single browser, which sometimes may be necessary (and we tackle this problem later), but we should design for graceful degradation. This is different from lowest common denominator design where we don't use any of the 'advanced' features just to be on the safe side. Graceful degradation is taking an approach that stands the greatest chance of still being displayable, even in browsers with many features missing. An example of this is shown in Figure 8.31, which shows the same design as shown in Figure 8.30, but this time with a few subtle changes.

Firstly, as seen in (a), we also specified a background colour in addition to an image. If the image is not displayed, perhaps we can still benefit from a more appropriate background colour than the (presumably) default white. However, if we also stick with black colour font for the title (as in b), then it still looks fine under all cases, even if we regain our background image support, as shown in (c). In all cases, the page is still readable.

(a) Looks ok   (b) Bit better   (c) That's fine

Figure 8.31 Lack of browser features gracefully handled.

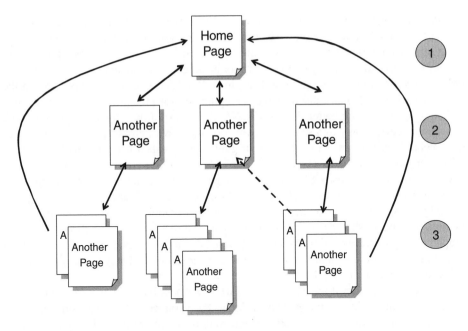

Figure 8.32 Applications should be shallow and easily navigated.

Navigational Efficiency   Designs should be navigationally shallow with typically three links of depth and a consistent navigation model on all pages, as shown in Figure 8.32. We can see that every page should have a one-click to get back home, an obvious point but surprisingly often overlooked. At any sublevel of the application, we should consider having the ability to easily move up and sideward (dotted line) in one click, where this makes sense, perhaps due to the likeliness of wanting to view very similar and related topics. For example, in a TV listings application, perhaps we have drilled down to a view of what's on a particular channel and then clicked on a programme to see the detail. Having got there, we could well imagine that our user didn't find something they were interested in and may well like to see the details of what else is on at the same time, but on a different channel. Rather than make the user go up and then back down, we could consider an immediate sideward link, such as 'What's on Channel 2 at this time?'

We should remember to use meaningful titles in the XHTML-MP title element, as this generally appears in the top of the browser and is valuable navigation context information as it helps to tell us where we are. We should make good use of this and change the titles to match the content, not just keep repeating the application or company name over and over again, which is an all too common mistake. We should then make page titles really count, not merely repeating what the XHTML-MP file title already states, which is redundant and a waste of valuable screen space.

Use text links only, avoiding clickable images unless it is obvious that we are expected to select an image, such as a group of icons on a home page. Beware of browsers that can't support any colouring of hyperlinks, like the Sharp GX10 implementation of the Openwave Browser, which only displays black links and ignores all style information. Also beware of browsers that put links on a new line, even where one wasn't intended.

Design Elegance   We should aim to keep colours simple and images simple, clear and relevant. As already noted, we should beware of using background colours and images when these are not supported. Our colour schemes should aim to use high-contrast design to aid visibility on small devices, especially with poor backlighting or under poor light conditions (which may mean overly bright as well as dim). We discuss device displays later in the book (see Chapter 10) and discover that there are several predominant types of display technology, some much better than others. We should avoid developing and testing on a feature-rich device, only to discover later that on other devices (perhaps a popular device) the colours look washed out or that colour guides (e.g. striped lists) are hard to discern. Wherever possible we should use standard link colours and be consistent across all pages. If we can, it is better to use CSS in place of images that are simply text blocks, as this will typically speed things up, although we should still be careful of variations in CSS support that might lead to excessive or unintended variation in our look-and-feel. The deeper we go in the hierarchy, we should consider simplifying page design for speed, such as eliminating graphics altogether. By the time the user has drilled down the design to get somewhere they want to go, they are no longer interested in what pages look like, they want to get 'in and out' speedily in order to get the job done.

It is likely to be safer if we try to use lists wherever possible, avoiding the use of tables to construct similar layout flows. This is due to the variance in table support that exists across browsers, with some browsers not able to draw a table without displaying borders, even if these are not described in the markup code. To display a table, rather than a list, a device probably has to perform floating-point calculations. On a very slow device, this will take time and thus the page will be slow to display, perhaps adding to the lag that the user experiences when fetching the page, decreasing user satisfaction. Tables do not render consistently on all devices, even some with the same browser. Some devices will display all columns with equal width, overriding the style information in the page. This can lead to particularly chronic layout failures that might even lead the user to conclude that an application failure has taken place. Of course, where data is naturally tabulated, then tables are ideal for formatting the information, so the idea is not to avoid them altogether, but generally to avoid relying on them for content formatting.

Browsers are Not All the Same   The fact that browers are not all the same is the bugbear of browser-application design for mobile environments: usually no two browsers are alike and they don't all fully support XHTML-MP, even if they claim to be 'XHTML-MP compatible'. This can make life difficult for the application designer who ultimately may have to consider designing an application that can dynamically adjust its presentation layer to suit the device in hand. Apart from some of the display differences we have already highlighted above, we need to pay particularly close attention to:

- Cookies

- Level of language support

- External CSS support

- Cache size

Cookie support can be very limited on some devices, perhaps with the ability to store only a few cookies at best. This can be mitigated by using the gateway to support cookies.

For an application that is being specifically designed to run through a known gateway, cookie support at the gateway can be exploited (if available). However, for applications for which we are unsure about the level of cookie support, we should aim to minimise their use, or possibly avoid them altogether. The usage of cookies and possible alternatives are discussed later in this chapter, when we look at something called *session management*.

In terms of language support, it is always worth consulting with a browser user guide first to understand which features are supported and which are not. We should not only be careful about understanding how many of the core XHTML-MP elements are supported, but we should also pay close attention to extensions, particularly if these involve HTML. We may, deliberately or inadvertently, revert back to using HTML elements, some of which are not supported by XHTML. In particular, the way attributes, especially relating to style (e.g. like 'width') are supported in XHTML is now different to HTML. A design that uses HTML might work in a backwards-compatible browser, but perhaps not in a strict XHTML implementation. This will give us problems later when we try supporting users with XHTML-only browsers. They will get errors on the pages that use HTML wherever it is not compatible with XHTML.

Language support goes side by side with CSS support, or WAP CSS (WCSS) support to be more precise. Not only should a designer be careful to check the browser guidelines on support for WCSS, but they should look out for chronic deficiencies altogether; like the inability to support an external CSS, where the designer is expected to embed style information in the header block of the XHTML-MP file instead.

Cache size is an issue if the user is positively relying on the cache for improving performance in order to get away with using rich graphical content. There is also an issue with caching of the external CSS; the designer should not assume that this support is automatically available, despite this being the intended wisdom behind it.

8.4 MANAGING DIFFERENT DEVICES

In our previous discussion it became apparent that not all devices are the same. This is a rather obvious point given the broad range of form factors, and something we discuss later in the book when we look at devices (see Chapter 10). However, we have also just seen that even on devices that may ostensibly be similar, there could be variations in browser implementation. This may preclude using a common design approach, or, if we were to go down that route, then we may end up gravitating to the lower common denominator. This may lead to design compromises that we are not willing to accept.

There are copious mobile device parameters to take into account: browser, language support, OS, user preferences, screen size, colour support, connection speed, graphics capabilities, audio capabilities. The catalogue of features keeps expanding. It is obvious that not every application design will work for every user in all situations.

If we look at a device like the Sony Ericsson P800 mobile phone, which is really a wireless PDA, then the integrated browser can access the Web using HTML, XHTML, cHTML and WAP. The screen, as shown in Figure 8.33 , is a quarter-VGA (videographics array) touch screen with 4096 colour depth. This means that when fully opened (the keypad is flipped out), the screen size is 208 × 320 pixels.

Figure 8.33 Sony Ericsson P800 (Reproduced by permission of Sony Ericsson.)

By contrast, one of the leading Java games devices (at the time of writing) is the Nokia 3510i, which has a comparatively small screen at 96 × 65 pixels (see Figure 8.34), although an equally impressive 4096 colour depth.

However, the 3510i has a WAP 1 browser, which means we have to code in WML. This is not too much of a problem if we wanted to use WML for our design, since the P800 supports it. However, a quick glance at the relative screen sizes in terms of resolution shows us that a common design approach may be problematic, as demonstrated in Figure 8.35.

It hardly needs explaining, but common formatting is going to be a problem. As an example of formatting problems, we might decide to use a header image like the picture portrait one seen in our previous examples of the WAP weblog. This image, which is 53 × 43 pixels, would appear quite large on the 3510i, taking most of the display space, as shown in Figure 8.35. On the P800, the image would perhaps be a bit small tucked away in the top right corner[19]. It seems as if we might benefit from using two different page designs here,

---

[19] The figure is not quite a true representation, as the P800 browser window does not occupy the entire screen space available.

Figure 8.34 Nokia 3510i. (Reproduced by permission of Nokia.)

or at least two different image sizes. The latter option would still necessitate two different WML pages, as the images would have to be stored as two separate files on the web server, thus requiring two separate file names. This would have to be reflected in the WML, so we would need two different WML pages to dish up these two different images.

This is perhaps a trivial example of formatting problems. If this were an application involving publication of articles, like the WAP weblog example, then the length of the articles (number of words) may have to be changed to adapt to the displays. Perhaps we could fit an entire article on the P800 display, which even with a bit of scrolling is still

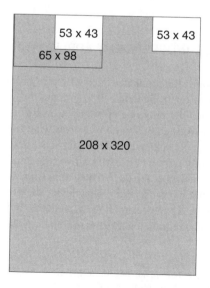

Figure 8.35 Pixel (not physical) size comparison 3510i with P800.

manageable. However, on the 3510i, we probably need to think about chopping the story up into smaller chunks and asking the reader to download one chunk at a time. Such an approach could be made adjustable to any device, simply by altering the 'chunk' size to suit the device's display size.

Depending on the application, we may decide that WML (WAP 1) is just too limited a language to express the rich content we would like to support. The P800 can support other languages, so it is likely that we would want to exploit one of these. Looking critically at the P800 capabilities, then we have the very advantageous pointing device courtesy of the touch screen and stylus. This means that we can implement exact positioning if we chose. The higher processor speed of the device also means that we are not too fussed about the processing overhead in displaying tables. Moreover, with full XHTML support, we could even consider image maps, should we feel that the size of images involved are not too problematic given whatever limitations the RF connection might impose. With a more conventional PDA, like a Palm Tungsten T, we may have a high-speed WiFi connection available and don't really mind about image sizes very much in terms of their loading times.

The problem we face is how to handle different device types if we chose to design different user interfaces to match device capabilities. One solution is simply to build N applications for N device types, where N could be, worst case, the total number of device types we are designing for, or could be a group of device families with similar display capabilities. From our consideration of J2EE, this would mean that whilst our business logic (EJBs on the backend) could definitely remain the same, we would have to implement different presentation logic for each application. Worst case, we end up with N different JSPs to control our view of the interface. Clearly, there is potentially a significant overhead in maintaining all these different applications, even if we are still able to use some component technologies within the servlets and JSPs (e.g. with Java beans[20]).

Another solution, which we will examine in the next section, is to build a framework that lets as much commonality as possible be implemented and allows dynamic content adjustment to the different device types.

8.5 BUILDING DEVICE-INDEPENDENT APPLICATIONS

The general approach towards the problem of designing for different devices can be seen in Figure 8.36. We can use a software element called a *redirector*, which is our entry point into the application. There should be only one entry point, as this is easier to publish and link to as the starting point. The redirector somehow senses what type of device is accessing it. Based on this information, it consults with a device database and extracts configuration data that gets used to subsequently redirect the user's browser to another resource that can better handle the display characteristics of the device they are using. This entire process is ideally automatic. The user does not have to specify what device they're using, nor do they have to manually link to the redirected page – somehow, as we shall find out, the browser can 'take them to it' automatically. The redirection process is all taking place at the presentation layer

---

[20] We didn't discuss Java beans, but they are not the same as EJBs. Java beans are like repeatable chunks of Java that we can reuse as plug-and-play components in our JSPs and servlets. They don't necessarily have anything to do with the J2EE platform and all its amazing capabilities that we have been discussing in this book.

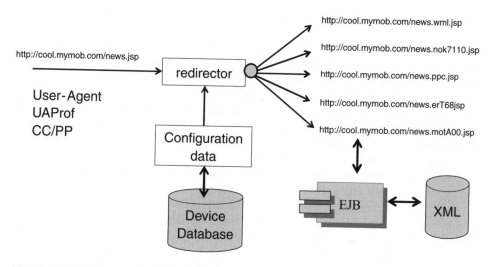

Figure 8.36 Device detection and application control.

level. The underlying business logic remains the same and is probably none-the-wiser of the process.

The use of a redirection scheme does not save us from being smart about how we design our pages. Earlier we discussed guidelines for XHTML-MP design. These guidelines were largely general and apply to other markup schemes. Moreover, these guidelines apply even if using a redirection approach. The reason is that we do not want to test for each and every browser/device combination, which in all likelihood is not feasible anyway, especially if we are required to gain access to new devices to conduct testing. This is very often not that easy to do, especially if we are designing applications that support devices as they come on to the market. Gaining access to pre-release devices for testing purposes is notoriously difficult, even for some bigger companies with a modicum of influence on device manufacturers. It is often engineering limitations that dictate that very few test devices are available before volume manufacturing commences and this problem is passed on to the application designers.

We should always aim to design pages to degrade gracefully in the absence of certain browser features, as discussed earlier. This is good design practice that will pay off in the end, as it means ultimately we will stand the best chance of attaining a favourable user experience, which is essential to the success of a mobile application.

If we wish to deploy the redirection (or 'resourcification') approach, then we need to consider how to achieve several mechanisms:

- Detecting device type (or browser type) and capturing device (or browser) capabilities

- Conveying the device capabilities to the server application

- Dynamically adjusting the presentation logic based on the above two information sources

Let's look at each of these areas in turn in the following sections, beginning with recognising the device type.

8.5.1 Detecting and Capturing Device or Browser Information

When discussing the details of HTTP, we understood that the request for service from the server consists of issuing a method, such as GET, along with associated fields (header fields) in the header of the request. One of those fields is called *user agent*.

The user agent request-header field contains information about the user agent originating the request. This has several uses. Firstly, it enables the web server to amass statistics about the devices and browsers accessing the server. Log files collect this information, which can then be analysed offline. The user agent field can also be used for tracing protocol breaches by 'rogue' devices. But, probably most importantly for our current purposes, it can be used for recognition of user agents for the purpose of automatically tailoring responses to match the capabilities of the user agent's display and other content handling functions.

User agents are expected to include this field with requests, although there is no particular convention for doing so. The field can contain multiple product tokens as well as comments identifying the agent and any subproducts, which form a significant part of the user agent. For example, in addition to identifying the browser itself, it could be used to identify that the host device has a particular audio codec available for decompressing sound. Perhaps the codec was downloaded by the user after purchasing the device, so is 'non standard' for the browser, thus we would not expect that our presentation layer would know of its existence unless we specifically informed it. Support for a particular codec may be useful for the server to know, so that it can include sound files optimised for the available codec. This will involve the presentation layer rewriting its output stream to ensure that the appropriate links (anchor tags - <a>) are included to reference the relevant sound files on disk (or on the audio streaming server). Possibly, the codec is the latest one available and offers an especially high degree of compression, making it ideal for conveying sound files across a relatively slow RF network.

It is useful to include information about particular capabilities of the device beyond its standard means. We can regard this as preference information. There are perhaps many other preferences that a user could specify, even regarding the user interface itself. This is not a problem as such. Once we have a mechanism in place to adapt our HTTP output stream per user agent, then there is no reason why we can't adapt it on a user-by-user basis; we could easily offer that degree of customisation of the output stream.

Within the WAP environment, the ability to determine device capabilities and preferences is not just limited to browsing; it also extends to WAP push. Later on, we shall see how WAP push is used to send multimedia messages that contain pictures, sounds and markup to glue them together. Clearly, capability information is useful in this context: there is little point in pushing an image to a device that cannot display it!

Our 'resourcification' problem has now extended to include not just capabilities, but preferences. This had already been highlighted by the standards community as an important feature in the desktop browser world and is the subject of a W3C project to define a more substantial means of identifying capabilities and preferences, a technique formally known as conveying *Capability and Preference Information* (CPI).

The W3C initiated a project called *Composite Capability/Preference Profile* (CC/PP) and the WAP forum adopted this and extended it into the WAP environment, sticking to the original project title of *User Agent Profile*, or *UAProf*.

The W3C has since gone on to expand its work in this area, recognising its importance across a wide array of browser-based applications and situations. In September 2001, the

W3C forwarded a draft proposal[21] for the formation of the Device Independence Activity (DIA), wherein they mentioned that the draft document:

> ... *celebrates the vision of a device independent Web. It describes device independence principles that can lead towards the achievement of greater device independence for Web content and applications.*

The document is worth reading as it highlights a much wider vision than the confines of our current discussion, which is still related to the more conventional browser paradigm. However, the W3C DIA project takes into account all possible access methods, such as accessing content via an audio-only connection. A mobile devices working group was established for a time and their output was fed into the DIA project, so the work of this group is fully expected to take into account mobile devices, which is our current topic of interest.

As we can imagine, the method of specifying capabilities and preferences needs to take into account a wide variety of device types and application scenarios, so inevitably it has moved way beyond simply identifying a suitable string to encode in the user agent header field.

The CC/PP project has worked to produce a definitive vehicle and framework for delivering capability information within the context of the Web. We shall briefly look at the CC/PP profiling method itself before considering its use within the WAP context. The latter is important as it tells us how the CC/PP information gets conveyed, handled and processed within a network where WAP protocols are being used.

CC/PP is based on RDF, the Resource Description Framework, which was designed by the W3C as a general purpose metadata description language. The origins and ethos of RDF are interesting. For a while, the Web has been attracting vast amounts of information to its cloud. However, the origins of the Web are in human interaction with information, notably the visual interaction paradigm of the browser we know quite well. However, it has long been recognised that machines themselves have difficulty in processing information on the Web. If we want to write a program to go look for information on a particular topic, then how do we identify the target information in a meaningful manner? For example, there may be lots of pages about cooking, but how would a computer know that it is looking at a recipe say? Moreover, even if it could find a recipe (perhaps searching for the keywords 'cooking recipe') how would we know what the ingredients are? In other words, we are articulating a problem to do with metadata, or semantics. We are not just interested in how to visually format information, but we are now interested in how to describe, or annotate, information – to add semantics.

RDF can be expressed using XML. No surprise there. But we are not interested in the details of RDF itself, as much as how to use it to support our CC/PP objectives, coming eventually to its use in WAP environments. A complete sample RDF file is given at the end of this chapter (Appendix A – 'Sample CC/PP RDF File'). Just to illustrate its usage, we can look at a few snippets of the XML, especially relating to screen size, as this was an example of a capability we were looking at earlier.

---

[21] http://www.w3.org/TR/2001/WD-di-princ-20010918/

```
<prf:ScreenSize>121x87</prf:ScreenSize>
<prf:Model>R999</prf:Model>
<prf:ScreenSizeChar>15x6</prf:ScreenSizeChar>
<prf:BitsPerPixel>2</prf:BitsPerPixel>
<prf:ColorCapable>No</prf:ColorCapable>
<prf:TextInputCapable>Yes</prf:TextInputCapable>
<prf:ImageCapable>Yes</prf:ImageCapable>
```

We should be used to the familiar XML tags by now, but the prf: bit seems new. This is referring to a namespace, which is an advanced topic. Simply put, although we can define our own tags in XML, like when we were defining our tags for Film-ML earlier in the chapter, the problem is how to avoid treading on someone else's toes. Perhaps the name we use is already taken by another vocabulary, especially if the name is somewhat generic, such as 'title'[22]. There is nothing wrong with mixing different XML vocabularies within one (host) document. However, this is when things are likely to conflict. To get around this problem, we refer to the namespace of a vocabulary, giving that space a moniker, like prf in our case. We can then prefix all tag names from that namespace, so that they are uniquely identifiable with the namespace, thus avoiding naming conflicts. In the case of CC/PP, prf is used here to refer to 'profile', which is the schema[23] used to define the allowable tags within the CC/PP profile.

Getting back to the XML snippet, we can see that the tags are quite self explanatory, as we would expect, as the objective of using XML is that it is also human-readable. The issue of screen size that we discussed earlier is covered with the tag <prf:ScreenSize>, so upon discovering the CC/PP profile for this device, we can extract its screen size by searching for this tag. Other tags in the snippet also relate to screen capabilities, including colour support and image display capability. There are many tags in the UAProf relating to various components of the device capability set. A sample of the structure of the UAProf, along with a few of the tags, is shown in Figure 8.37.

The issue still remains as to how this information gets conveyed from the client to the server. We also mentioned that it gets used in WAP Push, which is puzzling at first. If we are going to push information to a device, how do we know its capabilities in advance? Let's now look at these issues in the next subsection.

8.5.2 Conveying CC/PP Information

The suggested method for exchanging CC/PP information is captured in the W3C Technical Note[24] defining the CC/PP exchange protocol. This is a protocol that is based on the HTTP Extension Framework, which is a means to convey information via HTTP headers. The HTTP Extension Framework is a generic extension mechanism for HTTP/1.1, which is designed to interoperate with existing HTTP applications.

There are two ways of conveying the profile, both of which are supported by the WAP UAProf specification, though in a slightly modified form than we shall discuss here.

---

[22] It's always a good idea not to use names that are too generic. If we mean title of a film, then perhaps 'FilmTitle' is better, as it is more descriptive. However, there are arguments for sticking with just 'title', too. Namespaces can solve conflicts for us, however.

[23] Schema is like a DTD, which we referred to earlier in the chapter.

[24] http://www.w3.org/1999/06/NOTE-CCPPexchange-19990624

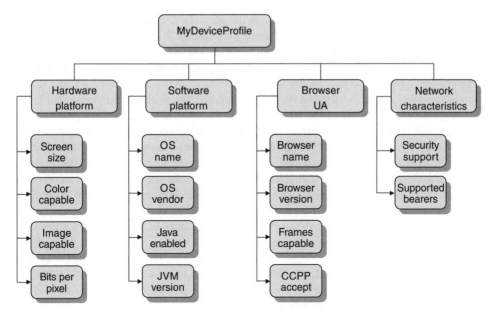

Figure 8.37 Structure of UAProf (partial).

The first method is to send a URI specifying where the CC/PP RDF file can be found on the Web. As we can imagine, this is a particularly efficient technique, as it means the amount of information we have to transmit from the user agent to the origin server is very little, which is particularly compatible with sending such information over relatively slow RF networks. The second method, as expected, is to insert the RDF file into the HTTP request header itself, thereby conveying in one go the entire capability set of the device. That said, it is possible to modularise the CC/PP (the specification allows for the description of capabilities and preferences in subunits called components). We can, if appropriate, just send a portion of the overall capabilities. This has several uses. Not only can it save on the amount of data to be transmitted, but it allows the user agent to convey subsets of information that may be pertinent, like the codec update we mentioned earlier.

The first method uses the *Profile* header field, the contents of which are a URI pointing to the RDF file.

Examples of profile headers are:

```
Profile: "http://www.aaa.com/hw","http://www.bbb.com/sw"
Profile: "http://www.aaa.com/hw","1-uKhJE/AEeeMzFSejs
YshHg==", "http://www.bbb.com/sw"
```

We can observe several features of the header. Firstly, there is no need to stick to one URI. Several can be used in conjunction with each other. This might be a good idea, perhaps to have one pointing to a repository of device capabilities, the other pointing to a repository of separately maintained browser capabilities. The profiles are listed in order of priority, with the first one listed taking precedence.

The next thing we may observe is that in the second example, the second entry in the field is somewhat strange; it certainly doesn't look like a URI that we are used to seeing. It is called a *Profile-Diff-Name*. This is actually a pointer to a profile contained within the HTTP request itself, which is found under a separate header field called *Profile-Diff*. The contents of a Profile-Diff header are the RDF itself. The strange looking string in the above example (second entry in the Profile header field) is actually a Base64 encoded MD5 (Message Digest 5) digest of the RDF file. That sounds like a complicated process, but we shall look at the actual mechanics of MD5 later in the book when we discuss how to authenticate HTTP sessions (see Chapter 9). All we need to know here is that this process takes the RDF file and produces a very compact fingerprint of it. The reason this is done is to allow any caching process (such as at a gateway, or the user agent itself) to perform lookups using the digest, which is a short process, rather than comparing entire RDF files to check for changes.

The Profile-Diff-Name entry in the Profile header is inserted into the list in a position commensurate with its priority with respect to the other entries. There can also be more than one Profile-Diff-Name, corresponding to more than one Profile-Diff header, which is why there is a single-digit prefix (in this case '1') to indicate which embedded RDF file is being referred to. Profile-Diff headers themselves are postfixed with the corresponding digit, like *Profile-Diff-1, Profile-Diff-2*, etc.

More than one Profile-Diff (i.e. RDF file) can be supported, as this may be a legitimate way of indicating capabilities. For example, the device itself may have an RDF, but sub-components on the device (e.g. audio codec) may have their own RDF, in which case we probably want to convey all of them, or as appropriate.

In the WAP UAProf specification[25], these headers are supported in both W-HTTP and WSP, but in the former case they are renamed to *x-wap-profile* and *x-wap-profile-diff*.

There is no need to refer to external profiles, particularly if they are not available or they cannot be relied upon. In which case, the entire profile is embedded in the request, in the manner we have just described. An example follows:

```
Profile: "1-P1GRkSjKK50aTWXXndFcSQ=="
Profile-Diff-1: <?xml version="1.0"?>
    <RDF xmlns="http://www.w3.org/TR/1999/
        PR-rdf-syntax-19990105#"
    xmlns:PRF="http://www.w3.org/TR/
        WD-profile-vocabulary#">
    <Bag>
     <Description about="HardwarePlatform">
     <Defaults>
        <Description PRF:Vendor="Nokia"
                     PRF:Model="2160"
                     PRF:Type="PDA"
                     PRF:ScreenSize="800x600x24"
                     PRF:CPU="PPC"
                     PRF:Keyboard="Yes"
                     PRF:Memory="16mB"
```

```
                            PRF:Bluetooth="YES"
                            PRF:Speaker="Yes" />
        </Defaults>
        <Modifications>
         <Description PRF:Memory="32mB" />
        </Modifications>
        </Description>
        <Description about="SoftwarePlatform">


      .....
</rdf>
```

There is also no reason why a Profile-Diff header field cannot be added by an intermediary network element like a gateway. Perhaps the gateway has additional information about the subscriber that is useful for the origin server to know. Such a technique also allows a consistent approach to be adopted with devices that do not support CC/PP exchange protocol and only support the user agent field. The gateway could add the profile information itself, based on what it knows about user agent fields and device capabilities and profile information, even about the user (e.g. from information stored in a mobile portal's personalisation database).

If we embed the profile in a WAP session using W-HTTP, then we keep the text description as it is, but when using WSP, we can apply tokenised compression to render the header into WBXML, which condenses it considerably, just as we saw with WML in our earlier discussion.

Within the CC/PP exchange protocol, we also have the ability to acknowledge that the profile has been accepted by the server. This is done by passing information in a header field called *Profile-Warning*. This header field contains a code number that indicates the success of the exchange, similar to the way status codes are used in HTTP. There can be one code for each RDF file that the server processed, or attempted to process, hopefully corresponding to the number that was referenced in the request. To make each warning clear, the name of the target RDF file (*warn* target) is mentioned along with each code.

```
<code>
Profile-Warning: 102 http://www.aaa.com/hw "Not used
                profile",
                202 www.w3.org "Content generation
                applied"

Profile-Warning: 101 http://www.aaa.com/hw "Used stale
                profile",

                102 http://www.bbb.com/sw "Not used
                profile",
                200 18.23.0.23:80 "Not applied"
                "Wed, 31 Mar 1999
08:49:37 GMT"
 </code>
```

This is a list of the currently defined warn-codes, each with a recommended warn-text in plain English, and a description of its meaning:

- *100 OK-* MAY[26] be included if the CC/PP repository replies with first-hand or fresh information. The warn-target indicates the absolute URI that addresses the CC/PP descriptions in the CC/PP repository.

- *101 Used stale profile-* MUST be included if the CC/PP repository replies with stale information. Whether the CC/PP description is stale or not is decided in accordance with the HTTP header information with which the CC/PP repository responds (i.e. when the HTTP/1.1 header includes the Warning header field whose warn-code is 110 or 111.). The warn-target indicates the absolute URI that addresses the CC/PP description in the CC/PP repository.

- *102 Not used profile-* MUST be included if the CC/PP description could not be obtained (e.g. the CC/PP repository is not available.).The warn-target indicates the absolute URI that addresses the CC/PP description in the CC/PP repository.

- *200 Not applied-* MUST be included if the server replies with the non-tailored content that is the only one representation in the server. The warn-target indicates the host which addresses the server.

- *201 Content selection applied-* MUST be included if the server replies with the content that is selected from one of the representations in the server. The warn-target indicates the host which addresses the server.

- *202 Content generation applied-* MUST be included if the server replies with the tailored content that is generated by the server. The warn-target indicates the host that addresses the server.

- *203 Transformation applied-* MUST be added by an intermediate proxy if it applies any transformation changing the content-coding (as specified in the Content-Encoding header) or media-type (as specified in the Content-Type header) of the response, or the entity-body of the response. The warn-target indicates the host which addresses the proxy.

By analysing these codes, the user agent can take corrective action if necessary. A similar heading is supported in WAP – *x-wap-profile-warning*.

When using CPI for WAP Push, there is no request from the user agent, so there needs to be an extra step in the protocol in order for the push server to know the target device's capabilities before pushing any content to it. This is provided by a standard HTTP request that gets made from the PPG to the device, remembering that in WAP, the device itself is capable of responding to HTTP requests; this being the mechanism by which data gets 'pushed' to the device (i.e. via HTTP POST requests). To first get profile information, the PPG issues a HTTP OPTIONS method, which is like a GET request, but without any resource actually being requested. As such, the device responds by sending back a response that only contains header information, but no body (as one wasn't requested). The HTTP header can contain the relevant profile headers (e.g. *x-wap-profile*) and so the PPG now has

---

[26] These codes are taken straight from the W3C document where the words MAY, SHOULD and MUST are used to indicate the desired level of conformance required from an implementer of the specification.

the CPI it needs before attempting to push content to the device, content that can now be suitably adapted to the device capabilities and to the user preferences.

A profile repository typically would be an HTTP server that provides UAProf CPI elements upon request. An origin server would go to the HTTP server to fetch the corresponding profile it found referenced in the Profile header field. UAProf profiles may reference data stored in repositories provided and operated by the subscriber, network operator, gateway operator, device manufacturer, service provider or any valid host on the Web. Specialist companies or agents may choose to track device capabilities and generate profiles on behalf of interested parties. Different ways of tracking profiles may emerge, some of which are as follows:

- Manufacturers and software vendors may provide static profile resources that describe the client devices and applications (e.g. browsers) that they produce. Such profiles will describe the hardware and standard software elements that exist on the client devices, as well as optional software elements that get loaded later.

- Network operators may provide additional profile information that includes critical information about network characteristics. Additionally, they may chose to operate gateways in such a way that they provide profile information that defines permitted service levels through the communications infrastructure. This facility may also be used to allow extra preference information to be stored and presented on behalf of the user based on service preferences they have made known to the operator, most likely via personal profile information associated with mobile portals.

- Subscribers may look to other entities to store preference information if they wish, perhaps to gain their own control over how they manage profiles. However, this would have to be a mechanism that the user agent supports. Just because a user agent can insert any profile URI in the header, doesn't mean to say that it will. Possibly some browsers will use constant header information preset at the time of installation. Even so, this could still be overcome on devices that offer sufficient programming access for a local proxy to be installed (on the device) to transform headers under the user's control.

- Service providers, including enterprise IT managers, could chose to manage their own profile repositories. Perhaps this is done with security in mind where a particular authentication technique is used, in which case the origin server needs to be made aware of it.

Independent of the content of the stored profile, policy controls may be imposed that limit what profile information is delivered to a particular requester. Any such limits, or methods employed to ensure compliance, are beyond the scope of this discussion. Also, whether or not it is a good idea to use profiles for the purposes of controlling policies and controlling application performance needs some reflection.

8.5.3 Dynamic Page Generation Schemes

Now that we have a mechanism for understanding device capabilities, we still need to solve the problem of adapting our presentation layer's output to suit each device according to its profile.

Before looking into the detail of any mechanisms, we need to recap on where we started this chapter of the book, which is the J2EE platform and its presentation layer capabilities. The platform supports several types of Java program, namely JSPs, servlets and EJBs. We identified that the JSPs and servlets were fantastically well optimised for writing output streams over HTTP. However, it was not made clear exactly how we should combine these elements to make best use of their combined capabilities. It is pretty clear that the EJBs always sit in the background and do the crunching and grinding of our business logic, talking to databases, XML streams, legacy systems and so on. But what about the servlets and JSPs; is there a particular usage of these program types that makes our lives easier, particularly now that we understand that what we want to be able to produce is a framework for dynamically adapting content?

In software engineering, we have the concept of *patterns*, which basically means a way of configuring software components to achieve a particular task. There is no real rocket science in this process. Largely, we take something that appears to work, refine it and then publish the idea as a pattern.

One J2EE pattern stands out from the crowd when producing mobile browser-based applications, and that is one called the 'Model View Controller' pattern, or sometimes MVC (and other names, too).

The MVC pattern consists of three kinds of main classes: Model, View and Controller. The Model represents the application's objective (core tasks) or its essential data. The View is the display of the model, and the Controller takes care of the user interface interaction with the user input.

We can see the MVC pattern on a J2EE platform shown in Figure 8.38. The EJB, the stalwart of the business-logic processing, is clearly the model in the pattern. The JSP, because of its superior support for page design (as noted earlier) is used to control the view – this is what produces the final output stream back to the user via HTTP. This leaves the servlet, which must be the controller. This also seems to be a good fit, as it naturally can handle HTTP input streams due to its intimate connection with the web engine (via the servlet API), but it also offers us the entire range of Java programming power to control how we want to handle the input stream. Comparing this to Figure 8.36, we can begin to sense that

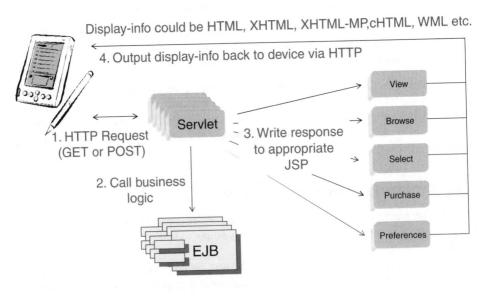

Figure 8.38 Model view controller design pattern.

the 'redirector' in the figure is going to be supported by a servlet, but let's dig a little deeper into the pattern and the adaptation problem we are trying to address.

Before continuing, we should clarify that the MVC pattern is generally applicable to J2EE application design for the Web. It is not confined to using servlets to select views based on device capability. In fact, this is not its origins at all. The MVC pattern can be used to develop scalable web applications that are relatively easy to maintain. The controller is actually used to determine which page to serve next according to where the user is in the page, or navigational flow of the entire application. In other words, we might want to progress from the 'login' page to the 'check my account' page. There may be a JSP to handle each function – LoginJSP and CheckAccountJSP. We can use the servlet to determine where in the design the user is, or what page they are requesting, and then divert to the appropriate view accordingly. This is the primary function of the model; to react to user interaction. However, we are currently considering extending this control function to achieve a finer level of control, which is to select the appropriate view based on user request and user device type (and capabilities, as per CC/PP).

We shall look at several methods for achieving the adaptation based on the MVC pattern. These methods are by no means exhaustive of all possibilities. Like with all software, there is more than one way to develop a solution, depending on what we are trying to achieve, what we know, what restrictions may be in place, and a plethora of other considerations – not always logical ones.

Reviewing Figure 8.38 for a moment, we can reflect on the possible means to achieve adaptation, introducing some of the programming techniques as we go along.

In Step 1 our servlet gets called by the user agent via a HTTP GET request. Within this get request we have three methods of passing parameters to the servlet. These methods are discussed in the Section 8.5 when we look at state management (or session management). But for now, we just assume that we are able to receive parameters from the user into our servlet.

These parameters may be used to determine what the user wants to do next. For example, if the user is currently examining their mobile airtime account details, then we may need to know the user's unique ID (account number). Let's assume that the user has already logged in to the application using a secure authentication technique (which we shall discuss in the next chapter). The user would have entered some username to login, and this is what we use to identify them in our system. Once the user logs in with this username, we may look up information about them, such as their account number. This is what we need in order to lookup any information from the accounts database, therefore we need to keep track of it for the duration of the user's session. The user may be looking at a particular part of their current phone bill and we have programmed a JSP that can produce a phone bill view, as shown in Figure 8.39.

The process of writing an output stream to produce a phone-bill view is clearly a repeatable one and we would end up using the same JSP to produce the view for each user and for each instance of each user's bill. It doesn't matter if we are looking at today's bill or yesterday's bill, the output will be same for a given device, and just the data itself will change. Let's say we have a JSP called MyBill.jsp to produce this view. HTTP itself is a stateless protocol. We saw that there was a single instruction to fetch data, called GET. We could have issued that same method for the same page 50 times and the server is none the wiser. There is nothing inherent in the protocol that enables discernment between the first and one-hundred-and-first requests. This is why we need to feed parameters into the

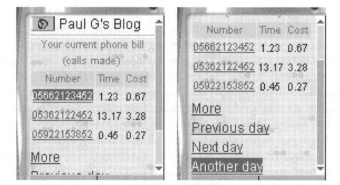

Figure 8.39 Viewing a phone bill via WAP.

servlet, so that by handing over control to the JSP, it knows that on this occasion, we are actually interested in viewing the bill from last Tuesday and that the bill in question is the one associated with Account ID 1234 (or whatever).

Let's imagine that somehow our servlet has access to the parameters that we can use to determine what to do next (i.e. state information). We can envisage a servlet called 'GetBill' that understands this is a billing enquiry. The servlet first extracts the parameters we have just been discussing. From these (however they are formatted) it understands that the billing request is for a particular date and account holder. If this were the first time that this particular user/device has accessed the application, then the servlet (actually the 'login' servlet) would extract the CC/PP header field information. This information could then be processed by some business logic to check the CC/PP RDF file. This process would consist of two parts. Firstly, we need to identify that we have valid profile information and that we can subsequently handle this device. To simplify our discussion, let's assume that we have divided our output views into three categories:

1. Small devices with XHTML support

2. Small devices with WML support

3. Large devices with XHTML support

Our business logic that processes the CC/PP information will attempt to put our user into one of these categories. If this is not possible based on exact profile information, then a default category is chosen, one that gives the best chance of graceful degradation. Once our user's device is categorised, this category information is stored somewhere so that it is accessible on subsequent page requests. On these requests, the profile headers are ignored by the servlet as the categorisation has already taken place. Either we find a way of subsequently passing the category information in the inbound requests, or we look for the category using a lookup table that stores categories against account ID; which will work because we have already stated that we are able to extract the account ID as a parameter in each request (even though we have not discussed how to do that yet , but we will do in the Section 8.5).

Of course, we may well have lots more categories − it is a design decision based on the nature of the user interface and how flexible it is across different devices. If it is very flexible, then we probably only need a few categories, like in our current example, with each category applicable to many devices with similar attributes. For very specific layout control and usability enhancements, we may have many categories, possibly with only one device per category.

The other function of the servlet is to actually call the relevant business logic to process the billing request. In this example, business logic is likely to involve talking to a billing database server via an EJB. The EJB would handle a request to the server via the JDBC API and extract the relevant billing data. At this point we need a place to store the data as it will be needed to build the output stream via the JSP. The actual mechanics of storing (or persisting) session data on a J2EE platform are beyond the scope of this book, but let's just assume that there are ways of passing data from one program to another, not forgetting that servlets and JSPs are just two types of program of course. Here we are assuming that in having handled the call to the EJB, the EJB is now left out of the process and control is handed back to the servlet, which subsequently invokes the JSP. In other words, the EJB does not itself act as an interlocutor between the servlet and JSP.

We shouldn't forget that all this process is taking place within the confines of the powerful J2EE platform, so we are automatically gaining all its 'goodness'; things that we have discussed earlier in the book, like the scalability across clusters thanks to RMI, and so on.

Moving on, and ignoring how data gets passed from the servlet to the JSP, we are interested now in how the JSP adapts its output to the device. Firstly, given that we only have three categories in our example, we could have actually considered having three very similar JSPs, something like 'MyBill_SmallXHTML.jsp', 'MyBill_SmallWML.jsp' and 'MyBill_LargeXHTML.jsp'. The servlet would then act as a redirector to the appropriate output page according to the category information that it extracted in the first instance, whether directly from the request or via a lookup table referenced by Account ID. We shall not discuss this approach as it is easy to understand without further explanation, and also because it is an unlikely approach due to its cumbersome nature, especially when large numbers of device categories are involved. We probably will have too many categories for this to be a viable option, as we would end up with too many variants of each JSP and the maintainability issue will become a dominant and inhibiting factor.

There are two ways we shall consider here that we can use to adapt the billing information:

1. *JSP tags* – this is where we write conditional JSP code such that the appropriate tags eventually get written to the output stream based on a filter that picks each tag according to what the designer wants and the capabilities of the device.

2. *XSLT* – this is where the billing information from the business logic gets stored as XML and we use the XSLT language to transform the XML to the particular vocabulary and structure we want for our output.

Adaptation using JSP Tags   The best way to discuss JSP tag libraries is via an actual example. Here we examine the WURFL project, which stands for *Wireless Universal Resource File*. This is an open-source initiative[27] that involves two concepts. First, the resource file itself is an XML configuration file that contains information about all the capabilities and features of a wide range of devices. This is in apparent conflict to the CC/PP idea where this configuration information is held in separate RDF files, but for now this doesn't matter

---

[27] http://wurfl.sourceforge.net/

in order to illustrate the second concept, the focus of our enquiry, which is the support of JSP tags.

As we have previously discussed, it is possible to write markup output directly into a JSP file, so we could write our XHTML, XHTML Basic, XHTML-MP or WML directly into the JSP. We also know that we can mix Java code in with the markup, but that starts to get messy and is also not easy for page designers who don't know Java. It is also not very maintainable. But with JSP tags, we can use tags, like the ones defined for us by WURFL, such as:

```
        :
<wurfl:if capability="xhtmldisplayaccesskey">
               :
        billing table markup goes here without
        numbering
                :
</wurfl:if>
        :
```

What this piece of code does is to only output the markup between the <wurf:if> tag pair, provided that the device accessing this page is XHTML-compliant and automatically displays numbers alongside anchor tags if they contain the accesskey attribute. The *accesskey attribute* can be used to allow a user to select a link on the page by pressing one of the digit keys on the keypad as a shortcut, rather than selecting the link with an arrow key and then activating the link. If the browser automatically displays the shortcut numbers for us, then we don't have to output them ourselves. To avoid having two sets of numbers, we detect this level of support and use the tags to select the piece of markup that either displays the numbers explicitly (e.g. using the image element <IMG> and possibly locally sourced WAP icons) or implicitly (the browser does it for us, so no need for the IMG tags).

What is actually happening on the J2EE server is that the tags are being mapped to Java classes that execute when each tag gets processed. In the case of the WURFL tag library, the code behind the tags is checking the user agent header field and then trying to match this against the XML resource file. Thanks to the powerful inbuilt XML processing capabilities of the J2EE platform, this process is efficient and scalable across our application. The types of capabilities that the WURFL identifies are grouped into feature categories:

- General product info (e.g. device brand name, device model name)

- WML UI features (e.g. table support, soft key support, etc.)

- cHTML UI features (e.g. display access keys, emoji support, etc.)

- XHTML UI features (e.g. honours background colours, supports forms within tables, etc.)

- Markup support (e.g. WML 1.1, 1.2, 1.3, XHTML Basic, XHTML-MP, cHTML 1.0, 2.0, 3.0, etc.)

- Cache features (e.g. cache can be disabled, cached items life time, etc.)

- Display features (e.g. resolution, columns, image sizes, etc.)

- Image formats (e.g. WBMP, GIF, JPEG, PNG, etc.)

- Miscellaneous (e.g. phone supports POST method, etc.)

- WAP Wireless Telephony Application support (e.g. supports voice call from browser, supports access to phone book from browser, etc.)

- Security features (e.g. supports HTTPS (SSL))

- Storage attributes (e.g. max URL length, number of bookmarks supported, etc.)

- Download fun support (e.g. ringtones, types of ringtones, wallpaper support, screensaver support, etc.)

- WAP Push attributes (e.g. various parameters to do with WAP Push support)

- MMS attributes (e.g. can send MMS, can receive MMS, built-in camera support, etc.)

- J2ME support (e.g. MIDP version, memory size limits, colour depth, API support, etc.)

- Sound formats supported (e.g. WAV files, midi monophonic, midi polyphonic, etc.)

The structure of the WURFL is interesting because it defines device features in such a way that they automatically inherit from a parent feature where applicable. This will let a new device inherit all the capabilities of the family of devices from which it appears to descend. For example, a new device that mentions the Openwave Browser (UP.Browser) in its user agent field will automatically inherit the capabilities of the Openwave Browser, even though the device is not currently entered in the resource file. This means that new devices do not default to the lowest level of support, which is not only undesirable, but is also unrealistic; newer devices should support the newer features, not the oldest ones.

For historical reasons[28], the WURFL itself is not CC/PP based, although it could be made so, depending on the need. What might be considered as value added for the WURFL project is its comprehensive repository of configuration information that is being constantly updated via an active and vigilant developer community. The fact that this is happening via the open-source community is also positive, as it means that support is likely to be strong enough that the accuracy and quality of the resource information is reliable.

Putting the particular strengths or weaknesses of WURFL aside for a moment, the principle of JSP tags is the important feature to grasp here. The tags are XML, so they are easy to learn and incorporate into the JSP file. They avoid the need to insert Java code except that we still need a way to get our actual user information into the JSP file, such as the billing records for the example we have been discussing. This will involve some coding.

If we didn't have the WURFL tags, we could use another tag library, including the possibility of developing our own. Returning to our earlier example where we only had three device categories, we mentioned how our servlet could identify the actual CC/PP information and then categorise our device: small devices with XHTML support; small devices with WML support; and large devices with XHTML support.

In this case, the JSP could use an inbuilt mechanism to accept the category parameter directly from the servlet, thus avoiding the need for JSP tags; we would use snippets of Java code to conditionally write out the chunks of mark up we require for each category, but still all within the one JSP file.

---

[28] WURFL existed before CC/PP had fully emerged.

The point about using JSP tags is that they are easy to understand and insert into our JSP file. The WURFL example shows us that it is possible to control markup output based on attributes rather than device types, which is a subtle but important point. Rather than have constructs like 'if p800 . . .' or 'if 3510i . . .', we would much rather have constructs like 'if screen > 300 . . .', or 'if java supported . . .'. With CC/PP and JSP tags we could do this; by mimicking the semantics of the RDF in our JSP. So, for example, we could envisage tags like the following made-up examples:

```
    :
<ccpp:if wtai.address="true">
    :
  Encode billed numbers to allow adding to address
  book
    :
</ccpp:if>
    :
```

When the JSP page runs, the Java class behind the <ccpp:if> tag will use the J2EE XML API, like JAXP, to search the embedded CC/PP RDF file for the following XML snippet (looking for the <prf:WtaiLibraries> tag>):

```
1. <prf:WtaiLibraries>
2.    <rdf:Bag>
3.      <rdf:li> WTA.Public.makeCall </rdf:li>
4.      <rdf:li> WTA.Public.sendDTMF </rdf:li>
5.      <rdf:li> WTA.Public.addPBEntry </rdf:li>
6.    </rdf:Bag>
7. </prf:WtaiLibraries>
```

Upon finding the <rdf:li>WTA.Public.addPBEntry</rdf:li> entry, the class can return true for the condition and this in turn will cause the markup between the tag pair to be included in the output stream.

Provided the necessary semantics are available in the CC/PP RDF file to identify the level of control we are seeking in the output stream (JSP file), there is no need to resort to resource files. All the information we need about a device's attributes is actually sent to the servlet or JSP via the HTTP header (Profile header field). This means that a highly maintainable process of device adaptation could be implemented, which is a useful achievement, especially given the complex sea of device possibilities that is expanding all the time.

Adaptation using XSL   It is possible to transform an XML document into another format using *XML Stylesheet Language* (XSL). This is a very powerful capability and can be used in a variety of ways to address the current problem being discussed, which is how to adapt content to a particular device, obviously assuming that the content is described using XML in the first place.

We shall introduce the concepts here, but a detailed exposition of XSL is beyond the scope of this book, as XSL is a complex set of programming technologies in its own right. We shall focus on the application of XSL to our problem, rather than its internal mechanics, recognising that there are actually several ways to apply XSL to the adaptation problem.

We have already learnt how XHTML can be delivered to a browser via the powerful J2EE presentation technologies, such as JSP and servlets. The browser can display XHTML because it understands the XHTML tags and can interpret them in terms of the implied layout semantics. For example, the existence of a <table> tag paid in XHTML tells the browser that a portion of screen space is going to be occupied by content in a tabular fashion. The browser's graphical rendering engine will interpret how to transpose the table contents to the screen. In terms of any style information, such as the colour of table cells, fonts and so forth, we already know that a separate file sent to the browser will contain the style information, which is the CSS.

However, some of our devices may not support XHTML at all, or support it but with limited capabilities in terms of screen size, colour depth and so on. The general proposition in this case, bearing in mind the availability of XSL, is to produce the page content in XML and then transform it to the desired final output form using XSL. From this, we can see where XSL gets its name. Just as CSS adds style information to XHTML, we can think of XSL as adding style information to XML. Clearly, the main difference is that XHTML has a defined set of tags and hence a constrained style sheet format (i.e. CSS), whereas the tags we use in our XML page descriptions could be anything, so we need a distinct style sheet format to match the chosen XML vocabulary. The general principle is shown in Figure 8.40.

The transformation process shown in Figure 8.40 gives us several problems to think about. Firstly, where is the XML coming from in the first place? Secondly, is the XML file purely the content, or does it contain the basis for driving the layout, or is the layout implied or contained in the XSL? Thirdly, what is driving the overall process in terms of selecting which of the final output forms is required? We shall examine some possible answers to these questions, but keeping in mind that there is no one approach to this solution and that the final approach will depend on a number of factors that should become apparent as we proceed.

In terms of the initial XML input, we are probably better limiting the XML to describe purely the content without any inference as to its intended layout. We do not have to produce

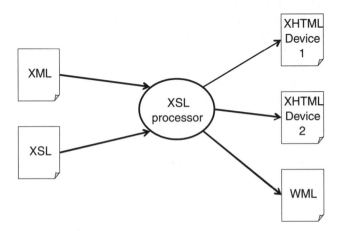

Figure 8.40 XSL transformation process.

XML natively from the database, although some XML database products are making their way into the marketplace. Adding XML tags to the native data can be done using the EJB that is responsible for collecting the data from the database. If we imagine an XML file to describe downloadable games, we might have something like:

```
<?xml version="1.0" encoding="UTF-8">
<gamelist>
<game>
 <title>Space Invaders</title>
 <type>arcade</type>
 <description>Classic space invaders game from the
    70s</description>
 <screenshot>images/spa.gif</screenshot>
 <jad-url>jads/spai1.jad</jad-url>
 <price>3.40</price>
</game>
 .

 .

 .
</gamelist>
```

We can see from this XML listing that the content is purely descriptive about the games; there is no layout information at all. It is up to the page designer as to how this content gets laid out, but this needs to be tempered by the layout restrictions of the device. The fact that the XML contains a reference to a screenshot image of the game clearly has display or layout implications. Not all devices will be able to display an image. Those that do may need to display the image differently, depending on factors such as the width of the screen size. For narrow displays, the image may need to be included in a vertical flow of information. For wider displays, it may be possible to show the image in a horizontal flow, such as to the right-hand side of the description.

What we need now is the means to go from the raw XML content to the final output form. What follows is a simplified explanation so that we can avoid having to go into any details that would require a working knowledge of XSL.

What we can imagine is that the page designer comes up with a set of layout templates to cover a range of devices. These layouts could themselves be described in XML, which we shall call *page profiles*. These would be generic 'families' of layouts, similar to out previous example when we had the notion of device categories, such as: small devices with XHTML support; small devices with WML support; and large devices with XHTML support.

We can think of these page profiles as being markup-independent descriptions of the page layout. For example, we could have a tag called to indicate the navigation area of the page. We could have a tag called <header> to indicate a page header area, such as might be used for logo and page titles. Within each file there can be variations on the basic layout according to the finer capabilities of a particular device, such as its ability to support a background image in a table. What is also added into the profiles is information to indicate what type of device capabilities are required in order to support the associated layout variations.

Now, what we are working towards is having an XSL file that can be applied to the XML content file in order to produce the desired output. The fascinating realisation is that XSL itself is an XML-based language (a language written in XML) and so could potentially be used to produce XSL or vice versa. We can take the CC/PP RDF files, which are XML, to identify the capabilities of the device requesting a page. Using the page profiles as a basis for the output, we can transform the RDF files into XSL files that are tuned for the device. This XSL file in turn is what gets applied to the XML content in order to produce the required output for the requesting (target) device.

This may sound a tad complicated, which the detailed implementation probably is, but its power is in its flexibility to handle all types of devices directly from the CC/PP semantics, which is ideally what we want. We can think of the process as follows:

1. Process the CC/PP file to produce an XSL file that in its structure will have code that looks for the custom layout elements defined in the page profiles. For example, part of the XSL output will be code that produces a navigation area. To do this, the XSL, when processing the page profile in the next step, will be looking for a tag pair called . Now, because our XSL has already been created by transforming the CC/PP file, in the case of a device that can support background images, this facility will be reflected in the output XSL. If, in the page profile, there is any tag element that specifies a background image, then this will be picked up by the XSL and propagated through to the output.

2. The XSL that was produced from the first step is now applied to the page profile to transform it to an XSL file that can be finally applied to the content. The XSL at this stage now contains the code necessary to produce the final output, such as the XHTML tags in the case of an XHTML-capable device. These tags are now suitably formatted in terms of say any inline style information required for the target device. All that's missing is the content, which comes via the next step.

3. We can think of the final XSL file as the final output file, such as an XHTML file, properly crafted for the target device, but just missing the content. This is now inserted by transforming the content XML into the final output, which really amounts to a merging of the content with the markup contained in the XSL template.

The above process sounds a bit convoluted. However, the reason for using XSL in the first place is that it is a powerful language for processing XML, unlike Java or any other language which is general purpose. XSL has been specifically designed for handling XML and has special constructs and subcomponents to assist with the process. We have not gone into these, but for completeness, XSL is actually made up of several components, including XSLT, which is the transformation language itself, and XPath, which is another language specifically for referring to elements within an XML file, recognising the tree-like structure of XML files (see Figure 8.15). The point is that the CC/PP input driving our device capabilities is an XML file. The content is increasingly going to be in XML. Therefore, in an XML-rich process, or pipeline, the ability to handle the XML using an optimised language built for the job makes sense.

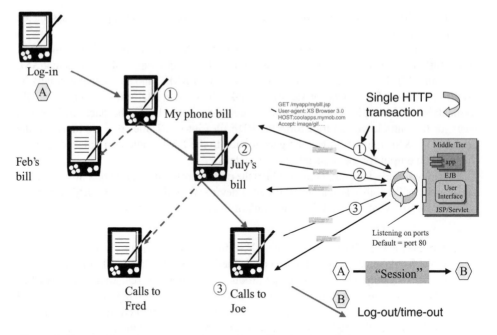

Figure 8.41 Tracking where we are in a web application is a challenge.

8.6 MANAGING SESSIONS

We have been looking at JSP and servlets for handling and generating HTTP streams. In the previous section, we identified a design pattern called MVC that enables a unified approach to browser applications, which allows us to handle requests, call business logic, and then hand off to the appropriate JSP to generate the output stream (the user interface).

What we omitted from the previous discussion was any detail about how to support state in our application, which is known as *session management*. This problem is easiest to identify by example.

Figure 8.41 illustrates the problem well. Here we have a sequence of page requests within our example billing application. Each page is requested by the user, initially by entering the application's homepage and then drilling down (and back up) the application by successive link selection. As we have been discussing, the billing application is most likely handled by one servlet or JSP on the server (ignore the issues relating to MVC patterns for now). For the sake of defining a *session*, we herein identify that a session is the traversal in time from the start of the application through to the end, either by logging out (or leaving) or timing out. This is shown as the journey from A to B in the figure.

It may seem an odd reflection, but how do we know where we are in the application? Or, more accurately, how does the servlet know whereabouts we are in the application in order to invoke the correct business logic and redirection to the appropriate JSP? The next diagram (Figure 8.42) perhaps illustrates the point better. Keep in mind that to invoke the servlet, our user agent has to issue a GET method on its URI, which we show here to be http://cool.mymob.com/myapp/myBill. This ultimately points to the myBill servlet that will get run by the J2EE web engine when it receives this request. However, within the application, we have several possibilities that the user can enjoy, such as showing billing

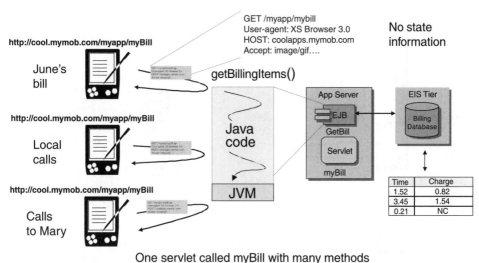

Figure 8.42 Session management.

items according to the month we made the calls (i.e. June, as in the figure), or the type of calls (i.e. local calls, as in the figure) and even to whom we made them (i.e. to Mary, as in the figure).

Each time, we end up calling the same servlet. For argument's sake, the servlet is also calling the same EJB ('GetBill'), actually accessing the method 'getBillingItems()'. Here the blank parentheses is the problem, as presumably we need to tell the EJB which billing items we are interested in, so that it can construct the appropriate query to run against the billing database using JDBC.

Before we elaborate on how to get these parameters from the user, let's look at the problem in a slightly different way, as shown in Figure 8.43.

In this example, we are showing more or less the same problem, except we are trying to access billing items for the month of June, but for lots of different users (at the same time of course). We are not bothered about how many users, as we know that our J2EE platform can handle it, but how do we know for each servlet request, who is who? Unless we know who is accessing the servlet, we will not know which billing records to query in the billing database. This is a big problem.

The point being made is that on each occasion, whether it's between users or between different views of the same billing records, the user agent issues the same GET request to the same URI (servlet). Apparently there appears to be no state information telling us either where we are in the application, or who is using it. Clearly, we need a mechanism for managing sessions on a per user basis, otherwise our application will not work.

This session management problem was not seen as an issue with the early origins of the Web. As we discussed earlier in the book, the aim of the browser paradigm was to publish documents and allow embedded hyperlinking to other documents. Within this paradigm, there is no apparent need for session management. A scientific paper on particle physics can be served up no matter who asks for it or what they did before, or what they want to do next. In other words, the need to maintain state was not required and not apparent; hence HTTP

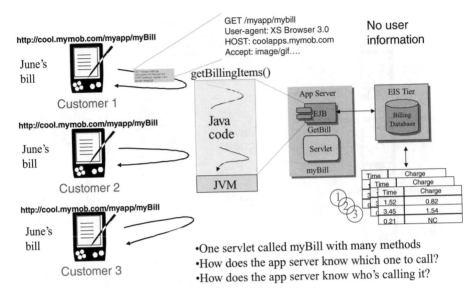

Figure 8.43 Session management between users.

remained a stateless protocol until the kinds of problems we have just been discussing eventually became apparent, especially with the dynamic (programmatic) generation of output, as opposed to static documents sitting on a web server hard disk.

8.6.1 Cookies to the Rescue

Fortunately, those clever people at Netscape, many moons ago, realised that session management was required – so they solved it. Their idea has subsequently become a public standard[29] [RFC 2109].

The solution is elegant and is important to understand if we want to go deeply into mobile application design using the browser paradigm, or possibly any mobile application that uses HTTP, whether it is browser based or not (as we can use HTTP from non-browser applications, as noted on several occasions in this book).

The cookie is simply a value that gets sent back and forth between the origin server and the user agent on a per-user, per-session, per-application (or path/URI) basis. In other words, we add a label to the incoming requests so that we can uniquely identify them.

The cookie process is shown clearly in Figure 8.44, which shows a response after the user agent has made a request from the server (GET method). Let's assume that this is the first request in a session, and we are using our billing example again. The server, noticing that this is the first request, issues a cookie to the user agent. This is a unique alphanumeric value added to the HTTP response as a header field called *Set-Cookie*.

As the figure shows, cookies have three parts. The first is the cookie name (e.g. 'JDSESSION'); the next is the value of the cookie, which can be any string of characters. Then there is the scope of the cookie, or the path on the origin server that this cookie

---

[29] http://www.ietf.org/rfc/rfc2109.txt

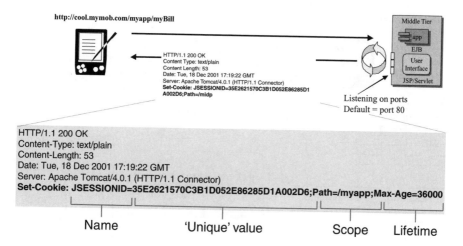

Figure 8.44 Adding cookies to the HTTP dialogue.

applies to. Finally, there is the age of the cookie, which is the length of time (in seconds) that it is valid for.

What the user agent is required to do is to store the cookie and its parameters in a cookie table (memory). As long as the cookie is still valid (Max-Age has not been reached), then the user agent must append the cookie to any requests for resources on the path stipulated in the cookie.

By passing a cookie back and forth, the server has a means of uniquely identifying a session, provided that a unique value was used for each unique user (session). On a J2EE server, the web engine creates a session object that is referenced using the same value as the cookie, as shown in Figure 8.45. This is done transparently on behalf of the JSPs and servlets being accessed by users. For example, if we take the billing application again, for each unique user that accesses the application (which we assume is only accessible via the path '/myapp') the web engine sets a cookie for each user and creates an associated session object. Whenever a JSP or servlet is subsequently accessed by a user, the relevant session object is made available to the JSP or servlet that is executing on the web engine. This is done without requiring the JSP or servlet programmer to track the cookie (and its reference) in software. A J2EE API is available, so that the programmer can set or get values from the session object.

Anything can be stored within the session object, as long as it's a valid Java programming entity; even objects can be stored. This is how we can keep track of a user's state. As shown in Figure 8.45, we could use the session object to store user information, such as a username, and for our billing example we can store state information about the current filtered view of the bill itself, such as 'month=July'. There really is no limitation on the way we structure information in the session object.

Supporting Sessions and Parameters via the URI   What remains now is to figure out how the user agent indicates to the server exactly what information it requires from a given servlet or JSP. For example, we know that the servlet 'MyBill' will produce a bill listing, but its URL is:

http://cool.mymob.com/myapp/mybill

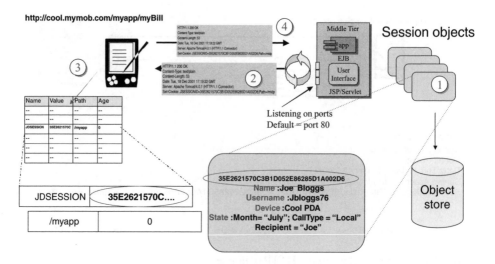

Figure 8.45 Session persistence using cookies.

How do we tell the server that what we actually want is the listing for July? There are two main ways of sending this parameter to the servlet, both of which apply equally to WSP (WAP 1) and W-HTTP (WAP 2):

1. By embedding the parameters within the URL itself by appending parameters to the servlet or JSP address (URI) as a query, like:

   http://cool.mymob.com/myapp/mybill?parameter1=this&parameter=that

2. By sending the parameters as fields in a form, so they get passed in the body as part of a POST request.

Let's consider the two methods and their implications for mobile application design and the foregoing discussion on cookies.

The first method is very straightforward and widely used. If, for example, we want to tell our MyBill servlet to get us the billing information for July, then we could construct a parameter called month and pass this as part of the URL:

   http://cool.mymob.com/myapp/mybill?month=july

This URI causes the user agent to send a GET request to the primary resource, which is located at the address to the left of the question mark. Everything to the right of the question mark is ignored as far as the GET request itself is concerned, except that the JSP and servlet are able to access this part of the string. As we might expect, the J2EE API for servlets and JSPs enables these parameters to be accessible in to the code on the page and can be used thereafter however the programmer chooses. For example, we can pull of the parameter called 'month' from the request string and ask the J2EE API for its value, which in our example would return 'july'. We can then take this parameter to run the appropriate query on the billing database.

```
http://cool.mymob.com/myapp/mybill?SESSION=ab34sje93nslk3&month=july&calls=local&filter=long
```
(a) Full URI with query

```
http://cool.mymob.com/myapp/mybill?SESSION=ab34sje93nslk3&month=july&calls=local&filter=long
```
(b) Query truncated by limited URI memory on device

```
http://cool.mymob.com/myapp/mybill?SESSION=ab34sj&month=july&calls=local&filter=long
```
(c) Full URI with query restored by shortening the session ID

```
http://cool.mymob.com/myapp/mybill?SESSION=ab34sje93nslk3j&m=7&cl=loc&f=lg
```
(d) Full URI with query restored, with full session ID, by abbreviating parameters

Figure 8.46 URI encoding truncation problems on mobile devices.

Using URI strings to encode request parameters is also useful for sending session values instead of using cookies. For example, we can just as well have the following URI:

```
http://cool.mymob.com/myapp/mybill?SESSION=ab34sje93nslk3&month=july
```

This has the same effect as using cookies, except that the value is not persisted in a cookie memory and via HTTP headers, but is persisted via the links in the pages returned by the server. If each link in the page has the session ID encoded into it, then whenever the user selects a link, the server will automatically get passed the session ID. Identifying a session is a large part of the session management problem and we now have an alternative method to using cookies.

Using the URI to encode session information, rather than cookies, has some important implications. Firstly, a user might have cookie support disabled. This might be because they perceive that cookies are evidence of having visited a particular site, which the user may not want to have used against them at a later date[30]. Secondly, cookies simply might not be supported, or only in a limited fashion. This is particularly true of mobile devices. Some devices do not support cookies at all, whilst others support only a few cookies. It is therefore desirable to either avoid their usage altogether, or to initiate a web application with URI session encoding and then move to cookies if a cookie gets returned from the browser.

There are some further problems with this approach on mobile devices. In particular, many mobile devices have very limited memory capacity and so very stringent memory-saving techniques tend to get deployed. One of these seen on some devices is the limited amount of memory available to process a URI. If the URI gets too long, it can end up getting truncated by the user agent, as shown in Figure 8.46. This can have catastrophic implications, as seen in (b) perhaps causing the application to stop functioning correctly. We can well imagine that if some parameters get chopped from the request, like 'filter=long' in the figure (b), then the application will not be able to determine correctly what the user is trying to do (i.e. look at long duration calls).

---

[30] Cookies can be used for more than just session management; they can be used to store anything. This can have surreptitious applications, but this is beyond the scope of this book and the user is referred to somewhere like Cookie Central (http://www.cookiecentral.com/).

(b) Output stream has the "rewritten URL"

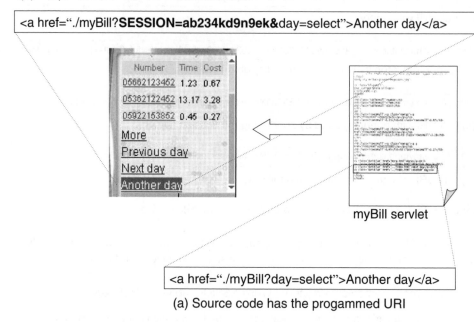

<a href="./myBill?**SESSION=ab234kd9n9ek**&day=select">Another day</a>

myBill servlet

<a href="./myBill?day=select">Another day</a>

(a) Source code has the progammed URI

Figure 8.47 URL rewriting by the J2EE server adds the session ID automatically.

There are several approaches to fixing this problem. Before looking at these, we should understand that the process of adding session IDs to the URI encoding is done automatically by the server itself, not by the programmer, as shown in Figure 8.47. The J2EE web engine simply rewrites URLs in the output stream to include the required session information.

To mitigate URI truncation issues by mobile devices, the first thing we can try is to shorten the session ID, which is typically a very long string on most J2EE servers. This may recover the situation some of the time, as shown in Figure 8.46(c). However, most J2EE servers generate the session ID automatically, and its length is not configurable. A notable exception is WebLogic Server, where 'WAP support' is a configurable option for web applications. If selected, this causes the session ID to be truncated[31], which is probably not a problem most of the time (unless it becomes ridiculously short of course).

Shortening the session ID may help us some of the time, but often we find that the rest of the URL dominates the string length anyway, so we may still have problems. One solution is to abbreviate the parameters, as shown in Figure 8.46(d). Abbreviating parameters can make the URL much shorter. In fact, with a set of very sparse abbreviations, combined with a degree of encoding, we can make the parameters very short indeed. However, the problem with this approach is that it can make matters very difficult when it comes to programming, testing and debugging, because the URI strings can become very obscure and make it

---

[31] In fact, the shortening of the session ID by WebLogic Server is via the removal of information that is needed to make sessions IDs work across a redundant cluster of servers, so it would seem that using this option with WAP devices means that seamless fail-over from a primary to a secondary server is not supported within a mobile browsing session.

difficult for the programmer to understand what's going on. Good for obfuscating, but bad for maintenance!

An alternative to using the URI to encode values is to embed them in the page requests by appending them to the request body. This entails using the POST method, so that form values can be submitted to the server. This can be done either by explicitly displaying a form, or by hiding it and simply using it to convey parameters, whether or not the user knows it. With an explicit form, we are limited to using the form interface as the navigation technique. For example, in the case of our billing application, we could ask the user to select a month from a drop-down menu on a form and then submit the form. The selection from the drop-down menu would get sent to the server in the POST message. However, often we do not want to use a form for navigation as it can become too cumbersome for the user as a form presentation is limited to certain widgets, like drop-down menus and selection boxes, and these can sometimes be hard to include in a page whilst maintaining a useable layout.

8.7 MMS AND SMIL

MMS uses its own method of presentation, which can include the use of WML, but more commonly involves the use of an alternative markup language called *Synchronised Multimedia Integration Language*, which is abbreviated SMIL and pronounced as 'smile'. Discussion of this technology is deferred until Chapter 13 on location-based services wherein we discuss the use of MMS within the LBS context (although the explanation of MMS technology is general and not specific to LBS).

APPENDIX A – SAMPLE CC/PP RDF FILE

```xml
<?xml version="1.0"?>
<RDF xmlns ="http://www.w3.org/1999/02/22-rdf-
  syntax-ns#"
xmlns:rdf ="http://www.w3.org/1999/02/22-rdf-
  syntax-ns#"
xmlns:prf="http://www.wapforum.org/profiles/UAPROF/
  ccppschema-20010430#">
<rdf:Description ID="MyDeviceProfile">
<prf:component>
<rdf:Description ID="HardwarePlatform">
<rdf:type resource="http://www.wapforum.org/profiles/
  UAPROF/ccppschema-20010430#HardwarePlatform"/>
<prf:BluetoothProfile>
<rdf:Bag>
<rdf:li>headset</rdf:li>
<rdf:li>dialup</rdf:li>
<rdf:li>lanaccess</rdf:li>
</rdf:Bag>
</prf:BluetoothProfile>
<prf:ScreenSize>121x87</prf:ScreenSize>
```

```
<prf:Model>R999</prf:Model>
<prf:InputCharSet>
<rdf:Bag>
<rdf:li>ISO-8859-1</rdf:li>
<rdf:li>US-ASCII</rdf:li>
<rdf:li>UTF-8</rdf:li>
<rdf:li>ISO-10646-UCS-2</rdf:li>
</rdf:Bag>
</prf:InputCharSet>
<prf:ScreenSizeChar>15x6</prf:ScreenSizeChar>
<prf:BitsPerPixel>2</prf:BitsPerPixel>
<prf:ColorCapable>No</prf:ColorCapable>
<prf:TextInputCapable>Yes</prf:TextInputCapable>
<prf:ImageCapable>Yes</prf:ImageCapable>
<prf:Keyboard>PhoneKeypad</prf:Keyboard>
<prf:NumberOfSoftKeys>0</prf:NumberOfSoftKeys>
<prf:Vendor>myprofileprovider</prf:Vendor>
<prf:OutputCharSet>
<rdf:Bag>
<rdf:li>ISO-8859-1</rdf:li>
<rdf:li>US-ASCII</rdf:li>
<rdf:li>UTF-8</rdf:li>
<rdf:li>ISO-10646-UCS-2</rdf:li>
</rdf:Bag>
</prf:OutputCharSet>
<prf:SoundOutputCapable>Yes</prf:SoundOutputCapable>
<prf:StandardFontProportional>Yes</
 prf:StandardFontProportional>
</rdf:Description>
</prf:component>
<prf:component>
<rdf:DescriptionID="SoftwarePlatform">
<rdf:type resource="http://www.wapforum.org/profiles/
 UAPROF/ccppschema-20010430#SoftwarePlatform"/>
<prf:AcceptDownloadableSoftware>No</prf:
 AcceptDownloadableSoftware>
</rdf:Description>
</prf:component>
<prf:component>
<rdf:Description ID="NetworkCharacteristics">
<rdf:type resource="http://www.wapforum.org/profiles/
 UAPROF/ccppschema-20010430#NetworkCharacteristics"/>
<prf:SecuritySupport>
<rdf:Bag>
<rdf:li>WTLS-1</rdf:li>
<rdf:li>WTLS-2</rdf:li>
```

```
<rdf:li>WTLS-3</rdf:li>
<rdf:li>signText</rdf:li>
</rdf:Bag>
</prf:SecuritySupport>
<prf:SupportedBearers>
<rdf:Bag>
<rdf:li>TwoWaySMS</rdf:li>
<rdf:li>CSD</rdf:li>
<rdf:li>GPRS</rdf:li>
</rdf:Bag>
</prf:SupportedBearers>
<prf:SupportedBluetoothVersion>1.1</prf:
 SupportedBluetoothVersion>
</rdf:Description>
</prf:component>
<prf:component>
<rdf:Description ID="BrowserUA">
<rdf:type resource="http://www.wapforum.org/profiles/
 UAPROF/ccppschema-20010430#BrowserUA"/>
<prf:BrowserName>Ericsson</prf:BrowserName>
<prf:CcppAccept>
<rdf:Bag>
<rdf:li>application/vnd.wap.wmlc</rdf:li>
<rdf:li>application/vnd.wap.wbxml</rdf:li>
<rdf:li>application/vnd.wap.wmlscriptc</rdf:li>
<rdf:li>application/vnd.wap.multipart.mixed</rdf:li>
<rdf:li>application/vnd.wap.multipart.
 form-data</rdf:li>
<rdf:li>text/vnd.wap.wml</rdf:li>
<rdf:li>text/vnd.wap.wmlscript</rdf:li>
<rdf:li>text/x-vCard</rdf:li>
<rdf:li>text/x-vCalendar</rdf:li>
<rdf:li>text/x-vMel</rdf:li>
<rdf:li>text/x-eMelody</rdf:li>
<rdf:li>image/vnd.wap.wbmp</rdf:li>
<rdf:li>image/gif</rdf:li>
</rdf:Bag>
</prf:CcppAccept>
<prf:CcppAccept-Charset>
<rdf:Bag>
<rdf:li>US-ASCII</rdf:li>
<rdf:li>ISO-8859-1</rdf:li>
<rdf:li>UTF-8</rdf:li>
<rdf:li>ISO-10646-UCS-2</rdf:li>
</rdf:Bag>
</prf:CcppAccept-Charset>
```

```
<prf:CcppAccept-Encoding>
<rdf:Bag>
<rdf:li>base64</rdf:li>
</rdf:Bag>
</prf:CcppAccept-Encoding>
<prf:FramesCapable>No</prf:FramesCapable>
<prf:TablesCapable>Yes</prf:TablesCapable>
</rdf:Description>
</prf:component>
<prf:component>
<rdf:Description ID="WapCharacteristics">
<rdf:type resource="http://www.wapforum.org/profiles/
 UAPROF/ccppschema-20010430#WapCharacteristics"/>
<prf:WapDeviceClass>C</prf:WapDeviceClass>
<prf:WapVersion>2.0</prf:WapVersion>
<prf:WmlVersion>
<rdf:Bag>
<rdf:li>2.0</rdf:li>
</rdf:Bag>
</prf:WmlVersion>
<prf:WmlDeckSize>3000</prf:WmlDeckSize>
<prf:WmlScriptVersion>
<rdf:Bag>
<rdf:li>1.2.1</rdf:li>
</rdf:Bag>
</prf:WmlScriptVersion>
<prf:WmlScriptLibraries>
<rdf:Bag>
<rdf:li>Lang</rdf:li>
<rdf:li>Float</rdf:li>
<rdf:li>String</rdf:li>
<rdf:li>URL</rdf:li>
<rdf:li>WMLBrowser</rdf:li>
<rdf:li>Dialogs</rdf:li>
</rdf:Bag>
</prf:WmlScriptLibraries>
<prf:WtaiLibraries>
<rdf:Bag>
<rdf:li>WTA.Public.makeCall</rdf:li>
<rdf:li>WTA.Public.sendDTMF</rdf:li>
<rdf:li>WTA.Public.addPBEntry</rdf:li>
</rdf:Bag>
</prf:WtaiLibraries>
</rdf:Description>
</prf:component>
<prf:component>
```

```
<rdf:Description ID="PushCharacteristics">
<rdf:type resource="http://www.wapforum.org/profiles/
 UAPROF/ccppschema-20010430#PushCharacteristics"/>
<prf:Push-Accept>
<rdf:Bag>
<rdf:li>application/wml+xml</rdf:li>
<rdf:li>text/html</rdf:li>
</rdf:Bag>
</prf:Push-Accept>
<prf:Push-Accept-Encoding>
<rdf:Bag>
<rdf:li>base64</rdf:li>
<rdf:li>quoted-printable</rdf:li>
</rdf:Bag>
</prf:Push-Accept-Encoding>
<prf:Push-Accept-AppID>
<rdf:Bag>
<rdf:li>x-wap-application:wml.ua</rdf:li>
<rdf:li>*</rdf:li>
</rdf:Bag>
</prf:Push-Accept-AppID>
<prf:Push-MsgSize>1400</prf:Push-MsgSize>
</rdf:Description>
</prf:component>
</rdf:Description>
</RDF>
```

9

Using J2EE for Mobile Services

In the previous chapters, we introduced J2EE as a platform for building mobile applications and then went on to establish the dominant communications paradigms for establishing interaction between our mobile device and the J2EE platform; namely HTTP and its wireless companions in the WAP family. We now want to look at how J2EE works in more detail and to understand exactly how its powerful infrastructure software services support our proposed communications paradigms. We shall proceed to look at devices in some detail in the following two chapters.

In this chapter we shall also look at how we make sure that the platform can be accessed securely, both in terms of implementing security within the J2EE environment and then ensuring that we can extend security policies to the devices, and ultimately, the users. It would be no use implementing a fancy means of making sure that only authorised users access a particular mobile service on the server if we then couldn't enforce this via the device, so we need to examine security within the entire mobile network context, revisiting our mobile network model.

In building our platform, we shall look at how harnessing the power of J2EE to support mobile devices is not enough. We also need to be able to support access to our mobile services platform by other participants who want to offer their services to the end-users whilst taking advantage of any common infrastructure built upon the J2EE platform. We shall take a look at ways in which this might be achieved and how such an arrangement would fit within an integrated services model, able to cope with all the key challenges of providing services that we have elucidated so far throughout the book. By integrated, we mean that whether an application is hosted locally by the operator or remotely via a third-party, the software infrastructure services (e.g. scalability, security, availability, adaptation, etc.) still function in a consistent manner across all services.

Next Generation Wireless Applications, Second Edition   Paul Golding
© 2008 Paul Golding

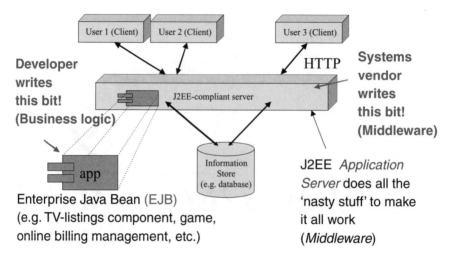

Figure 9.1 J2EE approach.

9.1 TECHNOLOGIES UNDERPINNING J2EE

As Figure 9.1 reminds us, J2EE is a platform that divides the gamut of software needed to implement a scalable robust solution into two parts. The first is infrastructure services that are likely to be commonly required for all applications. We call this infrastructure 'middleware' and it provides a host of powerful functions like distributed processing, transaction management, security management, multithreading and so on.

The other half of our software is the custom application itself, which we implement in units called Enterprise Java Beans (known as EJBs) where we construct the value-added business services that make the overall application do something useful. With J2EE, we have an approach that enables us to focus on 'our' bit whilst the application server does the rest (middleware). When introducing J2EE, we also learnt how the J2EE platform comes with a wide range of powerful system interfaces, such as the capability to connect with industrial strength database servers or mail servers. There is no need for us to implement any of these services. We utilise these services in our application by accessing the platform services via Application Programming Interfaces, or APIs. These interfaces are common to all J2EE implementations. Similarly, the low-level interfaces to enable the infrastructure services to function on behalf of our application are also common to all J2EE applications. This commonality is thanks to the J2EE specification being an open specification, accessible to anyone who would like to implement a J2EE platform, whether as a free-of-charge open-source initiative (e.g. Tomcat) or as part of an expensive enterprise solution portfolio (e.g. BEA's WebLogic Server and its companion products).

9.1.1 Containers – The J2EE 'Glue'

It is obvious that there needs to be some connecting mechanism between our EJB and the infrastructure services. If in actuality the EJB really only contains our application software and nothing else, then the infrastructure services will have a hard time enacting the middleware function because there is no relationship between it and the EJB. If we think about an example, then perhaps we can understand why this won't work.

Let's say our EJB provides a lookup function that provides a list of URIs where our mobile device can download ringtones. The bean will take an input value as a search parameter and then look for ringtones that match this topic, such as 'James Bond'[1]. Probably our ringtones are already organised as part of a Content Management System (CMS) for mobile content, which our bean is part of. Each ringtone has a description associated with it to identify its theme and title. More than likely these descriptions will be in descriptors that are structured as XML documents and probably stored in a database underlying the CMS. To access the database, we can use a J2EE API, which is a block of software already written to take care of the low-level interface tasks associated with connecting to a database server. This piece of software is called the Java Database Connector, or JDBC. We could imagine a command (method[2]) like 'gettones("James Bond")' that will return a record set from the database containing all the descriptors for ringtones that match the theme of 'James Bond'. We probably want to extract the exact titles of the matching tones in order to send them back to our user who made the request. We could use the J2EE API for processing XML (JAXP) to extract the content of the XML tags called <Title>. So far so good – all of the difficult code to process database requests and XML parsing is provided for us by the J2EE platform in the JDBC and JAXP programs.

Now, what if another user wants to do the same thing at the same time, but for a different theme, like 'Hip Hop'. Our bean is still running on the server to handle the first request. This is problem number one, because we want the bean to process the second request. Now let's make things a bit more challenging and imagine that actually we have 10 000 concurrent requests (this is a popular site for ringtones) and that the physical server (computer) we are using runs out of steam at 8000 requests. This is problem number two (scalability), after we have solved number one, of course (concurrency). We already know that J2EE has the capability to solve these problems – that's what it was designed for. But we are trying to understand conceptually how that process might work.

In our EJB, all we have written is the code to take the input topic, make the query and gather a collection of titles from the XML returned from the database. In other words, we have stuck to our 'business' and only written code to carry out the tasks we want it to do (hence 'business logic'). We know that computers run exceptionally fast and that via their operating systems they can appear to run many tasks simultaneously by swapping processes on and off of the central processor (e.g. Pentium). However, our EJB is not an operating system level task. It is a piece of software that runs within the J2EE runtime environment[3]. Creating EJBs and setting them running is the job of the J2EE application server. Furthermore, the J2EE platform can arrange for EJBs to be created[4] across a bank of servers, so that the processing load can be distributed. In some cases, this will involve the EJB having to talk to another EJB on a different server. This is a mechanism will shall look at in some detail in just a minute.

---

[1] 007 Secret Service Agent (character created by Ian Fleming).
[2] In Java, software commands, or functions (what used to be called subroutines) are called methods.
[3] We discuss what a Java Runtime Environment (JRE) is in more depth when we look at Java running on devices later in Chapter 10.
[4] An EJB is just a program. Creating an EJB means to bring the program to life, so that it is up and running on the system, ready to do its stuff, not just sitting on a disk.

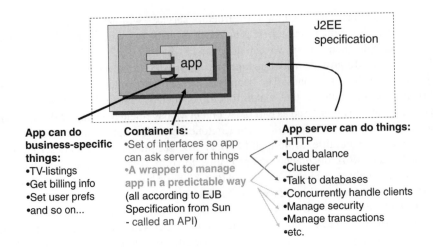

App can do business-specific things:
- TV-listings
- Get billing info
- Set user prefs
- and so on...

Container is:
- Set of interfaces so app can ask server for things
- A wrapper to manage app in a predictable way (all according to EJB Specification from Sun - called an API)

App server can do things:
- HTTP
- Load balance
- Cluster
- Talk to databases
- Concurrently handle clients
- Manage security
- Manage transactions
- etc.

Figure 9.2 EJB sits inside a wrapper.

Because we have not written any code to help create our EJB, or to enable it to talk to other EJBs on remote servers, we need to make sure that this code is available. Some of the code is system code that is part of the J2EE platform, but some of it actually needs to be bolted on to our program like a wrapper, so that the EJB is able to interact with the platform in a predictable manner and carry out these system tasks on our behalf, usually in concert with the application server's own processes.

The *wrapper* function is shown in Figure 9.2. We can see our EJB sitting in the middle and it contains, as explicitly programmed by the programmer, the business logic; the actual code we want to run to carry out the task required of this EJB, which in our example provides the ringtone-lookup function. The EJB sits inside a wrapper, which is housekeeping code that runs with our business logic as part of the EJB software process running within the J2EE platform. The wrapper provides the critical link between the EJB and the application server. At one level, it enables the EJB to access various interfaces, but another critical function is to enable the low-level infrastructure mechanics to operate, such as concurrency, program-lifecycle management (e.g. creation, suspension, destruction), transaction monitoring and distributed processing (inter-EJB communication across a network).

The wrapper function, whilst representative of how EJBs work, is not a function that is rigidly defined in the J2EE specification in terms of its physical representation on the server. Much of its implementation detail is left to the platform vendor to decide on how best to put into action. However, we shouldn't forget that the wrapper function is enabling a component model to be achieved whereby we can program to a certain specification and then rely upon being able to plug our EJB into a J2EE platform and expect it to work. In other words, the wrapper, whilst proprietary in implementation detail, is still used to realise an open interface so that at the black-box level it is consistent with what a J2EE server expects.

In the next section, we look at one of the critical functions of the wrapper code and a key underpinning technology for distributed processing within a J2EE environment (i.e. many J2EE application servers running in concert).

9.1.2 RMI – The EJB 'Glue'

One of the key design goals of the J2EE platform was to provide robust support for distributed processing. This is the ability to run an application across many servers. There are potentially several reasons why this is useful, perhaps foremost of which is the ability to achieve a scalable architecture. This is essential if we want to build a mobile services platform that can handle as many users as we want using a simple and consistent programming approach. We do not want a system that can only work effectively on one server, perhaps handling thousands of users, only to find that we have to throw it away and implement a new system to handle tens or hundreds of thousands of users across several servers. We would like to be able to take a single design approach to all these cases, an approach that simply scales when we need it to. This is a function that we would like the infrastructure services to support, not the EJB programmer.

Fortunately, this ambition has been fulfilled in the J2EE platform architecture. One of the key underpinning mechanisms for enabling this is something called *Remote Method Invocation*, or RMI. The fundamental building block for Java programming is something called a *method*. This is a unit of Java code that does a specific task. Associated tasks, common to one functional objective within our software, may be collected together into a collection of methods encapsulated into a *class*. For example, we could have a class called 'MyBill' that represents a user's mobile phone bill. It is common for classes to model real-world artefacts in this manner.

Within the class, we might have methods like 'UpgradeAccount', 'ChangeMyDetails', 'FetchCalls' and so on; all performing functions that are operations we are likely to want to carry out on our bill. These may not represent one-for-one correlations with tasks the end-user has in mind, like 'Pay my bill please', but will more likely represent tasks that the programmer perceives will enable appropriate end-user services to be realised, like paying the bill. We could argue that in a well-designed system, a lot of these methods and end-user tasks might well correlate directly, but that is not our concern here. A class is like a template – it is a definition of software behaviour achieved by specifying methods and then stating how they actually behave, which is done using the Java code that resides within the methods themselves. When a class is loaded, initialised and executed, it becomes an object. There could be many objects alive at any one time from one class. We could imagine lots of users accessing the bills and for each user we have an object dedicated to the task.

An EJB is a particular instance of a class. It is a special type of class that implies that along with the user-defined class comes all the wrapper code to enable J2EE infrastructure services to be accessed. The methods in a bean can be called from another bean, or, as we discussed in Chapter 8, from servlets and JSPs. The RMI mechanism is a means to do this between servers, as shown in Figure 9.3, which shows a servlet (or JSP, or another EJB) on one physical server (Server 1) apparently calling the method from a bean on the same physical server. This co-located software collaboration is the normal process in Java. However, what is actually happening is that somehow the server is conspiring with another server where the target EJB ('MyBill') actually resides (Server 2). As far as the servlet is concerned, the bean is on the same machine. The servlet programmer does not have to implement any software to go find the EJB elsewhere on the network and then figure out how to talk to it over the network: this is thanks to the RMI, an underpinning mechanism used by J2EE.

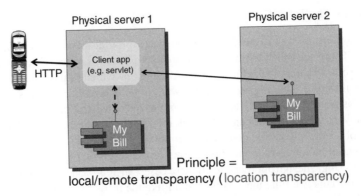

Figure 9.3 Remote Method Invocation (RMI). (Reproduced by permission of IXI Mobile.)

The mechanism of RMI enables what we call *location transparency* to be achieved. This is where the location of the software object being reference does not actually matter, nor is it known in advance by the referring object. Location transparency is built into the J2EE platform by default. In fact, this technique is not new and is not unique to the J2EE platform. An earlier attempt to create an open standard for managing objects in this way had already been undertaken and is called *CORBA*, which stands for Common Object Request Broker Architecture. Some vendors had already implemented systems using this standard, although J2EE has in many ways supplanted CORBA due to wider acceptance in the software and IT industry. To achieve compatibility with CORBA, especially under pressure from some vendors who participated in the J2EE specification process, the RMI mechanism is compatible with parts of CORBA. The full name for the mechanism is actually RMI-IIOP, where IIOP stands for Internet InterORB Protocol, which is an IP-based protocol to talk to calls methods on any system that implements CORBA.

9.1.3 Stubs and Skeletons – The Inner Workings of RMI

Having introduced such strange sounding names, like InterORB, the inner workings of RMI introduces us to yet more interesting names, this time called *stubs* and *skeletons*. Whilst this sounds like a horror movie, it is actually something quite elegant and magical from the wonderful world of J2EE.

As Figure 9.4 shows, the reality of location transparency is that our software receives help from a pair of software mechanisms that act as interlocutors for our two Java programs that wish to collaborate in a distributed manner. The idea is rather simple and a determined programmer could probably implement a rudimentary system from scratch[5], although this is not necessary thanks to J2EE.

Let's imagine again that we have a servlet that wishes to call an EJB, but unknown to our servlet (and the programmer) the target EJB resides on a machine somewhere else on the network, or possibly on lots of machines on the network, but we shall focus on just one. For our servlet to think it is communicating with the EJB locally, as per our original claim for location transparency, clearly there has to be an entity on the same machine as the servlet,

---

[5] For an illuminating example of how this might be done, see Monson-Haefel, R., *Enterprise JavaBeans*. O'Reilly Media, Inc (2001).

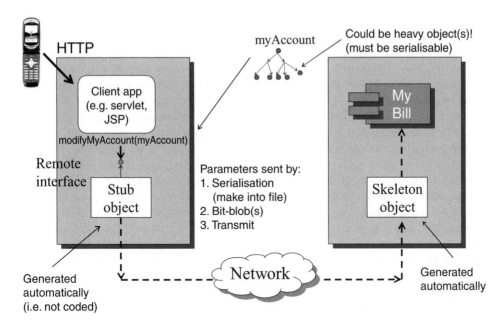

Figure 9.4 Stubs and skeletons collude. (Reproduced by permission of IXI Mobile.)

which is 'standing in' for the EJB. This is what the *stub object* does. It pretends to be the EJB 'MyBill' as far as the servlet is concerned, so that the servlet is able to pass its requests for accessing methods as usual (the servlet doesn't really know that anything different is happening than usual).

The stub object rather cleverly 'mimics' the interface of the EJB and can happily accept the appropriate method calls on its behalf. However, the stub is just a shell – it doesn't actually have the inner workings of any of these methods. If it did, then the whole exercise would be pointless of course, because effectively it then becomes the EJB anyway.

The stub object knows where to find a real 'MyBill' EJB on the network and uses a network protocol to route the method request from the servlet over to the EJB on the target machine. For now, we shall ignore the issue of how the stub 'knows' the methods of the EJB in order to successfully mimic them, and we also ignore how it knew where to find an actual implementation of the EJB in order to route the requests to the appropriate remote server.

In keeping with the J2EE ethos, it is not a requirement that the EJB explicitly program any special communications handlers to process the requests from a stub object. Remember that we don't require our EJB programmer to do anything special in order to benefit from the infrastructure services of the J2EE platform. The EJB has the user-defined methods implemented by its programmer and these can be called in the usual way via a co-located piece of software, which is a mechanism that is inherent in the way the Java programming language works. To complete the process that our stub object initiated, we have a mirror-like object on the target server called a *skeleton object*, courtesy of the J2EE server (i.e. not programmed by the EJB programmer). This accepts from the network the incoming method requests from the stub and then passes those method calls from the servlet by proxy, passing them on to the actual EJB. At this point, it is as if the stub and skeleton do not exist and the servlet has a direct connection to the EJB.

Once the EJB has received a networked request to run one of its methods, it takes its turn in execution on the server's processor and carries out its inner workings for the appropriate method. The results are then ready to be returned to the caller and this is done using a reverse process, passing them back to the skeleton who contacts the stub who passes them up to the servlet, which remains none the wiser about where and how the job got done. This is the magic of RMI on the J2EE platform!

There are some characteristics of the RMI process that are useful for the EJB programmer to understand as a user of the technology. One of these relates to the passing of any parameters that are needed by the EJB to carry out the required method. Because this process is happening by proxy, we need to send the parameters over the wire from the stub to the skeleton. If these parameters are large in size, then this can become a performance bottleneck, particularly if there is a high volume of RMI requests between lots of colluding objects; remembering that the mobile context of this process probably results in a large numbers of clients generating plenty of requests. In some cases, parameters can be software objects in themselves, which can reference yet further objects. Moving objects around poses several challenges. The technique employed is a process called *serialisation*, which involves converting an object to a self-contained flat file representation that will pass over a communications pathway. Not all objects will serialise, but for those that do, we have to keep in mind that choices made in the design in our software solution may have unforeseen consequences on performance when running on a distributed platform.

An interesting observation about the stub–skeleton collusion is that it is an entirely automatic process. The code to produce the stubs and skeletons is generated by the J2EE platform. The actual details of their construction and the low-level integration of the infrastructure services with our own user-defined (EJB) code is a matter of interpretation for the J2EE vendors. This is an issue beyond the scope of this book. For now, we should be happy that we have a means to build scalable mobile service delivery platforms based on a reliable and robust software infrastructure, like J2EE.

Before we move on from this topic to look at other facets of the J2EE platform in relation to mobile services platforms, let's briefly address some of the issues that we glossed over in our cursory glance at RMI. Firstly, when the stub object gets involved in the process, it already seemed to know something about the EJB it was masquerading as, in order to mimic its interface. This would be exceptionally clever if that were possible without some prior reference to it (real magic). In any particular implementation of a distributed system using J2EE, the collaborating servers are not acting aloof of each other. When a J2EE application is installed, it is effectively installed 'onto the platform', which means all the collaborating machines at once, not necessarily a particular machine. Installation processes are complex, but in effect we can think of each machine as receiving at least the templates[6] for all the EJBs, even if they don't intend to run them – this is at least required for our servlets and JSPs to gain meaningful references into the EJBs.

Another challenge is the invocation of an EJB. Just as with a desktop application, such as a word processor, even though it is installed on a particular PC, that doesn't meant to say it is running and ready to do work – we have to start it. The same applies to EJBs.

---

[6] In Java parlance, these templates are called *interfaces*, but that is not really important to grasp to understand the general principles.

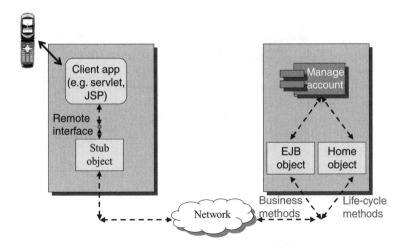

Figure 9.5 Skeleton a little more complicated than we thought. (Reproduced by permission
of IXI Mobile.)

Furthermore, if, as we have portrayed matters, we need a skeleton hanging around for each
EJB that we are running, we also need to start these; not forgetting that in a large multiuser
system, we are talking about large numbers of EJBs and their interlocutors running at the
same time[7]. However, we can't sensibly have a system where all of our EJBs have been
started, along with their proxy skeletons, for as many users as we are likely to encounter.
This would take up resource irrespective of the actual demand and dynamics of the working
system, which doesn't make sense.

To get around these problems, we have *life-cycle methods* for an EJB that enable us to
start them up (create) and shut them down (suspend or destroy). A special object exists to
carry out the life-cycle processes and this is called a *home object*, as shown in Figure 9.5.
Each EJB has at least one home object on the server where the EJB is going to run. We
can assume that the home object is always running and available to receive requests, unlike
the skeleton object or its associated EJB. All we have to do is first of all locate the home
object and then ask it to create an EJB for us. We can think of the home object as having
a locatable address on the network where our servlet can easily find it. This is the only
time that the servlet programmer has to be aware of what is actually going on in the
J2EE distributed environment. A small piece of code needs to be included at the start of
the servlet code that then gets called to find the home object by its name, something like
'findHomeObject("MyBill")'. This is a slightly simplified explanation, but fairly accurately
represents the process.

Ignoring the details of the software process that's executed to find the home object, once
the servlet has found it, requests can be made for the home object to create an EJB and its
skeleton. The home object then tells the servlet the address of the skeleton so that its stub
knows where to look for it. What actually gets created on the target server is not officially
called a skeleton (even though it has that function), but is called the *EJB object*. The reason it

---

[7] A technique called *instance pooling* can be used to avoid having one EJB and skeleton for each user, but that is
an advanced topic.

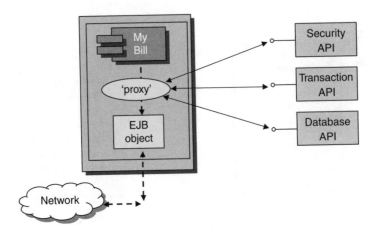

Figure 9.6 EJB object as a proxy.

gets called this it that it actually is the object we end up talking to once things are underway. We never ever get to talk to the actual EJB directly[8].

The fact that we do not talk to the EJB object directly actually offers the opportunity to implement a neat mechanism by which the J2EE server can now do all of the 'nasty middleware stuff' that we so desperately didn't want to program ourselves in our software, leaving it to the J2EE platform.

As shown in Figure 9.6, because the EJB object intercedes between the servlet (or any calling program in general) and the EJB being called, this gives the EJB object the chance to carry out infrastructure services for us, such as checking the security settings of the bean (i.e. are we allowed to use it), restoring the state of the bean from a previous use (i.e. persisting state between uses) and, if necessary, keeping track of all our activities as part of a transaction sequence that must be monitored. The alternative is that we could have programmed all this ourselves in the EJB, but that is going away from the ethos of J2EE. Moreover, it is not just a case of programming these proxy functions, but we still need all the infrastructure services to go with it, like the security management system, transaction monitor and so on. These are all built in to the J2EE platform.

In Figure 9.7, we can see the proxy process in a bit more detail. In Step 1, our servlet makes the method call to our EJB, which it does by referencing the local stub object. This reference was achieved during the 'start-up' process when the servlet first located the home object and invoked the EJB and its EJB object. Part of that process involves the J2EE platform passing a reference to the servlet that it can use as if the referred object were the EJB, whereas it is actually the local stub, which we are not showing in the diagram for the sake of simplicity. The EJB object, acting as our skeleton and proxy, receives the method call. It then accesses the infrastructure services for the EJB on our behalf. This could involve checking security permissions on the bean, monitoring transactions and many other

---

[8] To a certain extent, this is all semantics. The EJB object and the EJB may be hard to distinguish in reality depending on how the J2EE server has been implemented, but the abstraction is a useful way to specify J2EE and to understand it. Implementation of any technical mechanism can always deviate from the model by which we designed it, usually by virtue of optimisations that make sense when building it.

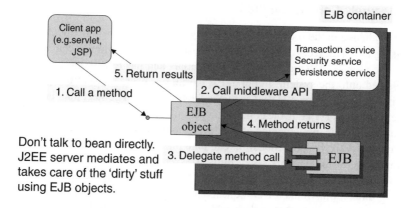

Figure 9.7 Proxy function of an EJB object as part of the process flow.

housekeeping functions. The key point is that the EJB object is doing this for us and we don't have to any programming. After infrastructure services have been satisfactorily completed, only then does the method call get passed to the EJB itself, as in Step 3. The method, if successful, returns the results (Step 4) and these get marshalled back to the servlet (Step 5) using the reverse process to Step 1.

9.2 MANAGING SECURITY

We have looked at some of the fundamentals of how the J2EE platform functions. In Chapter 8 we looked at the mechanisms for connecting mobile devices using HTTP, using JSPs or servlets to handle the inbound HTTP streams and to generate the outbound ones.

Returning to our goal of using J2EE technology to build a mobile services platform, we need to consider some of the implications of handling lots of users and lots of applications within a commercial context. We can envisage that different users will have different applications requirements and will therefore only subscribe to the ones of interest. It soon becomes apparent that we need a method for securing our applications. Some of our requirements might include:

- Being able to authenticate a user onto the services platform (i.e. that the user is a valid subscriber to the platform services).

- Being able to control subscriber access to particular applications (i.e. that a subscriber can access the applications they have paid to access and no others).

- Being able to protect one user's data from being illegally accessed by another user.

- Being able to centrally and easily manage security mechanisms in a maintainable and scalable fashion.

We shall now discuss how to address these issues in the sections that follow.

9.2.1 Securely Connecting the User

The first challenge in the security mechanism is to find a way to authenticate the user who is accessing the platform. All authentication systems follow a similar challenge–response pattern. The user is *challenged* for their credentials and has to *respond* correctly in order to be allowed onto the system. In our J2EE world, our users connect using HTTP, so the mechanism has to work over HTTP as this is the first point of contact that the user has with the system. Also, as HTTP is the common access method for all applications, it means that any security methods implemented at this level can be used across all applications.

As Figure 9.8 shows, authentication can take place at several levels in our system. We can authenticate at the network level, quite irrespective of who the user is and what the application is. This is typically done at the IP level by allowing certain IP addresses and preventing others, plus related techniques. This is something outside of the scope of the J2EE platform and more related to physical networking and how the LAN is set up, including the TCP/IP networking on the physical servers.

The next level of authentication is at the server level, which can still be done using IP address blocking, but at a different service access point on the machine. The way to think of this is that in the absence of network level authentication, the server itself still has the ability to prevent HTTP packets from reaching the web engine on the server. Of course, both of these levels can be used conjointly. A general level of filtering out of disallowed IP addresses can be done at the network level and then something more fine grain by the server. A good practice is to use both these methods.

Then we reach another layer of authentication at the application level, where individual applications can be assigned authentication access controls. The resolution of access control can be anything from restricting access to an entire domain (e.g. www.mymobileapps.com right down to an individual JSP or servlet. Irrespective of the access-control resolution, the best approach to achieve the application control is by adding authentication mechanisms to HTTP, as we shall discuss.

If we are going to use HTTP to authenticate our users, we need to achieve two things. The first is implementing a method to add an authentication mechanism to HTTP. Secondly,

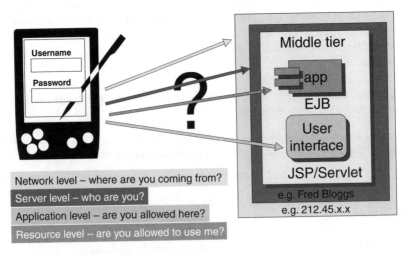

Figure 9.8 Different points to secure a service.

we need a method to connect this layer of authentication to the security mechanisms within the J2EE platform itself. There is no point in enabling HTTP requests to be authenticated if we cannot subsequently pass user credentials to our JSP or servlet in order for it to run and fulfil the successfully authenticated request. We also need to also ensure that these mechanisms are compatible with WAP (e.g. WSP), so that WAP 1 and WAP 2 devices can securely access our services platform via HTTP, either directly (e.g. W-HTTP) or via a proxy (e.g. WSP). Although seldom used, resource level control is an even finer grain access control mechanism to control access to individual EJBs and other elemental programs. Typically, this would be used to ensure that any resources with administrative-type privileges can't be accessed without appropriate rights and permissions.

9.2.2 HTTP Authentication – Basic

The principle of authenticating an HTTP connection is quite straightforward. The first step in authenticating a user is to challenge them, which means that the server has to ask the user agent to identify itself, or, more likely, its user.

When a servlet (or JSP) is requested to be run in response to an inbound HTTP request, the J2EE platform first checks if the resource is protected or not. It is considered protected if particular permissions are assigned to the resource other than allowing *everyone* to access it. Allowing everyone to access a resource amounts to no protection at all and therefore no authentication is applied, which is the so-called *anonymous user* mode. Assigning permissions to a resource in the first place is done using the J2EE security subsystem and has nothing to do with HTTP. The link between the two is how the J2EE server web engine responds to a HTTP request for a protected resource.

If a resource is protected, the web engine must first return an HTTP response to the user agent without running the required resource. This makes sense as we are not yet ready to send the content anyway, not until we have authenticated the user. In the HTTP response, instead of the usual '200 OK' status code in the header followed by the output from the requested resource, the status code '401 Unauthorized'[9] is sent back to the user agent, which tells the browser application that it must first authenticate if it wants to access the requested resource. On a desktop browser, this would cause the familiar 'logon' window to pop up, as shown in the lower right-hand side of Figure 9.9. It is up to the browser on the mobile device to display a similar logon prompt on the device user interface. However, before doing this, it checks the HTTP response header for another field that indicates how the server would like the user agent to authenticate, as there is more than one available method on most web servers.

The main two methods supported by HTTP 1.1 are called *basic authentication* and *digest authentication*. Both of these methods of authentication should be supported by WAP devices, either using W-HTTP or WSP. In the latter case, it is the duty of the WAP proxy to forward the authentication information in the header.

The particular field used to identify the type of authentication is labelled thus:

- 'www-authenticate:basic' for basic authentication

- 'www-authenticate:digest' for digest authentication

---

[9] Note that the spelling of 'authorized' with a 'z' is the American English spelling that is faithful to the relevant technical specifications.

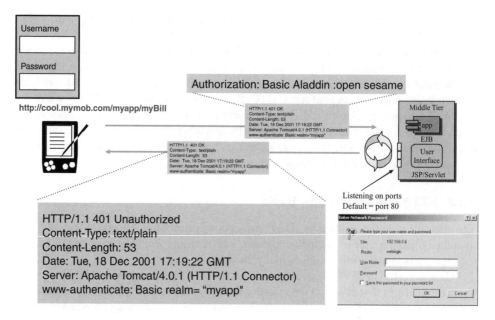

Figure 9.9 Authenticating a HTTP connection.

We shall look at both cases, beginning with basic in this section, which is shown in Figure 9.9 – we shall discuss digest more fully in Section 9.2.3. The response to a basic authentication challenge is very straightforward. After eliciting the username and password tokens from the user, the user agent writes out the same resource request (GET method) to its output HTTP stream, but includes an authentication header – 'Authorization:Basic *username:password*' – where the appropriate username and password information is inserted. For example, if the username is 'Daredevil' and the password is 'Maximus', then the header shall contain the string 'Authorization:Basic Daredevil:Maximus'. If the username is 'Aladdin' and the password is 'open sesame', then the string would read 'Authorization:Basic Aladdin:open sesame', as shown in Figure 9.9.

Figure 9.10 shows what happens at the J2EE server when we get a request with the 'authorization' header. In this example, we are considering a request to the resource with address http://cool.mymob.com/myapp/myBill, which is the address of a servlet on the server. HTTP is a stateless protocol, so the J2EE web engine doesn't know, or is not required to know, that this request is in response to an authentication challenge. It simply assumes this to be the case by detecting the 'authorization' header in the request. Because the servlet 'myBill', or possibly the entire application path – '/myapp' is protected (i.e. requires authentication), the web engine will check for the presence of the 'authorization' header. In our example, the web engine detects the appropriate header and so proceeds to extract the username and password token. However, notice that the token looks rather different from our 'Alladin:open sesame'. It has changed to something altogether more obscure. This is a result of the user agent encoding the token string using a technique called *Base64 encoding*. This is *not* an encryption technique; it is merely used for obfuscation of the string so that it is made harder for a human to read casually, were someone able to intercept it. This is not at all secure enough to defeat a determined and well-equipped interceptor, but we shall look at this weakness and its significance in a moment.

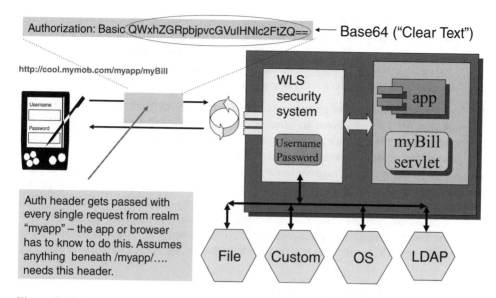

Figure 9.10 Processing a basic authentication request.

What happens to the token is generally the same on all J2EE implementations. The token is passed from the web server engine to the server's security subsystem and checked against a database of valid and authorised users. If the credentials match those on the server and the resource is tagged as being accessible by the authenticated user, then the web engine will run the servlet as normal and return its output stream in a standard HTTP response along with a '200 OK' status header. If the user does not check out, or they are authentic but don't have permission to access the 'myBill' servlet, then the 401 challenge is resent.

The method that the J2EE server uses to authenticate the user on the J2EE server is implementation dependent. The example in Figure 9.10 portrays several methods that are supported by BEA's WebLogic Server (WLS), so we shall mention them here as examples of possible authentication mechanisms. The WLS security mechanism can check the user credentials against a variety of possible user directories via four methods:

1. *Using a flat file stored on the server file system locally to the J2EE server installation –* this is locally encrypted so that a casual observer cannot interpret its contents. The problem with a flat file method is that it is not scalable, nor very maintainable. If we have a large mobile services portfolio with a substantial subscriber base, then this method will not suffice.

2. *Using the underlying security system inherent in the host-operating system –* such as Windows or a Unix variant. This is a powerful technique as it means that a consistent approach can be utilised where we do not have to leave the operating system security administration should we want to use it for our platform authentication method across all applications and servers.

3. *Referring to an external directory system –* use of the Lightweight Directory Access Protocol (LDAP) to send our authentication requests to an external directory somewhere in our network. This approach is likely to be a common one because as the opportunity of maintaining a central directory of all subscriber information is highly

attractive and more than likely already undertaken by many network operators. A centralised directory can be used to store and associate any information with a hierarchy of users, not just authentication details. In a mobile services scenario, we may want to use a directory to store subscriber information, like user address, account type, device type, and services tariff and so on.

4. *Adopting a customised approach via the J2EE JAAS API* – utilising a security API within the J2EE platform in order to implement a custom solution, or at least a custom interface to an existing (legacy) solution that is not catered for by any other access technique. We might well have our own favourite security system or directory equivalent, but not accessible via any of the other methods thus far mentioned. In order to continue using the available system and make our J2EE applications compatible with it, we can utilise the security API to create a custom authentication interface. This mechanism can be as elaborate as we like as we are able to utilise the full power of the Java programming language and J2EE APIs to implement it, perhaps involving an API like JDBC to talk to an external database that underlies the external user directory.

9.2.3 HTTP Authentication – Digest

The problem with basic authentication is that the username and password are both sent in clear text over the link, albeit obfuscated, but weakly so. To appreciate the vulnerabilities of this approach, we need to digress for a moment into the world of hacking (computer spying).

As shown in Figure 9.11, we can think of our hacker as the 'man in the middle', a person who is fully able to access the message flows across our network going in both directions. We shouldn't concern ourselves with how this is possible. It is a well-established assumption in

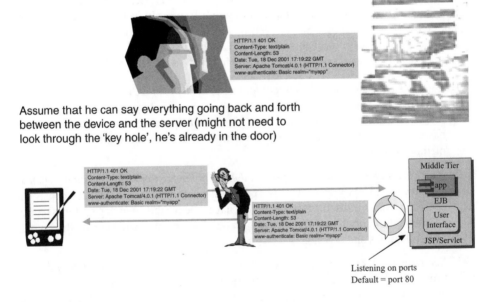

Figure 9.11 The despicable hacker.

the computer security world that not only is this possible, but that our default frame of mind should be to assume that it is likely or definitely going to take place. A word of caution from the security gurus is not to assume that the hacker is some 'evil spy' with a fantastical set of spying apparatus and gadgets from a Hollywood movie. It could well be the friendly cleaner or an employee. There are plenty of places that the hacker could be accessing the link, so anything is possible. Once we are comfortable (uncomfortable) with these gruesome facts of life, then we can go about the business of deploying techniques to address them.

It should be reasonably clear that if our hacker can snoop on our HTTP requests, then immediately they might gain access to critical user information by copying the 'authorization' header information from the inbound requests. Base64 encoding can be easily undone with the most rudimentary of computer programs. It is an openly available and widely known algorithm and is not even pretending to be an encryption algorithm, so there is no need to make it difficult to decode anyway, in fact the reverse is preferable; that it should be easy to decode for the purposes of computational efficiency.

With the 'man in the middle' threat firmly planted in our minds, let's proceed to explain an alternative to basic authentication that may deter or overcome this threat. We shall first explain the technique, called *digest authentication*, and then see how it addresses our concerns with basic authentication. However, we shall then proceed to see if there are cunning ways to defeat even this improved scheme, as impressive as it seems at first glance.

In general, there are really only two ways we can approach the problem of securing our authentication method. We can either encrypt the password information or we can encrypt the entire link so that nothing can be read going across it. Digest authentication is an attempt to encrypt the password.

In Figure 9.12 we can see the process of digest authentication, but it needs a little explaining to figure out what's going on. As with basic authentication, because our resource is protected, this results in our J2EE web engine issuing a '401 Unauthorized' header in the response. However, the 'www-authenticate' header has an altogether different structure

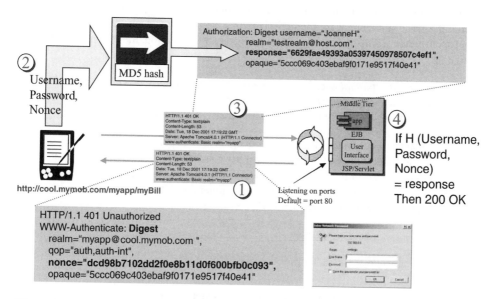

Figure 9.12 Digest authentication.

than we saw with the basic authentication. We need to focus on the field called the *nonce*. At first, this sounds like a rude word[10] (well, it does if you're British judging from the sniggers I get from training classes). However, it refers to a unique string of characters to be used as the basis of our authentication process.

What the user agent must do with the nonce is to put it through a mathematical process called an *MD5 hash*. If that sounds complicated, that's because it is, at least at the deep mathematical level. However, the concept is rather more straightforward. The function is what mathematicians call a 'one-way' (or 'trap door') function. This means that once we have submitted the inputs and generated an output (*hash*), it is extremely hard to reverse the process, even if we know all the input parameters except for the one we are looking for (i.e. the password).

Having calculated the hash, the user agent embeds it in a repeat request for the resource, using the 'Authorization:Digest' header, similar to basic authentication, except with different fields. One of the fields is the username, which the server clearly needs to know in order to authenticate our user. Unlike with basic authentication, we do not attach the password. Instead, we send the hash in the 'response' field (see Figure 9.12).

At the server end, the web engine detects a request for a protected resource, so the first thing it does is look for an 'Authorization' header, which it duly finds. Upon detecting that the header suggests a digest authentication response to a challenge, the web engine processes the fields by performing the same hash function that the user agent carried out. The web engine has the username, which is what it needs to look up the password from the underlying security apparatus, whatever they happen to be. Because the web engine generated the nonce, it then has all the same inputs available that the user agent had to generate the digest response field (the hash itself). Therefore, the web engine repeats the calculation and then compares the locally generated hash with the one embedded in the 'Authorization' header. If they are the same, then it is safe to assume that this request has been generated using the same password on file for the user, hence the user is now successfully authenticated.

In the digest process, notice that we have avoided transmitting the password altogether, unlike the basic authentication method. This is a much better proposition and evidently more secure. According to the proponents of digest authentication, the rationale behind the design is:

> *An improved, minimal security scheme for HTTP is proposed. This scheme does not require the use of patented or export restricted technology and is believed to provide the best effective security possible within those constraints.*[11]

At first glance, it seems that the digest process is remarkably effective. It is certainly an elegant idea and the mathematical basis for it gives a sense of robustness. However, those constraints mentioned in the justification above turn out to be its downfall. Upon careful scrutiny of the digest process it is actually very vulnerable after all. Let's now take a look at its weaknesses by putting on our hacker's caps.

---

[10] *Webster's Dictionary* definition – 'nonce':
Date: 13th century
1 : the one, particular, or present occasion, purpose, or use <for the *nonce*>
2 : the time being

[11] Taken from the specification that can be found at http://www.faqs.org/rfcs/rfc2617.html

1. Can replay the response and then gain access to the URI
2. Can intercept www-authenticate:Digest and replace with www-authenticate:Basic and then do the hash
3. Can make hash responses for common passwords from the dictionary
4. Spoof the server credentials and offer Basic challenge

Figure 9.13 Digest is perhaps not that good after all.

To uncover the weaknesses of the digest process, let's not forget that as a hacker we may have full access privileges to what's happening 'on the wire'. We are 'man in the middle' with full visibility of the message flows. Moreover, what we should have stated earlier, is that we are able to change messages to suit our purposes; this is what it means to have 'man in the middle' privileges (the life of a hacker[12] sounds so exciting doesn't it, but I hasten that I do not condone it in any way, but we do have to be aware of the nature of real world threats, especially if we need to design for (against) them).

Figure 9.13 lists the four major problems with digest authentication. We shall briefly mention them, but we do not need to ponder on the solutions too deeply. Soon we shall discuss a different approach that makes these problems irrelevant.

1. The first method of attack is to simply take the request packet and replay it to the server. A naïve implementation of a web engine will not notice any difference from a previous request for the resource. Indeed, a repeat request may well be a legitimate one. We can imagine that our spy has stolen the message and then whenever they want to access the protected resource (which could be any kind of sensitive material, like financial information), they simply send the request in to the server and siphon the response.

2. The second method of attack is to edit the authentication challenge from the server and simply change the string 'digest' to 'basic'. The user agent gets fooled into thinking a basic-authentication request has been made and gladly (sadly) offers up the username and password token in the subsequent request. Note in this case that the user agent must have actively requested the resource in the first place, otherwise it will not be expecting to process a response from the server, so there is restriction as to when our spy can carry out this attack. However, let's not be lulled into thinking that this constraint deters the hacker. The hacker is not going to be sitting on the wire with a 'stethoscope', actively 'listening' to our messages. They will most likely deploy a piece of electronic apparatus somewhere in the link to do all this automatically, so timing is of no consequence to them.

---

[12] The life of a hacker usually ends up abruptly terminated by a prison sentence, so this is not one to try at home folks. For the curious, I recommend reading Kevin D. Mitnick's interesting book *The Art of Deception:Controlling the Human Element of Security* by Wiley Publishing Inc, IN, USA (2002).

3. An alternative to this is a similar approach of offering a basic challenge, but using deception. The spy sets up a server that masquerades as the original server of interest. This is done by mimicking the user interface. The spy lures the user into accessing the protected resource, presumably because the user thinks that doing so is legitimate, possibly having been enticed onto it via an email message (or WAP Push message). Seeing that the user interface is no different from usual, the user is none the wiser and happily attempts to log in to the system, thereby giving their username and password to the hacker. This type of attack is called *spoofing*. It seems that mobile users are particularly vulnerable to this type of attack and should be vigilant (see sidebar entitled 'Mobile browsing securing threats'[13]).

Mobile browsing security threats

Mobile users should be careful about responding to messages (i.e. WAP Push) that take them to WAP sites. They might be fake messages.

When I started thinking about the vulnerabilities of authentication, it occurred to me that mobile sites are perhaps a lot less safe than their desktop counterparts.

One way to gain a password from a user is to spoof a server. Pretend to be a particular website and then ask your users to log in. Voila! If they bite, then you have their username and password. This has been done on numerous occasions with famous websites and is an ongoing threat. In fact, regular users of sites like eBay should learn to become vigilant against these types of spook attacks. With mobile sites it appears to be even easier.

Firstly, on most mobile browsers, to save real estate, the URL display box is not displayed, or there simply isn't one. This means a user typically has *no* idea what website they are actually on in terms of its web address – surfing on mobile sites is sometimes an eerie experience, like walking in the dark. If a user is directed to a spoof website, they would have little or no idea.

Secondly, due to sparse interfaces, it takes little effort to mimic a mobile website, perhaps just by copying a logo at the top of the screen.

WAP Push will soon start to become more widespread. It has been slow to catch on, but with the increasing number of picture-messaging phones, the necessary inclusion of the WAP browser means that more and more mobile users can access mobile sites, whether they know it or not. However, they don't need to know it to respond to a WAP Push message – the phone itself will take care of accessing the embedded URL in the message if the user chooses to respond.

The other weakness is with the WAP Push mechanism itself. It uses text messaging as the transport mechanism. With text messaging it is easy to change the sender's address in the message. In fact, many text-messaging bureaus offer this service to their bulk-messaging customers. This has some interesting and legitimate uses, but can also be malignant. It is easy to spoof the sender in order to make the message look legitimate, thus adding to the bait used to lure an unsuspecting user onto a spoof site.

Mobile users should be educated in the dangers of responding to WAP Push messages.

---

[13] Taken from the author's weblog: http://radio.weblogs.com/0114561/

4. The final approach in outsmarting digest authentication is to perform a dictionary attack. This simply involved repetitive entry of different password options from the dictionary until a matching hash is produced to the one that the user agent generated. This type of attack should theoretically be easy to thwart, simply by enforcing users to utilise non-dictionary strings for passwords. However, the reality of sensible password policy enforcement is often too embarrassing to mention because it frequently doesn't exist.

From what we can tell about the vulnerabilities of digest authentication, we can see that despite its apparent sophistication, it is not that secure. It is still worth implementing or deploying instead of basic authentication, as there are different levels of determination and sophistication that we might expect from our hacker. We shouldn't assume that we are inevitably going to confront the world's best hacker as our foe and therefore in the absence of a sufficiently strong technique we should simply not bother. This would be folly and it is better to always implement the stronger of any security option if we are able to do so. Casual hackers may well be deterred.

9.3 ENCRYPTING THE HTTP LINK

Ultimately, the best solution is to secure the entire pipe to make sure that nothing can be gleaned from our communications in the first place. If this is achievable, then any amount of hacking sophistication by being 'man in the middle' should be to little or no avail[14]. A protected link defeats all hacking attempts that we have thus far identified, except for some types of spoofing, which, as the sidebar[15] identifies, may be particularly troublesome for mobile users.

Figure 9.14 shows us what *securing the pipe* is about. Using encryption, we cipher all the information flowing between our device and server, with these two end points being the only the true endpoints of the ciphering. If the encrypted HTTP link runs from the user's device all the way to the server where the sensitive data is coming from or being requested by (e.g. a username and password), then this is call end-to-end encryption.

Ciphering results in the data becoming completely obfuscated to everyone, unless they have the means to decipher it, although that is a capability that we can use clever means to guard against, as we shall see in a moment. Only the intended recipient has the required deciphering ability. Because the information is obscured from everyone except the intended recipient, we don't really mind about a 'man in the middle' or even 'men in the middle'. Any amount of sucking data from the pipe in order to attempt reading it is no longer a concern. In fact, we could say that this is the real goal of an effective ciphering mechanism. We shouldn't really care about unintended recipients gaining our protected information, maliciously or otherwise. Our ciphering will take care of this threat for us.

As shown in Figure 9.14, we can think of ciphering as putting the data into virtual caskets using a lock and key. We lock the data up and send it. We hope that the protection is sufficiently robust that it cannot be compromised. The only way to unprotect it is to unlock

---

[14] In the sense that any technical means will be hard to implement. However, the greatest sources of risk are often found elsewhere, such as social engineering methods as described in fascinating detail in: Mitnick, K.D., *The Art of Deception:Controlling the Human Element of Security*. Wiley Publishing Inc, IN, USA (2002).
[15] Refer to the sidebar entitled 'Mobile browsing securing threats'.

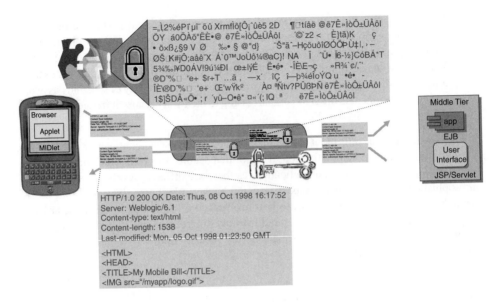

Figure 9.14 Securing the pipe.

it with a lawfully acquired key. The lock-and-key concept is a useful and obvious metaphor for understanding encryption, so we shall stick with it for our current discussion.

With the lock-and-key process, the mechanism for secure transmission is obvious. The sender locks the data with the key, sends the locked data and then requires the recipient to unlock it with their own key. A wonderfully simply process, but with a subtle, yet significant, defect. The challenge is *key management*.

In order for the recipient to unlock the data, they must have a copy of the key. This presents some challenges. The first one is distribution. How do we safely get a copy of the key to the intended recipient? If this were a house and two people shared the house, then giving them a copy of the key is easy. However, what about if someone were coming to stay in your house and they were travelling from the other side of the world to get there? How would we safely get the key to them? We could post it to them, but it might get lost or intercepted. In the latter case, it could be copied without us knowing. Alternatively, we could send the key via a courier, but we would have to know that we could trust the courier.

Let's say we could find a means to securely distribute a copy of the key, but what about making this available to more than one recipient? If we had lots of houses that we rented out to overseas visitors, then we'd have to take care of lots of keys. It would not be a good idea to have identical locks and issue everyone with the same key. This would mean that one visitor could open the house of another. Similarly, with a large website, we don't want to give a copy of the same key to each user, so we prefer to have unique keys. This poses a potentially significant key management problem, possibly involving thousands of keys. Not only does this make distribution that much more risky, but we now have to ensure that for each transaction with a user, we end up using the right copy of the key to cipher the conversation.

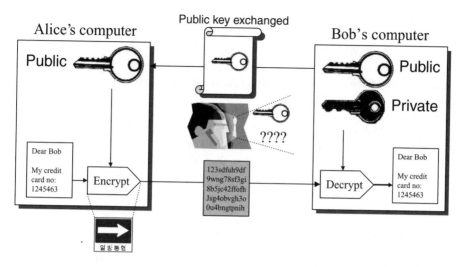

Figure 9.15 Public key cryptography.

Fortunately, some rather clever mathematicians have worked on these problems and came up with a genius solution called *public key cryptography*.

9.3.1 Public Key Cryptography

Perhaps the most famous example of *Public Key Cryptography* (PKC) is the *RSA*[16] *method*. The principles of this technique are shown in Figure 9.15, which we shall now describe.

With PKC, the genius idea is to have two keys. Both of these are generated (created) by the sender and are unique to the sender. In computer science terms, a key is actually a long number, long enough so that it would be too difficult to guess, but we shall return to this point later.

When the sender generates two keys (using a computer program on their device), one is designated the *private key* and the other the *public key*. Under no circumstances is the private key to be revealed to anyone – it is to be kept safely hidden away. However, the uniquely associated public key can be shown to anyone, which is actually what it is intended for.

As Figure 9.15 shows, let's imagine that Alice wants to send her credit card details to Bob in a secure manner. What Bob does is to give Alice his public key. Don't forget that this is not a problem, as long as Bob gives his public key and not his private key. Alice then uses Bob's public key to lock the data. This is where the power of PKC is exploited. A special mathematical *one-way function* is deployed that allows data to be locked with the public key. It is known as a one-way function for several reasons. Firstly, in the case of PKC, the public key itself cannot be used to unlock the data once locked. This seems remarkable and quite unlike our metaphor of the house key. Imagine a lock that we can secure with a key, but then cannot reopen with the same key: quite mind-boggling.

---

[16] RSA is the most common and well-known PKC method. It is named after its inventors: Rivest, Shamir and Adelman. Their patent on the method has now expired, so the algorithm is free for anyone to use without paying royalties.

The one-way public key is the magic of PKC. It means that Bob can freely distribute his public key without any fear of it falling into the 'wrong' hands. He really doesn't care who gets hold of it. With this approach, the key distribution problem is solved. The only way that Bob can open his encrypted message from Alice is to use his private key, which only he has possession of. Bob should never distribute his private key, nor does he have any reason to disclose it.

There is another clever trick that can be done with public and private key pairs. Securing a message with Bob's private key can only be undone using his public key. This may seem a strange thing to do. It seems as if everyone could open the message. However, that's the point. If we can open a message and read it, then we know that the message in question must have come from Bob and so we can use this process to authenticate Bob.

9.3.2 Using PKC to Secure Web Connections

Using PKC, we can secure web-based transactions using HTTP and WAP. We shall look at this process in general and then look at some of the issues concerning WAP, especially securing a WAP 1 connection.

The process of securing an HTTP session takes place at the transport level, which means that the entire HTTP communications link gets encrypted. In other words, every single HTTP message is encrypted from top to bottom and both the *user agent* and *origin server* are largely ignorant of the encryption process.

Recalling our description of Transmission Control Protocols (TCP) in Chapter 7, we may remember the idea of a socket. A socket is really a programming entity, like a handle used by a higher-layer application to pipe its data through (or receive) with the notion that it will squirt out somewhere on the other side; this being identified by the IP address and port number bound to the socket, as shown in Figure 9.16.

Ideally, what we want is the ability to secure the socket connection, so that information goes in one end (source socket), gets encrypted and comes out the other end decrypted (at the destination socket). This is what transport-layer encryption is all about, as shown in Figure 9.17.

In Figure 9.17, we can see that the sockets each have a key for ciphering the data that flows through them. Consequently, the connection is locked and secure all the way from one application to the other. Later in the book we discuss where and how this cryptography gets carried out on the device and make the point that a library of cryptography software routines

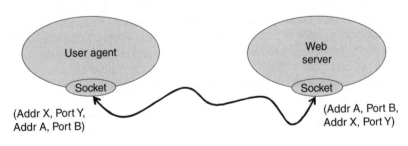

Figure 9.16 Sockets.

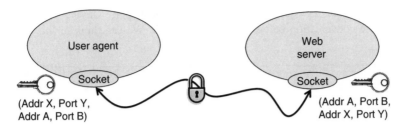

Figure 9.17 Secure sockets.

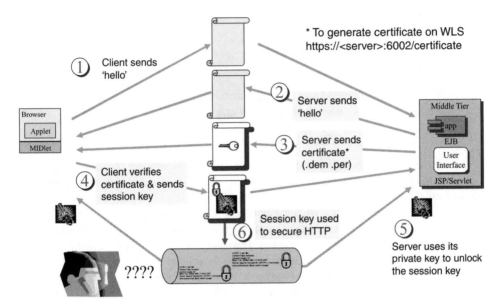

Figure 9.18 Using the SSL/TLS method to secure the Web.

is probably a standard inclusion these days on any mobile device, such is the importance of cryptography to an increasing number of mobile services[17].

The method of securing sockets that we shall look at here is shown in Figure 9.18 and is referred to by two names, the most apt for our discussion being *Secure Sockets Layer*, or SSL. This was a method of securing HTTP connections originally included into browser software made available by Netscape. It has now become a standard, although it has migrated to a new specification and name, known as *Transport Layer Security* (TLS), which has a wireless counterpart (in the WAP architecture) called *Wireless TLS*, or WTLS.

Stepping through the process shown in Figure 9.18, we can find out how SSL/TLS works, as they are both essentially the same in principle.

The first undertaking between both ends is a handshaking protocol. Just like hand-shakes between people, in data communications they server a similar function of greeting each other. Our mobile device and server exchange 'hello' messages. These

---

contain a request for an encrypted conversation to take place, laying down some parameters of the conversation before proceeding with it. We can think of this as follows:

Mobile:

'Hi there – I want to talk with you over a secure link. I prefer to use RSA to exchange keys, but can do Diffie-Helman, too – what say you? My opening phrase is "skipjack tuna".'

Server:

'Thanks for the request. I can talk to you securely, but prefer we use Diffie-Hellman key exchange. My opening phrase is "wiggly bigly".'

Apart from the legitimate key-exchange names, these examples are obviously contrived, but serve to show us how the SSL conversation is initiated. In a real system, the opening phrases are random data.

The juicy bit of the process is what happens next. The mobile receives the server's public key. This clearly allows the mobile to securely send information (i.e. HTTP requests) to the server, who can decode using its private key. However, there are a couple of subtleties that add a degree of complexity to this process, although the principle itself remains a sound one.

Authenticating the key is our first challenge: how can we check that the key we receive actually belongs to the server purporting to send it? If we think that we are about to transfer money from our bank account using MobileCash Bank (fictitious name), then how can we be sure that the public key is indeed theirs and not one generated by Joe Hacker (fictitious villain). In the latter case, we could be about to give our bank details to a thief.

The standards community have agreed upon a technique for embedding keys into a file called a *digital certificate*. We can think of these as having an electronic stamp on them from a trusted authority (Certificate Authority, or CA). The principle is that the certificate is not easy to obtain unless the applicant really is who they say they are. The CA should carry out checks to verify credentials and the legitimacy of the applicant.

Once we have a certificate containing our public key, the mobile can check the certificate first, before extracting the public key. The certificate will claim to belong to MobileCash Bank and we can check for a digital stamp from a trusted certificate authority. This digital 'stamping' itself is based on PKC. This is something we touched upon at the end of the last section. The CA 'signs' the certificate by encrypting it using the CA's own private key. The CA's public key is widely available[18]. If we can read the certificate by decrypting with the CA's public key, then we can trust that it was indeed issued by the relevant CA. Ultimately,

---

[18] In fact, in certain cases, the public keys for various CAs are already included in our browser software. In effect, this is a delegation of trust. By using these browsers we are agreeing that we trust the CAs whose keys we are going to use.

it is up to the users of the system to trust (or not) that the certificate has been issued with due diligence.

It is not enough that we rely upon the certificate to authenticate the server. The certificate simply authenticates the public key, but we still don't know who we are talking to until we ask for cast-iron proof. The way this is done is for the mobile to send a random code to the server and ask for it to return the code encrypted using the server's private key. If we are able to receive the same code by unlocking it with the corresponding public key, then we know that we are talking to the right server. This assumes that the server has never shared its private key and that the certificate it sent to the mobile checked out. To protect the random code from being spoofed by someone else, it is sent encrypted using the server's public key.

The details of this authentication process are more complicated still. As with the digest authentication technique used for HTTP session authentication, the server actually sends an encrypted hash of the original random code back to the mobile. The mobile can compute the same hash locally and compare. The hashing is really to protect the server from sending data that it doesn't want to send; the problem being that by using its private key, the data is effectively being 'signed' by the server. The server doesn't want to 'add its signature' to any old data and send it out, as this could be used surreptitiously elsewhere. With a strong hash (like MD5 used for digest authentication) the worst that a third party can do is to decrypt a message from the server that is garbled once decrypted.

The hashing involves the secret data exchange and the original public 'hello' messages, which we said were in fact random. This combined approach achieves several objectives, including making it difficult for the hacker to carry out replay attacks, which may be useful even if the contents of the requests remain unknown (encrypted). But this is an advanced topic not covered here.

Thus far, we have only covered Steps 1 to 3 in Figure 9.18, which is how we set up the secure conversation. This is all taking place at the socket level and eventually our sockets have a reliable, trustworthy and secure link. To indicate to the higher layer applications at either end that this is a secure channel, a special port number is actually reserved for SSL connections for web traffic, which is port 443. By assigning a separate port from the unencrypted connections (port 80), we can dedicate a resource just to handling this port, independent of any software that may already be serving port 80.

Having gone through all the elaborate procedures and fancy mathematics to establish a link using PKC, the next step (Step 4) in the SSL process seems a bit odd: the mobile generates a session key for a symmetric encryption session. This is back to the original idea where both sender and receiver have the same key. A moment of reflection reveals that this will work safely, given that the mobile can send the session key securely to the server using the server's public key to encrypt the session key. In other words, not only does PKC solve the key distribution problem in its own right, but it also solves the key distribution problem for any other encryption technique. The only question that remains is why bother to revert back to symmetric processing for the remainder of the HTTP session? The answer is that symmetric algorithms are a lot less resource intensive than asymmetric ones.

The processing requirements for PKC are not really a concern for desktop systems. After all, they have oodles of processing power and masses of memory to throw at problems, cost effectively too. Nonetheless, even in the immobile world, the processing requirement is still a problem – not for the client, but for the server handling thousands of sessions concurrently. Therefore, using a less intensive process is still worth it. We should not be misled into thinking that the lower resource requirements mean a weaker cipher. This is not the case

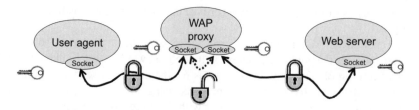

Figure 9.19 WAP proxy can't maintain a transparent connection.

at all. Symmetric ciphers are still very powerful and practically unbreakable with even an exceptional amount of applied resource.

For mobile devices, the use of asymmetric encryption for the entire session would definitely be out of the question, especially on devices with very limited processing resources. In fact, even the initial exchange using public keys can be a very slow process on a low-end device, so much so that the cipher experts have devised new ways of handling the maths involved in the calculations. *Elliptic curve cryptography* is one such example and is now a widely used technique for securing mobile browser sessions.

9.4 APPLYING SSL TO WIRELESS

In essence, the process of applying ciphering to a wireless browser session is the same as just described in the preceding section. In fact, the security layer between the mobile and the WAP gateway is the same process exactly. The problem is that the gateway introduces a discontinuity in the ciphering process that cannot easily be circumvented. As we noted when looking at IP protocols for mobile applications, the WTP and the WSP are not the same as their Internet counterparts TCP and HTTP.

The difference between WSP and HTTP is what causes the bigger problem. If nothing else, the contents of many HTTP messages are converted from WML to WBXML at the gateway[19]. This means that we cannot transparently pass data through our gateway; the WAP gateway process necessitates conversion. This implies that we have to decrypt the data at the gateway in order to carry out the conversion process. We then need to encrypt it again onto the opposing connection on the other side of the gateway. Transport-layer security only works if the higher layers require a transparent connection end-to-end (socket-to-socket). This is not the case with a WAP gateway, as shown in Figure 9.19.

For WAP 1, there is no way of getting around this encryption discontinuity at the WAP proxy. WSP and HTTP are not compatible, even though they are semantically similar and practically identical for the most part. The WAP stack and the HTTP stack are not the same for WAP 1, as Figure 9.20 clearly shows us.

This problem has caused a lot of controversy about WAP and was a contributory factor to its lack of uptake in the early part of its life. However, it is probably safe to say that the performance of early WAP devices was a much bigger problem and that, for the most part, the supposed 'insecurity' of WAP was more of a perceived problem than a real one for most applications. But how should we approach this problem?

---

[19] With WSP, the headers are also encoded, as we found out earlier in the book.

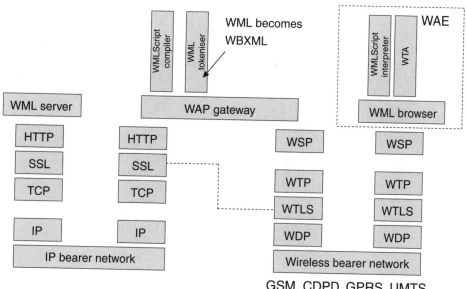

Figure 9.20 WAP stack not the same as HTTP stack.

Firstly, for applications that need to use WAP 1, the real impact of the discontinuity needs to be assessed. The point of vulnerability is the WAP gateway, so this is the only place in the end that an attack can be successfully carried out. Therefore, if we are confident that our WAP gateway is well-secured physically, then we may not really have much of a problem. If we still have doubts, then for applications that warrant it, we could deploy a private WAP gateway for those applications and ensure that its integrity is maintained without recourse to any third party, who we may not fully trust. In many enterprise applications, it was very easy to deploy a WAP gateway internally within the company firewall, either through the firewall itself or by private dial-in or data-access connectivity. Circuit-switched access from a WAP device is possible over standard dial-in servers that an enterprise may already have for remote workers. These can be configured to support fast connect times with mobiles on digital cellular networks, in the case where the dial-in lines are digital themselves (e.g. Integrated Services Digital Network, or ISDN). The benefit of using digital connections is that they are faster to establish and they are usually more reliable. Accessing ISDN modems requires ISDN V.110 or ISDN V.120 rate-adaption protocol support from the local GSM operator and the dial-in system, but this is widely supported.

The discontinuity of the WAP proxy introduces us to the concept of end-to-end encryption. A lot of the confusion with the security of WAP once arose due to a misunderstanding of what constitutes a secure connection. In digital cellular systems like GSM and UMTS, the air-interface (RF link) is encrypted. As it happens, the encryption is quite strong (in most cases[20]). If we return to our mobile network diagram established earlier in the book, we can see what the problem is with determining if a connection is secure or not.

---

[20] With GSM air-interface encryption, there are two types. One is strong and the other is weak. GSM operators in certain designated countries are only allowed to use the weak one so that they can be "spied upon by everyone else" (use of language my own).

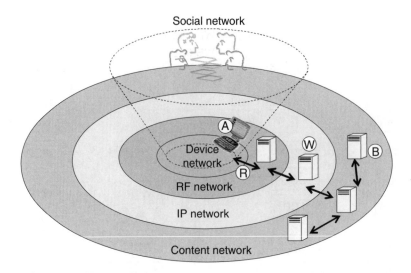

Figure 9.21 Parts of a mobile connection, not all that secure necessarily. (Reproduced by
permission of IXI Mobile.)

Figure 9.21 illustrates the problem. We are interested in passing data securely from our
user all the way to the application server that needs the secure information. This could be
password information, like that needed to complete a '401 Unauthorized' authentication
cycle in a HTTP session. It could also be information submitted in an XHTML-MP form,
which needs to be sent securely using HTTP POST method to the server, perhaps containing
credit card details.

If we are sending this information to the server designated (B) in the diagram, then we
need the link to be secure for the entire pathway from the device (A), to the server (B).
There should be no momentary break in the encryption, nor should there be any portion
of the pathway that is unencrypted. The confusion concerning securing WAP connections
over digital cellular networks came from the idea that the air-interface encryption provided
the necessary protection. Some mobile operators even mistakenly advised that using WTLS
was not necessary because the air-interface encryption was sufficiently robust. This is the
equivalent to only one segment (R) in the diagram being encrypted, but leaving everywhere
else unencrypted and, consequently, vulnerable.

Presented as it is in the figure, clearly, the notion that encryption of link R is sufficient
protection for a secure application is absurd. If the data is unencrypted everywhere else,
then that's where our hacker will attack, and with very little effort required. Ideally, the
encryption must begin on the device itself (A) and end at the server (B), and vice versa
going the other way. The next best thing is to deploy WTLS, which means we achieve
encryption at least from the device (A) to the gateway (W). If configured correctly, we can
also encrypt the path from the gateway (W) to the server (B), but the gateway itself becomes
a point of vulnerability, as previously discussed, but we might be able to live with that for
many situations in practice.

The ideal solution is true end-to-end encryption. Don't forget that in our network model,
the content network can include going out to an enterprise network from the mobile operator network, or, as we shall soon discuss, going out to a third-party network where our

application is physically hosted. In these cases, we still need to maintain the encryption to the source (destination) server, which is what our diagram indicates, but we may need to examine what this entails physically when we look at topics like *Web Services*, a method to extend our mobile delivery platform via the Internet to other platforms hosted elsewhere.

9.5 END-TO-END ENCRYPTION IN A MOBILE NETWORK

We have established that end-to-end encryption is the optimal solution for protecting confidential user information. Moreover, the use of encryption helps us to provide a secure channel for authentication; some of the encryption techniques even lend themselves to authenticating in a strong manner. With WAP 1, end-to-end encryption is simply not possible, though we can get quite close; matters can be made relatively secure if we can properly secure the gateway itself.

An important realisation is that with digital cellular networks, and with other wireless networks like 802.11, despite the ability to achieve OTA encryption, this does not allow us to be lax about application-level, or transport-level encryption end-to-end. Air interface protection is only one small part of the problem. In fact, if application-level or transport-level encryption is in place, then arguably we can dispense with air-interface encryption altogether (for data transfers, in any case, although we still need the encryption to protect other types of information such as voice traffic).

Given the issues with WAP 1, how can we achieve end-to-end encryption in a mobile application? Well, firstly we should point out that our current security consideration has been within the context of established software architecture paradigms, particularly the client–server approach based on the link–fetch–response, or browser, modality using HTTP or HTTP-like protocols.

As we shall find out when we look at devices in detail (see Chapter 10), there are several ways to develop an application for a mobile device. Browser-based presentation of a server application is only one method, but we can develop standalone applications too, which sometimes will want to communicate with a backend somewhere. However, even with these embedded applications, we may well end up choosing to use HTTP as the method of communicating with the backend, for all the reasons one would expect to do with the prevalence of the protocol and the infrastructure and tools supporting it, not least of which is the powerful J2EE platform itself.

Due to its prevalence, we can expect to find HTTP software libraries on many devices. Wherever there is HTTP, there is usually HTTPS (secured), which is the naming convention for using HTTP over an SSL (or TLS) link.

For embedded applications, there is no need to do anything that precludes connecting directly with an origin server. Similarly, as we have seen with WAP 2, our user agent is capable of connecting directly to the origin server using native HTTP and TCP protocols, without a proxy. However, in both cases, we may still chose to use, or may have to use, a proxy.

In the configuration of LANs hosting our origin servers, the network may be set up such that a proxy server is deployed as necessitated by the administration policy for HTTP access to networked resources. There are all kinds of reasons for this, which make it a common practice on many networks. We shall not discuss the reasons here, but we shall look at the implications for implementing end-to-end security using HTTP.

Even though the use of HTTP/TCP protocols is specified for WAP 2 browsers, we may still wish to utilise a WAP proxy. There are various reasons for this, some of them general, as found in standard HTTP networks, but some specific to the nature of optimising and supporting wireless access.

Firstly, the wireless profile of TCP is not going to be a standard configuration on most web servers, even though we noted that the profile consists of extensions, or options, to the standard TCP implementation. This means that a particularly comprehensive implementation might be compatible with features of W-TCP, like SACK. However, if we want to be sure about gaining the benefits from its optimised performance, then this is one reason to deploy a WAP proxy, as shown in Figure 9.22. Here we can see the two WAP stacks side-by-side, which is how they would exist in actual products if we choose to support them both – which is what most WAP browsers will do for some time to come. WAP 1 and WAP 2 stacks are not interoperable, so the only way to support both of them is to run a dual-stack solution.

We can see from the diagram, that when we are running a WAP 2 device it can go via the proxy using TCP or W-TCP. The proxy can handle this and proxy the connection over a standard TCP link on the server side. If we are running with encryption, then the encrypted data is still pushed around using TCP, so there is no problem in going from W-TCP to TCP across the proxy. We can think of this as simply how the encrypted packets get queued up and sent, and how they get resent if necessary. In a moment, we shall consider what happens at the proxy to the encrypted data itself. We have already learnt that with WAP 1 protocols, the data cannot continue its journey from here without being transformed, which means being unencrypted and re-encrypted so that the transformation can take place. This is not required with WAP 2, as we shall shortly learn.

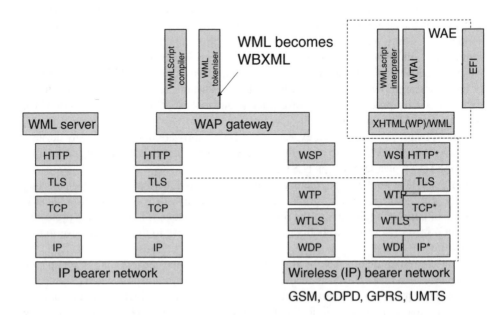

Figure 9.22 WAP Proxy can still be present for WAP 2.

There are other reasons why using a WAP proxy makes sense, even in a WAP 2 environment[21]. We discovered in the Chapter 8 that part of the mechanism of managing web sessions is to use the state-maintenance features of cookies. Yet, not all devices support cookies to the extent that we might like. Some devices have a very limited cookie memory and so extensive use of cookies across our applications portfolio could be problematic. Although there are solutions to mitigate this problem, like URL rewriting, we may still be better off sticking with cookies, especially as the use of cookies is part of the standard session management model for JSP and servlets in the J2EE platform.

As well as these issues, the WAP gateway/proxy is often a more fully fledged product than simply a proxy. It may include the WAP Push features for example. Some gateway products also include mechanisms for keeping records for billing and can even offer so-called 'real-time billing' capabilities, which is really a way of assigning different access metrics to different resources so that these can be billed for later. This is also known as *content-based billing* or *event-based billing*. If this feature is required, then a mechanism for it needs to be provided somewhere in our content network. We could implement it ourselves as part of our J2EE application, but then this type of feature is going to be a common requirement across all applications that need content-based billing in place. Therefore, we need a mechanism for enacting this across our applications, as a platform-wide feature accessible to all applications. Whilst we could envisage a variety of solutions to this problem, why not use an existing network entity, like the WAP gateway, if it can do it for us?

Fortunately, when it comes to securing our applications, WAP 2 has the capability to support end-to-end security, even with a proxy in the way.

WAP 2 clients support TLS 1.0, as do most web servers and J2EE application server web engines. In order to support end-to-end security, we use a technique called *tunnelling*. The name is fairly self-explanatory. What happens is that the proxy, under instruction of the client, passes through all its data unhindered to the server and vice versa.

To establish a TLS tunnel, the WAP client uses the HTTP method called *CONNECT*, as defined in RFC 2817. It requires that the HTTP proxy server should support the HTTP CONNECT method in the same manner (as defined in RFC 2817[22]. From the CONNECT method, the proxy knows simply to pass on the HTTP traffic to the server without touching it.

Any successful (i.e. 2xx) response to a CONNECT request indicates that the proxy has established a connection to the requested host and port, and has switched to tunnelling the current connection to that server connection. It may be the case that the proxy itself can only reach the requested origin server through another proxy. This may sound strange, but this could well be the case if we use a WAP proxy for all the reasons just noted, but end up accessing a content network with its own standard HTTP proxy (for the reasons we didn't discuss, but usually related to caching, security and access control, etc.)

In this case, the first proxy makes a CONNECT request of the next (cascaded) proxy, requesting a tunnel to the target server. Our WAP proxy should not respond to the WAP client with any 2xx status code unless or until, it either has a direct connection or a tunnel connection established to the requested resource.

We can see that the HTTP CONNECT method and the support for TLS at both end points of our service imply that end-to-end security is achievable.

[21] Our expectation is to run mixed WAP 1/WAP 2 environments for some time to come.
[22] http://www.faqs.org/rfcs/rfc2817.html)

10

Mobile Devices

10.1 INTRODUCTION

We now wish to look at the immediate human interface to the mobile network, that being principally through electronic mobile devices of one form or another. We wish to explore the possibilities for mobile devices in this chapter, investigating how they might work at a functional level. In Chapter 11, we shall look at the various ways of programming mobile devices. For the most part, we will not concern ourselves too much with device implementation issues involving the underlying physical electronics. Where relevant, it may be valuable to consider certain issues concerning the electronics, such as power consumption, memory capacity and so on, but only at a high level.

We have attempted to define and examine the mobile services landscape in general without confining how mobile connectivity might be realised. For example, it is just as feasible for a mobile service to be realised through WiFi as it is through a 3G connection, or both. It is just as feasible for mobile devices to talk directly to each other (P2P) as it is for them to connect to some remote server, most likely, as we have explained, via an IP network. However, we have not restricted ourselves to IP connectivity either. A P2P connection could use any low-level protocol to shift the bits physically from one application to another. Along the way, we have also understood that the mobile network is not just about people injecting and gathering meaning from the clouds of information that exist in our virtual information space. Software applications too, especially with the help of the Semantic Web (refer to Chapter 5), can participate in more ways than first anticipated, actively seeking out opportunities and information on the users' behalf, or, as strange as it currently seems, on their own behalf.

The myriad ways of handling information in the mobile network, and the rapidly developing device market (still thanks to Moore's Law) leads us to think, as already suggested, that an appropriate model for the device-end of the service delivery chain is to think in terms of a device network. This can be inter- or intra-personal. Attached to that network can be

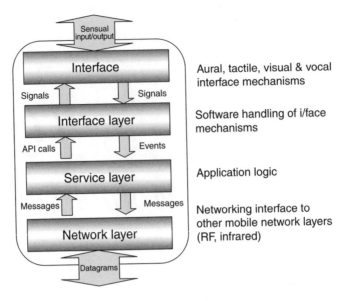

Figure 10.1 Abstract layered representation of device.

devices of various ergonomic propositions. A Bluetooth pen has no graphical user interface whatsoever, neither does a personal mobile gateway, but a wireless-enabled personal digital assistant (PDA) or a wireless-enabled laptop both have powerful graphical interface capabilities, the laptop especially so.

In developing our understanding of the device network, we need to take into account the possible device ergonomics, the various networking propositions, as well as the primary functions that each device, or a combination of devices, is trying to provide to the user in an effort to support mobile services as per our definition in Chapter 3. To approach this topic, a useful starting point is to construct an abstract high-level view of the device by representing its functions as a set of layers, as indicated in Figure 10.1, which we shall discuss in the sections that follow.

We can think of the software paradigm for a device as being centred upon a simple process. That is, to respond to either interface events or network events, to then react to those events with some kind of processing in order to carry out a useful *task* – and then send output back to the network or to the interface. This process, as a set of tasks, can be coordinated to make a useful mobile service. The granularity of this process may vary, such as a large amount of tasks being required to carry out a useful service, or perhaps just a single task. There is another type of event that is independent of the interface and the network, and that is a time-based event driven by the clock[1]. This software paradigm can be summarised visually, as shown in Figure 10.2. This simple model is the most basic operational view of a mobile device. Of course, we shall expand on this throughout this chapter and talk about how this model actually works and the various ways that it could be implemented.

---

[1] A mobile device will usually include a real-time clock on board that knows the time and keeps 'ticking' away and is able to notify the software services that a particular time has been reached.

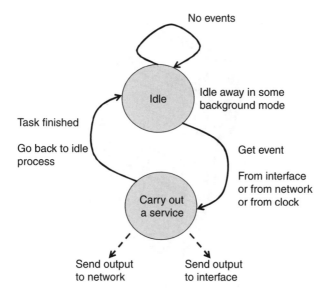

Figure 10.2 Basic (simplified) software model for mobile device.

10.2 INTERFACE ELEMENTS

Somewhere in the device network there needs to be a sensory interface between the device and the user. We shall not spend too long considering this layer, but it is useful to summarise its characteristics. The most common types of interface can be divided into four categories:

1. Tactile (pertaining to touch)

2. Aural (pertaining to hearing)

3. Vocal (pertaining to voice)

4. Visual (pertaining to sight)

We shall now discuss each of these interface types in the sections that follow.

10.2.1 Tactile Interface Elements

Tactile interface is the ability to interact with the device via the hands and fingers, such as through a keyboard or pointing device (stylus) in conjunction with a reactive touch display. These interface modes are designed for enacting commands and for inputting text, which could be in any language, or some kind of drawing input.

Keyboards can come in various modes, such as the alphanumeric 12-key keypad common to mobile phones, to fully-blown qwerty keyboards. There are also many versions of miniaturised keypads that are added to PDAs, as shown in Figure 10.3.

Figure 10.3 PDA showing built-in miniaturised keyboard.

For smaller devices, such as standard mobile phones, new and innovative keyboard designs, such as Fastap by Digit Wireless[2], enable a greater density of keys to be achieved than was previously thought possible, thanks to clever ergonomic design concepts (see sidebar – 'Fastap to a quicker message'[3]).

Fastap to a quicker message

Today I met with one of the guys from Digit Wireless. I previously blogged about their incredibly clever keyboard design, which they've trademarked as 'Fastap'.

Today I saw it, played with it, loved it! I not only played with the neat demo model, an anonymous phone mock-up with the Fastap keyboard, but touched and felt a production-ready unit under test by the operators (can't say who – sworn to secrecy).

Only when playing with the keys did I at last appreciate the genius of the design (see image below). Its simplicity was shaking; its inventor sure has ergonomic insight. The principle is that the letter keys are raised above the phone face, like 'hills', and the digit keys are sunk in between, like 'valleys'. That bit I understood already. Each raised key has enough space around it to more or less (a little less) occupy the space of a key on a qwerty keyboard, believe it or not! The finger does not clumsily hit the numeric keys by mistake, as these are sunk, or so they seem, as, in fact, there's no need for a key there at all. I loved that magic bit! By placing the finger on the sunken 'valley', over the desired digit key, the finger can't help but depressing some combination of the surrounding raised 'hill' keys and the software picks up the combination and knows how to map it to the desired digit. Genius!

---

[2] http://www.digitwireless.com/
[3] Taken directly from my weblog – http://radio.weblogs.com/0114561 – but note that this is my original blog address, which has since moved to blog.wirelesswanders.com

Intel's "Universal Communicator" prototype, featuring Digit Wireless' Fastap™ Technology

So what's my reaction? Is this another tech idea that has no real application? Absolutely not! This idea must surely come to fruition and I can feel that it has the potential to become commonplace. The reason is that we are not just talking about a better way to text. Actually, it is a better way to text, especially for those of us who cannot get to grips with predictive text (PT) input. I confess that I am a good user of PT, until I hit a word that's not in the dictionary, or I want to deliberately abbreviate. However, most people I know don't use PT, or find it somehow clumsy, although, to be fair, I think it is very phone dependent (T68i is one of the best implementations I've seen).

The reason that Fastap could be in demand is the gathering interest in Wireless Instant Messaging. Certainly, new initiatives, like Wireless Village (and the PAM forum entering Parlay) are generating momentum towards a big push for IM into the wireless space, but we seem to overlook the not-so-small problem of boringly slow typists – and that's talking about QWERTY users! I would like to use IM on my phone, but it's just too frustratingly cumbersome. Fastap could change that. If it does, then I think it will be a winner and could well assist IM take-off, along with email of course, which is the other mobile service still waiting to gather momentum, though it's problems are not just character entry, but often poor usability, full stop.

As probably already familiar to most of us, keyboards do not have to be mechanical devices at all. On most PDA devices available, a small virtual keyboard can be displayed on the screen. This virtual keyboard simulates both a numeric keypad and a qwerty keyboard.

The other emerging possibility is the use of light-projected keyboards, which using clever projection and optical scanning techniques, can project a keyboard image onto a flat surface and then sense where the fingers are on that image, as shown in Figure 10.4 .

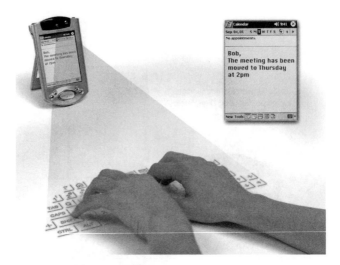

Figure 10.4 Light-projected keyboard attached to a PDA. (Reproduced by permission of
          Canesta.)

The technologies used by Canesta, Inc.[4], the inventors of the projection keyboard, are
able to sense where objects are in 3D space. This technology has wider applications for
mobile device input mechanisms. As Canesta states:

> *Electronic perception technology permits machines, consumer and electronic devices,*
> *or virtually any other class of modern product to perceive and react to objects and*
> *individuals in the nearby environment in real time, particularly through the medium*
> *of "sight," utilizing low-cost, high-performance, embedded sensors and software.*

This technology works by attempting to assign to each pixel of visual information a distance metric from the camera so that objects within the image can be assigned spatial depth.
This is done using radio waves in a similar fashion to how radar signals are used to detect the distance of remote objects, such as aircraft, although clearly at a different range
altogether.

In modern computing devices, we are now used to the familiar *point-and-click* interface
paradigm, as popularised by the windowing style of interface, an idea that actually grew
out of intent to provide an interface method that would correlate to an object-oriented
programming paradigm. This idea was first commercialised by the labs of Rank Xerox
Corporation[5].

The point-and-click interface has a variety of tactile possibilities, such as dragging,
bounding, re-sizing and other visually guided tactile moves that come from point-and-
clicking in a windowed world. These originally arose from the use of a mouse as the
primary point-and-click device. A standard desktop mouse is not a useful solution for
most mobile devices, so other options have arisen, most notably stylus pointers used

---

[4] http://www.canesta.com/ – the keyboard is now produced by Celluon, Inc. – http://www.celluon.com/
index.html
[5] The Xerox 8010 'Star' Workstation was the first commercial realisation of the ideas on object-oriented user
interfaces developed at Xerox Parc in the early 1970s.

with PDAs and tiny joysticks now available on some mobile phones, including 'smart phones'[6].

Tactile Feedback   It is also worth mentioning that tactile feedback or alerting is also possible, and is already widely used in mobile phones in the form of vibrating alerts. Such tactile feedback systems can be used in other ways, such as giving the sense of collision (e.g. racing games) or explosion (shoot-'em-up games).

10.2.2   Aural Interface Elements

The use of audio output is an obvious feature in a mobile phone where we need to be able to hear the other person, or people (e.g. in a conference call). It is possible to either have an audio output connector for a headset, or to incorporate a speaker. Speakers can be low power, such as the tiny speakers used in a handset for placing next to one's ear, or could be higher powered, such as a hands-free solution might predicate. Higher powered speakers will place a greater drain on the battery, so they are not common in devices with small battery capacity.

Of course, on laptops, speakers are common. However, on smaller form-factor devices, such as PDAs or phones, speakers may or may not be included. We have already noted that for certain applications, such as Push-to-Talk (PTT) mentioned in the book's prelude (Chapter 1), the use of a built-in speaker may improve the overall user experience.

There is no need to confine aural output to voice-generated signals. We can deploy aural output to support *text-to-speech* applications. This would be especially useful for accessibility for sight-impaired users, but can also be deployed for convenience or safety, such as reading of emails whilst travelling in a car. With new technologies, such as AT&T's Natural Voices, it is possible to achieve text-to-speech output that is increasingly natural sounding, as if read by a human rather than produced by a synthesis of artificial vocal sounds (phonemes).

Aural output can also be fed via a Bluetooth-connected ear piece, which is becoming increasingly commonplace.

10.2.3   Vocal Interface Elements

A microphone on a mobile device is necessary due to the telephony requirement, but there are other applications that might benefit from vocal input, such as recording voice notes, voice-control and speech-to-text conversion.

Increasingly, digital signal processing techniques can be used to assist with voice input. One possibility is noise reduction, so as to remove the unwanted background noise that might be experienced when say travelling in a vehicle or in a crowded place. All of us know the difficulties in trying to be heard (or in hearing) whilst using a mobile device in crowded room. DSP can be used to address such problems, but DSP always comes at a price, which is hardware and power-consumption differentials implied by the computationally intensive algorithms that the smart guys have invented to ameliorate such interface problems.

---

[6] There is no standard definition of a 'smart phone', but originally the term arose to explain a device with ostensibly a mobile-phone form factor, but with PDA-like capabilities. However, the use of the term has become blurred and we only mention it for completeness. Our consideration of devices incorporates all possibilities, no matter their marketing terms.

DSP can be applied in ways that are already used in immobile applications, but have not yet penetrated into the mobile device family. One such application is audio conferencing where a mobile device could be placed in the centre of a group of people and allows a full-duplex audio conference call to be placed. This would imply the need for a speakerphone and circumferential microphone pick-up technology. It also requires DSP to eliminate echo and to enhance audio perception and quality of signals generated by a group of talkers.

We could well imagine a useful combination of conferencing technology with the electronic perception technology mentioned earlier, as deployed in the light-projection keyboard. For example, a group of users could sit around a device and all take part in a simultaneous chat session alongside an associated audio call, courtesy of obliquely projected keyboards spread across an arc on the desk. This idea could also be used for games, such as multiple choice quizzes and so forth.

10.2.4 Visual Interface Elements

We are familiar with ever-richer graphical displays courtesy of Liquid Crystal Display (LCD) technologies that are improving all the time. It is commonplace now for new mobile devices with the smallest screens (e.g. mobile phones) to have colour displays. PDAs have increasingly better displays with better contrast[7] capabilities and higher resolution, all deliverable with achievable power consumption that enables portability.

Newer LCD technologies, such as the reflective variants, also enable better power consumption due to their ability to harness ambient light to enhance the display, rather than back lighting. Reflective LCD technology makes use of only the surrounding ambient light to illuminate the display. Beneath the surface of the display a rear polariser is combined with a reflector assembly to bounce back ambient light to aid screen visibility.

Transmissive LCD display technology is how a normal display operates. To view the screen, there must be a continuous back light, and this must be brighter than the ambient light to save the screen from looking washed out and subsequently becoming difficult to read. These displays are naturally very power hungry and operate best in darker lighting conditions.

Transflective LCD technology is a combination of reflective and transmissive types. The rear of the LCD's polariser is partially reflective and combined with a back light for use in all types of lighting conditions. The benefit here is the back light can be switched on only when there is insufficient outside lighting. Conversely, when there is enough ambient light it can shut off to conserve power. Transflective takes the best of both worlds and enables viewing in dark environments. Additionally the display won't 'wash out' when viewed in direct sunlight. However, the contrast rating is not as high as a purely transmissive display.

There is a high degree of association between display technology and tactile input methods, the most obvious being the manner in which a windowed graphics environment requires point-and-click navigation, manipulation and data entry. Therefore, a pointing device has

---

[7] Contrast is the ability to discern between different colours, especially in the presence of high ambient light levels. Clearly, displays should aim for high-contrast capabilities so that images can be clearly seen, including by visually impaired users.

to be able to move a cursor on the screen, or be able to click active areas in the parts of the windowing display responsive to tactile control.

With stylus pointing devices, the LCD screen needs to be sensitive to stylus position and pressure. This is achieved by a translucent overlay membrane that is able to act as a transducer from mechanical pressure to electronic signals that indicate whereabouts on the membrane the stylus is applying pressure.

There are a variety of types of touch technology available but the five major ones include analog resistive, capacitive, infrared, acoustic wave and near-field imaging. Of these, resistive are commonly used for small displays, such as PDAs.

Analog resistive touch technology is comprised of a glass overlay that fits exactly to the shape of a flat panel display. The exterior face of the glass is coated with a conductive, transparent layer. A clear, hard-coated plastic sheet is then suspended over the glass overlay. The interior face of the plastic sheet is also coated with a conductive layer. Between the glass and the plastic sheet there are thousands of tiny separator dots about one-thousandth of an inch thick. When a stylus applies pressure to the surface of the display, the two layers make contact and a controller (i.e. silicon chip processing device) instantly calculates the x and y coordinates. This accounts for resistive overlay's very high positional precision (i.e. fine resolution).

The combination of affordable, low-power touch-screen technology with equally affordable LCD displays has enabled the mobile device market to become established. The transition from monochrome to colour displays is now fully underway and only devices that are very much budget priced will continue to have monochrome displays, although, for some manufacturers of mobile phones, the inventory costs are altogether reduced by standardising on colour displays across an entire product range.

What is still an issue is the size of the display, especially where portability is a key concern. Clearly, the more portable something has to be, the smaller and lighter it also has to be. This invariably means a small display and this immediately presents usability challenges, ones that impose a different set of considerations when applying the windows design metaphor. The usability is made all the more challenging by the obvious tactile differences between a stylus and a mouse, plus the ergonomic differences, particularly the ability of a desktop mouse to accommodate more than one button and even more interface devices, such as a scroll wheel.

10.3 INTERFACE LAYER

The interface layer is what enables the electronic signals in our interface devices to be understood by the software services layer. Any interface device is supported by a silicon chip device that handles electronic signals, usually analog, to and from transducers and electro-mechanical or electrophysical components.

What we eventually want is for our services software to be able to drive interface devices or to respond to their outputs in response to stimuli from the outside world, usually due to human interaction. The controller chips that control the interface devices can be controlled using low-level digital commands issued from a microprocessor or similar, such as a microcontroller or general-purpose DSP device. This would usually be the same processor that runs our services layer. Each controller chip, such as a

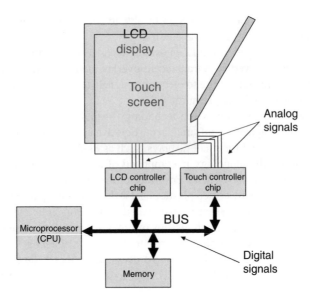

Figure 10.5 Microprocessor interface to interface peripherals.

controller for a LCD display, has its own specific set of commands that it will respond to via its microprocessor interface; this interface is usually referred to as a bus, as shown in Figure 10.5.

The microprocessor will run all, or most, of the custom software on our mobile device. There may well be an additional processor to carry out any DSP functions, such as speech processing. We shall be looking at possible mobile device architectures later, but for now our concern is with the handling of the interface peripherals. We have posited a layer in the software that is called the *interface layer*. This consists of is a set of device drivers. These are specialised low-level pieces of software that contain all the subroutines required to control the peripherals and to pass data to and from them.

The role of the device driver is not only to handle its associated peripheral, but also to present a simplified software interface to the service layer in order that the programmer of the service layer does not have to concern himself, or herself, with the details of peripheral operation. For example, the command to display a blue dot (pixel) on the LCD display may require a whole sequence of binary words to be written to the appropriate memory location on the LCD controller chip. However, the service layer would just like to issue a command like 'dot (120, 160, blue)' where 120 is the x coordinate and 160 is the y coordinate.

The device driver enables this abstraction of the interface to be achieved. However, it is usually the job of yet more software to make an even more powerful and abstract interface for the service layer. Perhaps the service layer would like to issue commands like 'rectangle' or 'button' or 'menu' and for the appropriate display widgets to be enacted on the LCD. We can imagine an entire chunk of software dedicated to providing a windowing interface paradigm to the higher software service layer.

We should remind ourselves that the set of possible peripherals for interfacing is increasing all the time, so we should not get overly fixated about displays. The inclusion of cameras is increasingly common on mobile devices of all form factors. A camera is based on a light-sensitive matrix and is like an LCD in reverse. Instead of writing to the controller chip to energise various pixel locations on the screen for subsequent lamination (reflective, transmissive, etc.), we grab values from the camera matrix to understand input light values.

A camera could involve us having to handle a continuous stream of images, just like with a video, so that we can subsequently grab images, either stills or sequences. We can imagine that for the microprocessor to keep processing an inbound stream of digital light values is a computationally intensive task. Therefore, the controller might be designed to grab images and write them directly to an area of memory on the bus without processor intervention, which is a technique called *Direct Memory Access* (DMA).

We do not want our service layer to be concerned with the low-level mechanics of controlling a camera, or any peripheral. So we would expect the device driver to do a lot of the work for us. Just like with the windowing display, we would also hope that we can find at our disposal a piece of software that can provide high-level interface primitives for the service layer to call upon, such as 'grab image', 'rotate image' and so on.

What we are describing here is the availability of toolkits or libraries of pre-written software that we can use to take care of housekeeping functions, such as windowing and image manipulation, leaving our service software to concentrate upon delivering or facilitating the actual mobile service we have in mind, like picture postcards (photo messaging) or videoconferencing.

10.3.1 Interfacing via the Network Layer

It is important to note that sometimes the human interface is not on the mobile device that runs the service layer. There could well be a separate interface device that communicates with the main device, or, more precisely, the software services layer, but does so via a networked connection. An example of this is the use of Bluetooth (BT) interface devices.

The most common device is probably the BT headset, in which the audio input (vocal input) is streamed across a BT connection to the processing device. Such a configuration could also be used for voice control of a device, not just for telephony.

Another example of a BT connected interface is the BT pen, an input device that enables ordinary handwriting, on either normal or special paper, which can be used to capture textual or graphical input for our services layer. This penned input could be used in raw form, as vector or bit-mapped graphical input, such as for taking freeform notes or making sketches, or can be used in conjunction with a handwriting-recognition process to convert the pen marks to character form, such as ASCII or UTF (two different standards for digitally encoding characters: American Standard Code for Information Interchange and Unicode Text Format, respectively).

As with any of these usages for computerised input peripherals, the interface handling is no longer carried out locally and the services layer does not talk to a local software process that handles interfaces. Instead, the interface dialogue takes place via the networking layer, such as through a BT protocol stack in the case of the examples we have just considered. This configuration is shown in Figure 10.6.

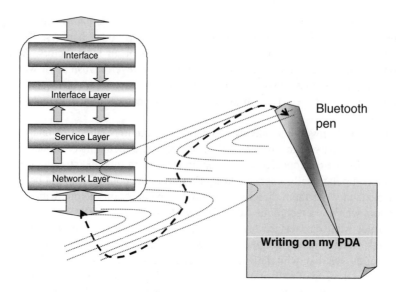

Figure 10.6 Textual and graphical input via a Bluetooth-connected pen.

10.4   SERVICE LAYER

The service layer is the 'business logic' of our mobile device application. A mobile service in its entirety can involve many interlocking functions, services and software spread across the entire mobile network. There are three modes of implementation:

1. Client–server (CS)

2. Peer-to-peer (P2P)

3. Standalone

These are purely modes of implementation, mostly reflecting the mode of collaboration with other software services in the network; they are not categories of application. For example, a particular application, such as a calendar tool, may well be able to operate in all three modes of operation depending on the context of use.

A user may wish to consult with their calendar, just to read appointments, survey the coming day's events, etc. This would seem not to require access to other services in the network, which makes it a standalone application. However, even this apparently innocuous requirement requires a lot of processing on the device, so we should not think of it as a trivial task just because we are not involving any networked activity at this stage.

For a moment, let's pause to consider what the processing requirements might be:

- Accessing a particular area of memory to read calendar values

- Processing the calendar values to transform them from an internal digital representation to a human readable one

- Creating a whole set of display widgets to display a calendar on the screen

- Rendering the graphics widgets on the LCD display, pixel by pixel

- Reacting to point-and-click stimuli on the touch screen

- Converting the point-and-click stimuli to appropriate calendar handling functions, such as 'get next month'

- Converting such functions as 'get next month' to a meaningful set of internal instructions to gather the appropriate data from memory, assembling it into a particular order

- Generating textual (character-based) output on the LCD screen and in the right place

We can see that there is a lot of processing potentially taking place just to display a calendar in the standalone mode.

The calendar application can also operate in other modes. We could find other peers to interact with and swap appointments. This would involve using the networking layer to communicate with them, and we shall describe the networking layer in the next section. We could also interface with software services running on the Internet, so as to share calendar spaces or to download appointments from event calendars, such as sports events, exhibitions, shows and so on. We may also want to connect with a calendar in our enterprise network and synchronise with it.

A great deal of what we want to do with a mobile device may involve common functions across a wide variety of applications. For example, accessing a calendar might be a common requirement, as might accessing an address book, or placing a call. There are many such functions that are potentially common to some or all applications. In would be advantageous to provide a set of common software building blocks to take care of these functions and for them to be available for the core software to call upon. This concept can be achieved in a number of ways and we shall be examining them throughout the rest of this chapter and in the Chapter 11 where we look at particular device software techniques, such as Java 2 Micro Edition (J2ME).

Later in this chapter, we will propose a possible generic device architecture that will take into account some of these common elements.

10.5 NETWORK LAYER

Now, we come to consider the final layer in our layered abstraction of a mobile device that we have been discussing (refer to Figure 10.1) – the networking layer. This layer is concerned with getting data on and off of the device, so it is a key enabler of our mobile service world. Part of the network layer is the actual hardware to achieve wireless networking. We have already introduced some wireless networking technologies, particularly those that enable us to connect to the IP network layer in our mobile network model, namely digital cellular technologies and wireless LAN technologies. In Chapter 12, we shall describe in detail the RF network and explain its functions and characteristics.

In terms of the implementation of a wireless connection on a mobile device, we can think of the wireless hardware as a separate entity or module, usually referred to as a *modem* (stands for <u>mod</u>ulator/<u>dem</u>odulator). An RF modem is a peripheral on the mobile device, just as the interfaces usually are (e.g. LCD display driver chip), and so we can generalise from the previous discussion about device drivers and software abstraction and apply the same principles to the task of interfacing to the modem. Typically, the modem will have a controller that facilitates the low-level software messaging interface with the microprocessor in our mobile device via the bus. This controller chip will have its own command set and protocols for receiving data via the bus to drive the functions of the modem. Therefore, on the main device processor, we shall need a device driver (software program) to handle the low-level control commands and data flows on the bus, whilst presenting these in more abstract terms to our service layer software. For example, the process to initiate a call via the modem might involve the writing of an elaborate software message to the modem's controller input on the bus. Rather than have to know how to do this, including the formatting of the low-level message, the service layer programmer would rather just be given a simple software command, like 'MakeCallTo(this number)'. It is the job of the device programmer to translate this higher level command to the low-level message format and also to take care of using the appropriate mechanisms to get the message to the controller's input on the bus.

Very often, a mobile device manufacturer will use a modem chipset from a third-party provider who specialises in modem design and manufacture. The chipset will come with datasheets that specify how to interface with the controller functions within the chipset. However, it is common these days for the chipset provider to also supply device drivers and libraries of software that make interfacing to the set even easier. Later on, we shall examine the various ways and levels that these libraries can be integrated into the device and how this eventually impacts on the development process for service layer software.

The software interfaces that our service software has to deal with may vary from device to device depending on the particular underlying software architectures that the device manufacturer has chosen; meaning the operating system and associated support for programming languages, such as Java. One can argue that much of the added value and skill in device manufacturing these days is in how effective the underlying software platform is in making it easy for the developer to quickly develop powerful applications for the device, being spared of low-level operational and implementation details. This will affect how much the device gets used to support various mobile services. In turn, this will affect device popularity and hence sales.

The software interfaces to the modem relate to how the modem is physically integrated into the mobile device. The modems may be fixed and part of the device circuitry, or they may be pluggable into expansion slots. In the latter case, the expansion slots will have their own controllers on the bus, such as a PCMCIA (Personal Computer Memory Card International Association) controller for PCMCIA peripherals. Industry standard expansion slot mechanisms have their own associated signalling protocols on the expansion bus. This is usually implemented by incorporating a controller device on the main processor bus that bridges to the expansion bus. This implies the need for a software handler to control the bridge.

As we shall see, the way that peripherals are handled by the service layer software depends ultimately on how the device manufacturer chooses to integrate these peripherals and make them available to the upper software layers (service layer). Most often, all the peripheral handling in a device is glued together by the operating system, which might be custom designed and proprietary to a particular device or family of devices, or it may be an industry-wide operating system licensed from a specialist supplier of mobile operating systems.

It may seem strange to the uninitiated, but typically an RF modem is like a computer within a computer as far as our mobile device is concerned. The inner workings of the modem may well involve its own microprocessor, memory, bus, peripheral connections and so on. Most digital wireless transmission techniques involve huge amounts of processing at the air-interface level, just to achieve the modulation and demodulation, RF interference mitigation and so on. Then, on top of the raw processing of bits in and out of the antenna,

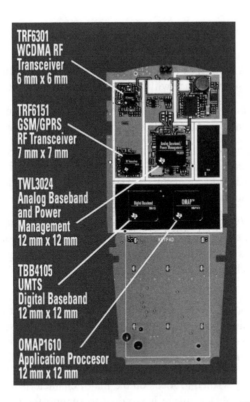

Figure 10.7 A reference design from Texas Instruments for a UMTS/GPRS device. (Reproduced by permission of Texas Instruments.)

we have the many protocol layers dictated by the particular RF standard in consideration, such as GPRS, UMTS or WiFi.

Due to the amazing achievements of silicon design, entire systems are implemented on a single chip, so many of the modem elements just described may well all be on the same piece of silicon, packaged into a single chip carrier[8]. This contributes enormously to lowering costs and power consumption. It is worth noting that there is an order of magnitude difference in power consumption between driving digital signals (by switching transistors on and off) across a silicon chip rather than across a printed-circuit bus (metallic tracks on some kind of insulating substrate). The electrical properties of printed-circuit tracks (e.g. capacitance and inductance) are such that electronic components have to drive output signals with a much greater push (energy) to get the electric currents flowing across the circuit tracks. Systems on silicon, therefore, usually offer a saving in power consumption.

It is useful to take a brief look at what these modems might look like at the functional level, so let's consider, for example, a wide-area RF modem that is both UMTS and GRPS capable (as we would commonly expect for wide-area devices until such time as UMTS subscriber penetration completely outstrips the need to maintain GPRS networks). Figure 10.7 presents a reference design from Texas Instruments, Inc., which shows various components required to implement a UMTS/GPRS modem.

---

[8] A chip carrier is the physical packaging that bonds the silicon chip to the input and output pins that are then soldered to the printed-circuit board hosting the device.

Let's take a brief look at the components in the reference design highlighted on this figure.

- *UMTS (WCDMA) RF transceiver* – takes care of the actual RF interface, processing the modulated signals in accordance with WCDMA principles (see Chapter 12).

- *GSM/GPRS RF transceiver* – takes care of the GSM/GPRS RF interface, processing the modulated signals in accordance with the GSM TDMA principles (see Chapter 12).

- *Analog baseband and power management* – this chip brings the RF signals down to their unmodulated frequencies and handles audio processing aspects in the case of voice and digital data stream interfaces in the case of data. It also includes power management functions for the entire chipset, including intelligent battery charging supervision.

- *UMTS digital baseband* – a good deal of digital signal processing is required to enable digital cellular transmission to reliably take place, such as error coding (convolution coding) that enables the hostile effects of the RF path to be mitigated. Ciphering is also carried out on this chip with the help of acceleration processors that offload some of the intense cipher calculations from the core processor on this chip.

- *Application processor* – this is an entire mobile device processor on a chip, as shown in Figure 10.8, including many of the peripherals we have discussed so far. Where some interfaces seem missing, this is because they are present on other devices in the chipset. For example, the touch screen interface is on the analog baseband chip.

In this section, we only wished to focus on the major characteristics of the network layer, and hence the issue of what software is required to support the reference design shall be deferred to our later discussion on generic device architectures (Chapter 11).

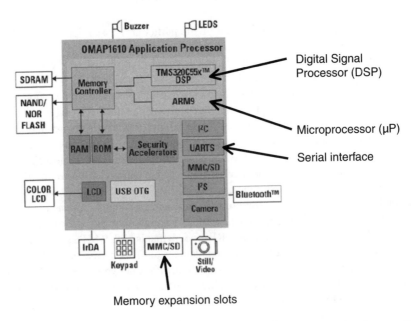

Figure 10.8 Application processor architecture from Texas Instruments' reference design.

In principle, the support of other wireless interfaces, such as Bluetooth and WiFi, is a similar affair. We expect to incorporate a similar modem chipset that presents an application interface to the core of our mobile device platform. The actual nature of the interface, how it is implemented and how it is presented to the core, is a matter for further elaboration later in this chapter.

Throughout the book, we have seen that there is usually an intimate relationship between mobile applications and networking protocols like HTTP; hence, we would expect there to be a means of providing a way for our service layer on the mobile device to easily gain access to such protocols. What we require is a software library on the device that can support HTTP without that much care about how this is implemented in the networking layer. Such a requirement seems sensible, and as we shall see, it turns out to be practical and increasingly available on many mobile device platforms.

10.6 ROLE OF DSP IN DIGITAL WIRELESS DEVICES

It is useful to review the basic architecture of a 2G mobile telephone, such as a GSM phone. This will enable us to understand what the basic building blocks are of a digital communications device, particularly the modem aspects such as the RF, analog and digital baseband modules, and to see how DSP processors are really at the heart of digital cellular (and wireless) devices.

In Figure 10.9, we can see in the basic components of a digital cellular device, such as a GSM phone. Just for completeness, we shall now briefly explain the functional blocks

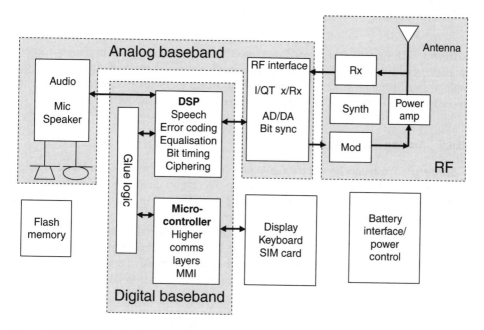

Figure 10.9 2G digital cellular phone architecture.

shown in the figure, though it is not really our intention to examine digital wireless modems in any more depth that this in this book.

10.6.1   Radio Frequency (RF)

Modulator/Power Amplifier   The raw bitstream that we wish to transmit is used to create a corresponding pattern in the RF transmission from the antenna that can be detected by a receiver at the base station. This pattern in the RF gets created by varying the frequency slightly about a central value, such that at certain points in time, corresponding to the bit rate, the RF signal has a certain characteristic that enables the receiver to detect the underlying bit values at corresponding bit intervals. The modulated signal is fed into a power amplifier that drives the signal into the antenna with sufficient power to radiate the required distance to the base station.

Synthesiser   To generate the underlying RF carrier that gets modulated we need to generate a signal at the required carrier frequency. We probably also need other signals at fractions of the RF carrier frequency to drive other parts of the signal processing chain in synchronicity with the modulation process. It is the job of the synthesiser to produce these reference fractional signals.

Receiver (Rx)   We need to be able to detect input RF signals from our antenna that are being received from the nearby base stations. It is the job of the receiver to detect signals within the required frequency band. It does this by applying filters to block out extraneous signal frequencies and by using a tuner to resonate with the required signal and detect it for further processing and eventual bit extraction.

Antenna   RF signals need antennas to convert from electrical signals to electromagnetic waves that can propagate through the air. The antenna also acts as a pick-up for signals that impinge upon it and resonate at that frequency to which it is sensitive. Antenna lengths are precisely chosen to facilitate the pick-up process at the desired frequency band.

10.6.2   Analog Baseband

IF Staging   We cannot handle the signal processing directly from the raw bit rate straight to RF, or vice versa. The problem is mainly in the receive chain where the RF signal is oscillating too fast for it to be converted into a digital signal that can be processed by a computing device at the raw rate. We therefore mix the signal down to an Intermediate Frequency (IF) where we can then deploy conversion. Generally, whenever signals get processed in electronics circuits, there is some kind of degradation in quality. There is a compromise between the chosen IF frequency and the reliability of conversion in terms of key metrics like signal quality and also component costs and power consumption.

Analog–Digital Conversion   The signal from the RF chain eventually needs to be converted into a digital representation whereby we use integer values to represent signal amplitudes. To do this conversion we use a device called an Analog to Digital (AD) converter. Such devices are selected according to how fast we want them to sample the analog signal and the accuracy of representing each sample as an integer. Of particular importance, is the range of integer values that we can accommodate, which is called resolution (or dynamic range, though not exactly). With AD converters, cost is related to resolution and sample speed.

Generally, AD converters are difficult to implement on the same silicon substrate as a large digital circuit, such as a microprocessor or DSP. This is because mixing analog circuit technology with digital usually requires stringent silicon-chip manufacturing conditions, silicon not always being the ideal substrate for analog electronics in any case.

10.6.3 Digital Baseband

Bit Synchronisation and Extraction   Having retrieved our input signal from the AD converter, we now need to extract the actual raw bit values. This may sound strange as the AD convertor is presenting us with raw bit values, so what do we mean by bit extraction. Usually the AD 'oversamples the input data stream. In other words, it is not just taking a digital reading of the analog signal at each point where it thinks the bit is, but at many points in between. This enables our receive process to use interpolation techniques in order to stand a better chance of finding the actual bit values that were transmitted.

We can see this in Figure 10.10 where a higher sampling rate enables us to better determine a point where there is a sharp inflexion in the signal. There is a fundamental minimum rate (called the *Nyquist rate*) that we have to sample at to catch basic detail in the signal at the rate we are expecting it to change, but if we exceed this sample rate we can do better at detecting the inflexions, particularly if our sampling is not ideally synchronised with the transmitted bit rate. Even so, we shall see that due to adverse propagation effects, even a perfectly synchronised system can still benefit from oversampling.

Equalisation   As just mentioned in the previous section, adverse propagation effects can affect our ability to extract the correct bits from the sampled data stream. This is because the RF signal seldom reaches our antenna directly from the base station transmitter. If we think for a moment, we should be able to confirm that for most of the time that we use a

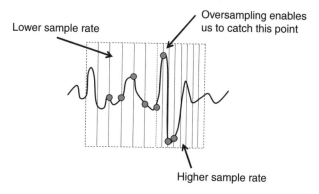

Figure 10.10 How sample rate affects signal quality.

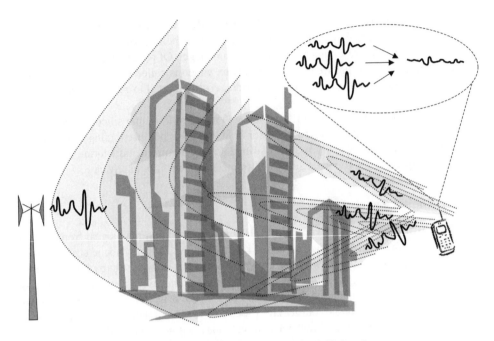

Figure 10.11 The effects of multipath propagation on the RF signal.

mobile phone we cannot see the base station, assuming we were close enough to do so with the naked eye and also assuming that we knew what one looked like (and it is not disguised as a tree, as shown in Chapter 12).

As Figure 10.11 shows, the RF signal bounces off objects on its way to the mobile device. However, because RF signals spread out as they propagate, we can end up with several instances of the signal bouncing around and arriving at the antenna. Under normal conditions, we can assume that these signals are basically copies of each other. However, they will actually have different power levels according to how far they have travelled and any path-specific attenuation experienced along the way, such as penetrating a building versus travelling unhindered over a field or over water.

Because these different signals take different paths, they are shifted in time with respect to each other. We can appreciate that the signal that has travelled a little further is a bit delayed. These multipath signals combine ('add' together) in our antenna and receiver. Due to the phase differences they can end up combining in ways that no longer resemble the original signal. Even without understanding the actual mathematics of it, we can perhaps appreciate what would happen if we listened to an audio signal, like a radio, but on several radio sets, each running slightly behind or ahead of the other, the phase difference constantly changing. It would sound garbled at times, and at other times not. This is what happens to the RF signal.

We can mitigate this multipath effect using a technique called *equalisation*. In fact, without equalisation, GSM would not be possible, it is that crucial to digital RF transmission. The way it works is simple in principle, but complex in practice and implementation; hence, why it accounts for a significant proportion of the computational complexity, and therefore cost and power consumption, in a mobile phone receiver.

We can imagine the sum path between the transmitter and the receiver as being modelled by several dominant pathways. If we are able to detect somehow the exact nature of these pathways, we could apply some kind of reverse process at the receiver. It turns out that this is possible so long as we have a means to measure the accuracy of our model each time we want to equalise a stream of bits in the receiver. We can do this by making sure that for a certain period of time we transmit a sequence of bits that is known in advance. When we receive these bits they will come out all scrambled due to the multipath effect. What we can do is to take the known bits and keep putting them through our model, adjusting the model through all its permutations until the output stream matches the messed up bits we actually received. We then know that we have a good model and we can use it in reverse to undo the multipath interference.

This sounds like a brute-force approach and in many ways it is. It therefore requires a lot of processing power. The actual process is slightly more complicated than I have explained here, but not quite so brute force in implementation. We can use elegant mathematical tricks to streamline the process, although overall it still remains a computationally intensive and power-hungry task.

Error Correction   Equalisation removes the effects of multipath propagation as much as possible, but it is not enough to ensure that we have an uncorrupted bitstream. The challenge when transmitting digital data steams is to achieve a very high level of data integrity that enables us to preserve what we are trying to transmit. Depending on the content, it may or may not be resilient to errors in the transmission. But let's say that we are transmitting financial information contained in a spreadsheet, even the corruption of a single bit of data could have grave consequences, as shown in the sidebar entitled 'Effect of bit changes on text'.

We therefore need very aggressive and robust error-correction capabilities when sending bits across the RF network. We can achieve this using a technique called *Error-Correction Correcting* (ECC).

Effect of bit changes on text

Imagine that we have a spread sheet with the discount at 1000.

As ASCII code, this is a bitstream of:

    0011 0001 0011 0000 0011 0000 0011 0000

Imagine that a single bit gets changed (shown underlined):

    0011 1001 0011 0000 0011 0000 0011 0000

Then we end up with a price of 9000 instead. A simple bit flip has cost us 8000!!!

If we were considering HTML codes instead of ASCII ones, imagine we have a price of £79 (79 pounds sterling), which in HTML is:

    &#163 &#55 &#57

which in Unicode Latin, as hexadecimal, is:

    26 23 31 36 33 26 23 35 35 26 23 35 37

---

which is (the first part in binary):

```
0010 0110 0010 0011 0011 0001 0011 0110 0011 0011
0010 0110 ........
```

Again, if we just flip one bit, as shown underlined:

```
0010 0110 0010 0011 0011 0001 0011 0110 0011 0010
0010 0110 ........
```

When rendered in a web browser the 79 pounds has now become 79 pence. If we were to upload this code to our server from our mobile device, as the latest price to be subsequently used in hundreds or thousands of online sales, then we'd lose a lot of money.

The way that error coding works is by adding additional (redundant) bits to the bit stream that we can use to tell us something about the other bits. Actually, that is probably a fairly loose way of describing the process, the problem being that error-correction principles are somewhat mathematical in nature and so difficult to describe here. They all rely upon a theorem called Shannon's Theorem that basically says we should always be able to devise a coding scheme to correct errors.

The error detecting and correcting capabilities of a particular coding scheme is correlated with its code rate and complexity. The code rate is the ratio of data bits to total bits transmitted in the code words, once we have added the error protection bits. A high code rate means information content is high and coding overhead is low. However, the fewer bits used for coding redundancy, the lesser error protection is achieved. A trade-off must be made between bandwidth availability and the amount of error protection required for the communication.

The most common type of coding is to use something called convolutional codes. We shall not look at what these are other than to note that the complexity is in the decoding process, not the encoding. This is an extra burden in the receiver processing chain in addition to the considerable burden already added by the need for equalisation. However, it turns out that decoding convolutional codes and performing equalisation can both use the same algorithm, called the Viterbi algorithm; therefore, we can benefit from using the same process for both parts of the chain. This becomes particularly agreeable if we use a special coprocessor to speed up Viterbi processing, as we save on power consumption and silicon costs by not needing such a fast main DSP processing core.

We need robust error-correcting techniques, especially for data applications, as we have already seen (see sidebar on 'Effect of bit changes on text'). However, what we have just discovered is that the more coding we use, the less bandwidth we have for the actual bitstream we want to protect. Previously we had understood that the presence of errors in the first place has something to do with the quality of the RF 'channel', i.e. the sum propagation path. What we can therefore postulate is a variable rate scheme that adapts to the channel conditions. When the channel is good, the amount of error-correction coding is reduced (i.e. the code rate goes up) and when it is bad, the error-coding is increased (i.e. the code rate goes down).

The adaptation of the error-correction code rate is something we may hear about, perhaps unwittingly, in the GPRS world where the coding schemes CS1, CS2, CS3 and CS4 get talked about a lot in the specifications (e.g. base station and handset specifications). We should clarify that we are not talking about classes here, like Class A, Class B, which are used to denote the level of GPRS service that can be achieved simultaneously with a GSM phone call. This is an altogether different concept. Coding schemes 1 to 4 refer to the amount of coding going on to protect the data stream. The handset can negotiate different coding schemes to be used in talking to base station depending on perceived channel conditions.

Voice Transcoding The underlying reason that digital cellular voice networks became possible when they did was due in no small part to the newly possible ability to digitise the voice and then *compress* it for transmission and decompress it at the receiving end. We discuss the significance of this technique with respect to its impact on digital cellular in Chapter 12 about the RF network. One of the most significant contributory factors in making the compression and decompression process possible was the availability of cost-effective DSP technology with low-power consumption. This continues to be a major factor in the success story of cellular technology.

Voice compression is referred to in digital cellular as voice transcoding. This is because the voice is first digitised into one format and then it is converted from that format to the compressed one, hence from *transformation* and *coding* we get the term transcoding. This should not be confused with an identical word that is used in the wireless developer world to refer to transforming web pages from one markup language to another automatically, so that they can be viewed successfully on mobile devices. The term *vocoder* is often used to refer to the apparatus (software and/or hardware) that carries out the transcoding function.

The underlying premise for voice compression is that within the range of sounds that the voice box (vocal chords) makes during speech, many of them are not critical in the hearing process in order for the voice to retain fidelity and still remain intelligible to the listener. We can think of the voice as being made of various components at different frequencies. We can throw away or diminish some of these components and it won't affect fidelity, although the more aggressively we dispense with some, or all, of the components, the greater the degradation in the speech quality.

Originally, voice compression was no different from any other type of audio compression. It was understood that the voice would remain intelligible if we filtered out the high-frequency components, as most of the vocal energy is in the lower frequencies. As we remarked earlier in our consideration of sampling RF signals, in order to catch faster inflexions in the signal, we have to sample at a greater frequency, notwithstanding the requirement to sample at least as fast as the Nyquist rate, which is basically twice the highest frequency component we want to reconstruct in our digital representation of the analog world. Hence, if we have voice components as high as 15 kHz (components moving as fast as 15 000 inflexions per second) then we need to sample at 30 kHz. Let's imagine that for each sample we use up to 8 bits to represent the level of the signal coming into our AD converter, so potentially we then have to transmit 240 000 (8 × 30 000) bits per second.

If we ignore the higher frequency components, as they are not required to make the speech intelligible, then we could sample at say 8 kHz and end up with 64 000 bits per second in our bit stream. This is already a substantial reduction in the number of bits we have to transmit. At this stage, we have not done any compression, just filtering or subsampling of the signal.

What is interesting to note is that most of the time the signal will not be changing rapidly, so we can deploy a technique to transmit the changes from one sample to the next, rather than the entire sample. We refer to this as Adaptive Differential Pulse-Coded Modulation (ADPCM), which in itself is a form of compression. It is also a form of compression that does not sacrifice quality as we are not throwing anything away by sending the difference values. We can reconstruct entirely the absolute values at the other end. Of course, we notice that this process is very susceptible to bit errors. If we failed to transmit the original reference value upon which we subsequently reconstruct the original samples, then we can imagine that we could easily change the signal altogether. This might cause loss of fidelity or even loss of intelligibility. Yet again, we see why error coding is so important; as it was for non-voice data sources (refer to the sidebar on 'Effect of bit changes on text').

ADPCM is actually the usual way of representing audio signals digitally, due to its compactness. However, using modern DSP techniques, we can do a lot better than this for voice. ADPCM is a technique that is generally applicable to audio. However, the voice has its own unique aural characteristics that enable it to be modelled in a way that we can attempt to synthesise.

We understand that vocalised words can be deconstructed into constituents, such as the production of phonemes. Phonemes can be thought of as instructions for articulating speech-sounds, and so a phoneme can be described in terms of the behaviour of the vocal apparatus that occurs when a speaker articulates his or her particular representation of the phoneme.

Phonemes can be divided into consonants and vowels. In the articulation of consonants, the flow of air from the lungs through the vocal apparatus is cut off or impeded. In the articulation of vowels, the flow of air from the lungs is not impeded, but the vocal organs are used to change the shape of the oral cavity and thus make different sounds for different vowels.

Amazing as it may seem, using DSP and mathematical models of how phonemes get translated into measurable signals, we can attempt to model a person's vocal apparatus in software. The reason this is useful is that instead of sending sound samples, we can transmit codes that represent phoneme production. It turns out that this is a much more efficient way of compressing speech signals. This type of approach is called parametric encoding, as we are now transmitting vocal parameters rather than sound samples.

Speech compression technology is changing all the time, as is the ability to deploy large amounts of intensive DSP processing at lower cost and lower power. This is why we have seen an evolution of speech compression offerings in the 2G cellular systems. The advent of multimedia communications in the Internet world has also resulted in greater interest in this area.

Speech compression is *lossy*, which means that some of the vocal excitement of the speaker does get lost in transmission. The more we compress, the lower the fidelity, but the lower the bandwidth required to carry the signal. Newer algorithms have sought to improve fidelity at lower bandwidths, rather than attempt to lower the bandwidth yet further. They are also adaptive, which means that, should we chose to, we can dynamically adjust quality and/or bandwidth to meet various requirements when running our voice network.

Encryption   It was understood early in the development of cellular radio, that eavesdropping would be an impediment to commercial success. First generation analog cellular systems were notorious for this problem, with many high profile cases of surreptitious behaviour, often involving famous people. Therefore, encryption of the signal travelling across the RF channel was deemed an essential requirement for 2G systems, and it remains so for 3G.

Encryption of the signal is a sophisticated task that involves intensive bit-level manipulation of the bitstream. This process is best carried out by a DSP as these devices have excellent bit-manipulation capabilities, as one would expect. Nevertheless, it is common to see the availability of ciphering accelerators. These are small dedicated digital circuits that are companions to DSP core processors. They get used to offload the processing overhead from the DSP, thus often lowering implementation costs and power consumption.

10.6.4 Digital Signal Processor (DSP)

As we have seen, the heart of all digital cellular phones is a DSP. It plays a key role in all of the digital baseband processing. The dominant DSP in this market is the TMS320C54x by Texas Instruments, Inc.,, which can be found in many modern cell phones. This processor is responsible for modulating and demodulating the data stream, coding and decoding to maintain the robustness of the transmission in the face of transmission bit errors, encrypting and decrypting for security, and compressing and decompressing the speech signal. In other words, it carries out all the baseband processing on one chip.

Carrying out lots of signal processing tasks on one chip means that it has to operate very fast. This is very much a real-time operation. Voice samples or data packets are arriving at a periodic rate, so each of the baseband tasks has to be carried out before the next set of samples or packets arrive for processing; not forgetting that this has to be done for both the transmit and receive processing chains concurrently.

A DSP carries out instructions that tell its internal processing core and mathematical engine how to process the incoming bit stream. The sophistication of the digital transmission algorithms, and the hard real-time constraints, mean that a DSP has to carry out millions of instructions per second (MIPS) in order to meet its deadlines.

In early 2G TDMA phones these functions could be accomplished with 30–50 DSP MIPS. As vocoders have become more complex and as data rates have risen in 2.5G phones, the total DSP load has risen past 100 MIPS. DSP technology has kept up with this requirement whilst, amazingly, also getting more efficient in terms of power consumption due to improved silicon fabrication techniques, such as the ability to operate the internal transistors at a lower voltage.

The CDMA standards require a somewhat different functional partitioning because of the data rates generated by spreading are too high for direct processing by a general purpose DSP device. While the DSP can still be used to process at the basic data rate (functions like forward error correction, encryption, or voice compression), custom silicon hardware operating under the control of the DSP must be used to process and modulate/demodulate the spread-spectrum signal.

10.6.5 Summary

Our discussion of basic mobile phone architectures has enlightened us about the sophistication of mobile devices as a necessary response to sophisticated RF networking requirements. Here we have only discussed digital cellular standards and implied architectures, the most fundamental of which we have introduced and examined their basic components. We discovered that a large amount of intensive processing is required to move data across an RF network, mostly due to the harsh conditions that our RF signals experience and our desire to maximise RF-spectrum capacity by compressing signals into their smallest form without compromising integrity and quality too far.

We have not yet discussed other DSP functions that require even greater processing power on top of what the RF modem already demands. Applications like voice recognition, voice-to-text, speech synthesis and video compression all require large amounts of DSP processing. We shall discuss these later in relation to their impact on device design and how we develop applications that require a DSP element. As we shall see, it is not necessarily the case that a maths-intensive application requires a DSP. Some newer microprocessor designs incorporate certain features to accelerate maths functions in a DSP-like manner. The distinction between a DSP and a general purpose processor is blurring in some cases.

10.7 SUGGESTING A GENERIC DEVICE ARCHITECTURE

Let us now develop our previous discussion to start expanding our understanding of device architectures, leading to an exploration of what possibilities are open to us in terms of developing software services to run on mobile devices as part of any mobile service offering.

We reiterate that there are three basic modes of application on a mobile device: CS, P2P and standalone.

We have posited that, no matter the mode, a basic layered model for a mobile device can be constructed consisting of; interfaces, interface software, services software, network connections. The guts of our service are enacted in the services layer. We may like to refer to this as the 'business logic' of the mobile application to draw a parallel with the software approach we witnessed with J2EE (refer to Chapter 8).

Thus far, we have only considered the layered model as a notional concept, but we have tried to convince ourselves that it is a sound and complete model. The issue remains as to how we actually deploy services using such a model (or architectural basis).

We can begin to approach the basis for software and applications deployment by adding some sound software patterns to our model. We shall tackle this process from two sides. First, we shall examine what framework we need to deploy software on a device, regardless of its mode or function. In other words, we shall introduce basic software practices for embedded devices. Second, we shall propose common functions that we would expect to find on any mobile device, or across a range of devices, things that are commonplace with little need to justify, such as a man–machine interface (MMI) for telephony, i.e. being able to dial a number, place a call, etc.

Let's take the first approach and look at a possible generic architecture for pulling our device together based on what we know so far from the layer model and its implications. A refined block diagram for our mobile device is shown in Figure 10.12.

We can still see our four layers in the diagram, but they are now set within a context of other companion or support software elements and hardware entities needed to support the layers and to glue them all together. We shall now discuss this refinement of our layered model in more detail.

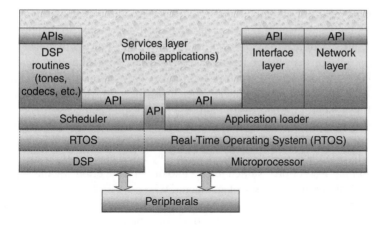

Figure 10.12 High-level generic architecture for mobile device.

10.7.1 Core Processor and Operating System

The workhorses of a mobile device are the microprocessor and DSP. Ultimately, our application software runs on these devices, although, as we have already noted earlier, the sum total of software running on a mobile device extends to include those elements running on the processors embedded within the various chipsets that bring the device to life.

Microprocessors are mostly serial devices, able to run only one logical unit[9] of software at a time. For example, the microprocessor may have to respond to an incoming message from the RF chipset indicating a call arrival. The processor would run a piece of software to activate the ringtone, display a call alert icon, possibly actuate a vibration alert motor, or whatever is required to handle this event. So we could imagine a software process that we could describe by the name 'handle call'. If the processor is running the 'handle call' process, then it is not running any other process. However, other processes may be vying for attention, such as a battery charge meter indicator process, an email alert process and so on. We don't have to think too hard to bring a long list of potentially useful and meaningful processes for our mobile device processor to run.

Due to its serial nature, the processor has to swap continually between vying processes, so that they all get a turn at being executed. If this process is done quickly enough, then the swapping is not noticeable and it is as if there are N virtual processors for N processes waiting to be executed.

When each process gets control of the processor, then we may have to implement some degree of resource control to make sure that processes do not violate the resources of other processes. For example, we wouldn't want the vibrate-alert process to start writing rubbish to the LCD display or to start sending messages via the RF data connection. We need some kind of process containment.

These types of challenge, such as scheduling of processes, swapping them in and out of the processor, protecting resources, and other mechanisms, are all the function of the operating system. In our case, we have the added challenge that certain processes need

---

[9] What constitutes a unit of software is not something we want to get into here, but I am thinking of software processes as units rather than lines of assembler code (or higher languages for that matter).

a deterministic means of gaining access to the processor and completing their designated tasks. For example, we would not want an email sending process to be able to hog the processor to the extent that the call-handling ('handle call') process is stalled and we end up missing a call. This is why we have the need for what is called a *Real-Time Operating System* (RTOS), the idea being that certain tasks have a very definite time-limit for initiation and completion and this is what we mean when using the term *real time*.

In real-time systems, the success of a process running on the processor is not just measured by its computational output, but is also related to how long it took to do it. Time really is 'of the essence', unlike in standard desktop operating systems, which generally deploy a 'best efforts' approach; processes finish when they finish and this is deemed acceptable most of the time. 'Best effort' is often turned into 'acceptable effort' by the application of brute force; plenty of processor speed and memory.

Clearly, what is meant by real time is a matter of degree. Even on a standard Windows desktop, an email polling process may well block our ability to type a letter, but not to such a degree that it becomes significant. However, more critical to our discussion is that in such a system, there is no means to control deterministically the scheduling and completion of processes, so the more real time that things become, the less likely they are to work on such a platform.

The concept of *hard* real time perhaps needs to be introduced. This is the notion that unless a process completes by a certain time, then its output is deemed a catastrophic (or complete) failure. Catastrophic need not mean that a nuclear power station is about to superheat into oblivion and irradiate everyone around it. In our case, we would probably deem that the inability to receive calls due to call-handling processes being blocked is catastrophic. Were it to happen often enough, the mobile service may be deemed to be a failure and would result in collapse of the service or even the business. That would be pretty catastrophic.

Soft real time is where the success of the process diminishes the more postponed it gets. We may consider that the reception of picture messages (via MMS) has a softer real-time constraint than call-handling. We might receive an MMS message indication whilst listening to our favourite MP3 track, or whilst watching an MPEG-4 video clip. If our device is too busy to download the picture-message payload from an MMS centre, then it can try again a bit later. Depending on whether we have already been notified of the availability of a message, and on its importance, the delay in being able to view it may or may not be problematic to us, but probably not catastrophic.

An RTOS is a specially written operating system that attempts to handle process timing issues as just described. There is no single, agreeable and sufficiently comprehensive definition of an RTOS, but we can elaborate on its characteristics in order to better understand what we require on our mobile device.

A hard RTOS must guarantee that a feasible schedule can be executed given sufficient computational capacity and once external factors are accounted for. External factors, in this case, are devices that may generate interrupts, including network interfaces that generate interrupts in response to network traffic, which in our case can be due to data or voice. In other words, if a system designer controls the environment of the system, such that events are within known bounds, the operating system itself will not be the cause of any delayed computations.

The types of mechanisms used to implement an RTOS are beyond the scope of this book, but the basic principle is the ability of the RTOS to gain a high degree of control over

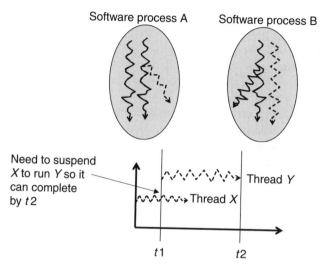

Figure 10.13 Concept of thread scheduling to achieve real-time deadlines.

the scheduling of software process threads[10]. The degree of control available enables one thread to be suspended and pre-empted by another thread, so that with sufficient processing resources and judicial scheduling, real-time constraints can be accommodated.

This concept is shown in Figure 10.13, where we show two software processes. We could imagine they are an MPEG-4 video media player and a Bluetooth pen handler. We can see that each software process has various threads of execution through its code. Thread X is the media player thread that fetches more encoded packets from an internet-hosted server via an RF network connection. Ordinarily, thread X would run to conclusion, but at time t1 the Bluetooth pen gets lifted form the page and wants to send a stream of vector pen strokes to the mobile device, to be handled by thread Y in software process B. Usability tests have shown that an upper limit of 500 milliseconds is tolerable for displaying the pen strokes. We therefore have until time t2 to complete the handling and processing of the pen input.

Because thread X was going to fetch a few more seconds of video from the server, we would not have been able to meet the deadline t2, so we use our RTOS to suspend thread X in favour of giving greater priority to thread Y. This is not a problem for our system as the delay in buffering the video will not affect playback, unless we did it so much that thread X never completed or was severely retarded.

Priority manipulation of program threads can benefit a variety of applications, including, for example, media players that are increasingly common on mobile devices (MP3, WAV, MPEG-2, etc.). The operation of a media player can be tied to the media rate required for proper playback (i.e. 44 kHz audio, 30 frames per/second video). So, within this constraint, a reader thread and a rendering thread can both be designed to wake up on a programmable timer, to buffer or render a single frame, and then go to sleep until the next timer trigger.

Timer-based pre-emption provides the tempo of execution that allows the priority to be assigned above normal user activities, but below critical system functions that must not be

___
[10] A thread being a particular path of execution through the software process that is active.

allowed to become starved of processor time. With well-chosen priorities, playback will occur consistently at the given media rate. A well-written media player will also take into account QoS, so that if it doesn't receive adequate Central Processing Unit (CPU) time, it can reduce its requirements by selectively dropping samples or follow an appropriate fallback strategy in an attempt to mask the problem from the user. This will then prevent it from starving other processes as well.

We may also wish to treat certain user events preferentially within the system. An example might be a continuous data stream from a Bluetooth pen as we attempt to jot down notes during a meeting. We would prefer that the mobile device appears responsive to our note taking, even whilst doing other tasks, such as possibly audio-recoding the meeting directly into a memory card (e.g. Multimedia Message Centre, or MMC), or playing back a video stream as indicated in the earlier example.

This works well when we increase the amount of concurrency within the application and when the event can always be handled in a small, predictable amount of time. The key concern here is the frequency at which these events can be generated. If they can't occur too frequently, it is safe to raise the priority of the thread responding to them. If they can occur too frequently, such as if we are writing a particularly long sentence and fairly quickly, then other activities will be starved under overload conditions if we keep assigning the pen-handler thread all of the available CPU time.

There are different programming strategies that can be employed to address these sorts of issue, such as dividing responsibility for events into different handling threads with different priorities, using timers to force suspension of priority threads and so on. In our earlier example of suspending the video-buffering thread to enable a more responsive handling of our BT pen stokes, we could imagine that we might better serve the video process by having two threads that fetch video, one that is a housekeeping thread to keep the buffer fed under normal circumstances, the other a higher priority remedial thread that runs if the buffer is nearing underflow, or has already emptied. Under this condition, we can suffer the infrequent delay in pen response in favour of keeping our video stream running.

The reader should be aware that these challenges exist and that different mobile devices will have different mechanisms available to assist our application with achieving the desired real-time performance.

We should also appreciate that under certain conditions, it may not be possible to have sufficient control over how the RTOS treats our application. Some of these issues shall arise further in the ongoing discussion in this chapter and in Chapter 11, particularly when we look at the pros and cons of different approaches to developing applications.

10.7.2 Digital Signal Processor

We previously discussed the importance of a DSP for the RF modem aspects of a mobile device, but we are also concerned with general DSP functions and how they might get integrated into our mobile devices and made accessible to the services layer. With the ever-increasing need for computationally intensive processes like video compression, voice synthesis and so on, the inclusion of a DSP on a mobile device is an obvious idea. We should caution that there is an alternative approach, as we shall see shortly, but for now the inclusion of a DSP does seem sensible.

If we consider a DSP by itself, then usually it is deployed to run a very narrow set of functions that are highly mathematical in nature, usually involving a good deal of repetitive tasks, most likely vector-based (e.g. matrix manipulation[11]). It is not intended to provide general-purpose computation, although, in theory at least, there is no reason why it shouldn't. However, this would generally not be a good idea because for each unit of processing power, measured in some general processing sense, a general-purpose device is cheaper than a DSP. Also, programming tools for general-purpose processors are more plentiful and more advanced, including, significantly, the existence of available operating systems to support the software processes.

Operating systems for DSPs are not that widespread and have limited features. For example, as a rule, we would not expect any support for graphical output devices, never mind support for graphical interface paradigms like windowing. On a DSP we usually expect some kind of basic task loader and scheduler, but nothing like the sophistication of a fully blown commercial RTOS.

If a mobile device has a DSP available for running software, it would usually be expected that the developer would develop for it almost independently of the general-purpose microprocessor device. Any collaboration between the two would usually have to be managed by judicious programming on both devices, most likely involving passing of messages between the two. There may or may not be existing support for this messaging process built in to the RTOS or any utility code that may come with the DSP.

The concept of a unified RTOS abstraction for a multiprocessor environment, supporting both DSP and general-purpose microprocessors, is not widespread. There are a variety of challenges, not least of which is devising a programming paradigm that would make sense for such an environment. Among other challenges, this would include grappling with the scheduling of tasks running on two processors. It is hard enough devising a sensible real-time scheduling scheme for tasks running on one processor, but when we have to manage parallelism, matters become significantly more challenging.

We should not be dismissive of the dual processor approach. After all, it works well in the case of the typical RF modem where there are generally two processors, one DSP and one microcontroller. A common paradigm, or metaphor, is to set things up in a way that the microcontroller is viewed as the master and the DSP as the slave. In other words, the microcontroller dictates the pace and probably is the instigator of what gets run by the DSP at any particular time. Many applications will fit the metaphor well, especially as the DSP functions are usually very deterministic, thoroughly enshrined in a rigid real-time framework that is repetitive and predictable. Data is injected into the input of the DSP, tasks get run one at a time, and then data gets squirted out; a predictable processing pipeline.

There is an alternative approach to the dual processor architecture. It is one that mitigates many of the problems we have just discussed, particularly how to provide a unified programming metaphor. Modern microprocessors are becoming more and more DSP-like by the inclusion of modified or additional circuits that are optimised for certain DSP (mathematical) operations. This idea has been pioneered on the desktop environment where DSP coprocessors were once expected to become popular, but never quite made it to the motherboard. They do exist of course, but in the form of embedded solutions (like a modem) inaccessible to the general programming environment.

---

[11] Many processes involving audio-visual streams can be reduced to matrix math problems. Hence, a processor that is optimised for processing matrices would be useful. Matrix manipulation typically involves repetitive arithmetic operations and a DSP is highly optimised for such processes.

Figure 10.14 JPEG image processing.

When examining the idea of including DSP structures within general-purpose processors, the decision by Intel Corporation to introduce Multimedia Extensions (MMX) technology is what comes to mind. This was initially an attempt to add some processing circuitry that could speed up graphics programming, or any vector-based processing. For a while, Intel named one of their Pentium chips with the MMX moniker, but later dropped any special reference to it after the circuitry became common place in subsequent processors in the Pentium family. From the Pentium III onwards, the enhancements got even better and became known as Streaming SIMD Extensions (SSE, and later SSE2 for the Pentium IV).

The technology used for MMX and SSE is a well-known processor architecture called Single Instruction Multiple Data (SIMD). Usually, processors process one data element with one software instruction, a processing style called Single Instruction Single Data, or SISD. In contrast, processors having the SIMD capability process more than one data element in one instruction. Let's say that we want to take an image using a camera phone for sending via MMS. We first wish to apply compression to the image using the familiar JPEG approach.

The processing chain for JPEG is shown in Figure 10.14.

To implement JPEG compression, we have to take the entire set of pixels in the image and apply mathematical processes, such as the Discrete Cosine Transform (DCT). The DCT is applied to 8×8 blocks of pixels at a time. This does not mean we have 64 operations to perform. On the contrary, the actual processing for each of the 64 pixels is more complex, entailing several operations per pixel, mainly successive multiplications and additions.

The resulting operation to compute the DCT for a single 8×8 block of data is 970 operations on a standard processor without MMX or SSE. When we talk about operations on a processor, we really mean clock cycles, which is the basic unit of time upon which all operations are synchronised. In the case of MMX architecture, according to Intel, this can be dramatically reduced to 280 clock cycles, and still further reduced to 250 clock cycles using SSE technology.

This four times' reduction in time (or four-fold increase in speed) is about what we expect as the SSE circuitry deploys four arithmetic units in parallel where previously we would have only had one, as shown in Figure 10.15. Therefore, we can construe that once a mathematical process is underway involving repetitive arithmetic manipulations, we would expect to be running fours times as fast using MMX or SSE.

Intel has developed a wireless version of MMX, called Wireless MMX, and this is part of the new breed of devices aimed at mobile applications or internet appliances. These devices are part of the X-Scale architecture. The Wireless MMX technology gives a large performance boost to many multimedia applications, such as motion video, combined graphics with video, image processing, audio synthesis, speech synthesis and compression, telephony, conferencing, and 2D and 3D graphics. Even without MMX, the Intel XScale core provides the ARM V5T Thumb instruction set and the ARM* V5E DSP extensions. To further enhance multimedia applications, the Intel XScale core includes additional Multiply-Accumulate functionality as the first generation of what Intel calls 'media processing technology'.

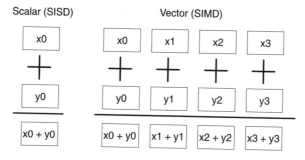

Figure 10.15 Parallel processing (SIMD) speeds things up.

It is fair to say that the distinction between a separate DSP processing element and an integrated one is blurred in terms of hardware instantiation. For example, the Wireless MMX technology is actually implemented as a coprocessor to the Intel X-Scale processor core, so it is like a separate processor in some regards. However, the degree of integration with the main processor is such that the software developer does not really have to concern themselves with a disjointed programming model. This is especially true with some of the newer software compiling technologies that are automatically able to map suitable code segments onto the appropriate instructions in the MMX or media co-processor.

10.7.3 Application Loader

In our basic device architecture, we have proposed that applications run on top of the RTOS. However, we did not discuss how the applications get loaded in the first place and how they get registered with the RTOS so that it knows where to find them and how to launch them. This is the job of the application loader. We shall not say much about this here other than to highlight that in a mobile context, we are going to be interested in being able to download applications to our device via the RF network, be that from the IP network or from a friend or colleague in a peer group, using P2P interaction. The ability to load OTA has become increasingly important, although historically it has not been present on that many devices, even though it is not a particularly challenging technical feat to achieve. Some of the new mobile-aware operating systems and related programming environments have started to address this issue.

We shall be looking at OTA provision in more depth later in Chapter 11, particularly in relation to the use of Java (J2ME) on a mobile device, although we shall discuss the general issues of OTA later in this chapter.

10.7.4 Application Programming Interfaces (APIs)

In our proposed basic device architecture, we have elaborated on some key attributes of a useful mobile device, namely:

- Core general-purpose microprocessor to run our software on
- An operating system, preferably real-time capable (RTOS), that runs on the target processor

- Provision of a capacity to handle complex mathematical tasks such as image processing, either by dedicated DSP devices or via 'DSP acceleration' techniques on the core processor
- A means to install and load our software onto the RTOS and start executing it

This just leaves us now with the job of writing our software to run in the services layer and being able to take advantage of the provisions just mentioned. The software might need to access a server (CS), or access another peer (P2P) or simply perform computations on the device for the user's benefit, without the need to network at all (i.e. standalone mode).

To write the device software, we mainly need two things:

1. Software development tools to write and test the code and then produce an installable application to run in the services layer
2. A known means to access the device resources from our code, like the RF modem

The first of these is mainly catered for by the existence of a compiler that can take our high-level language and then convert it to the instruction set understood by the target processor (whether core processor or DSP, or both).

The second matter is particularly important. This is about how to make our software work with the software already bundled with the device, like the software that controls the telephony features. Usually, the bundled software will come with published APIs, which tells us how to access and use it from our app.

We often think of computer programming as being about learning programming languages, data structures and data-handling algorithms. This is only one aspect of the discipline of programming. Among other requirements, a programmer needs to know how to use the APIs they are expected to make their program work with. In fact, it is often a more useful measure of a programmer's ability than what languages they know.

An API is a strange name, as it tends to imply something insubstantial. An interface doesn't sound like much. However, in the case of an API, this is not the case. It is better to think of APIs as software programs that come as pre-installed libraries of software functions on the operating system. The pre-installed routines can do things on behalf of our application, without us having to concern ourselves with the details. To access these library routines, the programmer has to know how to point to where they are on the device (somewhere in memory) and the names of the routines and their expected parameters. The supplier of the API usually publishes documentation saying how to use it. It is then just a matter of writing our application code and calling these API functions at the appropriate point in our code.

For example, let's take a look at the Pocket PC 2003 mobile device operating system. This OS can run on an Intel X-Scale processor, which reminds us that we can probably do powerful mathematical processing, if we need to, utilising Wireless MMX. In reconsidering our block diagram (refer to Figure 10.12.), we are now venturing that the network layer is accessible to us via an API, or possibly several APIs, perhaps one for each wireless networking device that we need to access. For example, Pocket PC provides an API to interrogate or configure the status of any Bluetooth modem that might be present on the mobile device.

In considering the use of the Microsoft APIs relating to BT, consider the following function:

BthSetMode

Use the **BthSetMode** function to set the Bluetooth mode of operation and reflect it in the control panel. This function is also used to persist that state across hardware insertion and reboot.

Syntax

```
int BthSetMode(
   DWORD dwMode
);
```

Parameters

dwMode
[in] Indicates the mode of operation that the Bluetooth radio should be set to. See BTH_RADIO_MODE for possible values.

Return Values

Returns ERROR_SUCCESS on success, or an error code describing the error on failure.

Requirements

Pocket PC Platforms: Pocket PC 2003 and later
OS Versions: Windows CE .NET 4.2 and later
Header: Declared in bthutil.h
Library: Use bthutil.lib

What this piece of documentation tells us is that there is a software routine called `Bth-SetMode` that enables us to set the operating mode of the BT hardware. For our current discussion, the key piece of information here is the `bthutil.h` and `btutil.lib` at the end. What this basically tells us is that there is a piece of software that already exists on a Pocket PC device, called `bthutil`, this being the software realisation of the API. This software, if we call it up from our program, will take care of configuring the Bluetooth modem for us at our application's behest. It knows how to talk to the BT modem to get things done. We simply ask it and it will make the configuration happen on our behalf. There is no need to concern ourselves with how this happens at the lowest level – this is the beauty of the API approach. In the case of the API example above, then the means to accessing its goodness is by inserting the piece of code indicated into our application, thus:

```
int BthSetMode(
   DWORD dwMode
);
```

For this discussion, we don't need to know about the anatomy of this function, but the principle of this is what we are trying to illustrate. In the case of Pocket PC 2003 and its sister Smart Phone OS, this API is provided courtesy of the operating system itself. The operating system has a built-in awareness of BT peripherals. For this to be so, clearly the BT peripheral must conform to an agreed interface specification for it to be accessible like this via the OS in an expected and standard manner. This would mean that any BT peripheral being inserted into a Pocket PC 2003 device would subsequently be accessible via the associated API.

In the case of the BT protocol set, it is quite diverse and complex. The API concept allows us to focus on what we want to do with BT at the application level, rather than how a BT modem works at the low-level interface. Let's say we wanted to write an application to implement a BT-chat application for use between a group of peers (an example of a P2P application), and we would like to focus on designing and programming the peer chatting paradigm, making a nice user interface, etc. We do not want to bother with the details of how we find other BT devices and make a connection with them. Using the API approach, we could simply issue a 'command' to sniff out other BT devices within range, and then using the results, subsequently establishing a paired communication pathway.

Looking at the diagram in Figure 10.16, we can see how the provision of a BT API works. Our software is the blob on the left, which is what we write and focus our creative

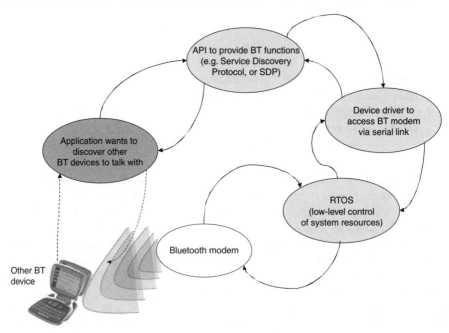

Figure 10.16 Using an API to sniff for other Bluetooth devices in the region. (Reproduced by permission of IXI Mobile.)

energies on. We can imagine that we are talking direct to the BT modem (or even better, transparently to other BT devices, as shown). All the other blobs are the API, device driver and RTOS working to make this happen.

What we can conclude from our discussion on APIs is that they are a useful way to enable the services layer to gain access to underlying resources on the mobile device. Different APIs can exist to take care of different functions. We have looked just briefly here at an API for a BT peripheral running on a Pocket PC device. This still fits within our layered model. We have a layer of network services, which can be any type of wireless standard, so long as an API is presented to the services layer so that mobile applications can be developed for the wireless modem concerned. In some cases, the API will be part of the Operating System (OS), as we have just examined, and in other cases, the provider of the peripheral chipset or device will provide the API themselves.

If the API comes from the peripheral vendor (either directly or via a third party), then it must be written to run on the host RTOS. Currently, the most portable way of doing this is to provide APIs written in the C programming language, this being the most common language for embedded systems. However, other languages for APIs are becoming increasingly popular; especially Java, but we defer this consideration until later.

Using this approach, the applications programmer responsible for programming a mobile application can rely upon being able to access powerful features of the mobile device without having to concern themselves with the interface details. For example, if a GSM modem is present on the device and the services layer wishes to send a text message (via SMS), then in the case of Pocket PC, an API exists for this:

```
HRESULT SmsSendMessage (
  const SMS_HANDLE smshHandle,
  const SMS_ADDRESS * const psmsaSMSCAddress,
  const SMS_ADDRESS * const psmsaDestinationAddress,
  const SYSTEMTIME * const pstValidityPeriod,
  const BYTE * const pbData,
  const DWORD dwDataSize,
  const BYTE * const pbProviderSpecificData,
  const DWORD dwProviderSpecificDataSize,
  const SMS_DATA_ENCODING smsdeDataEncoding,
  const DWORD dwOptions,
  SMS_MESSAGE_ID * psmsmidMessageID);
```

Again, we need not be concerned with the anatomy of this function call, but those who have tried it will tell us that interfacing to a GSM modem to perform text messaging is actually quite a messy affair at the low level, one that involves numerous serial interface messages. All that messiness has been removed from us courtesy of the API and we are left with one simple function call to take care of sending a message.

This principle is not unique to Pocket PC; we have simply taken examples from this particular operating system to illustrate the API concept. The API approach will exist on all mobile devices and it is how we expect our services layer application to interface with the rest of the mobile device. All the common mobile device operating systems support this approach. The issue that stands is whether or not there are common formats for the APIs themselves. Certainly, in the case of the different RTOS solutions, this is not the case, even for common functions like sending a text message.

10.8 MOVING TOWARDS A COMMERCIAL MOBILE PLATFORM

Thus far, we have been building our view of what a mobile device consists of. We have started out with a simplified layered model that captures the essence of what a mobile device does. Our first level of refinement was to propose an infrastructure for the layers, namely an operating system and a means of engaging with peripherals and the networking layer via defined software entities called APIs.

What we would like to do now is to further refine our architecture by the inclusion of certain utility applications that we might find useful on a mobile device and that are a sufficiently common requirement that we might expect to see them on most devices. Our further refined architecture is shown in Figure 10.17.

The figure shows that more detail has been added to the services layer. Essentially, we have added programs that will run in the services layer, some of which are useful utilities in their own right, accessible to the end-user, and some of which provide an additional framework for supporting and deploying our own service applications on the device. In this section, we shall examine what these applications and utilities are. We shall move towards demonstrating how this model is reflected in most commercial solutions available, and for various form factors of device. This will give us confidence that we have established a viable model and it will enable us to better grasp the various mobile platform solutions being offered.

The utilities for considering fall into four categories:

1. Communications utilities

2. Interface utilities

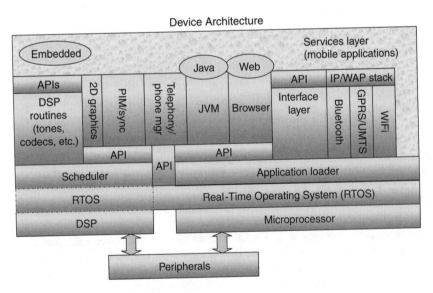

Figure 10.17 Typical mobile device functions.

3. Personal Information Management (PIM) utilities

4. Application support utilities

We will look at only two of these categories in depth here but this is a good point to state that each commercial mobile platform is different, although they have common traits, and it is these *traits* that we are trying to identify here. Particular approaches for each commercial solution can only be weighed up in light of real application requirements and associated commercial concerns.

10.8.1 Communications Utilities

IP/WAP Stack By this stage of the book, we should already understand the importance of being able to connect with an IP-based network, either using IP itself or the WAP refinement (1.x or 2.x). Therefore, it is useful to add a software library to the mobile device that enables any application to exchange IP or WAP messages without having to be concerned with the low-level programming required to do this.

Low-level IP and WAP programming would be concerned with message packet construction, error-handling, retransmission, packet assembly and so on. All this functionality is already included in the library, which is called a *stack*, as discussed in Chapter 5. The stack will support all the commonly used TCP/IP protocols, such as DNS, DHCP, etc. An IP stack would probably support the following:

- Transmission Control Protocol (TCP)

- User Datagram Protocol (UDP)

- IPv4/v6 support (IPv6 addressing as well as IPv4 addressing)

- Internet Control Message Protocol (ICMP)

- Point-to-Point Protocol (PPP)

- Domain Name System (DNS)

- Dynamic Host Control Protocol (DHCP) for IP address assignment

- HTTP 1.1-compliant client stack

This is likely to be a minimum set, although IPv6 addressing support may or may not be present. Additionally, we might expect the support for at least some of the secure protocol layers to support encryption and VPN connections, such as:

- Transport Layer Security (TLS) and Secure Sockets Layer (SSL); TLS module is effectively an enhancement of the SSL protocol.

- IPSec, which is the IP layer protocol used to secure host-to-host or firewall-to-firewall communication. IPSec provides tunnelling, authentication and encryption for both IPv4 and IPv6 so that secure connections can be made over the Internet, primarily for mobile enterprise access.

What the stack does is to enable the mobile application developer to concentrate on developing the core of their application without having to be concerned with implementing the IP protocols that it may require to connect to the outside world.

Accessing the stack from the service layer is done via an API that would be produced by the device manufacturer or, if applicable, by the operating system provider. Although TCP/IP is strictly speaking not an operating system function, it is common for such functionality to be bundled with the OS and be optimised to run on it. Often no distinction is made in referring to the actual OS – or the RTOS – and the utility libraries that get bundled with it; they all get referred to as the OS. These days, more and more utilities are included in the bundle, as this is how an OS vendor seeks to gain support (customers) for their particular product offering.

It is likely that in addition to providing an API to access the stack, a toolkit of some kind will also be provided, most likely as part of a Software Developer's Kit (SDK). At a minimum, this would provide documentation on how to access the API, how to program for it and utilise its services effectively. There may also be sample code demonstrating how to utilise some, or all, of the API services. Issues such as the programming language used to access the API will be discussed shortly.

Telephony Control/Phone Manager At least one of the devices in our possible device network will have integrated telephony; therefore, we expect some basic telephony control software to be made available on the device, as well as software to control the phone settings, such as the ringtone. The telephony manager will have its own UI to allow calls to be made and other such tasks. This will include the ability to respond to key presses, whether from a mechanical keypad or via a touch screen, as shown in Figure 10.18. The functions

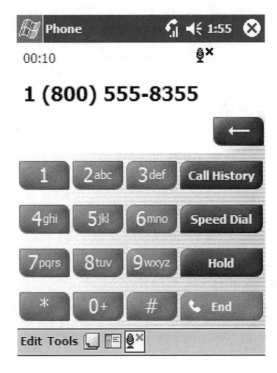

Figure 10.18 Phone dialler application on a Pocket PC. (Reproduced by permission of Microsoft.)

driving the display will also be included, such as showing the number being dialled, or the numbers being received and so on.

On devices that incorporate telephony, an array of telephony control and call management functions are to be expected, and most of these should be familiar to us. Certain telephony functions are expected to be present according to the requirements of the digital cellular standards like GSM and UMTS, such as features like phonebook, call diversion, called-ID display and so on.

Typical telephony features include:

- Dialling management (the ability to dial numbers)

- Phonebook store and manipulation

- Call barring

- Call records (last numbers dialled, received, missed, etc.)

- Call diversion management (divert always, on busy, etc.)

- Ringtone selection and volume control, including vibration alert where applicable

- Cellular network selection

- Call timers and possibly cost calculators

- Call waiting control

There may be other control features available, but this will depend on the appropriate telephony specification (e.g. GSM) and also on the device manufacturer's product design criteria, especially when it comes to incorporation of elective features.

Most of the functions that the telephony manager controls will be inbuilt features of the RF modem and we expect the software to interact with the control chip in the modem chipset. This arrangement is shown in Figure 10.19, which also helps us to understand how it might be implemented in a typical mobile device.

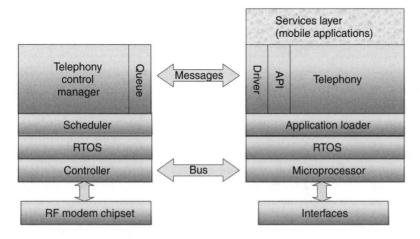

Figure 10.19 Telephony management architecture.

We can see from the figure that the telephony software runs as a process sitting on top of the RTOS that runs on the main processor in the mobile device. The microprocessor is physically connected to the controller chip in the modem chipset. The nature of the connection may either be on a shared bus or via some kind of dedicated communications pathway, like a serial link or a shared memory space. There is no preferred way of implementing the communication pathway as it is dictated by the chipset manufacturer and the interface options that they decide upon or which are possible in a particular device design.

It is possible that the telephony application could use low-level programming techniques to access the communication pathway directly, but this is unlikely. What is more probable is that a device driver is written that offloads this concern from the telephony software. The device driver will include all the necessary low-level software and start-up routines to initialise the communications pathway to the controller in the modem chipset.

The controller itself will run various software processes according to the tasks it is carrying out at any particular time. It is likely that it can carry out many tasks on behalf of the mobile device software and so it will run a message queue handler that can receive messages across the pathway from the microprocessor, interpret them and subsequently route them to the appropriate software process, such as the telephony control manager in this instance.

The device driver on the microprocessor side will be familiar with the low-level message formatting that the controller queue expects, including how to mark the message destination to be the telephony control manager.

On the microprocessor, the telephony software will access the device driver via an API, a predefined set of software function calls that can be successfully interpreted by the device driver and converted to the appropriate messages destined for the telephony control manager (via the controller message queue). With this arrangement, it is relatively straightforward to implement applications with rich graphical interfaces to control the telephony functions, such as call barring (see Figure 10.20), without the application developer having to know much about the interface details for the telephony controller and its low-level messaging protocol.

This low-level messaging process works both ways. A message from the telephony control manager, such as an incoming call alert, will be sent back to the microprocessor via its appropriate physical communications pathway, such as the bus or other options mentioned earlier.

What happens on the microprocessor is that the peripheral that handles the communications pathway will cause an alert to the RTOS to say that something has happened on the bus. This alert is called an interrupt. The RTOS is responsible for handling interrupts. It can interpret any interrupt from the processor, identify its source and then cause the appropriate software routine to be scheduled to run on the processor and subsequently handle the interrupt. In our example, the device driver for telephony handling being scheduled in response to the interrupt and the appropriate event-handling software routine within the driver will then be executed. This in turn will raise a software-generated event to the telephony software, which should then have the appropriate means to service the event. For example, in the case of an incoming call scenario, the event handler in the telephony software would take an action, such as updating the LCD display to indicate the presence of an incoming call, perhaps including the caller ID (number).

Increasingly, it is becoming blurred as to where in the mobile device the telephony management takes place. In the case of some chipsets, the ringtone generator is included

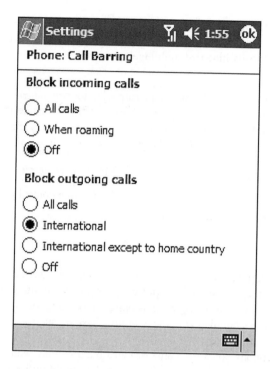

Figure 10.20 Call control via a Pocket PC (Mobile Edition) graphical interface. (Reproduced by permission of Microsoft.)

as part of the chipset. This can be driven by the core software on the mobile device, such as from the telephony software running on our microprocessor. Therefore, the device driver will also be expected to provide a low-level interface for sending messages back to the chipset controller message queue; in this case, to a different software process on the controller, one that is more than likely running on the integrated DSP, particularly if the tones can include sound samples (e.g. MP3 files or similar). Alternatively, the controller itself can take care of the entire call-handling function, switching on the appropriate ringtone automatically, without any intervention from the core microprocessor on the mobile device.

In terms of how these functions are implemented in software, it is sensible to enquire if the telephony device driver API is accessible to any mobile application running on the microprocessor. There is no reason why this shouldn't be so, but it is not that common in many devices. This is all related to how the device has been implemented and what the manufacturer decides in terms of providing third-party access to embedded device functionality.

We shall examine this issue in more depth later, but for now we might reflect upon the fact that most mobile telephones have been historically developed either using proprietary-embedded software solutions, or have, in any case, not been made open for other applications to be run on them other than the 'factory-fitted' ones. On the other hand, PDAs have been designed to run third-party applications from the outset, but they have not grown up with any history of integrated telephony functions. The issue of telephony APIs did not get addressed until relatively recently with the advent of the 'Smart Phone' concept, which is a PDA and mobile telephone combined (although there is no exact definition for a smart phone).

Specialised 'smart phone' or 'mobile aware' operating systems have since emerged, like Symbian. Efforts to produce standard telephony APIs have also emerged, like the JavaPhone API (albeit not specifically aimed at mobile telephony).

Phone Settings and Control   Whilst considering telephony management, we can also look at general phone settings and their control. Phone settings include things like adjusting display settings, interface language (i.e. display character set, such as English, French, German, Arabic, etc.) and similar housekeeping functions. These might include:

- Time and date settings
- Language
- Display contrast/brightness
- Power-saving modes

- PIN code protection
- Hands-free control
- Voice control
- Factory reset

Such functions would be present in any form factor of mobile device with a user interface, but not necessarily grouped like this, nor presented as 'phone' settings (given that the device might not be called phone, or recognised as being one). Some of these features would be common utility services bundled with the operating system, such as time and date settings, language, display control, etc.

As with the telephony software, some of these features may well be accessed through the modem chipset, such as voice-control dialling that might utilise an embedded DSP to execute voice recognition algorithms. We would face similar issues, as just discussed, in relation to how far these services are implemented on the main processor and what is the nature of the interface to them in terms of API provision. There is similarly a question as to how much of these services are deemed as operating system services, bundled with the RTOS, or as third-party add-ons. The architecture of a mobile device is no longer a simple issue (if it ever were)!

10.8.2   Personal Information Management (PIM) Utilities

Mobile devices are personal devices; therefore, they naturally take on some of the characteristics of other personal objects, like diaries. Early on in their history, PDAs were glorified diaries and address books, a more convenient and powerful version of the pocket book or Filofax. This is probably still the number one reason for buying one. However, email has become a very common personal communications tool and so there has been a growing interest to access email from a PDA, especially now that wireless connectivity is more widespread and affordable.

Mobile phones have always had the concept of an address book, or contacts list. This is an expected feature for a telephony device where number management is essential. Not so common is a diary feature, although many phones do incorporate one. On devices with limited data input abilities (i.e. with only a 10-digit keypad), the addition of extra features like a calendar has been slow to catch on. Perhaps it was felt that those people requiring calendars would most likely own a PDA.

Since the advent of 2.5G data solutions, like GPRS, some phones have started to appear with email clients. Email clients have existed on PDAs for some time, but in the earliest incarnations were limited to carry-and-go clients for desktop email solutions, like Microsoft's Outlook, which did not have any real-time email collection capabilities between the PDA email client and a remote email server. Emails could only be sent and received on the PDA during a synchronisation session in the cradle connected to the desktop PC. Such a solution was of limited use and even more so in the mobile context.

In newer 2.5G phones and wireless-enabled PDAs, email has become more important and is a standard offering. We have seen these two converge. Some mobile phones have expanded in functionality to take on PIM features, whilst some PDAs have taken on wireless connectivity and even telephony. It is interesting to consider the key features that we might expect on either type of device. Typical PIM features are:

- Diary with multiple viewing options (day, week, month)

- Contacts management with multiple fields, such as: name, number (mobile, home, office, fax), postal address (home, office), email address (multiple entries)

- To-do list (tasks)

- Notes

- Email client (POP3, IMAP4 support)

- PIM store (storing contacts, appointments and emails)

The existence of an email client usually implies the ability to connect with an IP network directly using TCP/IP, so the IP stack mentioned earlier needs to be implemented on the device. Most devices these days will be using an always-on wireless data connection and there is no longer the need to dial into ISP modem banks. Some older devices support this and even some of the newer ones still offer dial-up IP protocols as a fallback solution (and sometimes as the easiest route for corporate access, i.e. into an existing remote access server solution). Various devices using proprietary wireless networking, like some two-way paging networks, may not offer IP stack support and rely on gateways to go between a proprietary networking protocol and one of the IP email protocols like POP3.

It is possible to access email services without an email client. This can be done using web-based solutions, which is a common enough technique on the Internet, even without wireless access. Some operators offer email services via WAP, but often these solutions require a dedicated email account with the operator's email server network, so the solution is limited in terms of accessing other accounts. With some services, it is possible to implement POP3 collection from another account into the user's operator account, but this is still a limited solution because it requires that the user still sends emails from the operator account and not the user's regular account. As with any solution involving remote access to a server, once there is more than one way of accessing the mail (such as from a desktop and a mobile client), the problem of synchronisation arises. For example, sending an email via the mobile will probably mean that the message does not get stored in the 'sent items' folder on the desktop. Similarly, reading a message on the mobile probably does not flag the message as read on the desktop. There are many synchronisation issues that arise.

We do not wish to discuss particular solutions in too much depth in this section as we are simply trying to highlight the basic PIM functions and the value of their inclusion on a basic mobile device. We would not expect PIM functions on all devices, but the convergence is definitely well under way. The further level of convergence is with the mobile telephony itself. If a telephony function is present on the device, then we expect it to be fully integrated with the PIM functions. For example, we could dial a number directly from a contact entry, etc.

Text Messaging   In terms of the messaging capabilities of the PIM client suite, on a device with integrated telephony, we expect a text messaging client. Of course, all mobile phones on the common digital cellular networks already have text messaging support. The clients on devices with richer interfaces can be more user-friendly.

In terms of integrating the text-messaging client into the device's user interface, this raises an interesting point as to how we group functions on a device. There is one school of thought that says that we use a messaging metaphor and gather all types of message into one client. This used to be called *unified messaging* when the concept of handling all message types using one client was first proposed. This was originally conceived of as a means of unifying communications on the desktop, so that faxes, voicemail and emails would be gathered together into one client with a consistent messaging metaphor. For the wired world, there has always been an issue of how disparate messaging sources get integrated into one solution and this problem has been addressed by the production of dedicated infrastructure products to support a unified messaging paradigm. The challenge has been to create bridges between telephony and computing so that messages can be understood and processed by software, whether presented in a single client or not.

With digital cellular, this problem is not so acute. Voice mail messages have had visual indication for some time, courtesy of text message alerts. On mobile devices, text messaging has been present from very early on, so in one sense we have always had a unified approach to text messages and voicemail.

Text messaging and email seem natural cousins and it has made sense in some cases to integrate handling of these message types into one interface. It is usually done by taking the email metaphor and extending this to text messaging, which is reflected in the evolution of naming conventions. Text messages used to be available under a 'received' menu option, which has now become an 'inbox', whether or not it is also used for emails. Other naming parallels have been adopted.

We have the choice to adopt a unified messaging metaphor if we like. It is really a matter of software implementation on the handset where all the different messages types already converge.

The convergence process has been further provoked, or challenged, by the emergence of MMS on many mobile handsets. The same goes for WAP Push messages, yet another messaging type in the digital cellular world, and one rather peculiar to it.

It is really a matter of design choice as to how the mosaic of message types get handled by the mobile device in terms of their presentation and handling via a user interface. With modern interface techniques, such as XUL (Extensible User Interface Language) it is perfectly feasible to imagine giving users a choice of interface presentation. Some users are not keen on having all their message types grouped into one client, especially if it entails a single view of the message space through one inbox. This can become cumbersome to manage. With some users potentially receiving hundreds of texts and emails a day, the prospect of handling them all through one view is off putting. We could probably benefit from some serious usability studies in this area, but then why bother if we can offer users the option to configure the interface to operate how they would like it to? For those that want separate menus, the interface keeps them separate. For those that want to combine, they get to chose what they combine with what. This way, everyone's happy.

The type of unification that does make sense is a unified and consistent view of information like message addressing and contact identification. For example, if a user wants to send an email or a text message, they should be able to look up their target contacts from the same database. If a user receives a text message, then the sender's number should be checked against the contacts database to find a match. If matched, then the sender's name should be displayed instead of, or in addition to, their phone number. This is standard stuff on mobile phone text-messaging clients these days, but has taken a while to penetrate text-messaging solutions on PDAs (and even desktop PCs where text-messaging clients are possible and are increasingly being used[12]).

To do this implies that all the messaging clients have access to the PIM store, or, more likely to an API that provides PIM store management. We therefore would hope that a PIM API is present on our mobile device and not just the PIM client software. This leaves us free to develop new messaging applications based on the inbuilt PIM capabilities of the device. In fact, we would ideally like an API for the PIM client primitives, not just to access the PIM store. So, for example, we could use an API call to send an email message from our software via the PIM client, which means we can maintain an audit of messages in the client. This is potentially more powerful than just sending email messages (e.g. from a field service application) that are not recorded in the email client, or have to be assigned to a unique 'sent messages' folder with our custom application rather than the standard 'sent messages' folder.

---

[12] Text messaging is increasingly finding its way into the enterprise, especially with products that enable databases to be easily accessed and updated via a simple text message. For example, see http://www.xsonic.com

11

Mobile Application Paradigms

11.1 INTRODUCTION

In Chapter 10, we looked at the possible architectures for mobile devices. We identified a basic software and hardware architecture that provided us with the essential capabilities that a mobile service would need from a mobile device. The device elements are shown in Figure 11.1.

In this chapter, we will consider the following ideas and concepts:

- Application topologies – J2ME, browser, embedded

- Service topologies – client/server, peer-to-peer, standalone

- Device networking architectures – Personal Mobile Gateways (PMG)

- Device networking technologies – Bluetooth, WiFi and infrared (IrDa)

- Device types – PDA, Smart Phone, laptops

- Platform options – Pocket PC, Symbian, Linux, Palm OS, others

- J2ME approach

- Interfaces and usability

11.2 APPLICATION TOPOLOGIES

There are three basic approaches to developing mobile applications in terms of what runs on the mobile device itself, namely:

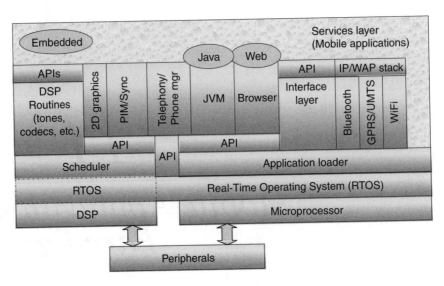

Figure 11.1 Typical (idealised) mobile device architecture.

1. *Java* – implement a Java program that runs in a Java Runtime Environment (JRE) and uses Java APIs to access mobile device resources.

2. *Browser* – provide access to a server-based application via a browser, where the browser accesses the mobile device resources for us, such as the network connection to the server.

3. *Embedded* – implement a program that runs natively on the mobile device real-time operating system (RTOS) and uses the device APIs to access the mobile device resources. (Java and embedded are similar approaches, but the Java approach is significant enough to warrant separate consideration.)

Using a combination of the above approaches is also possible, and we shall look at this later. Why are three methods necessary; can't we just make do with one? The short answer is that these three approaches have emerged and evolved to tackle the problem from slightly different angles. In fact, the Java and browser approach are really extensions of the embedded approach. Ultimately, any application on a mobile device has to run on the device's operating system and use the low-level resources of the device. This is the most generic solution. A browser is a particular embedded application, but a special kind of application that can subsequently enable an alternative general-purpose content-publishing model; namely, the page-based Web model that we have discussed and analysed so extensively in previous chapters. The Java model is also based on an embedded application, called the Java Virtual Machine (JVM), which subsequently enables a different type of general-purpose programming model whose appeal is that it presents consistent programming across a whole range of mobile devices, irrespective of the underlying operating system. This works by abstracting a generic programming environment and a generic set of APIs built on top of the specific underlying ones offered by the device operating system.

A more in-depth understanding of the options needs to be developed before answering the question 'which one should I use?' The first consideration is which one should I use for what? These programming paradigms are not the same and cannot be used interchangeably for all application scenarios. For example, if we hope to implement a P2P file-swapping application, so that one device can swap files directly with another, the browser option won't work. This means that we are excluded from using the browser approach and forced to use an embedded approach or a Java one. Now, if we attempt to access files on a device using the Java approach (J2ME in particular), then more than likely we cannot accomplish this because of the sand-box[1] Java environment that usually precludes access the underlying file system.

Our decision may also be influenced by the APIs available on the device. For file swapping, there may already be API services that are easily accessed via an embedded application. Protocols like Object Exchange (OBEX[2]) protocol may be included on the mobile device and exposed via an API[3]. OBEX is optimised for ad hoc wireless links and can be used to exchange all kinds of objects such as files, pictures, calendar entries (e.g. vCal files) and business cards (e.g. vCard files). OBEX is similar to HTTP but does not require the resources that an HTTP server requires, making OBEX perfect for devices with limited resources.

We might even be guided in our decision by our own technical competencies and understanding of the options and technologies. This is not unreasonable, though it may be inefficient. However, trade-offs are inevitable, even if we have good knowledge and proficiency in all approaches. Furthermore, there is no single way to implement a service. To a degree, the richness of device capabilities and software technologies combined with the increasingly powerful underlying processing power, means that we can probably afford to be more flexible in our approach to implementation. We don't have to agonise so much about evaluating the most resource-efficient method for a particular application.

As we remarked in the opening chapters, always take what already exists in terms of APIs, open standards and even open source, and learn to become a 'smart integrator'. The market can move quickly, so we don't want to spend a long time writing code that already exists, even if it only approximates what is required. As we have stated many times now, *usability* is a key design consideration. It is probably better to spend resources in this area rather than programming a low-level software function that could be integrated relatively easily. As we shall see, there is scope for rapid implementation and high degrees of personalisation across all of the approaches, whether browser, Java or embedded. Let us now look at the fundamental characteristics and attributes of each approach.

11.3 EMBEDDED APPLICATIONS

The *embedded* approach is without doubt the most flexible of all the options, so we could say that it is the 'king' of approaches. Anything that a particular device is capable of being programmed to do is theoretically possible using the embedded approach. There are fewer restrictions on accessing the underlying device resources, as compared to the other approaches. This is because the embedded approach provides the ability to get right down

---

[1] *Sand-box* means that the Java program is only allowed to access a subset of the underlying features on the mobile device.
[2] http://irda.org/
[3] There is an API for OBEX provided in Symbian OS and in Pocket PC, plus others.

to the guts of the device and programmatically access its low-level functions and hardware peripherals. In practice, the level of programming power available is determined by the provisions made available by the device manufacturer to third-party programmers, such as APIs, documentation and related programming tools. These considerations will become clearer as we progress.

Consider coming up with a new method of handwriting recognition. In theory, we could program the solution as an embedded application. Using a browser approach, we would not have sufficient access to the device resources. Although, to be fair, the browser paradigm is not intended to allow for this degree of flexibility – the programming and interface model is deliberately simplified for good reasons. Functional extensions to the browser might be possible[4], but this would involve a hybrid approach that still requires that the extended functions be programmed as an embedded application.

An embedded application runs on the main device processor on top of the RTOS, which manages how the application runs and how it gains access to the device resources. Via software pre-installed on the device, an embedded application has a plethora of APIs that it could use to harness the services of peripherals and other applications that may already be installed on the device as part of its native functions, such as a telephony manager, PIM and so on. For example, we already suggested in Chapter 10 that an email client might be a standard feature on certain mobile devices. If so, its functions might well be accessible via an API. An embedded application could utilise this API whereas a browser certainly could not. A Java application may or may not be able to access the API depending upon factors we shall discuss later to do with how the Java environment is implemented on the device.

11.3.1 What Do We Need to Develop an Embedded Application?

Developing embedded mobile applications is not that different to writing applications for desktop platforms. We need similar tools and resources, such as:

- Compiler for the chosen platform

- Debugger for the chosen platform

- Software Development Kit (SDK) for the chosen platform and possibly for a particular variant within a platform family

- Workstation (e.g. Windows PC) to run the tools on

In addition to the above requirements, we will also need:

- An actual target mobile device on which to try out our solution on, or possibly several devices to test on various, intended mobile products

- A live network connection for testing the application within a mobile network

This book is not about how to write software, nor about the development process. Therefore, we shall not discuss the need for a requirements specification, test plan and so on. We will

---

[4] The External Functionality Interface (EFI) specifications in WAP 2 provide means of enabling WAP applications to access 'external functionality' (embedded applications running on the same device) in a uniform way through the EFI Application Interface (EFI AI).

allude to the important aspects of the software process where necessary in order to emphasise certain mobile considerations.

We have not yet mentioned the choice of language to develop the applications. Usually, most SDKs specify a set of APIs and a toolset that support C or C++, this being a language that is particularly suited to embedded solutions for a variety of reasons, not least of which is its support for low-level manipulation of data. However, we should not think that C is the only option. First, we need to dispel some misconceptions about the use of portable languages like Java, as while we assign it to a category of application development in its own right, this is only in respect to a particular Java approach called J2ME, which we explain later. However, Java can also be used for general-purpose embedded programming, although it is rare to find this on mobile devices. Language support depends mostly on the device operating system and the programming tools available for it. Ultimately, the device vendor determines the language support and it is common for either the mobile operating system vendor or the device vendor (or both) to provide the tools required for the embedded development process.

11.3.2   C and C++ Are Not the Only Choices

Visual Basic .NET and C#   As we mentioned, C or C++ has the most widespread language support, but there are some notable exceptions. In the case of Microsoft's Windows CE RTOS, there is support for Visual Basic .NET or a relatively new language called C# in line with the general support for these two languages across all the Microsoft platforms using a programming framework called .Net. Even though Windows CE (sometimes abbreviated to WinCE) has been built from the ground up as a brand new operating system for mobile devices, its programming model and language support has been made deliberately familiar to Microsoft Window's programmers. This obviously makes it easier to program.

The reason that Microsoft developed a new operating system from scratch is that the microprocessor families for mobile devices are very different to the Pentium family that Microsoft has intimately supported for desktop and laptop PCs. In fact, there is no single processor family for mobile devices, but a collection of families including MIPS, ARM, SH3/4 and the PowerPC. This led to Microsoft developing a Common Executable Format (CEF) to include in Windows CE. The idea was to compile programs in Visual Basic (VB) or Visual C++, the original languages supported by early versions of CE, to a common format. This is then run on a virtual machine within the CE environment.

The CEF idea appears to have been the inspiration for Microsoft's latest overhaul of the Windows platform, the programming framework called .NET ('dot net') Framework, which uses a Common Language Runtime (CLR). This is a major revamp of the Microsoft Windows platform, largely to make it easier to develop applications that are internet-centric. The Microsoft .NET Framework is a software component that can be added to or is included with Microsoft Windows operating system. It provides a large body of pre-coded solutions to common program requirements, and manages the execution of programs written specifically for the framework. The pre-coded solutions that form the framework's class library cover a large range of programming. The functions of the class library are used by programmers who include them with their own code to produce applications.

Programs written for the .NET Framework execute in a software environment that manages the program's runtime requirements. This runtime environment, which is also a part of the .NET Framework, is known as the *Common Language Runtime* (CLR). The CLR provides the appearance of an application virtual machine, so that programmers need not consider the capabilities of the specific CPU that will execute the program. The CLR also provides other important services such as security mechanisms, memory management and exception handling. The Compact Framework is the mobile edition of the .NET Framework and has been ported to enable a common programming model and toolset approach across all Microsoft platforms.

It is still possible to develop using C++ (using the Embedded Visual C++ toolset) but new applications in classic VB are no longer supported. Programming in the newer languages enables access to a much richer set of APIs, so it is both tempting and sensible (advisable) to migrate to the .NET toolset. Furthermore, programs that are written using the newer languages benefit from advanced OS features like program management (PM). With PM, the OS prevents managed applications from freezing up the device or acting maliciously[5]. It is also possible to load these applications OTA, a technique that we shall discuss later because it is very important within the mobile context.

11.3.3 'Native' Java Support

RIM's BlackBerry   Shortly, we shall see that any mobile device can potentially support programming applications in Java via the porting of the JRE to the device as an embedded application itself in the services layer.

However, on some devices, Java is the only way to program applications in the service layer. This is because the whole OS is built from the ground up either using Java or to offer intimate Java support. This is not a surprising development given the increased capacity of newer mobile devices with greater processing power and memory yet still within an acceptable price/performance curve (including power consumption).

A notable 'native Java' device is the RIM BlackBerry, such as the BlackBerry 6210 GSM/GPRS device, as shown in Figure 11.2.

Except for restricting the language choice to Java, the development process for the Black-Berry is similar to the general embedded approach. We still have to use a developer toolset, in this case called the Java Developer Environment (JDE) and we still access system resources using APIs. The APIs on the BlackBerry are a combination of standard JRE ones (as applicable to the type of JRE running on the BlackBerry) and extension APIs provided by the company Research in Motion, Ltd, otherwise known as RIM (and not a standard part of Java).

It is important to understand that Java is not just a language; it is also a set of APIs and an associated runtime environment, which together form the Java *platform*. In the case of implementing the Java platform on the BlackBerry, RIM has implemented their own API set that suits the architecture of their device. This is what we would expect when looking back at our proposed generic device architecture shown in Figure 11.1. All the APIs would have to be made available in the Java platform so that they are made accessible to the Java programming language when writing applications to run in the services layer of the device. To implement a powerful application that exploits the potential of the device, we have to use the custom Java API set. This is going against the Java portability ethos, which is to use a consistent set of Java APIs to standardise an abstraction of device capabilities, thus

---

[5] 'Acting maliciously' means doing something on the device that would be disruptive to its overall behaviour, such as one program corrupting the data of another.

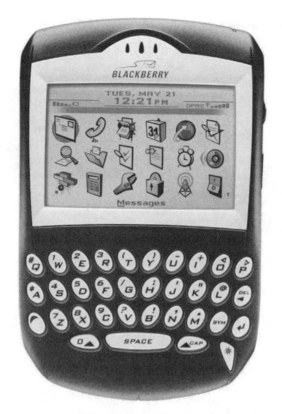

Figure 11.2 RIM BlackBerry 6210. (Reproduced by permission of RIM.)

confining development to this set in order to achieve portability objectives. We shall discuss this issue in more detail below.

If we consider the 2D graphics capabilities of the device, the API for this is made available in Java, as we would hope. As we shall see later, the various standard Java platforms have their own concepts of graphical user interface (GUI) widgets and related visual metaphors and they provide APIs accordingly. This is supported on the BlackBerry, but additionally, RIM has devised a custom GUI metaphor for the BlackBerry, including the BlackBerry track wheel for tactile response. The API for this is in the extended set, not the standard Java set. Similarly, to access some of the radio resources in the network layer, proprietary RIM Java APIs are provided, such as a mechanism to enquire about the radio signal strength.

In using custom Java APIs, there seems to be an apparent contradiction of using a language and platform in a way that seems to go against its intended benefit of application portability. This is only partly true, as a large amount of the Java code would probably remain portable, but the main intended benefit here is to provide a mobile development platform that is attractive to Java developers, simply because there are now so many of them that providing a Java solution makes sense. We only wanted to look at the programming support for applications on the BlackBerry, not its detailed architecture and features. For more information, see the RIM website[6].

---

[6] http://www.rim.net

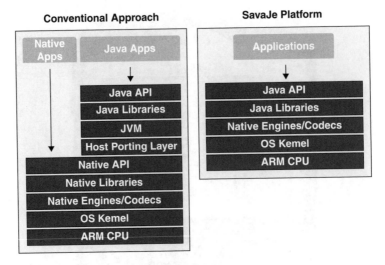

Figure 11.3 Comparison between java and SavaJE.

SavaJE OS   SavaJe[7] OS presents the entire embedded API set on the device in Java. However, the support for the Java platform is more comprehensive than with the BlackBerry. A larger subset of the standard Java platform is supported and the extended API set is quite extensive, including the provision of a large number of applications that are ready to use straight 'out of the box'.

The inner kernel of the OS is not written in Java, but the programming model presented through the API set is pure Java, even for library functions that have been developed to run natively on the kernel (i.e. not in Java). This hybrid approach has a performance advantage due to the ability of the kernel to run native applications quicker than Java ones; therefore, the computationally intensive tasks, like video compression, are all written in C and executed natively. The essential difference between a conventional approach to supporting Java on a mobile devices and the SavaJE approach is shown in Figure 11.3.

One of the most notable features of SavaJe is its support for the Java API called JSR 209, which is an advanced graphics library for mobile Java devices, including support for the popular Java graphics libraries called Swing and Java 2D.

In 2007, Sun Microsystems, Inc., bought the intellectual property assets of SavaJe Technologies. These assets are now used in the JavaFX Mobile product, which we cover next.

JavaFX Mobile   We should briefly mention the JavaFX platform from Sun Microsystems, Inc., as it is a notable development of the Java approach. Sun acquired SavaJe and this appears to be the basis for JavaFX Mobile, although how much of the system comes from the SavaJe OS is undocumented. As we have discussed, the SavaJe concept was to push Java further down into the OS, moving telephony applications such as messaging and call management into the JVM to provide a more stable and flexible interface. Like SavaJe, JavaFX Mobile is a complete stack run on a device that enables the programming of the device to be done in high-performance Java. However, there has been no

---

[7] Pronounced *savij*

attempt to create the OS kernel, which is Linux. This follows a number of other vendors who have taken the Linux route for mobile platforms, including Motorola with their LJ platform.

JavaFX Script is actually a scripting language (called JavaFX for short) whereas JavaFX Mobile is a complete platform for the mobile. This means that the JavaFX Mobile comes with all the operating system and telephony libraries required to implement the telephony and messaging functions found on any standard mobile phone. The JavaFX scripting language is yet another flavour of Java. It is specifically designed to optimise the creative process of building graphically rich and engaging UIs leveraging some key Java graphics APIs, namely Java Swing, Java 2D and Java 3D. The language is declarative, which means that the structure of a program written in JavaFX should represent what is being described, which is the UI. In other words, for each component of the UI, there will be a portion of code that represents that component. This makes writing UI programs easier and more intuitive and maintainable for both developers and content authors.

Just like its cousins, JavaFX is still a network-centric language, making it easy to create networked applications, so we should still expect the flexibility of creating compelling UIs that can be customised and adapted by live connections to the Web 2.0 platform. JavaFX is expected to be available on all digital end-points, or 'tops' as Sun prefers to call them, starting with phonetop and then moving to set-top (box) and desktop. Without doubt, an interesting area of convergence is between the mobile device and the set-top box (STB). Currently, to create interesting applications that run across the STB and mobile, two sets of very different middleware are involved, such as the ODP and something like it for the STB. They use separate, different and proprietary UI creation tools. JavaFX promises to solve this problem with a single programming approach across all devices. It is important to understand the commonality is in the programming approach, not in the program itself. It is still unlikely to support a write-once, run-anywhere goal, mostly because different device families require different approaches. However, a great deal of commonality should be possible. For example, if writing an email application, then many of the interface components and operations are going to be the same, just displayed differently.

11.4 EMBEDDED DEVELOPMENT TOOLS

Developing a mobile service is not just a question of choosing which programming language to use. The available API set will also influence the decision. As with most computing platforms, the richness of APIs on mobile devices is continually evolving, as is the architectural model that we developed in Chapter 10. New and exciting services are possible on an increasingly wide range of devices.

For certain applications, some platforms are better than others. For example, for secure enterprise connectivity, we might want devices with keyboards or built-in enterprise productivity tools, like a word processor. Perhaps a Windows Mobile device or something like the BlackBerry would be suitable. Developing a secure enterprise application on a BlackBerry will almost certainly dictate an embedded approach, particularly if we want to push data to the device asynchronously of user requests. The BlackBerry has an API to do this and a programming model on the device that supports the concept of listening for push messages. The OS manages this service for the service layer applications. Whatever the reason to use

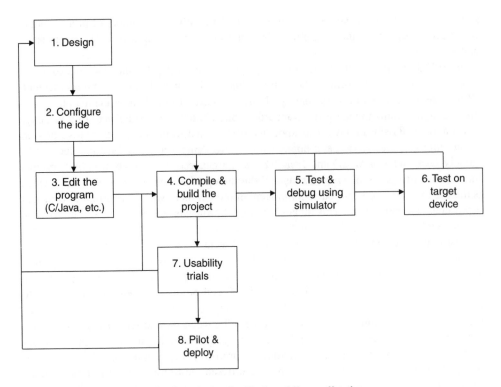

Figure 11.4 Typical process flow for embedded mobile application.

an embedded approach, the development process is usually the same for all devices and is summarised in Figure 11.4.

Let's briefly examine the development process and outline its constituent parts:

- Design
- Configuring the Integrated Development Environment (IDE)
- Editing the program(s)
- Compiling and building the project
- Testing and debugging via a simulator (emulator)
- Testing on a target mobile device(s)
- Conducting formal usability testing
- Pilot trials and deployment

11.4.1 Design

Remember, this book is not about how to develop software, so we do little justice to this subtask by devoting only a few lines to it here. Software designers will have their own preferred approach. The things to be aware of in the mobile context are the constraints of the mobile environment, such as:

- Potentially intermittent existence of a networked connection

- Bandwidth variability of the networked connection

- The existence of mobile push mechanisms and the different platform-specific techniques for handling push

- Constrained and varied user interface possibilities

- Limited memory capacity

- Limited processor performance

- Wide variety of devices and likelihood of differences in running an application, even on a supposedly 'standard' device model (e.g. a particular Java platform)

- Restricted ability to update software once deployed

- Minimising unnecessary drain on battery

- Optimising the communications paradigm at the application level in order to conserve bandwidth (and implied costs)

With all these considerations, we may wonder about the existence of a standard design approach. Unfortunately there isn't one, as there is so much room for variation. We emphasise the need for careful consideration of the service requirements and, having done that, a thorough review of the device possibilities, especially APIs, and the implications for the target user community if we go with a certain set of devices.

We cannot assume that modern mobile devices are free of the constraints listed above. Consider memory allocation as an example. If an application exceeds the available SRAM[8], then the application might start to use flash memory, which can be comparatively slow, and performance might suffer. This implies that we have to design applications carefully to avoid exceeding the available memory. For example, we should not allocate large arrays, and we should use techniques to create arrays of an exact size when possible. It is also possible for Java applications on a BlackBerry to leak memory, even though Java has a facility called *garbage collection* to remove unwanted processes from memory, such as after they have been used and done with. However, an object[9] cannot be garbage collected if references to it still exist, so the programmer must take care to release objects when finished with them, especially when objects are shared between different areas of code. Similarly, network connections should be closed when finished with. These are all good programming habits generally, but especially for mobile applications.

I recall visiting a GPRS applications testing lab run by Ericsson and made available for third parties to test their applications using a GPRS network simulator to evaluate application performance under different conditions. I was told that over 80% of the applications failed to work when the RF network coverage became intermittent. This was simply because there were no time-outs in the software to bring things back under control if the network failed to

---

[8] SRAM stands for Static Random Access Memory, which is the volatile memory space available for running an application
[9] A piece of code in Java is called an *object*.

respond within a certain time period. On some devices, this failure could lock up the device altogether rendering it useless until switched off (and then on again).

There are so many strategies that can be deployed to reduce code size and to improve execution speed, some of them general and some of them specific to the device and the toolset supporting it (like the compiler). In Java, a *long* data type is a 64-bit integer. BlackBerry handhelds, like many others, use a 32-bit processor, so operations run two to four times faster if we use an *int* instead of a *long*. We will also use less memory by using *ints* instead of *longs*.

Another example of the interaction between design decisions and the target execution environment is related to system design rather than just coding technique. Perhaps our overall system design leads us towards the use of XML for mobile message passing. This seems reasonable, as many software processes increasingly utilise XML vocabularies. However, we might not find an XML-processing API on the target device, thus forcing us to implement our own. This will cost memory and execution time. It will also cost bandwidth, which may or may not be a problem, but usually is. In a very constrained environment, the use of XML may not be appropriate, such as where the most reliable transport mechanism is text messaging, where the message payload size is very limited (160 characters, or bytes).

There are so many low-level details to take into account, or that can be taken into account, when developing software for a particular target device, which means that the only real option is to take time to consult the device documentation in the associated SDK; a laborious yet necessary process. We really need to get our hands dirty when it comes to appropriate design optimisations for mobile devices. Some general principles are applicable to all designs, but this whole area of optimisation is the subject of an entire book[10] in its own right (not one that I intend to write).

11.4.2 Configuring the IDE/Program Editing/Compilation and Build

The IDE is one in which we can write code, compile it and test it within one user interface. Sometimes, we are able to get the entire toolset from the platform vendor, such as the JDE from BlackBerry or Visual Studio .NET from Microsoft. Other times, we can use a third-party toolset. The crucial matter is that the toolset we use should directly support the device, or device family (or software platform) we wish to program.

If an IDE is aware of the particular APIs that we are coding to, it is able to provide direct support for programming to these APIs; for example, syntax checking and checking that we are passing the right parameters to the API. We would also probably like to have built-in documentation support (help system). Another aspect is the user interface programming. Many development tools support visual construction of the user interface using drag-and-drop interface widgets, such as menus, option boxes, buttons and so forth. We would, therefore, prefer to be able to build interfaces to suit the particular target platform we are coding for.

Some devices are supported by third-party toolsets. Where this is so, it is probably a good idea to invest in a well-established IDE product because it may pay off in the long run by enabling better productivity. Toolset specialists naturally put more effort into fine tuning of the tools, whereas a toolset in an SDK is often designed with time-to-market in mind and can be very basic, perhaps not supporting advanced features like visual construction of interfaces.

---

[10] See http://www.smallmemory.com for such a book.

11.4.3 Testing and Debugging with a Simulator

Testing and debugging with a simulator is a crucial part of the design flow that is not generally found or required during desktop development. Once we have written our software and compiled it, we need to test the program. The obvious thing to do is to load the packaged program onto the mobile device and then try it out for real. However, initially this turns out not to be very useful.

In the first place, we may not have a device. New devices aren't always available, even though there might be an SDK to support them. In this case, the solution to testing without a device is to use a simulator.

A *simulator* provides a graphical view of the target device, including all its physical interfaces, within a desktop-executable environment. An example of the BlackBerry simulator is shown in Figure 11.5. We can see from the figure that the entire device is visually displayed, including the interfaces such as the keyboard and buttons. The graphical display of the device is faithfully reproduced. It behaves identically to the actual device. Furthermore, the simulator will run our developed mobile application and behave in the same way as it would on the device.

A simulator can be a very valuable productivity and debugging tool, even when we have the target device available. It is far quicker to load an application onto a simulator for testing than to load it onto a device where extra steps are required, including physically uploading the program to the device. In some cases, the simulation can take place in lock-step with execution of the code within the debugging environment of the IDE. For example, we could step through the code in the IDE, watching where are in the code whilst viewing the results in the simulator. This powerful feature has many benefits during debugging.

Figure 11.5 BlackBerry desktop simulator. (Reproduced by permission of RIM.)

There is a difference between a simulator and an emulator. The difference can be an important one and so we need to understand it, especially in the context of testing our application. A simulator attempts to mimic the general behaviour of the target device by using features from the software environment of the desktop platform to replace equivalent features on the device. For example, if the desktop platform supports a windowing graphical toolkit, then this will be used to mimic the one on the device as closely as possible. Similarly, if the desktop platform has in-built HTTP networking functions, then these will be used to substitute the ones on the device. Overall, we can appreciate that simulation is trying to map one feature set onto another on a different platform in order to reach a useful approximation of the target device, as shown in Figure 11.6. In the end, only testing on the device itself will enable complete testing to be carried out.

An alternative approach to simulation is *emulation*. This is not so common because it is a complex task to achieve. Emulation does not map software functions at the higher layers of articulation and execution. Instead, it attempts to model faithfully and accurately the actual device hardware and low-level software structures, and runs this model on the desktop platform as a complete self-contained process. In other words, the emulator is a virtual device.

If we think about emulation technology, then we will understand why it is a complex process and can be difficult to implement. With emulation, we are proposing to take the actual compiled software image that we would want to load onto a physical device and to load it onto a desktop computer instead. The emulator will attempt to run the application image as if it were the actual target device – we have now moved beyond the level of mimicking APIs at a functional level. When the software image runs, it is expecting to run natively on the target processor and to gain access to chunks of code that are the images of the APIs themselves just as they would normally be stored and accessible on the target device. These chunks of code will exist in the emulation environment and when they get

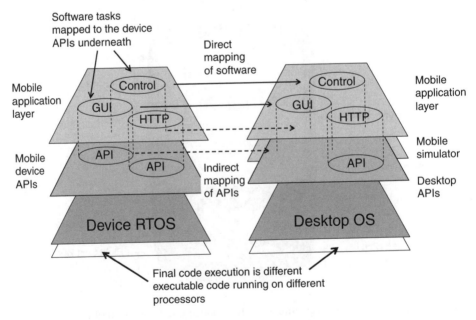

Figure 11.6 Mapping mobile software to a device simulator.

called they will want to interact with real hardware, like an LCD controller chip or a serial port. The emulator responds to the required interaction by faithfully reproducing the low-level hardware and device driver environment that will be present on an actual device. Thus, an emulator must succeed in creating a virtual device and virtual mobile OS that runs on the host system, as shown in Figure 11.7.

In the case of the Microsoft Windows CE 4.3 .NET emulator, it provides emulated access to the following hardware elements:

Emulated hardware	Description
Parallel port	Provides a parallel port that you can map directly to an LPT port on your development workstation.
Serial port	Provides two serial ports that you can map to communications (COM) ports on your development workstation.
Ethernet support (optional)	Allows the device emulator to support a single Ethernet card using shared IP. When Ethernet support is enabled, the device emulator uses the Ethernet card in your development workstation to emulate a DEC 21040 Ethernet card. This means that your development workstation does not need a specific driver for Windows CE .NET to emulate a physical Windows CE–based target device.
Display	Allows you to specify the emulated display size, from 80 × 64 pixels to 1024 × 768 pixels. You can specify a colour depth of 8, 16 or 32 bits per pixel.
Keyboard and mouse	Supports a standardised keyboard and a PS/2 mouse.
Audio	Provides basic support for audio, including audio input, audio output, microphone input, and line output as well as support for full duplex audio, which enables VoIP.

Emulating the general-purpose device features is useful, but emulating the RF networking and telephony features would be even more useful. It is not clear the extent to which this is available, but it ought to be possible to emulate access to other hardware elements. For example, it would be useful to emulate a serial connection directly to an actual Bluetooth modem attached to the desktop PC and then to interact with other devices in a device network via a real Bluetooth link. Such emulators can be built, but their complexity usually rules them out for general release as develop tools. Such test bench systems often do exist, but are usually limited to device manufacturers' labs.

The challenge that remains both with simulators and with emulators is to how to reproduce RF networking and telephony features. However, once we reach this stage in the

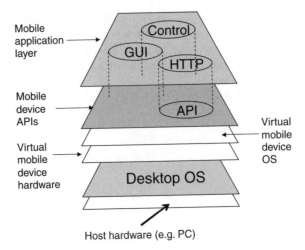

Figure 11.7 Mobile device emulation on a desktop PC.

development process, then we probably have to move to the real device to complete our testing scheme.

11.4.4 Testing on the Target Device

Eventually, we will have to move to the target device to test our application, including connectivity with other devices (in the device network) or with the RF network.

Even at this stage of testing, it is feasible that we can enhance our test strategy by the inclusion of test software within the application, accessible via a test mode. This might include software to help us gain access to internal messages and operational states of various software elements. This information could be written to a file on the device, made accessible via the screen, or any manner that we choose, such as via a serial link.

The use of on-target debugging is supported by some mobile device operating systems, such as the Symbian OS. Symbian ships with a debugger (GDB GNU Debugger), which supports on-target debugging of mobile applications. This is a particularly powerful feature as it means we can fully test our application within its intended environment, particularly the RF networking aspect, and gain access to useful debugging information. A special piece of code (called a stub[11]) is installed on the target Symbian OS phone and communicates, over a serial link, with the debugger running on a host system (i.e. a desktop machine locally networked to the device), as shown in Figure 11.8. On the host the command-line interface is supported. Nokia includes an application debugger in its CodeWarrior IDE toolset for Symbian OS, called MetroTRK. Nokia has now switched to a new IDE for future releases of Symbian OS and its Series 60 devices, called Carbide.

This is one of the advantages of the embedded software approach. We can interface at a very low level with the device to enable debugging monitors and host communications to take place. This is because we have access to the device resources to do this. With other

---

[11] The stub here is not the same as the remote method invocation stubs we looked at in J2EE in Chapter 9; these are totally different.

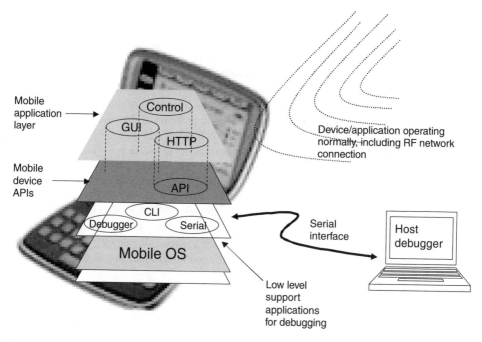

Figure 11.8 On-target application debugging via host PC interface. (Reproduced by permission of IXI Mobile.)

approaches, such as the sand-boxed Java approach we shall look at soon, this is not so easy, and may even be impossible in many cases.

One of the problems with testing software on a mobile device is attempting to produce a range of test scenarios that represent the expected variation in usage and operational context. This is particularly so for the RF networking. A useful feature would be the ability to put the device itself into some kind of engineering mode that can facilitate some of the test cases required. For example, the RF modem could deliberately throttle bandwidth or simulate radio signal variation.

Engineering modes are available on mobile devices and have been in existence for a long time, but these are usually undocumented features that are ordinarily only accessible to authorised test personnel working for the device manufacturer. These modes usually enable access to special test features and measurement and reporting information from the RF modem subsystem. Engineering test modes can often include operations outside the bounds of normal device interaction with the RF network, which is why they are usually unavailable to developers.

If we have implemented a solution that targets many different devices, then it is essential to test on all possible target devices. This usually presents the problem of having all the devices available. This can sometimes be addressed by participating in development forums that provide a testing service, usually at a fee. Some of the operators provide such facilities, although the test devices are often in short supply in relation to the size of the developer community being supported. Otherwise, we face the prospect of getting hold of all the devices, assuming that they are available. There is no easy solution to this problem.

11.4.5 Conducting Usability Tests

Something that I learnt very early on in mobile applications design is that the UI design can be very misleading. What appears to the designer to be a perfectly sensible UI implementation often baffles a user unfamiliar with the application. Fortunately, there is an easy solution to this problem, which is to let the users have a go! It sounds obvious, because it is. Nevertheless, it is remarkable how often designers overlook, dismiss or ignore *formal* usability testing.

The topic of usability for mobile devices is slowly being addressed by the usability gurus. An instructive study was conducted by Jakob Nielsen who is somewhat of a sage in the general field of usability, particularly web interfaces. To quote from one of Nielsen's Alertbox newsletter on usability (extracted from an article called Why you Only Need to 'Test with 5 Users'[12]):

> *Some people think that usability is very costly and complex and that user tests should be reserved for the rare web design project with a huge budget and a lavish time schedule. Not true. Elaborate usability tests are a waste of resources. The best results come from testing no more than 5 users and running as many small tests as you can afford.*

Usability testing is straightforward. The idea is to assign tasks that the users are usually expected to carry out whilst using the application. For example, if the application is a mobile email service, then we would set tasks like:

- Compose and send an email to 'Joe Bloggs' whose name is in the address book

- Add an appointment to the diary for one day next week

- Find a message from 'Fred Smith' about opportunities in wholesale coffee

In other words, we set typical tasks that we expect end-users ordinarily to carry out with the finished application. Then we conduct surveys to get user assessment of how easy it was to carry out each of the tasks. This should be done without any unnecessary intervention in the process. If a user struggles to carry out a task, then it's useful information that we need to know. As tempting as it may be to intervene, we should not give hints on how to do the tasks. After all, we won't be standing over the user's shoulder in the field.

The curve in Figure 11.9 shows us a nominal metric for how effective usability testing can be in relation to the number of testers. As Nielsen points out:

> *The most striking truth of the curve is that **zero users give zero insights**. As soon as you collect data from a **single test user**, your insights shoot up and you have already learned almost a third of all there is to know about the usability of the design. The difference between zero and even a little bit of data is astounding.*

Nielsen's comments are much needed. Whilst it is obvious that zero testers will tell us absolutely nothing about usability, this is the shocking reality of many mobile projects, with operators often the worst offenders. I have personally tested many commercially available mobile applications and found that the usability was inadequate, clearly not tested. Until recently, the topic of formal usability testing was noticeably absent from most formal

---

[12] http://www.useit.com/alertbox/20000319.html

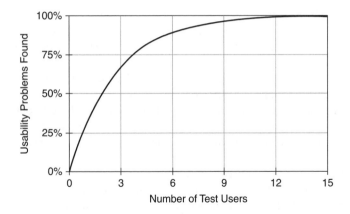

Figure 11.9 Curve to show usability effectiveness dependent on users.

software processes. The tidal wave of websites that have surged into view has caused a new focus on this important topic. Usability in mobile applications is even more important than with the Web, as the tolerance for poor UI design is often much narrower.

11.4.6 Pilot Trials and Deployment

At the end of the development process we need to deploy the application. For embedded applications, this can take a variety of forms:

- Web download and manual install via desktop PC

- OTA installation

- Pre-installation in Read-Only Memory (ROM) or other non-volatile memory

- Packaged distribution with device, or separate, such as on CD-ROM

In the future, there may be other distribution formats, such as removable memory devices, like MMC cards, which is more akin to the games cartridge distribution model. The method of distribution may be dictated, depending on whether or not there is any intervention in the process from a mobile network operator. Indeed, we could well have chosen to deploy our application via an affiliated scheme with the operators. Such a scheme is likely to have its own preferred distribution requirements. Let's briefly examine each option.

Web Download via PC   This is the most common method for distributing any software solution today. It is already commonly used for PDA applications, so there is a legacy of experience with distributing mobile applications in this manner. The actual loading mechanism onto the device is via a cable, either Universal Serial Bus (USB) or serial, sometimes via a cradle. For PCs with Bluetooth connectivity (e.g. some laptops, or via a USB-BT dongle) it is possible to establish the serial connection wirelessly.

The advantage of the direct installation method, where available, is that it is relatively quick and less prone to errors than alternatives, such as OTA. Using the Web for distribution has obvious benefits, not least of which is its prevalence. The existence of various marketplaces to offer software solutions is also an advantage. There is an increase in marketplaces for mobile applications, especially games, with some of the big names like Yahoo offering a mobile games arcade in addition to what many of the mobile operators are already doing.

The disadvantage of downloading from the Web is that the device needs to be connected to the PC. In some cases this is not practical as the user may not even have a PC. In may also be the case that users' expectations are such that once they have a connected mobile device, the need to use a wired connection (i.e. via a desktop PC) may seem at odds with the underlying mobile service concept. Thus, manual installation via a web download may frustrate users.

At this point we might sensibly enquire as to why we have to bother with a web download via the PC if we do indeed have a connection available via an RF network. There are two reasons. The first is that on some devices there simply is no mechanism for installing applications other than via a desktop PC. The second reason for downloading via a PC is the potential size of the file. Mobile applications could easily be several hundred kilobytes. Over a very slow RF link, this may be too much to expect a user to download. Even if it were possible, then we would probably still want to ensure that we had a purpose-designed mechanism for handling software installations via this method. For example, we would want to ensure that chunking of downloads is possible with error detection for each chunk. Should the link become disrupted, then we would not have to start again the whole download process, something that until recently was required on desktop implementation until some downloads became truly massive. This prompts us to also consider the need for application compression, but this topic is discussed in the next section.

Over-The-Air (OTA)   There is nothing new about remote loading of software over a communications link. This technique is used all the time in embedded applications. Interestingly, it has been used in cellular base stations for some time, to avoid the cost of deploying service engineers for upgrades[13]. In other words, the techniques for remote software installation are not rocket science and are well understood. They are not that complex either. Remote installation is not even new for wireless, the most obvious example being satellites.

However, it has taken some time for the concept to arrive on consumer mobile devices. It has really been advanced by the recent developments in mobile Java (more on this later), where initially some recommendations were made for OTA provisioning of Java applications on the device, but recently the technique has become formally part of the specification[14].

There are a number of technical issues that need addressing relating to OTA provisioning:

- An agreed and available mechanism on the device to cope with an OTA installation

- An agreed file format, if applicable, for encapsulating downloads

- A compression method(s)

---

[13] Note that the software downloaded to a base station is done over the landline connection to it, like a 2-Mbps leased line or something similar. This is worth stating in case we thought that base stations were somehow upgraded wirelessly – due to their obvious wireless nature, but we shouldn't forget that they usually have a wired connection too, which is the backhaul to the rest of the cellular network.

[14] The so-called MIDP 1.0 version of Java did not have a formal OTA policy, just a recommended practice, whereas OTA has been fully integrated into the MIDP 2.0 specification onwards.

- A suitable file download protocol

- A network protocol for establishing the OTA link with the application source

Let's look at each of these issues in turn to see what the details might entail.

In terms of a mechanism for coping with OTA, then this clearly implies that the device itself has a software application that is able to receive application files and then install them. We don't really need anything new because the existing application loading mechanism could be used; it simply receives the installation file from an RF modem connection rather than a serial port, both being serial streams in any case, so the adaptation is not that challenging.

However, there are some subtle differences, which may seem trivial, but can end up being significant if not addressed. It may have escaped our attention that when installing software via a cable and PC, the PC itself is running part of the overall installation application. One of its functions is to provide backups, this being particularly important before we install a new application. If something goes wrong during the installation, then we would like a means to get back to a previous known working state. The PC can provide this assistance for us. Potentially, the PC installation could even back up the entire memory image from the device so that we could reliably get back to a known working condition whenever things go wrong.

A backing up procedure over a wireless connection may not be feasible, though in general it is probably something we are going to see emerge with faster RF connections, not just for applications but for user data, too[15]. If backing up over the network is not possible, then we probably want to take a more conservative approach towards installing new applications. For example, we may want to provide local backup via spare memory on the device. Applications may contain APIs or shared files that are already on the device and need updating for the new application to run properly. It would be a good idea for shared files to be backed up first. This would enable a previous state to be retrieved should the new application not install properly or should any of the currently installed applications stop working.

SAR (or WTP-SAR) stands for Segmentation and Reassembly. It's an optional feature of the Wireless Transaction Protocol (WTP) within WAP[16]. SAR defines a means for a WAP gateway to break a large message (e.g. an application we're downloading) into smaller chunks (the segmentation, or 'chunking') and for the phone to piece it back together (the reassembly), as shown in Figure 11.10 and discussed fully in Chapter 7.

Not all mobile devices use SAR, but many will access web servers through a WAP gateway. Nokia WAP phones with Java games support use SAR, whilst Motorola devices apparently use features of HTTP 1.1 to retrieve small chunks of a file one at a time and then reassemble it, as we shall discuss shortly.

The sender can exercise flow control by changing the size of the segments depending on the characteristics of the network. Selective retransmission allows for a receiver to request

---

[15] Data backup services for address books and text messages are already becoming increasingly common.

[16] We should note that SAR here is not to be confused with how it gets mentioned in the context of IP datagram communication, where packet sizes can be made small for 'small packet' networks. Even using packet-based IP communications, we can still lose the link, no matter how well the IP layer is functioning. If anything goes wrong in the download session, we want to be able to recover from the problem without restarting the entire download process; hence we need SAR at the transaction level, not the packet level. (Refer to Chapter 7 for more insight into this issue.)

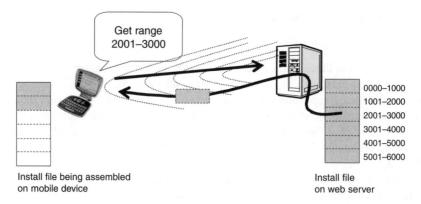

Figure 11.10 Segmentation and Reassembly Process (SAR) for downloading applications.
        (Reproduced by permission of IXI Mobile.)

one or multiple lost packets. Without this, in the event of packet damage or loss, the sender
would be forced to retransmit the entire message, which may include packets that have been
successfully received. This could rapidly become very tedious for the user.

One problem that arises with WAP gateways concerns MIME (Multipurpose Internet
Mail Extension) types returned by the web server. If a user requests a JAD (Java Application
Descriptor) or JAR file (Java Archive – both of which are Java program files) and the server
returns the wrong MIME type, the gateway and subsequently the phone will handle the
transfer incorrectly. This reminds us that our mobile device application is only one piece
of the puzzle. There are many other participating elements in the mobile network that need
to be considered, which is why wide systems knowledge is important (a good reason for
reading this book).

An optimisation introduced in the HTTP 1.1 specification is something called *byte range*
operations. We've all experienced trying to download software from a website, only to have
the connection fail with only a few kilobytes left to go, out of 10 Megabytes! At that point,
we might be forced to initiate the download again, hoping for the best, maybe giving up after
a few tries. Using HTTP 1.1, the client software application can request just the last few
kilobytes of the resource instead of asking for the entire resource again. This significantly
enhances usability and general user satisfaction for web downloads. This technique can also
be used for downloading applications, not just data files. When requesting a byte range, a
client makes a request as normal, but includes a *Range Header* field specifying the byte
range the resource is to return, as shown in Figure 11.10.

The client may also specify multiple byte ranges within a single request if it so desires,
causing a queue of requests to be aggregated into one response. In this case, the server
returns the resource as a *multipart/byteranges* media type. This seems to go against the
ethos of ranged requests, but it might be a useful mode to deploy dynamically during a
download, should the client detect that the RF network connection is robust enough to
support it.

The use of byte ranging is not limited to recovery of failed transfers, or for OTA provisioning of mobile applications. Certain clients may wish to limit the number of bytes
downloaded prior to committing a full request. A client with limited memory, disk space or

Table 11.1   Possible compression gains for OTA provisioning of CAB files

	CAB file size	AirSetup installer size	Compression level
Sample CAB file #1	757 KB	188 KB	24%
Sample CAB file #2	2.69 MB	1.09 MB	40%
Sample CAB file #3	901 KB	424 KB	47%
Sample CAB file #4	1.47 MB	0.64 MB	43%

Source: SPB Software Warehouse.

bandwidth can request the first so-many bytes of a resource to let the user decide whether to finish the download. This will depend on whether or not the nature of the proposed download allows for this pre-emptive interaction with the application prior to finishing the download.

Web servers are not required to implement byte range operations, but it is a recommended part of the protocol and widely implemented in any case. Clearly, if it is not implemented, then the fallback is to attempt downloading the entire application. This may suggest that it is a good idea to examine carefully the application file format for OTA, taking into account compression optimisations for OTA and any inherent chunking of the installation files and process.

To deploy an OTA application, we possibly need a new file format for the transfer process, a file format that is optimised for OTA deployment and can be used to encapsulate the installation file.

In some cases, we should even consider an implementation for installation files, one which is optimised or built for OTA provisioning from the ground up, rather than an afterthought.

In the case of an encapsulated file format, it should simultaneously support OTA-optimised compression and software chunking. Compression is usually a feature of software installation files, but this is not always the case; nor can we assume that the default approach is the best method, as it may not have been optimised for OTA provisioning. General-purpose compression techniques applied to a binary file will never be as good as ones specifically designed for a particular purpose, such as compressing application files for OTA provisioning. Table 11.1 gives us some food for thought. It is a table reporting possible improvements on compressed CAB[17] installation files for Microsoft's Pocket PC devices. These improvements are the claims of SPB Software House[18] when using their AirSetup installer technology, a third-party OTA solution.

Irrespective of the file format required, we need a protocol for managing the download process. We would like to incorporate the ability to establish that the download has been successful. This will enable us to support our end-users in a reliable fashion. Furthermore, the ability to determine that an application has been successfully downloaded and installed provides an opportunity to complete a financial transaction with the user, assuming that we want to bill them for the downloaded application.

---

[17] CAB is short for cabinet, which is a metaphoric name to indicate many files placed into one cabinet (one place) for download purposes.
[18] http://www.spbsoftwarehouse.com/products/airsetup/index.html

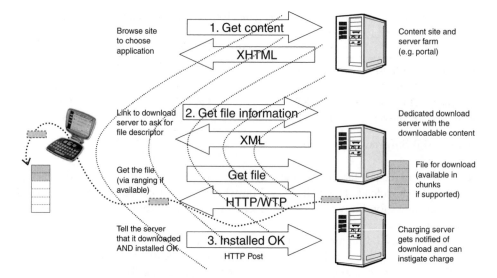

Figure 11.11 Possible content download mechanism as per OMA guidelines. (Reproduced
         by permission of IXI Mobile.)

As we shall see, a mechanism for doing this has already been proposed by the Java
Community Process whilst formulating the specification for the Mobile Information Device
Profile (MIDP) version 2.0, a version of Java aimed at mobile devices, which we shall
discuss later. Using the MIDP mechanism as a model, the Open Mobile Alliance (OMA)
has proposed a similar method for the downloading of any content to mobile devices[19]. We
emphasise here that this mechanism is intended for all types of content, such as multimedia
clips, ringtones and so forth. However, it is a generally applicable model that can be used
for any type of file, so it would be sensible to presume that this method be adopted for
downloading embedded applications. Even if such a provision does not exist natively on a
device, as part of the underlying operating system bundle, there is no reason why it should
not be installed on the device first and then used to fetch the target application(s).

OMA download mechanism   The OMA download mechanism that is summarised in Figure
11.11 consists of three main steps, which are described in detail as follows:

Step 1. **Browsing (discovery process)** – The method for perusing downloadable content
        is presumed to be via a browser-based presentation layer. The browser picks up
        web (WAP) pages from a server and the user eventually selects an embedded link
        to take them to the next step, which is to download (and/or view) a descriptor
        file about the proposed target download application (or file).

Step 2. **Evaluation (content negotiation)** – Prior to downloading the required applica-
        tion install file, we would like to be able to evaluate the application in terms of its
        resource requirements. To do this, we firstly fetch an object descriptor file. This
        is done using HTTP, so any sever capable of delivering the file would suffice,
        not necessarily the same one used for the browsing in Step 1. The reliance upon

[19] OMA-Download-OTA-v1_0-2002121920030221-C,   Version   19-Dec-200221-Feb-2003   available   from
http://www.openmobilealliance.org

HTTP for this process does not preclude non-browser solutions. Any mobile device application that can handle HTTP would be able to fetch the descriptor information from the server, so we could even fetch the descriptor from within another application, such as an embedded one. This may be useful for software upgrades.

The Download Descriptor, or DD (as named by the OMA), is written in an XML vocabulary and contains various information about the application attributes, such as file size and file type. The mobile device can use this information to ensure that the proposed download is compatible with the device and its available resources at that time (e.g. memory space).

An example of a DD is shown here:

```
<media xmlns="http://www.openmobilealliance.org/
xmlns/dd">
        <type>image/gif</type>
        <objectURI>http:/download.example.com/myapp.app
        </objectURI>
        <size>100</size>
        <installNotifyURI>http:/download.example.com/
        gotis.asp?id=ab35612</installNotifyURI>
</media>
```

If we examine the anatomy of the example DD just given, we can understand some key elements of the download process:

(a) *type* – this is the MIME type for the resource. This may or may not be useful, but indicates something about the file type to the client, which it may find useful or simply ignore. It could be used to confirm that the file type is as expected.

(b) *objectURI* – this is the web address of the downloadable resource and is where the mobile device can now go to get the file via HTTP or WTP accordingly.

(c) *size* – this is the size of the resource in bytes. Note that this only indicates the size of the download file, it does not mention how much space is actually required by the resource once it is installed. After decompression and installation, the size of an application may be bigger (this does not appear to have been considered by the OMA).

(d) *installNotifyURI* – after the resource has been downloaded, a status code can be sent back to the server via this address, which can be any server, not restricted to the same one from where the download just came. The advantage of this process is that the downloading process can be state-monitored, so we should always know what state the download is in and what went wrong with it, should it go wrong. This 'download success' acknowledgement mechanism is a unique addition to the downloading of content. It is not included in the HTTP protocol or associated WAP protocols. The codes used to indicate the download status are to be submitted via HTTP post to this web address. Some of the codes suggested by the OMA download specification are shown in Table 11.2.

Table 11.2   Status codes to indicate content download success

900 Success	Indicates to the service that the media object was downloaded and installed successfully.
901 Insufficient Memory	Indicates to the service that the device could not accept the media object as it did not have enough storage space on the device. This event may occur before or after the retrieval of the media object.
902 User Cancelled	Indicates that the user does not want to go through with the download operation. The event may occur after the analyses of the Download Descriptor, or instead of the sending of the Installation Notification (i.e. the user cancelled the download while the retrieval/installation of the media object was in process).
903 Loss of Service	Indicates that the client device lost network service while retrieving the media object.
905 Attribute Mismatch	Indicates that the media object does not match the attributes defined in the Download Descriptor, and that the device therefore rejects the media object.
906 Invalid Descriptor	Indicates that the device could not interpret the Download Descriptor. This typically means a syntactic error.

Step 3. **Installation/notification** – After downloading the file, the mobile device attempts to make use of the downloaded resource. The actually downloading process has already been discussed in terms of the possibilities for compression, file encapsulation and chunking. We are now assuming that we are at the state where a useable resource file is now available on the mobile device. Clearly, we need the installation process to be initiated. The program that initiated the download, which may have been a dedicated OTA-handling application or another embedded application, would also take care of the installation. In either case, the program will call the appropriate API for installing and registering applications with the operating system, bearing in mind that we have already shown how there needs to be a part of the device platform that can load applications.

An aspect of the OTA process is that the installation process must have a means to report to the download server (or the designated install-notify server) via the installNotifyURI that we said is included in the DD. There seem to be several ways of managing this process on the device itself, and we mention two of them here.

The first option, as shown in Figure 11.12, makes the assumption that a separate download manager exists to control the installation of the downloaded application and that it does this via an API call. After a series of events to process the installation, there would be an API call-back to the manager to notify it that the installation has finished and to indicate how successful it has been. The download manager can then interpret the call-back information and use this to initiate the HTTP POST of an appropriate download status code to the installNotifyURI.

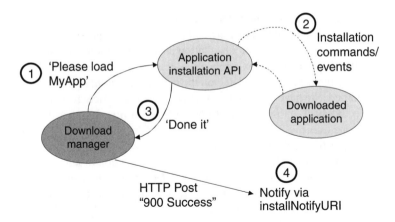

Figure 11.12 Download manager manages both the installation and the reporting back to the download server.

The second option is to let the actual downloaded application initiate the communication back to the installNotifyURI, either directly, or via an API call to an appropriate application capable of carrying out an HTTP POST (such as a browser on the device that has its inner workings exposed via APIs, like Pocket Internet Explorer on Windows CE). This would simply require that the installed application be passed the installNotifyURI as a parameter, so that is knows where to respond.

The key difference between the two approaches is that the second approach enables a greater degree of flexibility in reporting back meaningful status information. For example, a user could be asked to fill out a licence acceptance form for the application before a status code is returned. This may seem a bit obtuse; as presumably we could use the application itself to communicate anything we want back to a web server in order to implement charging or some other process in relation to the download and installation operation. This is entirely true. We should understand that the need for a separate download manager and status reporting function is largely dictated by the need to support a wide range of downloadable content. For example, a ringtone could be downloaded using this mechanism. A ringtone does not have the ability to communicate back to a server, so we can see the need for a separate reporting mechanism in this case.

The final observation about the installNotifyURI is that is should contain a unique identifier that can be used to identify each download uniquely. This is so that installations can be associated uniquely with users. This is desirable for the purposes of charging and support.

We might wonder if there are other ways of managing the OTA process. In fact, there are probably many ways it can be carried out. We have discussed a procedure that is generic in nature and has been supported by a large number of mobile industry players via the OMA initiative. The procedure is deliberately modelled on a similar process for OTA provisioning of MIDP Java applications, which is something we shall discuss shortly.

What we could argue is that the OMA process is aimed at single-file downloads, or content that is not that complex, such as a single ringtone file or a photo. The process of application installation is obviously more complex than this. Potentially, a very complex application could involve a considerable amount of files being installed or updated on the device, including some system files being modified. We have discussed that altering critical system resources may require selective back-up procedures to be undertaken first. It is feasible that this can be done over the RF network. Such a process would involve an intelligent OTA manager on the device that could record device state information, file changes and other device alterations, almost like a transaction monitor that in the event of failure could undo all the changes. Clearly, certain changes might preclude backing up over the network should they impact on the networking mechanisms themselves, like critical device drivers and so forth. The OTA manager should have the ability to detect such implications and flag them to the user, giving the option to back-up files to a memory device instead of over the network, which may subsequently become unreachable if the installation process fails.

Even the process of backing-up would need to be made more suitable for the OTA environment than perhaps we are used to in the wired world where everything gets backed up prior to starting the installation. We would not want to ask the user to sit through 5 MB of back-up across the network only to discover that the first files we download for installation are corrupted, or that one of them is not compatible with the device. There may be opportunities for incremental approaches, gradually backing up and replacing files so that any download failures result in a more graceful degradation or quicker recovery. We may still be able to benefit from some of the updates to our device system files, even though the entire installation process was unable to complete; therefore, we should apply 'undo' operations in an intelligent manner.

Looking at the approaches to remote software installation and maintenance that have been developed over many years in other fields, it is entirely feasible that robust OTA mechanisms could be implemented. This is a major necessity for next generation devices, simply because if we are to achieve mass penetration of services and devices, we need to remove as much as possible any technical support activity from the users. Ideally, the users should not even be aware of what an application is, except in terms of its service offering expressed via a user interface. Low-level concerns with installation, code signing and so forth should not be a required part of the user's sphere of concern. This seems entirely possible given the level of technological resource that we can now deploy and apply towards this objective in the mobile device world.

Packaged Distribution   Returning to our options for installing software onto our mobile device, the final option is to simply distribute the application physically to the user, usually through a CD. This may be useful for over-the-counter sales, such as in a mobile device shop. It also has its place in the usual distribution channels and in some promotional channels, such as magazine covers. In future, we may also find that pluggable non-volatile memory devices, such as MMC cards, are used to distribute applications.

The distribution of applications in this manner brings us to a related topic, namely the digital signature of an application by its creators. This will become increasingly important in the pursuit of guarding against surreptitious and malicious practices, such as viruses.

We are often prepared to delegate our choices on a basis of trust. We will do something because we trust the people with whom we are dealing. Software is no exception. If a vendor claims that their software is free from viruses, then we have the option to trust them, or not. The problem is verifying that we are in fact dealing with a vendor whom we trust. For example, if we download an application, or load it from a CD, then we want to be sure that it is from a particular vendor purporting to be the creators of the application. We are willing to trust their software to run on our device, so long as we can be sure that the software has come from them.

The ability to determine authenticity is not unique to mobile applications and has been tackled already using digital certificates and code signing and similar method can be applied to mobile applications.

Trials   Before we engage in full-scale deployment, a trial of our application is a useful undertaking and probably worth the investment in time and effort. This is not to be confused with usability testing, which, as we have already discussed, is actually a part of the development phase, *not* the deployment phase. However, we should expect further usability issues to be uncovered during a trial.

A trial is often confused for what we sometimes call a beta test, but they are different. A *trial* is really about rating the acceptance of our application according to its intended usage and market. We want to see how users get on with it. It is confirmation of our concept and its realisation, *not* testing of the software per se. *Beta testing* is about uncovering the inevitable latent bugs and flaws in the system that we could not expose during testing. However, it is perfectly valid for these two activities to be combined, but the point is that end-users should be aware of what's going on.

11.5 BROWSER-BASED APPLICATIONS

The second method of accessing a mobile service is to use a *browser*. The advantages of the browser approach are well known and we have already discussed them in some depth in our earlier discussions on HTTP and CS architectures, along with HTML and its derivatives. Today, we have the wireless-optimised variants called WAP. What we are interested in here is a discussion of the characteristics of a browser-based approach, especially in contrast to the embedded and Java platform applications discussed in this chapter.

One of the advantages of using a browser is that there is no need for application distribution. This is the beauty of the Web. We could have a hundred applications and we don't need a hundred installation files on our device, just the single browser. What we do need is the means to access those applications, which is through a URL, or address, where we can find each of them. Finding applications via their URLs is perhaps the most cumbersome aspect of the mobile Web. We have two problems:

1. Application discovery

2. Physical entry of the address (URL) into the device

Application discovery is about finding the application in the first place. In the mobile context, we have limited options for how we discover applications, simply because we tend to browse infrequently and in general, there are fewer ways to find out about a site we might be interested in surfing, although *WAP Push* could affect this, as we shall see.

Address entry is the second problem. An application is accessed via its URL, but these are lengthy and can be cumbersome to enter on a mobile device, especially small-form factor devices with fiddly keypads. However, this problem can be made easier. In some browsers, it is possible to type only the unique portion of the URL and arrive at the site by auto-completion of the address, a feature found on some desktop browsers (although still sadly lacking on many mobile devices). We could simply type 'trains', where we mean http://www.trains.com or http://wap.trains.com, there being a trend for some site designers to use 'wap' as the sub-domain in place of the conventional 'www'. Since the introduction of the new .mobi (also known as DotMobi) top-level domain (TLD), the auto-complete might also arrive at http://www.trains.mobi, or simply http://trains.mobi. That aside, a well-implemented website should be able to redirect automatically to the mobile-compatible pages from a single entry point (home page), so there is really no need for the user to be aware of 'wap' subdomains or what the TLD is for the destination. This redirection is done by detecting the browser-type accessing the home page and then steering it to an appropriate page to match its content-handling abilities (made known to the server in the HTTP headers).

Of course, once a site has been found, users can bookmark the site. Sadly, on a many devices, managing bookmarks can be cumbersome and is almost an abuse of the word 'shortcut'. On some devices, the route to enacting a shortcut can be circuitous and hard to fathom. What makes most sense is to include the ability to display a shortcuts page as a personal home page, so that useful sites are only one click away from starting the browser. Such conveniences are notably missing from most mobile browser implementations.

There have been some suggestions proposed for making services easier to access via browsers, especially to overcome the entry of cumbersome addresses. Bango.net Ltd[20], for example, pioneered the use of unique numbers to access services, to circumvent the addressing problem. With Bango.net, the service assumed that the user can at least access the Bango.net web page in the first place, which itself would be bookmarked. Thereafter, entering the Bango number for a particular service takes the user straight to it. So, instead of entering http://www.usefulservice.com to access a useful service not previously known by the user, they could simply enter the Bango number, like '1224' to navigate straight to the same destination on the mobile Web. It suffers from the problem of ensuring that the Bango service itself is sufficiently widely known and publicised that the Bango numbers become widespread and subsequently meaningful to a large user population. The people at Bango have since launched a site called wap.com that enables users to keep track of the mobile sites they have visited, and to recommend sites, etc.

This shorthand method of entering accessing browser-based services is a useful principle, but it could be enhanced by its adoption as an internet standard or by some other industry body, instead of a commercially owned and controlled addressing space. Of course, for Bango, the control of an address space is useful commercially as it compels users to use their service, which is exploitable in various ways and also enables premium 'short code' addresses to be rented at a premium.

---

[20] http://www.bango.net

Figure 11.13 A special access key to make browsing easier.

If there were a standard way of using short codes and of registering them, then it would be possible to add a new key to devices that becomes the 'master shortcut' key. Instead of dialling a number to access a phone line, we could 'dial' a number to access an application. We could use a reserved character – such as '@' to indicate the 'access code' for applications, just like we have country codes or area codes (see Figure 11.13). Dialling '@1221' will take us to the website application for code 1221.

It is relatively easy to implement this as a standard by adopting a convention for translating domain names to numbers, or vice versa. For example, if we introduce a new high-level domain name, called .mob, then we can allow the use of numbers as the domain, by convention. For example, if we have an application registered at http://www.usefulapp.com, then we can register for an alias, like http://www.1221.mobi, which would be accessible via standard domain mechanisms (Domain Name System, or DNS). Then, all that we require is that mobile device manufacturers implement the access shortcut in software on the devices. So if we now dial '@1221', the device is programmed not to attempt dialling this number, but to open the browser at page http://www.1221.mob, which is a simple procedure to implement in the device software.

The purpose of having the '@' character is that users would become familiar with this symbol, which is already strongly symbolic of the Internet. After some time, users would instinctively know that dialling '@1221' (or any number) means they will be accessing an application rather than making a voice call. On billboards and in adverts, we simply add the line 'dial @12213 for more information on your mobile phone', or however we want to say it; simple and effective. Over time, the prefix will become widely understood.

We could consider extensions to the idea that make sub-sites accessible from a single domain. For example, '@12213.1' and '@12213.2' would translate to http://www.12213.mob/1 and http://www.12213.mob/2 and so on. More likely, as we would want to use this for targeted advertising, we would use a parameter, like http://www.12213.mob?param=1 and http://www.12213.mob?param=2. This would be useful for location-specific adverts that do not need to rely on location-finding processes in the mobile network. So an advert for train times could have different post-fixes that map to different timetables on the website.

11.5.1   Limited Local Processing

A browser-based application has limited local processing capabilities compared to embedded applications and Java platform applications. The browser has two main functions:

1. Display web pages from the web server

2. Accept user input in forms for sending back to the web server

The Web 1.0 idea is really one of page rendering, with very little local processing of the fetched information apart from rendering it to the screen. It was not intended to implement an actual spreadsheet or a word processor as a browser application, only to display statically the information that might come out of such an application.

The extent to which the browser can carry out local processing tasks is still very limited on mobile devices, unlike with desktop counterparts where the Web 2.0 browser has evolved to include powerful scripting capabilities (e.g. JavaScript) and the ability to host embedded applications within its framework, something that originated with the Netscape plug-ins. A well-known example of a plug-in is Macromedia's Flash player.

Therefore, the browser framework still does not support a general-purpose programming model like the embedded approach. The browser itself is an embedded application, written in a general-purpose programming language like C++, and runs natively on the underlying mobile operating system. It interprets text files that contain markup languages of one type or another – there is no such concept as running compiled code, like assembly code that has been compiled from C. It is possible for browsers to enable compiled applications to run, using the OS as the execution environment, with the browser as the display and communications framework. The conceptual link is already available via the *object* tag in XHTML-MP, which is a mechanism to refer to content in a web page that requires an external program (helper application) to handle it. This is primarily aimed at handling multimedia content types, but it could be used to launch a custom application written to run natively on the mobile operating system. In practical terms, this may not be possible on many devices, especially where there is no support to install custom applications in the first instance. Where Java is supported on a device, this mechanism may be used to launch a Java application, though in some environments, such as the MIDP Java platform, this mechanism is not directly supported by the MIDP specification[21].

In the specification for WAP 2.0, there is no support required for browser-based scripting, unlike the original WAP 1.0 where WMLScript was proposed and supported. This is a limited capability scripting language that enables some processing to be done locally, but this is constrained to handling data in the published pages and rudimentary checking of data submitted via user input forms. Like all browser scripting approaches, the scope of execution of a script is also constrained to running only when the container page is displayed in the browser, so general-purpose scripting is not possible; therefore, we should not think that browser scripting provides a container for general-purpose scripting on a device. In fact, the scripting only has access to APIs that enable access to the page (document) components, not to any other resource on the device. JavaScript support is available only on a few mobile browsers.

---

[21] There is no API to handle the communications between the browser and the Java MIDP program.

11.5.2 Requires an Available Network Connection (Caching)

An essential requirement for the browser approach is the availability of a network connection from the browser to the server. If we do not have a network connection at a particular time, then we have a problem with access to the required application – this is the weakness of the browser approach for mobiles. Without an RF connection, we are unable to fetch pages into the browser; in effect, we have no application. As much as we would like ubiquitous network coverage, there will always be circumstances where radio coverage is intermittent or even non-existent, for whatever reason (see Chapter 12 about the RF network). The absence of a network connection not only affects retrieving information, but also submitting information back to the network. This would be problematic for any application that required gathering information from the user. In both cases, the use of temporary storage may help us. This is a technique called caching. The principle of caching is that a copy of everything we fetch from a server is stored in a local memory space on the mobile device. If the RF connection is lost, then we can view pages previously fetched by fetching them from the cache, as shown Figure 11.14.

Caching in web browsers is a technique originally used for performance enhancement to speed-up page access, but in mobile scenarios, it can have an even more useful purpose, which is to enable information to be displayed when the connection is absent. In this situation, we are obviously constrained to viewing information that is already in the cache. This is clearly a problem if the information becomes out of date. It is also possible to implement a cache in the uplink direction, so that forms can still be filled out and submitted, even without an RF connection. In this case, an embedded cache-management program will watch out for an RF connection to be re-established and then upload the form as soon as a connection with the server is present.

Caching uploaded forms is only workable when no immediate feedback is expected, or required, from the server. In this case, it can be implemented as a transparent feature whereby the application does not need to be designed specifically to work with a cache. As far as the user is concerned, the connection appears to be present and the form appears to have been submitted as usual. In reality, there is a delay before the form actually gets processed.

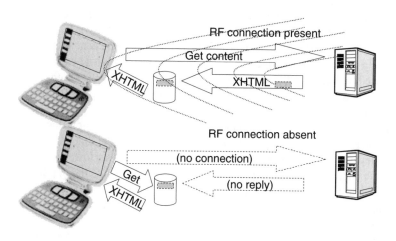

Figure 11.14 Caching enables page to be fetched if already accessed before. (Reproduced by permission of IXI Mobile.)

However, it is unusual, and probably undesirable, to process forms without offering some kind of response to the user, so the use of a form cache is limited. It would be the job of the cache-management program to supplant this response with its own response in order to assure the user that the form has been submitted.

Clearly, this method of caching is only appropriate for certain application scenarios, namely wherever the submission of a form by itself is a meaningful operation without being part of an atomic sequence of page and information exchanges that require server contact. Caching might be suitable for a number of job reporting applications in the field where an agent is required to submit an onsite report. Whenever this is the case, then this caching process has value. It effectively enables a browser-based solution to operate in standalone mode for a period, in order to maintain a useful level of service to its user.

11.5.3 User Interface Constraints

Browser-based applications use markup languages to allow the annotation of data for the purposes of directing what the data looks like in a suitable browser. We have already discussed this process and its variations for wireless devices at length throughout the book (e.g. see Chapter 9).

Markup languages, like XHTML, use a set of layout primitives (e.g. paragraphs, tables, inline images) for presenting data, the construction of which dictates the level of design freedom that the application designer has in determining the richness of the user interface. Additionally, the browser paradigm itself places certain constraints on the user interface. There is no explicit concept of persistence in the user interface, other than how much of the browser area we choose to dedicate to an aspect of the interface that we wish to remain constant across pages. We might want to emulate a persistent display in order to provide certain user interface elements that would be useful to constantly display throughout the user's engagement with the service, such as menus. However, it is difficult to fix menus or buttons to a locked screen position relative to the browser screen. Display primitives are relative to the page layout, which means that they will scroll with the page. This is not ideal for mobile UI design.

Persistence can be supported in conventional browsers using framesets. However, the markup languages for display-limited devices, like XHTML Basic, do not support frame sets. Moreover, even if they did, then the ability to utilise the persistent area is still limited by the relatively crude display widgets. For example, the use of pull-down (overlay) menus is not supported by XHTML Basic or XHTML MP. Defining a lowest common denominator graphical display experience, not a rich one, has influenced the design ethos of XHTML Basic and MP.

The increasing use of colour bitmapped displays, even on mass-produced phones, suggests that the lowest common denominator is getting better all the time, notwithstanding the legacy of devices that remain in circulation with older displays. However, even on devices with a much enhanced display capability, the use of a browser as the application-delivery mechanism will probably still impose user interface limitations that don't exist with embedded applications where it is usually possible to access the richer low-level graphical user interface APIs on the device.

The exception to the limited capabilities of the markup languages is when a browser is able to support a plug-in feature that provides a richer display capability to the content provider. A notable example of this is the Macromedia Flash player, which is very prevalent in the desktop browser world and now also available as Flash Lite on certain mobile device platforms, such as Windows Mobile and some Symbian devices. Flash allows vector-based graphics to be displayed within the browser; thus, giving pixel-level control to the design, which is a level of design freedom far greater than XHTML-MP can provide. However, Flash is not available on most mobile devices, so it cannot be relied upon for general (wide audience) usage, and so XHTML MP is still required for widest device coverage.

11.6 JAVA PLATFORM APPLICATIONS

So far, we have looked at embedded applications and browser-based applications. We are now going to explore the remaining option, the Java platform, which in some sense might be the best of both worlds, especially in its newest variants called Java 2 Micro Edition (J2ME), which is going to be the focus of our attention. We shall now examine the use of Java for mobile devices, highlighting the key attributes of this approach whilst comparing with the alternatives.

To understand the use of Java for mobile devices, we need to first summarise what is meant by Java. What is this technology is all about? We have already looked in some detail at Java 2 Enterprise Edition (J2EE); but, we did so from a systems perspective, focusing on its ability to support highly scaleable applications based on the CS architecture. We did not examine the Java programming environment itself, only its features within the J2EE environment, like RMI and JMS. It seems obvious that J2EE is not a suitable technology for implementing applications that run on mobile devices. We don't need clustering, distributed processing and heavy-duty database access on a mobile device!

Before exploring the general concepts behind Java, we might sensibly question why bother at all with running Java on mobile devices. Certainly, most of the heavy-duty features in J2EE are not required, so why not use a different technology altogether; indeed, what does it mean to support Java on a small device? To answer this question, we need first to look at the origins of Java in terms of its unique software engineering approach that popularised its appeal when first introduced into the computer world back in the glory days of the early Internet boom epoch.

11.7 THE JAVA ETHOS – A TALE OF TWO PARTS

Let's remind ourselves of how a computer works. Computers have more or less stuck to an architectural theme that was popularised by the PC (what we used to call the IBM-compatible PC).

Figure 11.15 shows the basic computer architecture that revolves around the central processing unit (CPU), like a Pentium or Power PC, which can crunch numbers loaded from memory and output results back to memory across a bunch of printed-circuit wires called a bus[22]. Certain memory locations are special because they also enable the CPU to interface with peripherals, like a graphical display. Writing certain values to the display

---

[22] The Von Neumann architecture introduced the idea of a stored program in the same writable memory in which the data was stored.

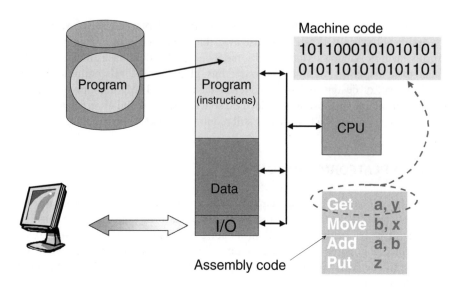

Figure 11.15 Basic computer architecture.

memory will affect what appears on each pixel on the screen and so in this way a software program can write a whole sequence of pixel values in order to control what the screen display. The operating system (OS) is a program that runs on the CPU and can run utility code to manage how this area of memory gets updated and in doing so also provide a more meaningful interface, like a windowing one, to other programs running on the CPU. These applications, instead of writing to the display memory, will write values to a portion of memory monitored by the window-utility part of the OS, which in turn writes to the display memory in a way that manages a windowed look-and-feel.

For the CPU to know what to do with memory values in order to carry out useful work, it has to follow a set of instructions, which is what we call a program (or an *executable*). A CPU can only run one set of instructions at a time and hence one program at a time. However, the OS can swap programs in and out of memory and give them turns running on the CPU. In this way, if the programs are swapped often enough and given long enough time to run on the CPU, then we create the effect of many programs running at once. This all takes place without the user perceiving any swapping, because the CPU runs so fast that it can do the swapping and the time-shared processing without the user noticing.

The CPU executes instructions one at a time whilst stepping through a program. How do we, as programmers, tell the CPU what to do? A CPU is a machine that is exceptionally good at manipulating binary numbers. It turns out that this capability is useful for automating a good deal of processes once we can articulate them as numerical problems, which is possible for almost all types of problem. A CPU contains number-crunching devices[23], like an adder, a subtractor, a multiplier, and also the means to get the operands from memory and store them back to memory. There are other kinds of operators beside basic arithmetic, such as bit shifting, that can be used to manipulate numbers usefully, some of them not familiar to us from mathematics unless we have studied binary logic.

---

[23] The number-crunching bit is called an Arithmetic Logic Unit (ALU)

From one step to another, the CPU knows which operator to use and which operands to load from memory by following a program. Each step in the program is called an *instruction* and these are also stored in memory, but in patterns of binary digits (bits). Fortunately, we do not have to remember these bit patterns as programmers (though we used to have to learn them once upon a time); instead, we can use labels, called *mnemonics*, such as LOAD, ADD, STORE and MOV. Using these mnemonics to program the CPU is called assembly language programming.

We naturally want to express a problem and a solution in unconstrained natural language, such as English, or some other higher-level symbolic form, like arithmetic, algebraic and calculus notations, not assembly mnemonics. To bridge the gap, computer scientists invented what we call high-level programming languages that are sufficiently readable to be understood by humans (those who are suitably trained) and are sufficiently constrained by a formal syntax that they can be mapped to the mnemonics by a special program called a compiler. An assembler simply converts the mnemonics into their corresponding bit-pattern instructions. This hierarchy of symbols, from high-level language representation to binary patterns is shown in Figure 11.16.

Different CPU designs require different instructions and sets of mnemonics. This means that a compiler has to be made for each target CPU. Therefore, if we make a program, like a tax calculator, and want to run it on different device with varying CPUs, we have to compile it for each type of CPU. Ideally, we would prefer to write a program only once and then be able to run it on any suitable device, thereby facilitating portable software program. This is what the Java language concept was devised for. With Java, the process of symbolic conversion is modified so that the high-level language, called Java, is compiled first to an intermediate form that we can think of as a set of virtual mnemonics, called *bytecode*.

Instead of a physical CPU with a hard-coded set of mnemonics, we now imagine that we have a *virtual machine* that runs the bytecode mnemonics. This virtual processor is called the Java Virtual Machine (JVM) and we can think of unplugging the actual CPU and plugging in the JVM, as shown in Figure 11.17.

Before we look at how the JVM works, we need to consider the wider process of writing a software program to run on a particular machine, not just the CPU. There are other aspects to the process to take into account besides variation in the CPU and its corresponding

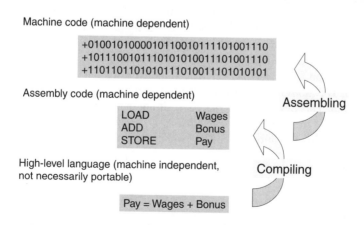

Machine code (machine dependent)

+01001010000101100101111101001110
+10111001011101010100111101001110
+11011011010101110100111101010101

Assembly code (machine dependent)

LOAD	Wages
ADD	Bonus
STORE	Pay

Assembling

High-level language (machine independent, not necessarily portable)

Pay = Wages + Bonus

Compiling

Figure 11.16 The hierarchy of computer languages from high-level to low-level symbols.

HttpConnection c = (HttpConnection)

Connector.open("http://coolapp.mymobile.com/mybill");

c.setRequestProperty("Cookie", "JDSESSION=2e5fb45n3");

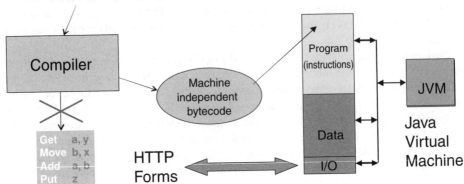

Figure 11.17 The Java language concept – from Java to bytecode, not assembly code.

mnemonics. In Chapter 10, we explored how a useful computing environment needs to includes APIs, which are utility programs that enable our main program to call upon powerful pre-written chunks of code that can carry out routine tasks, such as supporting a windowed-GUI. Java has an identical concept so that Java programs can continue to benefit from APIs that make general-purpose programming a lot easier and more productive. The APIs in Java are implemented using the same bytecode as our main application; therefore, they will run wherever a JVM is available. This is an essential and powerful part of the Java concept that extends it beyond just a programming language. We have the language, some accompanying APIs and the JVM to run them. In the official parlance, we have the Java language and we have the Java Runtime Environment (JRE), which is the JVM and the associated APIs and, of course, the ability for all these elements to interact coherently.

The Java APIs are blocks of code already available in bytecode format and which are pre-installed[24] on the target machine in order to be readily accessible by the JVM to call on their services whenever the main program requires one of their code routines. This concept is shown in Figure 11.18. Here we can see that the program in Java is still compiled to bytecode to run on the JVM. The code snippet in the figure shows an important keyword in the Java language called *Import*, which is a link that points to an API we wish to use in our main program. Once we have referenced the API, we can use its software routines, which are called *methods* in Java parlance. Whenever we refer to a method that is not in our code, the JVM detects this fact and looks for it sitting in one of the APIs. The JVM uses a convention[25] to find the required *pre-compiled* APIs, which are called *class files*, and loads them into memory for execution by the JVM whenever it needs to run one of the referenced methods.

---

[24] Pre-installed usually means that the APIs are already sitting in non-volatile memory, which would be a disk on a desktop PC or in a ROM or Flash memory on a mobile device.
[25] There is usually a system-wide pointer, called *Classpath*, that the CPU can access to find where to look for referenced class files (APIs).

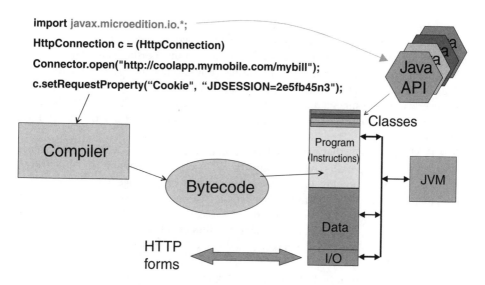

Figure 11.18 The Java API concept.

Of course, we don't swap the real CPU for a virtual one (the JVM) as previously suggested. The JVM is actually an embedded software program itself, written using mnemonics compatible with the target CPU. Once the JVM is running on the CPU, it is as if it takes over from the CPU to start execution of the bytecode rather than native mnemonics, like a translator or interpreter. What the JVM does is convert the incoming bytecode from our program in memory to the native mnemonics, mapping the bytecode to the native code (see Figure 11.19). Clearly, this is an added processing overhead not present in program running directly on the CPU, so we expect Java programs to run slightly slower than natively

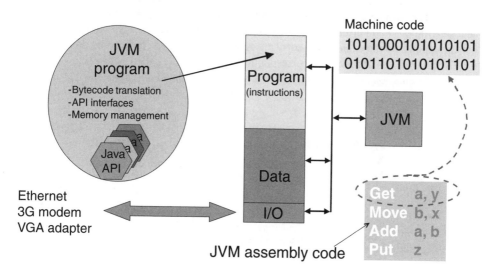

Figure 11.19 The JVM is a program written for the target CPU.

programmed ones. Moreover, in the translation from bytecode to native code, there is not always a convenient one-to-one mapping from bytecode commands to mnemonics, so there might be additional overhead in the execution itself after the translation has taken place.

Techniques can be used to mitigate performance losses due to bytecode interpretation, such as the use of a Just in Time (JIT) compiler that takes chunks of bytecode on-the-fly and converts to native mnemonics, rather like a second compiler to compile the bytecode. Other optimisations exist and later we shall look at some of those relevant to wireless device operation.

The JVM is a program that achieves the bytecode translation to native code and manages the loading of reference APIs. Additionally, the JVM performs other housekeeping tasks to keep things running smoothly. One of these tasks is memory management, or cleaning up unused memory after our program finishes with it, a process called *garbage collection*. We don't need to know the details here, but in the process of executing a Java program, memory gets assigned to store data associated with the creation of software entities called *objects*.

For example, we might create an object called *VideoPlayer* that plays a video clip that we download over the RF network. We can appreciate that there is a chunk of bytecode in our program, or possibly in an API, that contains all the instructions for the video player. This chunk of code gets loaded into the volatile memory when it is needed to play a video clip. We have to ensure that the code is deleted from the volatile memory space once we have finished with it. Otherwise, the memory will become full up with code that is not actively running (or being referenced) and eventually the JVM will run out of memory, which is called *memory leakage*. Garbage collection solves this problem. In addition to memory management, the JVM also checks the bytecode inside class files to ensure data integrity. This is to detect file corruption and thereby avoid running any code that might subsequently go haywire when being executed.

If we consider the process of developing a program in Java and then running it, we can divide the lifecycle into two distinct phases – design time and run time, as shown in Figure 11.20. At design time, the programmer is writing the software in the Java language. A small snippet is shown Figure 11.21 along with the output screen showing what happens when we run it. The Java language is written using keywords that are expressed in English, Latin symbols (like { }) and arithmetic operators. A Java file is saved to a text file with a name that corresponds to the name of the software object (user-defined class) being defined in the file. This file then is compiled to a corresponding bytecode file, called a class file, which is a binary representation of the meaning of the Java code translated into bytecode instructions.

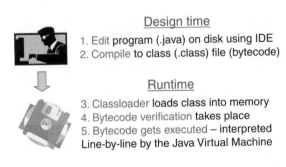

Design time
1. Edit program (.java) on disk using IDE
2. Compile to class (.class) file (bytecode)

Runtime
3. Classloader loads class into memory
4. Bytecode verification takes place
5. Bytecode gets executed – interpreted Line-by-line by the Java Virtual Machine

Figure 11.20 Java development life cycle.

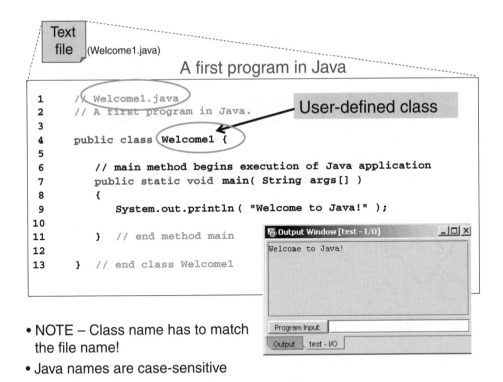

Figure 11.21 An example Java program.

The class file is installed onto a mobile device that already contains the JRE, which includes the JVM and the associated APIs. When the program is invoked, the JVM will load our class file into volatile memory and run it, interpreting the bytecode into the native CPU instructions for the target CPU. Prior to execution, the bytecode integrity is checked to verify that the class file is not corrupt in any way.

11.8 JAVA 2 MICRO EDITION – 'WIRELESS JAVA'

We have just looked at some Java fundamentals. We learnt that Java is both a programming language and a runtime environment, including APIs. Earlier in the book, we learnt about a very powerful technology for implementing scaleable backend systems called J2EE. Most of the goodness of J2EE was in its huge set of powerful software mechanisms to enable interfacing to myriad enterprise systems, such as databases and directories. All these powerful utilities in J2EE are implemented in Java and work using the underlying processes just described, such as using a JVM, Java APIs, class loading and so on.

Figure 11.22 shows that with a JVM available on a target machine, we should be able to run Java. This is highly attractive to developers as it means they can focus their efforts on mastering one language that will allow them to create applications for any type of environment, especially for mobile devices where there are huge numbers ready for running all those interesting mobile services we want to go on to develop. However, as the figure

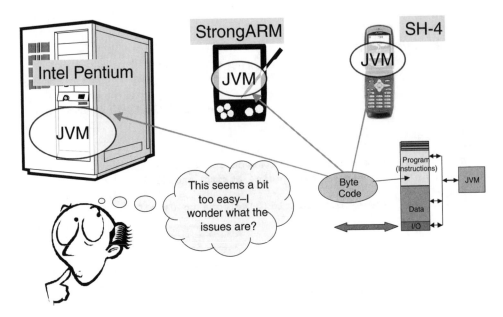

Figure 11.22 Will Java really run everywhere the same?

shows, our thoughtful developer is rather curious that this all seems a bit too easy. If, like our curious developer, we ponder for a moment on the implications of what is shown in Figure 11.22, we might start to realise some of the problems here, especially when taking into account the differences between the devices shown in the figure.

Two issues might come to mind. Firstly, the devices themselves are very different. Servers, like the ones needed for J2EE web-server clusters in a mobile SDP probably have massive amounts of memory, both volatile and non-volatile, and huge displays capable of rich colours with high resolutions, which are big enough to display many applications (or windows) at once. In other words, the first set of issues relate to the physical characteristics of the devices in each category. Whilst device capabilities are changing all the time, Figure 11.23 presents a snapshot of some typical device characteristics for servers, PDAs and mass-manufactured cell phones. The key differences are in processing power (CPU), memory size, and connectivity and display characteristics.

The second set of issues is to do with the types of applications that we might want to run on each type of device. Clearly, on the servers that are supporting our SDP, we need all those wonderful J2EE interface capabilities to hook up with database servers, directories, legacy mainframes, messaging systems, mail servers and so on. But, on a mass-consumption handset, such interfaces would be redundant. We can hardly imagine installing[26] an enterprise-class relational database application on our handset, never mind finding something useful to do with it.

And what about display capabilities? On a server, or desktop, it is normal to offer a rich user interface based on the Windows metaphor that is now so pervasive in human-computer interfaces. To make life easier for the Java programmer, one of the APIs (called Swing) that

---

[26] A particular instance of a well-known relational database server requires over 1 GB of disk space just to install it!

• Slow 32-bit (<200 MHz)
• Not so much memory (8 MB)
• Linear low-resolution display
• Variable network connection
• Mainly personal device

• Very fast 32-bit (>1 GHz)
• Lots of memory ('infinite')
• Nonlinear rich display
• Fast and reliable network connection
• Distributed architecture

• Slow 16-bit (<50 MHz)
• Not so much memory (1 MB)
• Linear low-resolution display
  possibly character-based
• Low-bandwidth network connection
• Personal device

Figure 11.23 Physical differences across device categories – showing that with Java, one size really does fit all! (Reproduced by permission of IXI Mobile.)

was made available by the inventors of Java is a rich set of code to handle all the windowing, leaving the programmer free to concentrate on what the windows are to be used for, rather than constructing and maintaining them (a non-trivial programming task).

Again, on a small mobile device with a tiny display, the use of windows is not possible or practical.

These differences between devices might lead us to think of chopping out the APIs that we don't need and creating a slimmer version of the JRE for resource-limited devices. This is exactly the step that Sun took in 1999 to revise the Java strategy, giving us three versions:

• Java 2[27] Enterprise Edition (J2EE)

• Java 2 Standard Edition (J2SE)

• Java 2 Micro Edition (J2ME)

Figure 11.24 shows the family of Java editions. On the left side of the diagram, we see the already familiar J2EE version of the Java platform. It has all the powerful enterprise APIs that we have already discussed in earlier chapters. Java 2 Standard Edition (J2SE) is closest to the original version of Java and we can consider it a very feature-rich implementation suitable for general-purpose programming on desktop PCs. Both the J2EE and J2SE editions can run on the same design of JVM, which is also available in the C programming language as the CVM, so that those who want to port the JVM to their own environment can do so by a suitable compiler to the target native code (mnemonics).

The Java 2 Micro Edition (J2ME) is targeted at a range of consumer and embedded electronic devices with constrained resources and it is this version that we shall concentrate on for the remainder of our discussion about Java in this chapter. Because the J2ME was

---

[27] The '2' was not coincidental with this split, but rather with an earlier decision to release a significant upgrade to the APIs.

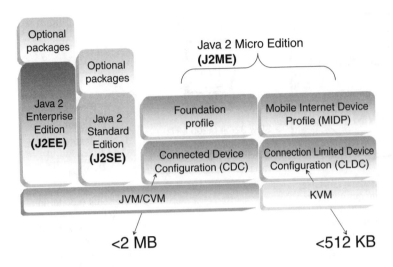

Figure 11.24 The family of Java editions.

intended to support a wide range of consumer devices in various markets (not just mobile electronics), the J2ME is further subdivided into configurations targeted at particular hardware families (or types). We can think of a configuration as a core set of Java APIs and an associated virtual machine specification, which are grouped together in a way that is optimised as far as possible for a particular group of target devices and possible application scenarios. More specifically, a Java configuration specifies which:

- Java programming language features are supported

- Java virtual machine features are supported

- Java libraries and APIs are supported

The configuration aimed at the lower end of mobile devices (e.g. mass-produced phones) is the Connected Limited Device Configuration CLDC, which aims at 160–512 KB memory available for storing the JRE. Other considerations are that the environment will probably be battery powered most of the time, most likely have a very slow CPU and also have an intermittent network connection. This is clearly appropriate for the typical mobile device, remembering that we are talking about mass-produced mobile phones, not just PDA-like devices, which can easily include much heftier processors and memory.

To make the CLDC available for resource-constrained devices, the JVM functionality has been cut down and it now becomes the Kilo Virtual Machine, or KVM[28]; the 'K' standing for 'kilo' to indicate device memory spaces in kilobytes. Sitting on top of the CLDC is the MIDP, which specifies an API subset appropriate for mobile phones. Potentially, we could simply use the CLDC core API set as it is, so why the further need for a specific profile stating the designated API support intended for mobile devices? The answer is that the profile is more akin to an industry agreement that states how supporters agree to implement

---

[28] The KVM technology came from a research system called Spotless developed at Sun Microsystems Laboratories. More information on Spotless is available in the Sun Labs Technical Report 'The Spotless System: Implementing a Java system for the Palm Connected Organizer' (Sun Labs Technical Report SMLI TR-99-73).

and support the CLDC on devices they will make and sell in their industry. With a particular set of devices and application scenarios in mind, the profile defines appropriate APIs for the target market. All hardware supporters of the profile agree to implement these APIs on their devices running on top of the KVM. The MIDP profile has the backing of a large group of mobile phone operators, phone manufacturers and affiliated technology and service companies, such as toolset vendors. In this way, developers can be confident that if they develop an application that conforms to the MIDP, then it should run on any device that the manufacturer claims to support MIDP.

Key J2ME definitions (taken from Sun):

Configuration – A J2ME device configuration defines a minimum platform for a 'horizontal' category or grouping of devices, each with similar requirements on total memory budget and processing power. A configuration defines the Java language and virtual machine features and minimum class libraries that a device manufacturer or a content provider can expect to be available on all devices of the same category.

   Profile – A J2ME device profile is layered on top of (and thus extends) a configuration. A profile addresses the specific demands of a certain 'vertical' market segment or device family. The main goal of a profile is to guarantee interoperability within a certain vertical device family or domain by defining a standard Java platform for that market. Profiles typically include class libraries that are far more domain-specific than the class libraries provided in a configuration.

11.9 USING MIDP TO DEVELOP MOBILE APPLICATIONS

Now that the principle of J2ME has been described, we can return to our main consideration for this chapter, which is the ways and means of developing mobile applications, not just the technical processes underlying the technology. What we need to do is understand what it means to use MIDP to develop mobile applications, especially compared to the other two methods we have already identified and discussed; namely the embedded approach and the browser approach.

   In essence, the Java approach is very similar to the embedded one. We write the application in a high-level programming language and follow most of the same production cycle for developing embedded applications. However, the key advantage is that the MIDP program, called a MIDlet, will run on any device in the market that is MIDP compliant. If we had written an embedded application for a Motorola device using the Motorola SDK, then the program would only run on the Motorola device. If the same device supports MIDP, then we can run our MIDlet on the Motorola device as well as any other MIDP device.

   There is a huge array of devices in the market, especially in the mass-manufactured mobile phone category, which accounts for hundreds of millions of devices. If we want to develop applications to run on as many devices as possible, MIDP is clearly an advantage. Native embedded solutions will still be developed for utility applications, like the WAP browser for example, or a media player. In fact, the set of APIs supported by the MIDP are too restrictive to allow such applications to be developed. In order to ensure wide device coverage, the MIDP API subset is deliberately narrow.

If we attempted to use the embedded approach for a particular end-user application, such as a game, then in order to get a large number of users we would have to develop the game for many different types of device. Each device probably has a unique programming environment, such as a specific operating system and associated system APIs. This would be prohibitively expensive and cumbersome, as there are potentially hundreds of devices on the market, taking into account the legacy of devices already in circulation. At any one time, there are probably at least several hundred different devices on sale in any one market.

The MIDP solution addresses this problem quite elegantly, as shown in Figure 11.25. This factor is its main selling point. By standardising on a single language and associated API set, independent of the underlying operating system and processor, developers can now write an application only once and then make it available for a wide set of devices. What MIDP offers is:

- A standard API set that is guaranteed to be available on all MIDP-compliant devices

- The possibility to use a single toolset (IDE) to develop any application

- A single way of packaging and distributing applications (more on this later)

- A standard network communications model (IP based) that is also guaranteed to be available on all MIDP-compliant devices (and can utilise HTTP should we want to)

The unified programming environment of MIDP is a huge step forward for the mobile data future, especially ubiquitous access. There's little point in building ubiquitous RF network coverage if we can't deploy applications in a ubiquitous fashion. Naturally, there are still some limitations to the MIDP approach, and we briefly mention some of them here, although realistically they are not serious problems and don't really present roadblocks to the potential success of MIDP.

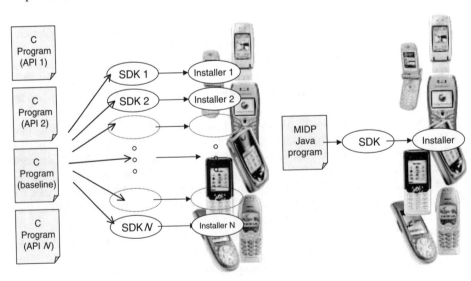

Figure 11.25 Embedded headache versus MIDP 'walk in the park'. (Reproduced by permission of Nokia and Sony Ericsson.)

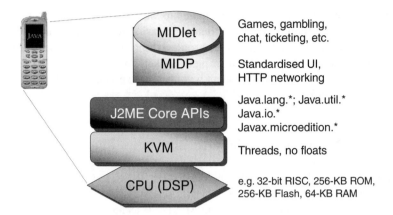

Figure 11.26 The MIDP (MIDlet) software 'stack'.

To provide a unified programming environment for MIDP meant that we had to agree on a single set of APIs to define the profile. This means that all devices will behave the same as far as the programming model is concerned, even though one device might be more capable than another in terms of its underlying OS and embedded API set. For example, a particular device OS might offer powerful system-level APIs for cryptography, whereas MIDP (1.0 in any case) does not offer any cryptographic APIs. A key restriction of J2ME is that it is not possible to access the underlying APIs, unlike J2SE, which supports the Java Native Interface (JNI) to allow underlying native code to be called from within the JVM. This is a deliberate limitation of the CLDC core set of Java functionality, remembering that the MIDP is built upon the CLDC and has to inherit its limitations, as shown in Figure 11.26.

Figure 11.26 shows the microcosm of a MIDP program (MIDlet). The MIDlet is restricted to the API set available in the MIDP, unless we use an extension set of APIs provided by the handset manufacturer and already pre-installed on the device. This is possible, although it is moving away from a standard profile and we are therefore making our development process more customised; which is possibly going to limit our application's appeal if it only works well on one device that supports the extension APIs (assuming we rely heavily upon them to meet our application's requirements).

As an example, if we consider the Motorola A008 device, which was probably the first to market with MIDP 1.0 support, the user interface of the device supported a limited windowed appearance. This user interface metaphor is not present in the basic graphics API for MIDP 1.0, but it is possible to implement using the lower-level graphics APIs in the CLDC. Therefore, on the A008, by sticking strictly to the default MIDP implementation, we would produce an application that has a look-and-feel inconsistent with the A008 user interface, which is a limitation. However, Motorola made an extension API available to support the limited windowing capability of the device, thereby making it possible to avoid the inconsistency. The downside is that we end up with a particular instance of our program having to be supported specifically for the A008, should we choose to use the extension.

The idea of offering extension APIs may seem like spoiling the wonderfully standard world of MIDP. Well, that's what the world of software is like! Everybody has their own idea of what is useful and no two mobile devices are the same, so differentiation emerges despite standard approaches, or even because of them. With the first version of MIDP, there were perhaps a lot of features missing compared to what some of the latest phones could support. Not surprisingly, quite a few proprietary extension APIs cropped up. Vodafone, for example, almost created an entire customer profile to replace MIDP in order to get the best from a set of new devices they wanted to bring to market. The good news is that many of the API enhancements have become redundant with the later versions of MIDP, an altogether more feature-rich platform.

11.10 WHAT DOES MIDP 2.0 OFFER?

At the time of writing this second edition, MIDP 2.0 is the widest supported J2ME profile for mobile devices, although MIDP 3.0 is nearly ready. We shall talk about MIDP 3.0 in a moment, but first let's look at what using MIDP 2.0 technology can do for us. Where possible, in keeping with the ethos of this chapter, we shall compare and contrast the available capabilities with other means of developing mobile applications, so as to keep our discussion relevant to the theme of mobile service delivery, not J2ME per se, as there are plenty of books in this topic[29].

The MIDP 2.0 specification is based initially on the MIDP 1.0 specification and provides backward compatibility with MIDP 1.0 to the extent that MIDlets written for MIDP 1.0 can execute in MIDP 2.0 environments. The momentum behind MIDP had been gathering in the marketplace, so backwards compatibility was an important consideration. This is a problem particularly pertinent to the mobile devices market because upgrades in the field are very unusual, unlike the desktop PC environment, where they are the norm.

As already discussed, the MIDP is designed to operate on top of the CLDC. While the MIDP 2.0 specification was designed assuming only CLDC 1.0 features, it will also work on top of CLDC 1.1. However, it is probable that most MIDP 2.0 implementations will be anchored in CLDC 1.1. As we shall see, MIDP 3.0 allows for the possibility of J2ME devices running CDC on top of a JVM (not KVM).

Mobile Information Devices (MIDs) capable of running MIDP Java will be in abundance and the number of expected applications will be vast. Therefore, rather than trying to provide capabilities for all eventualities, the MIDP 2.0 specification defines a subset of useful APIs, attending to those functional areas that were considered essential to achieve broad portability goals for a range of typical mobile application scenarios[30].

MIDP 2.0 capabilities include:

- Application delivery and billing

- Application life cycle (i.e. defining the semantics of a MIDP application and how it is controlled)

---

[29] Enrique Ortiz, C. and Giguère, E., Mobile Information Device Profiles for Java 2 Micro Edition: Professional Developer's Guide. John Wiley & Sons, Inc., NY (2001).
[30] How do we know what's typical? The MIDP was arrived at by industry consensus though the Java Community Process, so the final specification is not without considerable input from those with a view of the market. Whether it is the correct, or typical, view remains to be seen.

- Application code-signing model and privileged domains security model to control deployments within a sanitised scope of behaviour

- End-to-end transactional security (HTTPs, as discussed in Chapter 9)

- MIDlet Push registration (a server push model, like WAP Push, but for MIDlets)

- IP-based networking

- Persistent storage

- Sound support

- Timers

- User interface (UI) (including general display and input, as well as some unique requirements for games)

Let us examine each of these capabilities in turn, summarising the key attributes, keeping in mind our concern for understanding the array of applications options open to us on the device, not the specifics of Java programming (MIDP, or otherwise).

11.10.1 Application Packaging and Delivery

As discussed earlier in this chapter, an effective means to deploy applications is essential, so it is no surprise that the MIDP expert group focused quite heavily on this aspect. MIDP supports application loading via any networking route onto the device. Most importantly of all, it supports an OTA download mechanism. This is actually contained within the MIDP 2.0 specification, unlike MIDP 1.0, where the OTA mechanism was a recommendation only. The MIDP OTA mechanism should look familiar to us, as it is similar to the OMA method we described earlier in the chapter. OMA OTA was based on the MIDP OTA model. Therefore, as with OMA, the MIDP method uses HTTP and is practically identical to the OMA except that the Object Descriptor is a file type specific to Java, called a Java Application Descriptor (JAD). The downloadable executable Java file is the MIDlet itself, which is encapsulated in a file type called a Java Archive (JAR), which is like a compressed ZIP file. As we have already described the OMA method, we defer the explanation of the MIDP OTA mechanism to the end of this chapter.

What we want to focus on here is the application packaging, especially the new developments for MIDP 2.0, such as code signing. MIDP 2.0 introduces the idea of *trusted* MIDlets. This is a means to authenticate that a MIDlet is from a particular vendor and thereby enable the user to benefit from its features without fear of any malicious misappropriation of device resources. One fear that operators had was a MIDlet accessing a data connection, or sending text messages, without the user being aware. This might incur costs that the user should be made aware of. There was also a fear of viruses of other types of 'malware', but thus far mobile viruses have proven not to be a widespread problem.

MIDP introduces the concept of code-signing using digital certificates. The Java community, motivated to promote open standards and support existing ones wherever possible, elected to incorporate an existing digital certificate method for code signing. Recalling our discussions earlier in the book about HTTP, public key cryptography, or PKC, is a powerful method for supporting digital certificates, the principle being that with certain unique one-way key associations, we have a means of achieving irrefutable digital identification[31]. The mechanisms defined in the MIDP 2.0 specification allow signing and authentication of MIDlets based on the X.509 Public Key Infrastructure (PKI) specification[32]. In Chapter 8, we looked at the principles of PKC. However, the means of issuing certificates, sharing keys, authenticating whom they belong to and so on is what PKI is all about.

We should note that this mechanism is open to any application developer to adopt for code signing, so we could similarly use such a method for distribution of embedded applications for mobile devices, whether or not we are using CLDC, MIDP or even Java. However, the use of code signing has been standardised for MIDP 2.0. It also extends to include the concept of permissions-based access to APIs and MIDP functions, as we shall discuss next when examining the API support in MIDP 2.0.

11.10.2 API Summary

We are not going to look at how to use the MIDP 2.0 APIs in a program, as that is beyond the scope of this book. Here we shall look at the capabilities of the MIDP 2.0 APIs and their usefulness within our current evaluation of mobile application paradigms.

The good news is that the API support of MIDP 2.0 is constant across all MIDP-compliant devices, at least in terms of the core API features that are deemed mandatory by the MIDP specification. Of course, vendors can offer APIs as optional packages, should they wish to do so. We have already argued that this might be a good thing in order to exploit particular device capabilities, or it might be a bad thing if it leads to lots of versions of the same program to cater for different API extensions (thus taking away from the portability advantage of J2ME and MIDP). There is no easy answer. It all depends on a particular project's objectives.

The API set is aimed at general-purpose programming, not system programming. The difference is that with system programming, we are concerned with implementation of applications that are to be incorporated into the device ecosystem itself, almost at the OS level, and not run as an end-user application on top of the system. For example, a telephony management program (e.g. dialler and associated phone book) is an essential system program on a mobile device that supports telephony. Another example would be an intelligent battery charge management program. These types of program are characterised by the need to gain low-level access to system resources on the device, such as right down to the device driver level (see earlier discussions to understand this in the context of a mobile device architecture). System level applications would almost certainly require a native embedded approach.

11.10.3 User Interface APIs

The MIDP user interface capability is contained within two APIs: the high-level and the low-level graphics APIs. The high-level API is designed for applications where a workable and basic level of interactivity is required to be guaranteed on all devices. Interaction at this level is at its most basic and is more akin to the forms capabilities and limited layout options of

---

[31] In Chapter 8 we discussed PKC and mentioned that if we could open an encrypted message using someone's public key, then we know that they must have at one time secured it with the corresponding private key, which provides a means for digital identification.
[32] [RFC 2459] – Internet X.509 Public Key Infrastructure (http://www.ietf.org/rfc/rfc2459)

XHTML-MP in a browser, than a fully blown graphics API with pixel-level control of the UI. The high-level graphics widgets are typically text boxes, buttons, selection lists and so on.

For applications using the high-level graphics API, portability across devices is important. To achieve this portability, the high-level API employs a high level of abstraction by providing very little developer control over the look-and-feel. In fact, the look-and-feel is left to the MIDP implementation on the device itself, and so typically will differ from device to device, although the widgets will all have a similar morphology. This very high-level abstraction is further evident in the following ways:

- The actual drawing to the device display is performed entirely by the underlying MIDP implementation. In other words, if we consider a button for example, then how it looks on the screen is dependent on the system code that the device manufacturer has written to support a button, notwithstanding the actual capabilities of the display itself. The MIDlets do not define the visual appearance (e.g. shape, actual colour, font, etc.) of any of the components in this mode.

- Navigation, scrolling and other interactivity mechanisms are entirely encapsulated by the MIDP implementation. The MIDlet application is not aware of these interactions and has no control over their manifestation on the device.

- Applications cannot access concrete specific devices, like a particular key on the keypad. When using the high-level API, it is assumed that the underlying implementation will do the necessary adaptation from interface widgets to the device's hardware and native UI style.

The low-level API, on the other hand, provides very little abstraction and allows for detailed control of the display via software. This API allows exact positioning and management of graphical elements, as well as access to low-level input events. Some applications also need to access special, device-specific features. A typical example of such an application would be a game.

Using the low-level API, a MIDlet can:

- Control entirely over what is drawn on the display (confined to the portion of the display given over to the KVM, or MIDlet, by the device operating system)

- Listen for programmable primitive events like key presses and releases

- Access specific keys and other input devices, where available

Because of the variation in display sizes, colour depths and variations in keypad design, the freedom to program to the limit of the device capability clearly means that applications utilising the low-level graphics API are going to struggle to be portable; the low-level API tends to push the programmer towards device-specific interface design. For example, a program with interface elements sized to 300 pixels across is not going to work too well on a device with only 200 pixels width. This might not be a problem depending on what the programmer is trying to achieve, and there are ways of writing low-level graphics software that can allow for display variations, but the problem of adapting the low-level graphics code to various mobile devices is the bugbear of graphics-rich MIDP programming, such as gaming.

If the application does not use low-level API features, it is guaranteed to be portable. It is recommended that applications use only the platform-independent part of the low-level API whenever possible. This means that the applications should not directly assume the existence of any keys other than those pre-defined in the API, and they should not depend on a specific screen size. For something like a game, it is possible to use a generic game-key mapping technique rather than 'hard code' the keys (see sidebar 'Generic game action keys'). This is clearly advisable to enable portability. Similarly, assumptions about screen size should be avoided where possible, so that the design approach is adaptable to the available display characteristics. Alternatively, the MIDP designer has to take care to code for different devices, but this will add complexity and increase the cost of developing an application, as well as making it more burdensome to move to new devices as they come out in the marketplace.

Generic game action keys

Earlier we discussed the Java language concept and introduced the unit of software called a *class*. As we now know, some classes come already written for us in the Java implementation, and that is what we have been discussing – they are called APIs. In MIDP, there is a class called 'Canvas', which enables us to control the screen, but also to get events from the keypad.

Rather than figure out on a device-by-device basis what keys map to what, the Canvas class defines a number of constants for commonly used keys. In particular, it defines abstract game actions (UP, DOWN, LEFT, RIGHT, FIRE, GAME_A, GAME_B, GAME_C and GAME_D) whose key-code mappings can be determined at runtime.[33]

What this means is that as the MIDlet developer, we deal in our software in terms of keys like 'UP' and 'RIGHT', not '2' or '6', or even other codes relating to input devices like joysticks.

Games keys in MIDP

(Reproduced by permission of Nokia.)

MIDP Canvasclass defines keys "LEFT" and "RIGHT"

Don't explicitly use "1" and "3"
-- won't work on a novel keypad design, like 3650.

---

[33] This means that only once the MIDlet is running on the target device can we tell which actual keys on the device are mapped to the abstract key names. Therefore, let's say that we wanted to provide on-screen help instructions to convey which key is the 'jump' button in our game (which let's say we assigned to GAME_A), then we could not hard-code this instruction in the software, such as 'press the 1 key to jump' because we would not know in advance which key is going to be assigned.

Making assumptions about specific keys can be problematic and, in the worse case, render our application unusable, even in ways that we don't expect. We shouldn't be making assumptions when writing software! For example, an operator in the United Kingdom running a MIDP games arcade from their portal found that their arcade became disrupted by the introduction of the Nokia 3650, which is a wonderful phone, but it has a circular keypad. Had we hard-coded the keys, perhaps quite sensibly, '1' and '3' for left and right, then on the circular pad we clearly have a problem, as the above illustration testifies!

In addition to the high- and low-level APIs for graphics, MIDP 2.0 introduces a new graphics API for gaming. This API supports graphical metaphors that work well for games where common graphics operations are to be expected, such as complex and movable background images, animation and layering of the scene to aid the illusion of perspective and movement. For example, in a game with a complex scenery, the effect of motion can be produced by constructing the scene from layers and then animating (panning) them at different speeds to get a feeling of depth. Foreground elements, such as action characters can be animated as single entities (*sprites*) rather than pixel-by-pixel.

Some of these gaming concepts are shown in Figure 11.27, which is a deconstruction of an arcade game programmed as a MIDlet. We can clearly see logical layers in the graphics, which if they were treatable as programmable entities would make our programming lives a lot easier. This also enables the device manufacture to take advantage of any underlying hardware that is capable of handling graphics in this manner directly, such as graphics accelerator chips.

There are two movable sprites shown in the figure that represent the characters in the fight scene. We could have different fixed sprite shapes to represent their major action poises in the game. We are showing a zoomed image on the right of the 'jumping position' sprite for one of the characters. We can move this sprite as it is (i.e. a predefined and self-contained block of pixels) wherever we like on the canvas simply by calling the relevant Java method in the API. The sprite will be defined by a complex polygonal shape, as we have shown (though this distinct boundary is not visible in the game, it is just for illustration purposes). If the entire sprite were the large rectangle, then we can nominate the background colour of

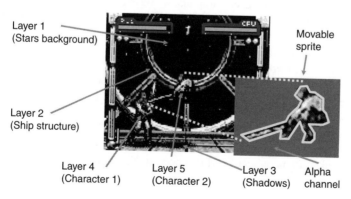

Figure 11.27 Game graphics layering, sprites and alpha channels.

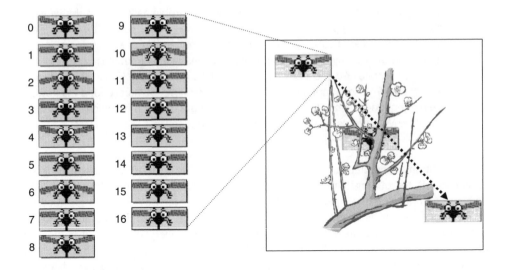

Figure 11.28 Sprite animation is achieved by sub-frames in the sprite image object .

the sprite to be transparent. This is called alpha channelling and is how we achieve layering with visual continuity between overlapping layers.

With a sprite, we can move its container frame around in order to create panoramic movement in the game. To achieve animation of a sprite local to its frame, we can move between sprite frames within a composite image object divided into sub-images of equal size (and equal to the frame size), as shown in Figure 11.28.

11.10.4 Networking API

The underlying CLDC platform has a powerful class of connection methods and, quite sensibly, a particular focus of the MIDP interpretation was provision of an HTTP API. Significantly, and something that was lacking in MIDP 1.0, we now have a secure connection option via an API that enables us to establish HTTPS communications pathways (i.e. HTTP over TLS). This is in line with everything that we have discussed in the book thus far to do with security concerns for various mobile service scenarios. As we had already established, the only viable solution for secure transactions is the provision of an end-to-end encrypted pipe, and this is now possible with MIDlets.

It is also possible to work at a lower layer in the TCP/IP stack, with the ability to form datagram streams from our MIDlet. This would be useful for media streaming applications and implementing any high-level IP protocols not included in the API, such as XMPP for Instant Messaging, or SIP for IMS (see Chapter 14).

A new addition to the MIDP 2.0 specification is the provision of an API to connect with a logical serial port on the device, whether that manifests as an actual physical port, such as an RS232 link[34], or an onboard port to a peripheral, like an infrared modem, or remains

---

[34] An RS232 link could be used to connect a mobile device with a PC, such as via a cradle. It may be useful for a MIDlet to gain access to this link, despite the OTA ethos of MIDlets.

a logical serial port (emulated by underlying system software, such as a Bluetooth driver). Presumably we could use a serial link to talk with a Bluetooth peripheral, although the Java Community is working on a separate CLDC-based API set (JSR-82) specifically to handle Bluetooth peripherals.

11.10.5 Securing the APIs

With MIDP 1.0, all the APIs were available all of the time to the main MIDlet program. MIDP 1.0 also adopted a very tight sandbox approach whereby the MIDlet could not access system resources directly, only indirectly via the basic API set, if supported by the device vendor. For many types of mobile application, it would be useful to access other resources on a device via extension API sets, whether vendor-specific, or via the expected evolution of the MIDP itself or via other APIs that the Java Community Process might make available for vendors to optionally include in their Java implementations. We can think of MIDP 2.0 as enabling access to the device from the KVM, but in a sanitised and controllable manner. Whilst this initially seems a bit odd, the idea is nothing new. In the desktop world, it is unusual that all programs running on a computer are allowed by the OS to access all its resources for any user. The concept of restricted access rights has been a necessary and essential part of OS architectures for some time. This concept has now been carried across into the mobile world in order to allow various usage and security policies to be implemented. Let's say that a particular device has an API to allow location information to be accessible to MIDlets. This API may be something that the mobile operator introduced to a device and want to control its usage. With MIDP 2.0, this API could be restricted to be only accessible to MIDlets provided by the operator.

11.10.6 Push Mechanism

An exciting new feature of MIDP 2.0 is the *push registry*. This is a means to allow an otherwise dormant MIDlet (i.e. one not currently running on the device) to start running following an event on the phone, namely the receipt of a push message (such as a text message) or the occurrence of a timer event. Any of the MIDlets installed on a device can register with the push registry to ask for notifications of particular events of interest. An API is used to carry out the registration process. There are many interesting applications for this technology. An obvious example is the implementation of an email or PIM client that can receive alerts of changes to a central mailbox or PIM database, thereby allowing real-time synchronisation to occur. For example, if an email message arrives at the central mailbox, then some kind of server mechanism could push a text alert to the mailbox user's mobile device. Using the push registry, the alert could activate a particular MIDlet, even if the MIDlet was not currently running on the device. This would allow a constantly synchronised PIM application to be implemented using MIDP technology. Not only could email be notified, but also changes to a central calendar or contacts database. The technology is also useful for CRM applications, such as maintaining real-time inventory information for field sales representatives. It is also useful for various types of media applications, as we shall examine in Chapter 15.

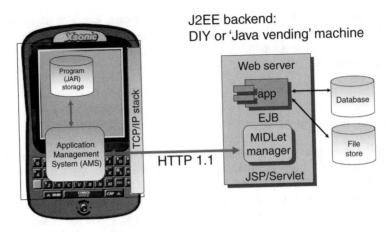

Figure 11.29 Basic MIDlet download architecture.

11.11 MIDP OTA DOWNLOAD MECHANISM

The mechanism for downloading a MIDlet to a device takes place using HTTP or WSP, as shown in Figure 11.29. The objective of the process is to download MIDlets, in the form of JAR files, from a central source, such as a database or file server, placing them into the JAR memory space in the device. As previously mentioned, a portion of device memory belongs to the JRE and it used for storing MIDlets. The file format is a compressed form of the compiled class file, using the ZIP file format.

The Java Application Management System (AMS) is responsible for the downloading and the process, which also includes installation of the MIDlet class file on the device. The AMS is mandatory for MIDP-compliant devices and gives the user the ability to download and run MIDlets. The Java application manager will provide a means for the user to run any MIDlet in the future.

Figure 11.29 shows us the server component responsible for managing MIDlets as a centralised resource, which might well be a J2EE platform. This 'MIDlet manager' or 'vending machine' could be a bespoke application, or an off-the-shelf *service delivery platform* (SDP) aimed at content management, including games 'vending' (assuming the MIDlets are games[35]). There are many vending servers on the market. An example is the Integrated Mobile Marketplace from July Systems, Inc.

As Figure 11.30 shows, the first thing that the application manager does is to download a JAD file. The JAD describes the essential attributes of a uniquely associated JAR file that contains the MIDlet. One of the functions of the JAD is to inform the user and the device about the size of the JAR file. The application manager should then indicate to the user whether there is enough spare memory on the device to download and install the referenced MIDlet. The JAD also contains the URL that points out where to download the JAR file. If the user decides to download the JAR file, the application manager uses HTTP GET

---

[35] The OTA download principles are the same for games or any other type of MIDlet. An SDP that specifically caters for games will have additional features that enable payment to be made and extra levels of gaming difficulty to be charged for, etc.

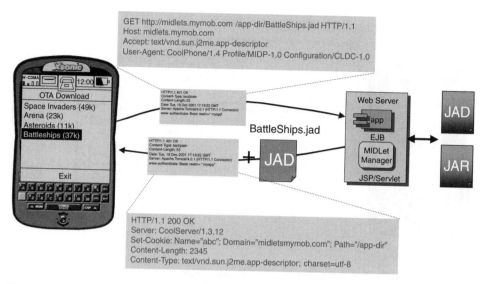

Figure 11.30 Downloading the JAD file first.

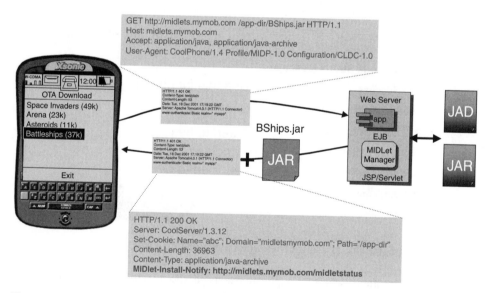

Figure 11.31 Downloading and installing the JAR.

to fetch the MIDlet file, as shown in Figure 11.31. We have already noted in Chapter 7 on IP protocols and earlier on in the current chapter that the download process can use segmentation and reassembly to enable a successful download in the face of data errors on the link.

Figure 11.31 shows the HTTP header that the 'vending' server returns as part of the JAR download response. Within the header, the field *MIDlet-Install-Notify* contains a reference

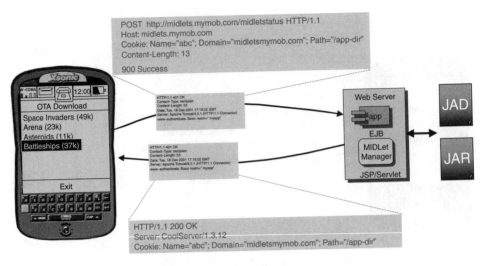

Figure 11.32 Reporting the status of the JAR installation attempt.

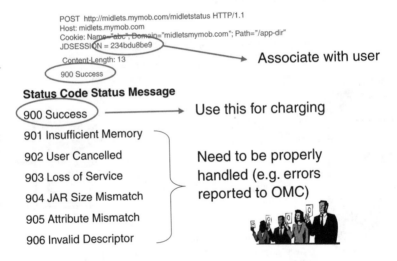

Figure 11.33 Importance of status codes when downloading a MIDlet.

to a URL. This location is where the application manager should send (POST) a status code[36] back to the server using HTTP POST, as shown in Figure 11.32. Some of the possible status codes for the download process are shown in Figure 11.33.

Notice that a cookie value is part of the POST response from the application manager, reflecting a cookie set earlier by the 'vending' server during the download process. This cookie enables the 'vending' server to identify uniquely a particular download session and its user. This allows the association of download status reports with their users. This was the recommended practice for MIDP 1.0, but MIDP 2.0 states that a cookie is not mandatory

---

[36] Note that these codes are similar in principle to the status codes used by HTTP, although they are entirely different codes with independent meanings.

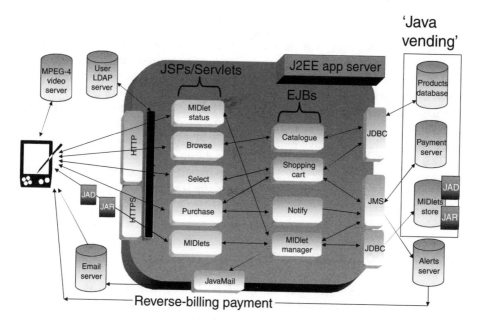

Figure 11.34 Possible architecture of J2EE system with MIDlet download.

(nor recommended) for this process. Instead, the URL used to post the status code should be rewritten to include unique session information, as per the techniques discussed in Chapter 7 on IP protocols (e.g. http://www.midletsmymob.com/midletstat? session=a23sdn32n9fhu).

It is an important part of the overall mobile service to report the download status codes back to the server. The common assumption is that the value of the status codes lies in being able to detect a successful download in order to charge the user. However, this only makes use of one of the codes. A fault-management system should log these codes and make them accessible to application-support staff and help-desk staff. It is likely that throughout the course of potentially thousands of downloads, some errors will occur during the download process, including human errors, such as trying to download a file to a device with insufficient spare memory[37]. In such a situation, if a user consequently places a call to a help desk, ideally the help-desk operator should have meaningful access to an interpretation of the status codes. This will greatly assist with customer care initiatives by helping to identify and eliminate the common errors.

Earlier in the book, we looked at an example of a J2EE system for managing a video-streaming application. Figure 11.34 reminds us of this application in its basic form, but this time we have added a MIDlet download facility to enable MIDP-capable devices to download a MPEG4 video player in order to use the video dukebox service. This would be useful for users with devices that don't have an MPEG-4 player.

Notice that a text-messaging server facilitates payment for the service via reverse-billed text messaging. Otherwise, the application is the same as before. The ability to first offer a

---

[37] On modern devices, it is expected that the AMS itself will prevent the download of a MIDlet that is too large for the available memory space.

MIDlet to the potential user shows how new services can be offered to users without being concerned about whether their device can handle the information or media format, such as MPEG-4 video in this case. With the use of MIDlets, it is entirely feasible to provide the user with the application needed to enjoy the service. Furthermore, this automated process can be carried out remotely OTA. On RF networks with high-speed access, this download process would not incur excessive delays. This may add to the success of OTA provisioning in the future, allowing for applications that are more complex to be downloaded. However, the sensitivity (intolerance) of users to download delays should not be underestimated. It is certainly the case that in the wired world, the success of application downloads declines with size. This is usually because the user gets frustrated with a large download taking too long and subsequently cancels it, or they are loathe to retry if it all goes wrong. Segmentation and reassembly can ameliorate the latter case, but care is still required to avoid overly large file sizes.

11.12 WHAT DOES MIDP 3.0 OFFER?

At the time of writing the second edition of this book, the MIDP specification has reached version 3.0 for draft consideration. Therefore, it is useful to examine the major enhancements that MIDP 3.0 offers over and above those of MIDP 2.0 that we have so far been discussing. Some of the changes might not make it to the final release, but it is likely that most of them will, so this section should be considered to be a fairly accurate description and appraisal of the changes. The nine main proposed changes for consideration are as follows:

1. *MIDlet suites and LIBlets* – It is now possible to package a suite of MIDlets for download to a device. The really powerful feature of suites is their ability to use and include LIBlets. These are, as the name suggests, libraries of code and not complete standalone programs with user interfaces like a normal MIDlet. We can think of these as helper applications that provide functions needed by a MIDlet or group of MIDlets and that can remain on the device and be shared, thus reducing the amount of space required to store MIDlets. The concept of library code is well understood in programming circles and has a number of benefits, such as facilitating better code design and maintenance. MIDlet suites can be downloaded in a single JAR file for download. LIBlets can also be downloaded if a particular MIDlet requires a LIBlet that is not currently residing on the device.

2. *External and portable RMS binaries* – Any MIDlet can store persistent data as records in the memory space allocated to the J2ME environment. However, a limitation is the inability to share this data with other MIDlets or modify it outside of the originating MIDlet. With MIDP 3.0, RMS data can be referenced and accessed externally, including over the network. This adds a powerful dimension to the MIDlet-programming model.

3. *Improved OTA mechanism* – Various improvements have been made to the OTA model, including more reporting codes, giving greater robustness to the user experience. CDC support (not just CLDC) for higher end phones – MIDP 2.0 was strictly meant to run on the CLDC configuration of Java, which mostly means the KVM. However, as

processors continue to increase in power and mobile phones become increasingly capable of handling complex computing tasks, some devices will inevitably support the CDC. This greatly enriches the possibilities for using Java on a mobile device. However, so that the portability and ubiquity goals of MIDP are still realised, CDC devices should still run MIDlets, which means that the MIDP 3.0 specification has insisted that this be the case. This is more of a porting issue than a feature of MIDP, but it is still an interesting development.

4. *MIDlet concurrency* – Explicit support for MIDlet concurrency has been added, including the ability to keep a MIDlet running when not in focus. However, no scheduling technique or algorithm has been specified and this is left to the handset vendor to determine. Some guidelines are given, such as the obvious need to avoid processing starvation for a background MIDlet and the need to provide a responsive user experience generally, especially to the MIDlet with the focus. MIDlet concurrency has been possible on some MIDP 2.0 phones, but only in a limited sense whereby the background MIDlet is effectively suspended whilst another MIDlet grabs the focus. MIDP 3.0 goes beyond this model and greatly improves the richness of the MIDP environment for supporting mobile services. After all, it is unlikely that a user would want to use an IM application that went to sleep when he or she decided to play a game, both being MIDlets. MIDP 3.0 allows such application multitasking to take place. Let's hope that vendors implement this in a usable fashion.

5. *MIDlet security* – More flexibility has been added to the concept of trusted MIDlet security, allowing finer grain access to a single resource or function via an API instead of all or nothing (boolean) permissions. It is also possible to control permissions on how different MIDlets can share data, share LIBlets and communicate with each other via the Inter-MIDlet Communications (see Point 8). Additional security policy mechanisms have also been added to enable trusted MIDlets to be verified against an operator domain via root certificates stored in the SIM or USIM (or other smart card device).

6. *Screen-saver MIDlets* – MIDlets can now be marked as being screen-saver programs. This means that the MIDlet is executed or given the focus when the screen-saver period starts. This is great for dynamic content apps fed by the network, such as ticker-tape type of applications and widgets. The MIDP 3 has defined the behaviour of MIDlets when in screen-saver mode.

7. *IP version 6* – As with most IP-based solutions these days, IPv6 support has been added to the MIDP 3.0 platform.

8. *Inter-MIDlet Communications (IMC)* – This is a great addition with an obvious function – it allows concurrent MIDlets to communicate. This is done via a low-level socket-like interface. This shouldn't be confused with a MIDlet calling code from a LIBlet, which should be thought of more as one big program that just comes in parts. IMC is two separate programs (which themselves might call LIBlet code) sending messages to each other.

9. *Mobile media API* – This is now mandatory in MIDP 3.0, which is an obvious step. Media support is so crucial for many mobile services.

11.13   ON-DEVICE PORTALS

11.13.1   Introduction

The mobile browser is an easy way to deliver services to the mobile. It is probably still the most reliable in terms of device penetration because nearly every mobile on the market now has a browser. However, the browser has several limitations that have led to alternative approaches. Two limitations in particular are:

1. The richness of the user interface of a web page is limited compared to using an embedded programming technology, like Java.

2. Apart from the recent advances with AJAX, it is not possible to pre-fetch data into the browser, especially if a particular website isn't actually being surfed. This is particularly problematic for rich-media solutions where the user would like to access many media clips. Jumping between pages whilst opening and closing media clips is an extremely poor user experience in mobile browsers (and I would argue in desktop browsers – it's no coincidence that iTunes uses an installable client, not a browser).

We cannot dismiss the browser altogether. Its universality is still important for reaching as many users as possible with a particular service. However, in terms of a service strategy, we really have to consider that the browser approach is the lowest common denominator, notwithstanding that some browsers on some devices can support relatively rich pages.

The above limitations have led to the emergence of On-Device Portals (ODPs) for the purpose of offering users a rich media experience, but with the singular goal of offering the users multimedia content, such as video, picture and audio files, plus related assets like ringtones. ODPs are increasingly prevalent and seem to be emerging as a class of mobile application in their own right, hence why we should include them in the closing sections of this chapter.

ODPs are great for supporting rich-media content, but they are mostly designed for this purpose alone. For more general-purpose programming where we would like to consider the mobile as just another end-point for digital services, like the desktop is today, the problem is that the API capabilities of browsers are very limited or non-existent. For example, from a web page in the browser it is generally not possible to access the phone's address book and other key functions on the device. This is particularly challenging as we move towards greater convergence between services. For example, if I wanted to write an application that mixed some telephony with instant messaging and video services, this would be extremely difficult, if not impossible, in a browser. Java in its MIDP form can provide a solution to many of these problems. However, it still has some limitations in that as far as the user is concerned, their phone is divided into two parts. There's the phone itself, with all its applications, interface components and services and then there's the Java environment, almost always accessed via an application button that seemingly takes the user into a separate applications space that is hemmed in and doesn't seem to interact with the rest of the phone in any integrated sense.

What if we could provide the programmer with an environment on the mobile phone that allowed network-centric (i.e. Web 2.0 platform) rich-media applications that integrate well with the native phone environment, giving the user a rich and integrated experience?

Well, isn't this what we get with the truly native embedded approaches, like Symbian? Yes it is. However, a native Symbian application will only run on a Symbian device. Moreover, the programming style is distinct to the Symbian platform. If we wanted to use web-style programming and ideas, then we're stuck. In particular, if we still have convergence in mind, then we would like a programming approach that works across a variety of key digital end-points (devices) including mobile, set-top box and desktop.

To address these sorts of issues, a variety of solutions have appeared on the market, which can be thought of as 'phonetop' solutions, which is supposed to represent an idea of making the mobile a more generic end-point, especially one of the key triad: desktop, set-top (box) and now phonetop. In the next sections, we shall look at ODPs and other possible phonetop technologies.

11.13.2 ODPs

The ODP approach is singularly concerned with giving users access to multimedia content, but with the best user experience possible. We think of content as really being accessible via a catalogue. These catalogues are usually accessed via a WAP site, usually called a portal. When we talk of great user experience, we mean something that is graphically rich and very responsive (i.e. quick). It looks nice, it's easy to use and they can get what they want very quickly. This is from the users' perspective. However, content providers themselves are often looking to provide a great experience as part of the image they would like to convey with their content and brand. We should keep in mind that an ODP could potentially be used by anyone who wants to promote content. They are not limited to operator use only. For media companies interested in this approach, Chapter 15 explores strategies for mobilising media.

A typical ODP architecture is shown in Figure 11.35. There is always a client that is downloaded to the handset (or it could potentially be pre-installed). We can think of the client as a very high-end browser. It is capable of rendering a self-contained set of 'pages' that are graphically rich. The richness is achieved by using an embedded programming environment, which is typically J2ME based. There is then a middleware server that sits in the network and controls the packaging and distribution of the 'pages'. In other words, the architecture is similar in many ways to a standard browser approach, except that an entire 'site' can be downloaded to the phone in one go and it can utilise the native graphics libraries of the Java environment, which are usually richer and more flexible than the browser. Included in the 'site' is not just the layout of the interface, but the content itself. For example, along with a page to show the top ten music hits for the week, the actual audio files (or most likely preview clips) can be downloaded, too. Thus, when a user hits the button to play a preview, it happens instantly.

The entire look-and-feel of the ODP can be customised at any time. It is often branded to match the desired look-and-feel of the portal owner, which is usually linked in some way to their existing branding. A layout tool controls the look-and-feel. This is either a desktop or web-based application that enables an author(s) to create the ODP site itself, including the layout, colour schemes, interactivity and so on. The tool is also used to create the links between the layout and the content, both of which are then uploaded to the middleware platform. The platform controls the packaging of the layout and content information into a

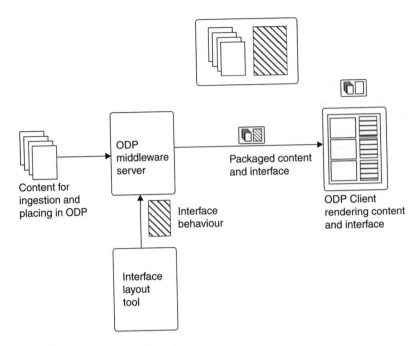

Figure 11.35 Typical ODP architecture.

format that can be distributed to the devices. The package can be pulled down by the device or pushed to it. Incremental updates are also possible and very often the ODP vendor has designed the packaging to allow partial updates so that only the changes are pushed to the handset; thus, optimising the use of bandwidth.

Because ODP clients are built with an embedded programming environment, they are usually able to access any underlying features of the device that are exposed via APIs to the program. One example is the API to access messaging on the phone, including texts. This is a mechanism that can be exploited in order to implement charging using premium-rate texts (i.e. reverse-billed messaging). Critically, it is possible to access the media player on the device in order to allow media files to be played within the ODP client. This can also include media streaming; thus, enabling ODPs to be used for Mobile TV applications. Another possibility is accessing the camera in order to allow the user to take pictures and upload them to a service. This could be used to support rich features like blogs as part of the overall portal experience.

The middleware layer in the ODP architecture is typically web-based and has a set of APIs to allow links with other services. For example, it is often the case that the service provider already has a powerful content management platform to manage all the content. The ODP middleware could therefore interface with the content platform rather than ingest the content directly. The ODP approach isn't always used for entertainment content catalogues. It has found use in various networks as a means to distribute mobile versions of magazines or for rich news services. However, the disadvantage with the ODP approach is that it isn't standardised. Most solutions use entirely proprietary technology to implement the ODP, although the underlying protocols and software technologies are standard IP and Java (or similar).

Web page   JavaScript

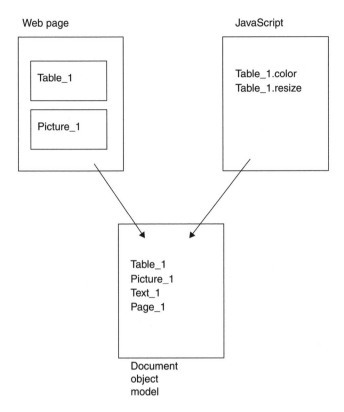

Figure 11.36 Document Object Model (DOM) inside browser.

11.13.3   Alternative Application Paradigms – Opera Platform[38]

Opera Platform (OP) was proposed as a client technology that runs on the device. It doesn't involve any custom network resources, like the middleware layer used for most ODP solutions. However, it adopted a web-based approach to programming whereby pages can be requested from a web server, but these pages can also access local resources on the phone to create a seamless user experience between networked and device resources.

Despite its web orientation, OP wasn't just a browser for generally accessing web pages, although it was built using Opera's popular mobile web-browser technology. It was a way to build applications that run on the device, so it is a more of a networked programming technology than a browsing technology. However, to build an interface within the framework, the developer would have used web-like programming techniques, scripting and markup languages, notably XHTML, CSS and JavaScript. Not surprisingly, these elements are based on the Opera browser, which is already available on a number of mobile devices, mostly S60 Symbian and Windows Mobile. It is through the extended power of JavaScript that new levels of functionality are exposed to the programmer. To do this, the OP first had to be

---

[38] Note that this product was never released to the market, but it is included here because its unique approach offers much food for thought for future approaches to mobile application development, especially ones that are exploitative of the Web 2.0 platform.

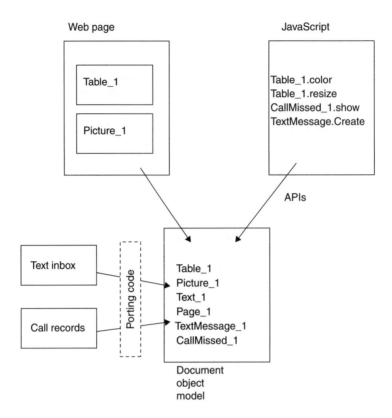

Figure 11.37 Extending the DOM in Opera Platform to include phone functions.

ported to the target device. This meant implementing the underlying code libraries to enable the extended APIs within the JavaScript environment via what's known as the Document Object Model, or DOM. In a JavaScript program resident in a browser, the page inside the browser is accessible programmatically via the DOM. The reason we need the DOM is because the page itself isn't written in JavaScript. It is still written using XHTML and so all of the UI components declared in the XHTML sit outside of the JavaScript program. Therefore, in order to access them from the JavaScript, there is a set of APIs to allow access to the DOM. This is shown graphically in Figure 11.36.

The OP extended the role of the DOM to include a model of some of the phone functions so that they can be made available to the JavaScript engine; thus, enable a mix and match of graphical elements (XHTML) with phone functions, as shown in Figure 11.37. There is no doubt that this was an elegant attempt to combine the world of web programming with the functions of the phone.

As an example, within the JavaScript framework available in an average browser, there is no API to expose the text-messaging capabilities of the device or functions such as battery strength, call records and so on. These telephony and messaging functions are present in the OP via an extended set of JavaScript APIs to access the extended DOM. In order to extend the DOM, some porting code is required to enable the DOM to model the phone functions; hence, integrating OP into a device would have been more complex than embedding a standard browser and would have required extensive cooperation from a licensee vendor. This is perhaps why it wasn't successful.

As a key part of the OP approach, the entire home screen of the mobile could be implemented as an OP application. This enabled a great user experience to be provided directly to the home screen incorporating live data from the network, such as weather forecasts, financial information, TV programme gossip and so on. From the outset, the user experience is fully integrated into the phone's look-and-feel and other operations, thus enabling an intuitive and seamless experience. This also allowed content teasers to be displayed right on the home screen where users are likely to 'bump into' them, which generally is more likely to lead to content being accessed than if it is hidden away underneath a separate button, or on a web page. In everyday use of a mobile, it is almost impossible not to look at the home screen, such as to check calls, messages, etc. As the home screen is essentially a web page that can respond to phone events, it is possible to fetch network content in response to activities on the phone, such as receiving a new message. For example, if a message is received from User A, then we can make a web call and bring up content from the server related to User A, such as their latest photoblog entry.

This approach of putting dynamic content on the home screen was not unique to OP. Other vendors, such as Motorola's Screen3 technology, which allows content to be pushed to the home screen, have done it. This technology is currently unique to Motorola devices and uses a Java MIDlet that is given special permission to access the home screen via an API, something that other vendors have now started to implement, such as Sony Ericsson. Content can be viewed via the home screen but for more details and to make any content purchases, the user is driven to an appropriate website via the embedded browser in the device. It is worth mentioning that the concept of dynamic home screens is possible with some ODP solutions, although these obviously depend upon the appropriate APIs being available on the device to allow access to the home screen in the first place.

12

The RF Network

Computer networks are now able to utilise a variety of wireless connection technologies. In this chapter we are interested in wireless connectivity that facilities portable or mobile computing, although our main interest is in wide-area (cellular) solutions. An increasingly diverse range of wireless solutions provides users with potentially ubiquitous access to services, including the Internet. As already discussed throughout this book, the evolution of internet-orientated software technology, like J2EE, XML and Web Services are helping to make the Internet become an incredibly powerful delivery mechanism for services. The IP network, including the Web 2.0 platform and underlying web-centric services, clearly has much strength as a means to deliver software services, an assertion amply discussed in this book. It is time to look at how we achieve the wireless freedom needed to make these services mobile.

Continuous wireless access to the Internet is now possible on a wide scale thanks to cellular networks. Providing a cellular connection with IP compatibility has been possible since the 2G networks, like GSM. However, the 2G infrastructure and its technical roots grew from the need to talk, not the need to exchange data. These early data connections were not packet-orientated, so they were very inefficient for carrying IP data. Attempts were made to overlay a packet access method on the 2G networks and this became the basis for 2.5G, otherwise known as General Radio Packet Service, or GPRS. For IP-based traffic, 2.5G networks are very limited, so along came *3G networks*. These are wide-area solutions built and designed to carry both voice and IP data. Devices can communicate over distances of many kilometres whilst maintaining useful battery life and portability. In many countries, roaming devices will always be within range of a 3G *base station*, the access points to the 3G network. The 3G network itself can carry data back and forth between the devices and the Internet, or any other packet data network.

If we think of an area being 'illuminated' with RF coverage from 3G base stations, then underlying that umbrella coverage will be pockets of local-area wireless networking

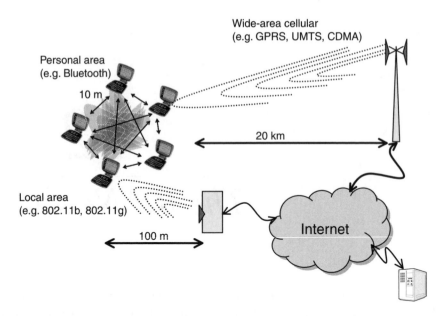

Figure 12.1 Heterogeneous and ubiquitous wireless access. (Reproduced by permission of
IXI Mobile.)

coverage, like 802.11b (WiFi) or 802.16 (WiMax). The lower range of these localised
islands means a much faster connection speed than with current 3G standards, shifting
greater quantities of data within similar power and portability constraints as 3G. These
technologies complement each other and can run side by side, as shown in Figure 12.1,
even on the same mobile device. At an even greater level of intimacy, devices can directly
talk to each other within the vicinity of a few metres, as offered by personal networking
technologies like Bluetooth, as well as WiFi. Since the first edition of this book, the so-
called 3.5G networks and devices have started to appear on the market, which are called
High-Speed Packet Access (HSPA) networks, fulfilling the dream of a mobile broadband
service, although the uses for such high speeds have yet to be fully appreciated.

   In this chapter, we shall take a look at how 2.5, 3 and 3.5G networks function. We shall
also look beyond the RF physical layer and examine what other assets exist in the operator's
infrastructure that can help us to build mobile services, such as access to messaging and voice
platforms. In the next chapter we focus on one capability of the RF network in particularly,
namely is its ability to locate mobiles, which can be exploited to deliver location-aware
services.

12.1 THE ESSENCE OF CELLULAR NETWORKS

At its simplest, the RF network layer of our mobile applications network is fundamentally
a big switch. It can route information, voice or data, from fixed information appliances to
moving ones, as shown in Figure 12.2, or from one moving appliance to another. Wired-data
communications connections are on the fixed network side, whilst on the mobile side, the
connections are made using RF.

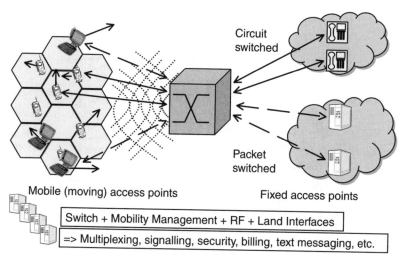

Circuit switched

Packet switched

Mobile (moving) access points   Fixed access points

Switch + Mobility Management + RF + Land Interfaces

=> Multiplexing, signalling, security, billing, text messaging, etc.

Figure 12.2 Essence of a mobile network. (Reproduced by permission of IXI Mobile.)

The degree of movement that we are interested in supporting should ideally cover the scope of the mobile applications and services we have been considering thus far. With a mobile service, we would like user to be able to have:

> The ability to interact successfully, confidently and easily with interesting and readily available content, people or devices *whilst freely moving anywhere we are likely to go* in conducting our usual day-to-day business and social lives.

The freedom to move anywhere (ubiquity) is our objective. We have seen that allowing voice calls to be placed anywhere at anytime has amply proven itself as a liberating service for a large number of people. Those 'crazy' inventors of cellular telephony have vindicated themselves and their audacity. Now, the extent to which ubiquity is important is no longer a matter for debate, but how to achieve ubiquity often is. Some pundits argue that WiFi hotspot coverage is sufficient for many applications, whereas others argue that the contiguous coverage of wide-area cellular systems trumps WiFi in most cases. New voices with new ideas, like WiMax, have yet more opinions. In this book, we are not interested in the debate, but we are interested in what the technologies can do for us.

Being faithful to our mobile services objective above, our philosophy and approach towards building applications should encompass any access method if it enables successful end-user engagement with mobile services. It also seems entirely possible that mobile services will develop from both ends of the spectrum (no pun intended) and then often meet in the middle; WiFi applications and 3G applications will converge, overlap and augment each other. This is already happening with WiFi/3G phones and services like UMA, which enables a 3G phone to make cheap calls from inside the home via a WiFi router that can route the traffic back to the mobile network over a fixed wire-line broadband connection. A similar expectation is now arising around the evolution of WiMAX from a fixed wireless solution to a mobile one.

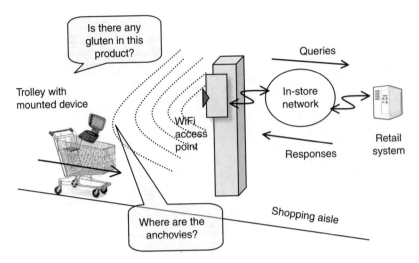

Figure 12.3 A wireless shopping assistant system using WiFi. (Reproduced by permission of IXI Mobile.)

12.1.1 RF Network Convergence

These two access methods, WiFi and 3G, are bound to converge, rather than collide. The attributes of WiFi services that might make them attractive will emerge and become familiar to users, such as its high speed and apparent cost-effectiveness for large data transfers. Similarly, users will also come to understand the meritorious attributes of 3G that are due to emerge, such as its ability to determine user location. It often goes unnoticed by the pundits that it is entirely possible to have the best of both these worlds. Dual-mode devices are not only technically possible; they are both economical and entirely desirable.

Let us ponder for a moment on the idea that each technology has its place. As an example, imagine that we are interested in providing wireless shopping-assistant services to mobile terminals whilst inside a supermarket[1]. As the shopper moves around the shop, we know their exact location (to within a particular aisle) and we can offer all kinds of services accordingly. We could build such a system using the browser paradigm and many of the solutions and technologies discussed in this book, even though the freedom to move anywhere in this case starts and ends at the supermarket door. A cost-effective solution for this system would be to use WiFi to connect the terminals to the retail backend system, as shown in Figure 12.3. With WiFi access points at regular intervals inside the shop, it is possible to locate shoppers quite accurately[2]. The best that a wide-area solution could do is place the shopper inside the shop, but nothing finer in terms of spatial resolution.

Many solutions like this will probably emerge; WiFi services that are limited in coverage to a particular place, or collection of similar places (e.g. all stores in a chain), although not necessarily for the reason just mentioned (i.e. fine-resolution location finding). However, we live in a changing world, a world of convergence, to use that often-ambiguous expression.

---

[1] There are various ideas like this under review by major supermarket chains, including providing services via terminals attached to shopping trolleys.

[2] Indoor positioning is an area of intense research, being carefully scrutinised by retailers. For a useful summary, see 'Retrieving position from indoor WLANS through GWLC', by G. Papazafeiropoulos *et al.*, which can be found at http://www.polos.org/reports.htm

Despite its overuse (abuse), it does have some value and relevancy here. For example, no sooner have we defined and built a service that works in our supermarket, when along comes a bright spark who suggests extending it into the host shopping mall itself. Perhaps someone has identified that the commercial interests of the supermarket converge with serving the general interests of shoppers in the mall. We can think of this as a convergence of opportunity, in this case influenced heavily by common geography (co-location) and common consumer interest (shopping). However, the original technological solution scoped for the supermarket might need a rethink in terms of how to extend it into the shopping mall. At this point, perhaps the economics suddenly become very different and the coverage area and footfall[3] no longer lend themselves so readily to the original solution.

Our degree of movement (mobility) changes, potentially dramatically, between the relatively constrained and partially predictable movement within a shop, to the wider less-perceivable movement in the shopping mall. After the success of the shopping mall service, someone might then make the connection with trying to lure potential shoppers in the retailer's den; an altogether different technical challenge and more than likely delving into 3G territories.

On the other hand, a critical examination of the dynamics of the in-shop service might suggest that such a service can only work in conjunction with supplementary services, like an associated and already established Internet shopping channel. Perhaps the implementation of electronic coupon redemption on phones has become more widespread than anticipated, forcing our in-shop service to support such a scheme. This is convergence at the service level. One service, like mobile electronic coupons, ought to work with another, like wireless shopping-trolley assistants. The availability of the coupons may be a key driver that attracts shoppers, so the discovery of available coupons cannot be limited to shoppers already inside the shop. Coupons need to be discoverable and available outside of the shop, which means another degree of mobile freedom is required in our application.

There are yet other areas of possible convergence. Cost effective delivery of the in-shop service requires enough customers to reach critical mass, so it is probably only viable to offer such a service on customer's *existing* personal devices, or at least in conjunction with them. However, newer devices, with better displays, the ability to watch streaming-video adverts, receive images, etc., will make them more suitable for the shopping service. In other words, we require technological convergence. Perhaps it is also possible to reward shoppers with credits on their mobile phones in proportion to the amount spent in the shop, so we have a further convergence; this time between the service fulfilment ends of the two systems, perhaps executed using Web Services.

Our shopping assistant service now seems to require accessibility from many different place and devices and to a variety of back ends. From these considerations, possible system architectures begin to emerge, at least at its highest level. We end up with a system (see Figure 12.4) that looks remarkably like the RF network model we proposed earlier (see Figure 12.2). However, the access methods are heterogeneous. If a single operator is able to offer both access methods (a likely development, already seen in places like the United States), then a common core network (routing function, authentication) will already be in place. However, if there is no common service provider (an equally likely scenario in many cases), then we need a different means of service convergence, such as Web Services across the Internet.

---

[3] Footfall is a retail term for the number of feet (i.e. people) who walk into the shop (or shopping area).

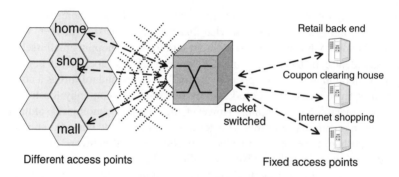

Different access points

Fixed access points

Figure 12.4 Useful service requires many access points.

In our shopping example, mobility has two dimensions, or, more accurately, two levels of resolution: fine-scale movement within locations and the more coarse-scale movement between those locations. We are going to need to support this type of mobility in the future. However, this challenge is not without other considerations. We have to take into account not just the freedom of movement, but also the qualitative criteria that the opening part of our mobile services definition suggests: 'The ability to interact *successfully, confidently and easily . . .*'

Would requiring the user to own several devices, one for each location (in-store, in-mall and out-of-mall), facilitate an ability to interact easily? It seems unlikely. Multimode devices do seem inevitable in future mobile service architectures.

The foregoing discussion has highlighted the likely emergence of multimode RF services and devices. However, for the reasons that we mentioned earlier regarding the importance of other capabilities of the cellular network, we will be focusing on cellular networks during the rest of this chapter. We need to expand on the notion of an RF network as a giant switch, as shown in Figure 12.2, in order that we can more fully appreciate its capabilities. Before doing so, we ought not to forget that we may need to support mobile application topologies that do not require a wide-area network, such as P2P topologies and the close range interdevice communications within the device network itself (personal networking). The RF networks we can use to support this are mainly Bluetooth and WiFi, but used in ad hoc mode, they largely function as cable replacements. Given such an obvious function, all that remains is to understand the technical details of such connections. However, such details are not within the scope of our enquiry.

Looking again at Figure 12.2, the core of our network appears to be little more than the switching (routing) function and a means to support the RF connections themselves. However, this is an oversimplified view. Not only is it simplified in terms of how these components of the network actually work, but more critically it overlooks a good deal of mobile network assets and a host of other support functions that are key to many types of mobile service. These network assets have been included since day one of 2G networks, or slowly accrued over long periods of network evolution. Digital cellular networks are now very mature and most of them in their second decade of operation. Subsequently, they have a wealth of assets that feature very strongly in the anatomy of the network, beyond an RF connection and a switch. As Figure 12.5 shows, cellular networks are no longer just wire-line replacements with an ability to support roaming. Text messaging and multimedia messaging

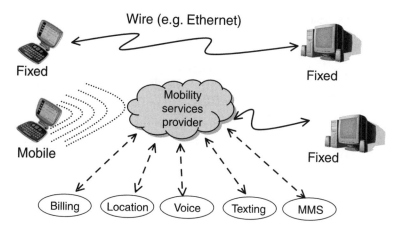

Figure 12.5 Mobile network is not just a wire-line replacement. (Reproduced by permission of IXI Mobile.)

are prime examples of important mobile network assets that have become important service platforms and services in their own right.

Before we discuss these network assets, first let's look at how the RF and switching function is enabled. What is a cellular network?

12.2 THE RADIO PART

Some books tend to place a lot of emphasis and time on discussing the radio part of a cellular network. This is perhaps because the RF is what naturally feels like the essence of a mobile application or a mobile network. However, from an applications perspective, understanding the actual RF part is perhaps the lowest priority in understanding the entire network of networks that we have been traversing throughout this book. As we pointed out in Chapter 4, the overriding challenge for the RF component is for it to be as transparent as can be, meaning that it is actually as 'wire-like' as possible in terms of its information handling attributes. To an extent, one could treat the RF component as such and ignore its characteristics and vagaries. As we shall see, that would probably be a folly approach to mobile service design. Perhaps the most crucial parts of the RF network to understand are its boundaries with the IP network. This often seems to be a source of confusion; how do we get from the IP world to the RF one?

We do not wish to belittle the RF network. As we shall come to appreciate, the processes and mathematics that underpins RF transmission technology are by no means trivial. The information theorists and communications experts have invented something quite incredible. However, given a functioning RF network, the greater challenges in application system design are seldom to do with the RF. They are probably more likely to do with the configuration of the network assets (like the Short Message Service Centre, or SMSC), as well as how to connect with them.

That said, an appreciation of the RF is useful, especially the ways in which its behaviour can affect application performance and, ultimately our approach to design. When considering the RF network, we shall place emphasis on understanding the essential characteristics

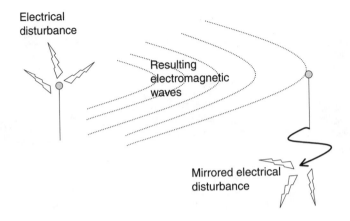

Figure 12.6 Electronic disturbance (excitation) felt at a distance.

of a cellular network, rather than a rote description of the different cellular standards, such as GSM, GPRS, CDMA, UMTS, etc. There are plenty of good books on these networks and their details.

12.2.1 Basic RF

To cut a long story short, and to avoid giving a physics history lesson, certain electrical disturbances in a metallic conducting rod (antenna) can produce an effect at a considerable distance from the point at which they occurred, as shown in Figure 12.6. The conveyance of the effect is by electromagnetic energy that travels outward from the disturbance source as invisible waves moving at the speed of light. We all know this effect well by tuning into a radio station on our transistor radio; we benefit from the effect in the receiver by converting it into sound energy that comes out of the loudspeakers. I make special mention of the word 'transistor', because this is possibly a much more profound scientific invention, as this is what makes modern RF communications possible. This is particularly so when millions of them are combined to make a silicon chip that can process RF signals using digital signal processing (DSP) techniques. We made mention of the importance of DSP in Chapter 10 on devices.

   Thinking of the radio is probably a good place to start. We probably already appreciate that the radio effect can take place within a band of possible frequencies (the frequency being the rate at which the disturbance takes place, such as 900 million times per second, or 900 Mega[4] Hertz[5], or 900 MHz). We divide RF frequencies into bands[6] that we use for transmitting information from one source to its destination or destinations (see Figure 12.7).

   From basic experience, and perhaps common sense, we know that occupying a band with the output of a particular station prevents its use by another station. We can easily imagine

---

[4] 'Mega' is from the Greek word *megas*, which means large, but in scientific measurement refers to 'times one million'.

[5] 'Hertz' was someone's name (Heinrich Hertz) and is a unit of measurement that indicates one cycle in a periodic event, like a repeated RF disturbance (excitement).

[6] Although we tune our radio into a precise frequency to get a particular radio station, the information transmission process actually requires a small band of frequencies either side of the main frequency that we have tuned to.

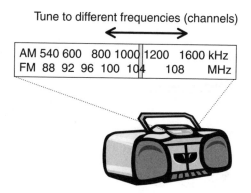

Figure 12.7 Radio spectrum divided into channels.

what happens if two radio stations attempt to transmit on the same band ('frequency'[7]). Chaos would ensue and we would hear garbled noises, if we were indeed able to hear anything meaningful at all. Interference is the name given to this basic limitation of RF transmission. What radio engineers mostly do when designing radio transmission systems, especially cellular radio ones, is to invent schemes for maximising capacity (i.e. the number of stations in our case) whilst mitigating interference; in effect, reducing interference and maximising capacity are the same thing. We could say that this is the number one challenge – the RF *information capacity* challenge.

Those familiar with Citizens' Band (CB) radio (or any two-way radio system), already know one obvious solution to interference. We simply avoid more than one user transmitting on the same frequency at the same time. We divide the spectrum up and allow each user (or radio station say) to use their own band, or *channel*, exclusively. This is how we achieve more than one user accessing the RF spectrum, or multiple concurrent access, which is why we call it Frequency-Division Multiple Access (FDMA).

Exclusive use of a given band is either permanent, like with broadcast radio stations, or temporary, like with two-way radio. With two-way radio, we generally use a system on a first come, first served basis. If I'm already on Channel 8 (say), then you can't use it till I finish. This procedure is what we call a *protocol*, and we need *channel reservation* protocols like this in order for a communications system to function in its entirety. Clearly, we don't remember checking to see who's on our channel when we place a cellular voice call. In this instance, the reservation protocol is automatic, carried out by software running in the device and in the RF network.

The use of some bands (most of them) in any given country requires a licence from the government in order to transmit. This requirements came about in order to stop chaos from ensuing in a 'free for all' situation, particularly because the chaos would affect emergency service transmissions with possibly onerous consequences. However, some bands are unlicenced. The chaotic situation can be managed using appropriate RF-transmission techniques, such as the spread spectrum method we shall discuss shortly. It is because of these technologies, or the ability to implement them cost effectively with today's electronics, that

---

[7] We say 'frequency' because although RF communications does take place across bands of frequencies, the band is usually referred to by its centre frequency, which is why we hear radio stations as being on a particular frequency, like '99.8 FM'.

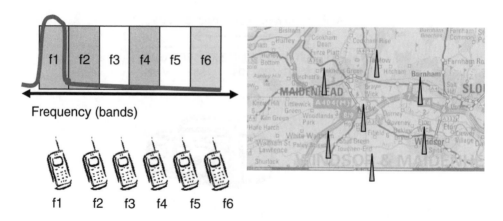

Figure 12.8 Assigning bands to each device.

some people, such as in the United States, are contesting the need to have a licence. Their motivation, in the main, seems to be about removing restrictions that might stifle innovation and competition, or about removing restrictions in principle on the unfettered use of a public common, such as the RF spectrum.

12.2.2 Building an RF Network

Given what we know so far, if we want to use radio signals to support a mobile phone network, then our starting point might be to assign a unique frequency to each mobile, as shown on the left-hand side of Figure 12.8. Here we can also see that each mobile actually transmits on a small range (band) of adjacent frequencies. In practice, due to the physical limits of RF circuit components, it is very difficult to confine the transmitter to a distinct, vertically edged band. As the figure shows, for our device on band f1, there is some residual 'spill' into adjacent bands, so there is a tiny bit of interference experienced by f2 in this case. This interference is called *adjacent channel interference* (for obvious reasons).

To enable our mobiles to communicate with the fixed network, we can install base stations in our service area, as shown on the right-hand side of Figure 12.8. Thinking of each base station as being like a broadcast radio station, we can assign a frequency for each base station to handle. What we end up with is a grid of base stations, each transmitting on their own frequency, so we end up with the situation shown on the right-hand side of Figure 12.9. Radio waves propagate like ripples on a pond after a stone causes a disturbance. In the figure, we can see these 'ripples' propagating out from the base on frequency f6. Clearly, each base station is also causing a similar radiation of waves from the centre, but the hexagonal cells are an idealised way of representing each region served by a particular frequency. We can see that f6 radiates way beyond its immediate vicinity, the exact distance related to the power or the transmission, as measured in watts.

In the geographical region immediately next to f6, it would be folly to use the same frequency (f6 again) as clearly the signals would interfere with each other. This is why adjacent cells transmit on a different band. However, as we can see from the figure, the waves spread out a long way from the base station. Just as light from a torch weakens over distance, the reach of the RF signal causes them to weaken until the signal eventually

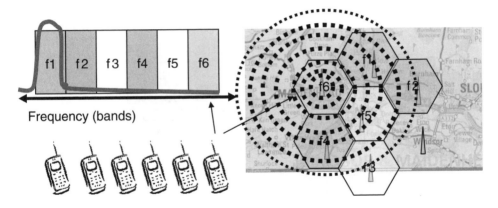

Figure 12.9 Allocating frequencies to regions.

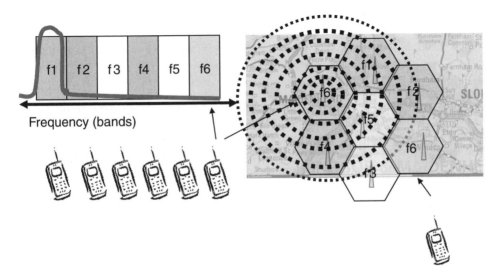

Figure 12.10 Frequency reuse at a safe distance (notice f6 is used again).

becomes quite weak. If we go far enough away from the base station transmitting on f6, we would eventually be able to use the same frequency again, without any fear of detrimental levels of interference on the same band (*co-channel interference*). There would still be a residual level of co-channel interference, but we can design our network in a way that seeks to minimise it[8].

Moving far enough away from the first station using f6, we can install a second station on f6 and use that frequency all over again, which is called *frequency reuse* and is how we arrive at a patchwork of cells across a wide area, like a country, and is the inspiration behind the name 'cellular' network. The frequency reuse concept is shown in Figure 12.10.

---

[8] Co-channel interference in a cellular network is one of the limiting factors that affects capacity and performance of the network.

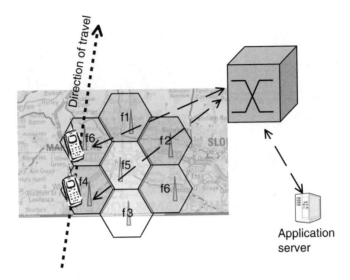

Figure 12.11 Mobile network switches (hands over) from one cell to another as the mobile moves through the network.

Our system would be extremely limited if we didn't have frequency reuse because after using our six bands our system would run out of bands and therefore capacity.

The basic operational scenario of our cellular system is that as a device moves through the service area, it swaps from one frequency to another, called *hand over*, to keep a continuous transmission active. Cleary this has implications:

- Devices have to support operating on more than one frequency and the ability to swap frequencies as required.

- The network has to support a mobile staying in touch from one cell to another, which means it has to hand over (i.e. from one cell to the next), as shown in Figure 12.11.

- If the mobile moves into an area where there is already a mobile transmitting on that band, then it cannot transmit on that frequency at the same time, otherwise it will cause potentially catastrophic co-channel interference.

With sophisticated software, we can track mobiles and facilitate handover. This is not a problem. It is clearly important to track where mobiles are anyway, because if we want to initiate a communication session with one, then the cellular network needs to know where it is in order to route information to the right cell. Tracking mobiles for routing purposes is part of something called *mobility management*.

Reflecting on what we have achieved thus far, we now have a basic design for a wide-area mobile device network, using cells to allocate frequencies and by preventing devices from transmitting on the same frequency within the same area (FDMA). However, if we look carefully at our cellular layout as shown in Figure 12.11, the system has some obvious and severe limitations; namely a lack of capacity. If we can only handle one device per frequency per cell, then the people of Maidenhead (the town on our map) are not going to be very

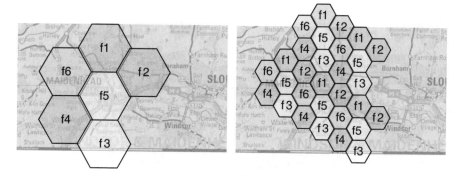

Figure 12.12 Smaller cells means more mobiles means higher capacity network.

excited about cellular technology, as our service can apparently support only a handful of users. Presumably, the population of Maidenhead is more than a handful of people!

There are two ways to solve this problem:

1. Make the cell sizes smaller

2. Find a way for several devices to share the same frequency in a cell

Shrinking the cell size seems a sensible proposition. In actual cellular systems, this turns out to be the main way of increasing capacity, everything else being constant (which, as we shall see, it probably is once we have built the network).

We can cope with a higher density of mobiles (i.e. more customers) if we have a denser cell pattern, as shown in Figure 12.12. Clearly, the higher capacity does not come without a cost, namely the costs associated with installing more base stations and connecting them back (the *back haul*) into the core of the mobile network (*core network*), such as the switch.

Smaller cells bring some added benefits to the end-users. The amount of power needed to send an RF signal back to the base station is much less, simply because it is much closer. This means lower power consumption for the device, which in turn means longer battery life. In fact, cell size reduction is one of the major reasons that cellular handsets have improved so much in terms of improved battery life. The other reason for better battery life is the dramatic improvements in the electronics inside the handsets, mainly the silicon chips running off lower electricity levels (voltage), which means less energy consumption[9]. This also significantly benefits the standby time of the device as the lower operating voltage causes a reduction in draining of the battery whilst in standby mode (from reduced *current drain*).

As mentioned, the second way of increasing capacity is to find a means to share the same frequency between several mobiles. This seems to contradict our earlier concern about interference, but it turns out that there are some neat tricks to get round that problem, one of them almost mechanical in nature (TDMA), the other very mathematical, almost magical (CDMA). We introduce these two methods in detail in the next sections.

---

[9] Integrated circuits (silicon chips) are made of lots of transistors. These are tiny switches. Turning a transistor on requires energy in proportion to the voltage level it has to rise to. For lower voltages, less energy is therefore required; hence, the battery drains that little bit less.

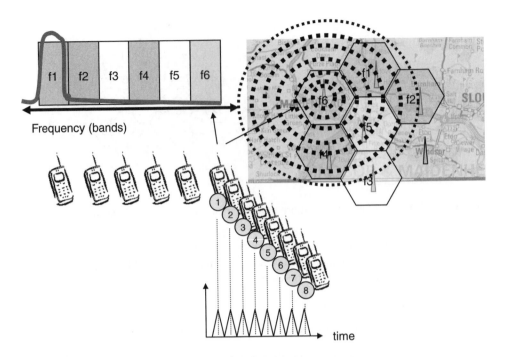

Figure 12.13 Mobiles take turns to transmit on the same frequency.

12.2.3 Increasing Capacity using TDMA

The principle of Time Division Multiple Access, or TDMA, is simple, as shown in Figure 12.13. The figure shows that eight mobiles are all transmitting on f6 in the cell over Maidenhead, but not at exactly the same time. They take turns to transmit in quick succession in a continuous cycle: 1, 2, 3, 4, 5, 6, 7, 8, 1, 2, 3, . . .

The cyclic nature of sharing the same frequency seems strange at first. If we think again of the radio station analogy, it would seem as if we would be listening to our favourite station, but as a stream of quick bursts; an audio stream punctuated mostly by silence. If that were the case, then clearly such a proposition is unworkable. However, with digital cellular, this is not what happens. Because everything takes place digitally, the digitised speech can be compressed using digital processing techniques. We can imagine the mobile device as having a digital recording function (which in effect, it does have) and then passing the recorded digital samples through a compression algorithm, as shown in Figure 12.14. The way to think of this process is chopping the recorded sound into lots of tiny sound files. Then, a compression program compresses each file down to a smaller size. In our example, we use a compression technique that manages to squash any given sound file into one-eighth times smaller. For example, if we are recording in one-second chunks, each input file might be 24 KB in size. Our compression process takes these files and makes them only 3 KB each (which is 24 divided by 8). Our device continually takes input files, squashes them and then puts them into a queue in the mobile device's modem, ready for transmission.

What have we achieved with this compression process? Well, each compressed file now only needs one-eighth of the transmission time across the RF connection. If the mobile

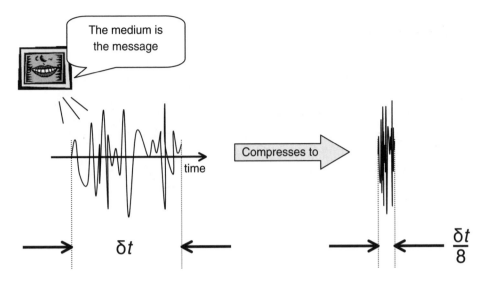

Figure 12.14 Speech compression means we take less time to transmit a given sample.

transmits each compressed sound file and then waits its turn while seven other mobiles transmit (in succession), by the time its transmission slot comes round again, it will be ready with the next file. As long as the recording and compression process keeps up with the arrival of each transmission slot, the whole process will work smoothly. Well, almost. What we need is a reverse process at the receiving end that can take the compressed files, uncompress them and then *seamlessly* stitch them back together into a continuous audio output stream. Using digital processing, this process is easy. After all, if dinosaurs can be made to appear in films[10], then voice files can be stitched back together again.

In summary, TDMA is the process of taking it in turns to transmit on the same frequency band, and this method is used for GSM and GPRS. The principles of TDMA have been established for a long while and are not the reserve of radio systems; the technique applies equally to wired (or even optical) transmission systems. TDMA allows digital information streams to share a common channel; the streams do not have to contain speech. There is also nothing magical about the eight slots in our example either; I chose eight, as this is the number of subchannels per frequency originally supported by GSM[11], which is probably the world's most prevalent and successful digital cellular system.

With GSM, a better voice compression technique emerged after the design, construction and delivery of live systems. Consequently, GSM evolved to support 16 mobile devices per channel (frequency), but had to stick with the original eight slots arrangement to maintain backwards compatibility. However, this was not a major problem. Simply by transmitting every other turn, two mobiles could share the same transmission slot (sometime called a *burst*) by taking turns. Effectively, that means a pair of devices alternately use the same slot,

---

[10] An altogether different proposition, I know, but once we can get information into the digital domain and apply enough processing power, we can do some incredible things.

[11] Why GSM uses eight bursts per channel is a deeper technical consideration to do with a variety of parameters.

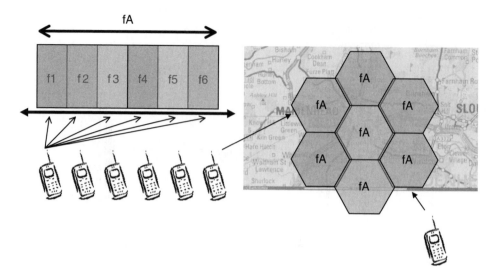

Figure 12.15 With CDMA, mobiles blast out on all frequencies.

but that's not a problem, so long as they remain synchronised and don't try transmitting at the same time.

12.2.4 Increasing Capacity using CDMA

With Code Division Multiple Access, or CDMA, things get a bit confusing very quickly, so we have to resort to a degree of poetic licence in explaining the essence of the technique, just so that you are aware that the following description is not a literal explanation of the technique. Having earlier said that we can't transmit on the same frequency, we've already discovered that TDMA does provide a way out. The benefit is that more mobiles can use a single frequency within a given cell, which is good for capacity and good for the users.

TDMA doesn't really allow mobiles to use the same frequency at the same time. Taking turns means they never transmit at the same time[12], thus avoiding the dreaded interference that we said was a fundamental limitation on RF communications. However, CDMA seems to defy these principles and violate common sense. With CDMA, the mobiles do indeed transmit at the same time on the same 'frequency' (or band).

The inverted commas around the word frequency give the clue about the reality of CDMA, so let's proceed with an explanation of how it works.

As Figure 12.15 shows, the CDMA set up is a strange one at first glance. Each mobile within a cell transmits across the entire range of available frequencies. There is no division of the spectrum into bands, like our radio station analogy. It is perhaps an odd thing to consider in the first place, that instead of tuning in to a particular frequency on the dial,

---

[12] TDMA mobiles have to be careful to ensure that they don't end up transmitting at the same time, so synchronisation is important. However, this isn't always easy. An added problem is that the signal from a distant mobile will be arrive late at the base station, with the chance that it can still be arriving when a nearby mobile transmits on the next slot. To avoid these problems, there is a guard time at the end of the slot during which mobiles should not transmit.

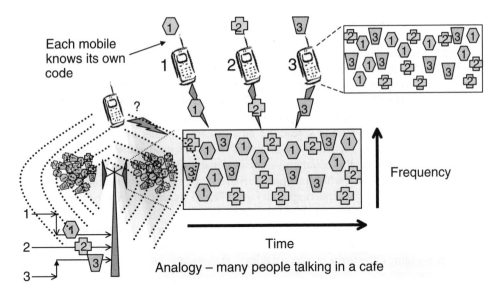

Each mobile knows its own code

Frequency

Time

Analogy – many people talking in a cafe

Figure 12.16 How CDMA works.

our radio station is now transmitting everywhere on the dial 'at the same time' (almost). This is a by-product of the transmission technique called *spreading* or *spread spectrum*, so-called because quite literally we spread the signal across the entire spectrum. This is not a book about RF communications theory, so we spare ourselves the mathematical details of spreading[13], but we can take a closer look to figure out what's going on without having to consider the maths. To reiterate, our explanation does have a hint of poetic licence, but it serves us well enough for our purposes in this book.

Figure 12.16 gives an indication of what is happening in a CDMA system. We have here three information streams going into the base station. The transmitter takes each stream and transmits it in an apparently scrambled fashion so that at any particular point in time the signal can be found transmitting somewhere[14] in the frequency spectrum. Returning to our radio set again, we can think of this as our radio station jumping around, apparently randomly, all over the dial on our radio. This analogy is to help us avoid the mathematical explanation, but allows an appreciation of the inner workings of CDMA. Each input stream is scrambled, so on the output of the base station we appear to have a complete mess, with everything mixed up. To an extent, this is true, but it's a bit like mixing lots of coloured beads in a bowl. The big picture is messy, but using the colours, we can extract each set again and recover from the chaos. Knowing the colour in this case is like knowing where to go on the radio dial in order to pick up the signal for that moment in time. To know where

[13] Ultimately, all communications theory is mathematical, but with CDMA in particular, a mathematical explanation is the only satisfactory way of explaining how it works because in a mathematical sense (or in the language of mathematics), there is nothing odd or counter-intuitive about the explanation. When we start thinking of what's going on physically, it become a little less clear.

[14] The amount of information in speech (or data) that we wish to transmit does not warrant a frequency band as wide as that used by CDMA, which is why we can think of the narrower and required amount of frequency as jumping around somewhere within the entire spectrum that gets used as a result of the spreading process.

to go on the dial, we use the spreading code that was used to scramble the signal in the first place.

With the scrambling of any particular input stream, the output becomes a pseudo-random mess. However, it is not truly random. In fact, if it were, then our signal would no longer contain useful information because complete randomness is meaningless; it conveys no information[15]. Deeply embedded in the apparent randomness is actually our information stream. It can be extracted using a special code to reverse the pseudo-random spreading, the same code that was used to spread the input stream. As long as our receiver has the right code, as shown in Figure 12.16, it is as if it can see its signal clearly amongst the mess made up from the other 'randomly' spread signals in the mix. A receiver without the right code just sees a mess, which is actually an advantage as it offers a degree of immunity to spying, similar to what ciphering achieves[16]. This is hardly surprising given that the historical context of CDMA was clandestine military use.

A popular analogy for CDMA is imagining lots of people in a room who are all talking different languages. Overall, the cacophony of the crowd is a noisy and unintelligible mess. However, standing close to the person speaking your language, suddenly everything is clear and makes sense.

The language analogy serves us well for elaborating on some of the challenges of running a CDMA network. For example, if there are too many people in the room, then the overall noise level goes up and we probably have to either move closer to the person we are listening to, or, ask them to talk louder. If people are unable to move, then talking louder is the only option. The downside to talking louder is that others nearby get drowned out and they too have to move closer. For those unable to move, the net effect is that the range of the speaker reduces and eventually we shall no longer be able to hear them clearly enough to understand what they are saying. In CDMA cells, this is the equivalent to the cell actually shrinking and those on the fringes being 'pushed' out. When the problem goes away, the cell size can grow again, a process called *cell breathing*, as shown in Figure 12.17. One solution is for an adjacent cell to take over the dialogue with the muted mobile.

The codes used to scramble the signals in CDMA have special properties. The result is that the signals spread in a way (*orthogonally*) that they are least likely to interfere with each other in practice, so that we can still extract them. As long as each mobile knows the code of the signal it wants to receive, then it can extract the wanted signal with only a tiny residual of interference from the other signals, due to their orthogonal properties. Using the café analogy, the more unintelligible the other languages, the less they can cause interference. However, some languages, like modern Punjabi for instance, utilise many English words. Therefore, in the case of an English conversation, a nearby Punjabi speaker would occasionally break through into the conversation because the English words would be detectable by the unintentional listener. However, a nearby Chinese speaker would not cause any such interference because the sounds of the Chinese words would be very different (orthogonal) from anything the English speaker is able to detect.

---

[15] Ironically, this is what we are trying to approach with CDMA. The more random we make our signal, the less it can be 'understood' by a receiver that doesn't know what to look for. No matter how hard we try, a receiver is always contaminated by unwanted signals entering its input. The less these contaminants are understood the better, as they can't interfere with the informational coherence of the wanted signal. Something that has no meaning can hardly interfere with something that does have meaning.

[16] Although we should be clear that spreading is not the same as ciphering.

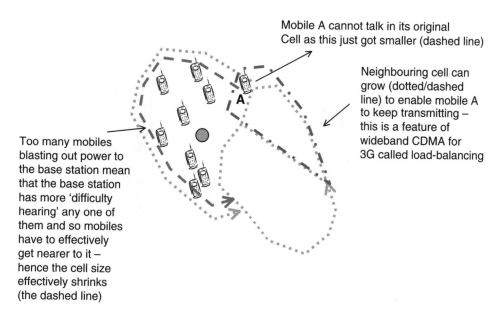

Mobile A cannot talk in its original Cell as this just got smaller (dashed line)

Neighbouring cell can grow (dotted/dashed line) to enable mobile A to keep transmitting – this is a feature of wideband CDMA for 3G called load-balancing

Too many mobiles blasting out power to the base station mean that the base station has more 'difficulty hearing' any one of them and so mobiles have to effectively get nearer to it – hence the cell size effectively shrinks (the dashed line)

Figure 12.17 Cell breathing in CDMA.

CDMA is the technique used for 3G systems like UMTS. Theoretically, it offers the highest possible capacity[17]. Its mathematical complexity requires a lot of computer processing power to extract the signal, but advances in silicon technology have made such techniques accessible to consumer products. Costs are kept low enough and power consumption is manageable within the capabilities of current portable battery technology[18].

CDMA is also the technique used for 802.11b, the very popular wireless replacement for Ethernet (local-area networking).

12.3 THE HARSHER REALITY OF CELLULAR SYSTEMS

Our brief tour of cellular systems has shown us that despite the problem of RF transmission being susceptible to interference, we can find different ways around the problem in order to gain enough capacity for lots of users to benefit within a realistic population density. However, unlike the entirely controllable and predictable world of using cables and wires, the wireless environment still offers a degree of hostility to our attempts to beam information to our roaming mobiles, especially from having to keep up with their every move.

---

[17] Compared with TDMA, but there are other theoretical multiple-access techniques that promise even better capacity than CDMA, such as Spatial Division Multiple Access (SDMA).
[18] There are always signs of breakthroughs in battery technology, like fuel cells, but otherwise battery technology has advanced the least out of all the electronics used in mobile devices, so we are still very much reliant upon improvements in silicon technology (Moore's Law).

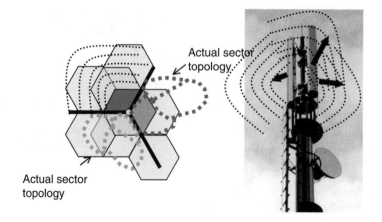

Figure 12.18 Cells are divided into sectors and the edges are fuzzy.

The reality of modern digital cellular communications is that we are constantly operating on the limit of the available implementation technology. Therefore, under certain conditions, wireless systems start to degrade. There are three main difficulties:

1. Unpredictable terrain

2. Unpredictable mobile location

3. Unpredictable traffic loads

A brief examination of these challenges will highlight the difficulties in meeting our original challenge for the RF network, which was to achieve transparency, or to make it as 'wire-like' as possible.

The simple problem with cellular systems is with the RF waves not reaching the target directly, either from the mobile to the base station (uplink) or vice versa (downlink). RF signals do have their limitations, bound by certain physical laws. Despite seeming counter-intuitive, RF signals can go round corners (diffraction) and can go through materials (refraction), like buildings. They can also bounce off certain materials (reflection). However, there are limits to this and the signals eventually end up dispersed and possibly unable to get to certain points. Powerful signal processing techniques are used to mitigate these effects in the margins, but nothing can be done about the laws of physics once the limits are reached, so RF systems do fail (as do wired systems if you push them to the limit[19]).

In an actual implementation of a cellular system, rather than the base station sitting in the middle of the cell, as we originally intimated, it often sits on the edge and beams out in several directions, as shown in Figure 12.18. In GSM, a typical cell site would utilise three directional antennae equally spaced at 120 degrees, each area of RF coverage called a *sector*. It is unlikely that the cell sites sit on a nice rectilinear grid suggested by the idealised hexagonal pattern. This is because cell coverage is related to where the users need service, following the patterns of population distribution, both static and migratory patterns.

---

[19] After all, Ethernet was stuck at 10 Mbps for years before the engineers figured out fancier transmission techniques (made possible by better electronics). However, there is still a point at which a signal going down a wire will no longer properly travel down it.

In terms of static populations, we would expect to find tightly packed clusters of cells in densely populated areas, particularly busy commercial zones, like the business districts of cities and so forth. In terms of migratory patterns, we also expect to find many cells along major commuter routes, both road and rail, and associated terminals, like railway stations and car service stations. Within the areas of coverage, the actual position of the cell sites is determined by a range of factors. Their idealised positions are first calculated using sophisticated cellular planning tools that attempt to arrive at optimal locations to achieve the best frequency (or code) reuse patterns, maximising capacity whilst minimising interference. The planning tools take into account the terrain of the land as much as possible. This is done using three-dimensional models, the accuracy of which is not always certain, particularly if it involves the modelling of buildings and so forth. However, the actual site locations will then be chosen according to restrictions as to where they can be built, initially governed by finding suitable physical locations in the first place. Site location is influenced by the cost and feasibility of getting backhaul communications lines to the sites in order to connect the base stations back into the core network of the cellular system (i.e. the cellular switch and so forth). Planning restrictions also affect site location.

The net result of the irregular distribution of sites, combined with the irregular terrain, is an irregular cell pattern, as also shown in Figure 12.18. In fact, the true shape of a cell is difficult to know without a detailed survey using sophisticated RF-measuring equipment and painstaking coverage detection, typically by driving around with equipment, hoping to find the extent of the coverage. Despite all the sophistication of the RF-planning tools, a degree of drive testing is usually required in a network, especially to fine-tune the coverage. In terms of understanding usage patterns that determine cell locations and density, this is well understood thanks to the experience of running 2G systems, although these patterns are mostly related to voice service, not data.

In terms of design challenges in a cellular network, the constant movement of mobiles is not a particularly nice problem to design a communications system to handle. This imposes various limitations. Firstly, given the cellular pattern that we need to handle capacity[20], roaming mobiles need to switch from one cell to another, possibly during an active session (voice call or data connection). This is not an easy situation to cope with, so it opens the door for possible performance disruption and even erroneous operation, such as connections not making it from one cell to the next (being *dropped*). Secondly, intuitively we can sense that there must be a lost opportunity in having to beam out an RF signal over a wide sector when the target is actually sitting in a very narrow part of the signal's beam. The signal that is spilling out everywhere else can only be 'wasted' signal, a source of interference for whatever else lies in its beam. This problem could be considered as the limiting factor to capacity within a given cell radius. The net effect is that we still will experience capacity problems when too many mobiles vie for the RF resource, even with the more gracefully degrading CDMA. Overcrowding of the cell is not always easy to plan for as it is related to unpredictable numbers of mobiles trying to make a call.

12.3.1 Data-Rate Variation

In understanding the actual limitations on the capacity of a cell, we start with a theoretical best performance that our RF network is designed to cope with at its peak and then work

---

[20] CDMA still has cells, but the mosaic is staggered around different code sets, not frequencies.

our way down from there. So using GSM slots to send packets of data rather than voice (i.e. *General Packet Radio System, GPRS*) we start with a theoretical best performance of using all eight slots[21] for one mobile on one frequency in a cell with ideal radio performance and we get a maximum data rate of 160 Kbps. However, a user will seldom get, if ever, this kind of performance. Firstly, the statistical likeliness of getting all eight slots is remote[22], never mind that most operators are highly unlikely to configure any system to use eight slots because they will always reserve a minimum number of slots for voice calls. Moreover, most mobiles cannot handle eight slots. Typically, they can handle four to receive data (downlink) and two to transmit (uplink). The reason for this is that we begin to defeat the processing advantage we gained with the slot mechanism. If a mobile only transmits, or receives, on only one slot per frame, then it has ample time to do all the processing before the next slot. If we now ask it to process twice, or four times as much, it begins to run out of steam using cost-effective processors. More challenging is the greater rate of energy consumption, worsened to the extent that an eight-slot system would most likely not have a very useful battery life.

If the RF conditions begin to get more hostile due to interference, poor signal strength or other artefacts then we have to use some (or more) of the available traffic capacity to carry redundant information that helps combat errors in the transmission. As we noted before, most data transmissions are extremely sensitive to errors – a single corrupted character in a bank statement can make a huge and unacceptable difference. To protect against errors, *error-control coding*[23] is used and this means that our theoretical 20 Kbps capacity on one timeslot can be reduced to 8 Kbps.

The vagaries of cellular RF networks inevitably lead to variable levels of data-rate performance, possibly suboptimal for some applications at least some of time. Therefore, our mobile service should be able to cope with the various levels of performance whilst staying within the bounds of an acceptable end-user experience (not forgetting our mobile services definition is qualified with words like 'successfully, confidently and easily').

Many of the performance optimisations we discussed when looking at WAP 1 and WAP 2 begin to make sense when we take on board these very real limitations in the RF network. Some scenarios are particularly challenging, like 3G operators that allow roaming to GRPS (or other 2.5G systems) wherever 3G coverage is poor or non-existent. Dropping from a 200 Kbps link speed to something like 16 Kbps is a dramatic discontinuity. How is the application going to cope? We need strategies to overcome these problems, or else we probably shouldn't be launching certain services at all. WAP 1 suffered this on some networks (though not all), where the RF performance overall was not able to carry the service with an acceptable level of performance.

We should expect variable data-rate performance and then design our application to cope with it. Figure 12.19 gives us a feel for what may occur in practice. We can think of the coverage area as being a constantly changing patchwork of data rates.

---

[21] With GPRS, the mobiles can try to use as many of the eight time slots as possible when transmitting a burst of data, such as sending an email. Mobiles are not limited to one time slot and can use up to the maximum assigned to GPRS, provided they are not being used by another device.
[22] This can be properly estimated using statistical estimation techniques and queuing theory, but we can safely use words like 'remote' in this context.
[23] There are four levels of stringency in the error-coding scheme for GRPS, called Class 1 to 4 (CS-1,CS-2, CS-3, CS-4), which give us 8, 12, 14.4 or 20 Kbps throughput per slot.

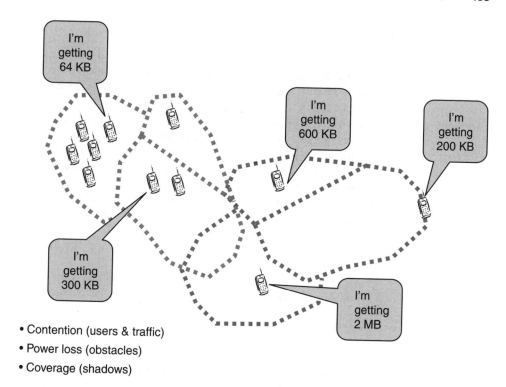

- Contention (users & traffic)
- Power loss (obstacles)
- Coverage (shadows)

Figure 12.19 Don't think that we get the 'advertised' speed everywhere.

The core services of cellular networks, such as voice, are already designed to cope with this variance in QoS. A voice compression algorithm like Adaptive Multirate (AMR) is able to adjust dynamically the level of compression according to the prevailing network conditions.

Compression techniques for audio and video signals are what we call *lossy*. This means that in order to achieve compression, one of the underlying principles is that some of the source information is not going to make much difference to the user's perception of quality if omitted, so it gets removed to achieve the compression. In an image, this is easy to conceptualise. The fine detail in a collection of trees is difficult for the eye to perceive. Similarly, detail in an image with *motion blur*[24] is also almost imperceptible. Converted into a digital data stream, this fine detail would require a lot of network capacity to transmit. Getting rid of some of it (the finer details) saves on capacity and makes little difference to the result in the eyes or ears of the recipient.

The advantage of a lossy compression technique is that more and more detail can be thrown out in order to achieve greater levels of compression, provided we are prepared to accept a reduction in quality consequently. Obviously, we can keep removing more detail and subsequently producing a coarser output, but eventually quality will suffer and can even reach a point of becoming intolerable. In the case of the AMR voice codec[25], voice

---

[24] Motion blur occurs whilst panning the camera across a scene, such as following a football in a soccer match. As the grass whizzes by on the screen, it is blurred and so detail gets lost anyhow.
[25] Codec stands for coder–decoder, which normally is taken to mean compressor/decompressor.

Table 12.1   Difference service levels according to data rate.

Data-rate available	Level of service for multimedia-based, location-based service
2 Mbps	Full-quality video clips for local cinema can be streamed to the user. Product advertisements can be viewed (e.g. for a nearby car under offer)
300 Kbps	The buyer can hold an online videoconference with a sales assistant specialist and view product videos before buying the product
56 Kbps	High-quality colour animated advertisements sent to user, possibly with some audio streaming available. Images could be accessed when looking at restaurants, attractions, etc.
40–600 variable	For user drill-downs into a tourist application, the system will attempt to return highest quality data (images, audio, etc.) and switch back to low-grade if not possible
28 Kbps	Ordinary base-line, text-driven and map-driven service with basic advertisements
SMS alerts	Push adverts; buddy notifications still get sent. User can make limited requests that are better fulfilled when full service becomes available again

reproduction quality is traded against capacity. If we have many mobiles vying for a limited resource in an overcrowded cell, we can back off the demand for capacity by simply lowering the voice quality on some, or all, of the mobiles.

This adaptation principle should also be present in our application design. If we are not used to such a notion, then at first it might be difficult to think about how to design for it. Perhaps for some applications, it is not necessary, but let's examine some possible scenarios.

Table 12.1 gives us an idea of the range of service levels for a particular application, depending on the available data rates. The transition from one data rate to another is not just a matter of service variance within a particular RF technology. With a multistandard device, the access method itself may change, such as a transition between GPRS/UMTS and WiFi; the potential contrast in speeds could not be starker (see Figure 12.20). It is feasible that a GPRS user could be accessing services on a train at a measly 40 Kbps, only to jump to 1.5 Mbps (or higher) on a WiFi access point in the station itself.

12.4   MOBILE BROADBAND NETWORKS

Since the design of the UMTS air interface was underway, it was clear that the promised 3G data rates of 2 Mbps would not be enough for data-hungry services like music downloading and media streaming. The challenge in a mobile network is that the resources are shared in each cell, so 2 Mbps is not what each user is likely to see in a cell with multiple users accessing data services. In other words, capacity is still an issue. This is even more so for operators wishing to offer flat-rate data tariffs, which is looking more and more likely as the only way to attract users onto data services like web browsing and media streaming. It has not been possible to shake user sensitivity to metered services. If anything, users have

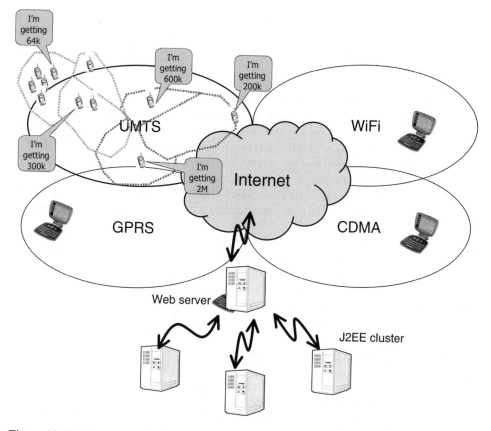

Figure 12.20 Data rate is also affected by service type. (Reproduced by permission of IXI Mobile.)

learned to adapt to a mobile life centred on texting and then letting rip with rich media services online when they get to their PCs. This limitation has led to the need for another access method beyond 3G, often referred to generically as *mobile broadband*, or 3.5G. At this stage, we shall look at two technologies: HSPA and WiMAX.

12.4.1 HSPA

High-Speed Packet Access, or HSPA, technology is the set of RF network technologies marking the migration path from UMTS. Most of the operators who have already deployed UMTS are proposing to migrate to HSPA technology. It comes in two flavours: HSDPA and HSUPA, where the 'D' is for Downlink and the 'U' for Uplink. The initial release of HSPA is HSDPA, promising a 5–10 times speed increase over current 3G deployments, thus brining it into the realms of fixed-line broadband rates (at their lower end). HSUPA is to be made available somewhere in late 2007 and offers similar improvements to the uplink speed. This is particularly important for services where users would like to upload their content – the so-called User-Generated Content (UGC) boom triggered by services like YouTube.

By the end of 2007, the expected data rate of HSDPA is somewhere in the region of an upper limit of 7.2 Mbps. HSUPA will be somewhere around an upper limit of 5.7 Mbps. Again, these are still aggregate speeds, representing the total bandwidth available for sharing in a single cell.

HSPA is a set of techniques on the air interface to allow a higher overall throughput for a number of concurrent data users. Of course, to start with there is simply a higher data rate, taking advantage of better silicon technology. But this is achieved by allowing a much wider set of modulation and error-coding schemes than previously supported. As we discussed earlier, sending an RF signal is a messy business. Interference on the radio channel tends to distort the signal and causes the data rate to be lowered. It is possible to adjust the modulation and coding scheme to take into account the radio conditions. This is not a new idea and has been possible since GPRS (2.5G). However, what's new with HSPA is the extremely fast rate at which the transmission scheme can be adjusted and modified on an ongoing basis.

The HSPA interface is divided into 2 millisecond blocks of transmission. The modulation and coding scheme can be switched from one block to another. This enables the most efficient adaptation to the channel conditions. For example, if the channel allows a 1 Mbps connection, then this is switched on during the block. If things gets worse, then for the next block it can be reduced to 500 Kbps, but immediately ramped up again the moment things improve. Moreover, with HSPA, the codes can be shared amongst the users, even to the extent that for a particular block, all the available codes could be assigned to a particular user in order to give them the entire throughput for the cell at that particular instant because good channel conditions apply. We can almost think of HSPA as a continually variable pipe that can be assigned to any user at any time and passed on rapidly enough to enable all users to maintain a useful throughput. The scheduling of the pipe is influenced not only by user demands, but also by the actual quality of the channel that each user could achieve. There is little use in assigning resources to a data-hungry user if they are in a very interference prone situation, unable to benefit from the resources allocated to them; the resources would simply be wasted. Using the *fast scheduling* capability of HSPA, resources can be rapidly assigned to another user who can better utilise them. This principle is shown in Figure 12.21.

HSPA requires new devices because of the new processing required to handle the various schemes used to improve the data rate. There are 12 different categories of handset defined

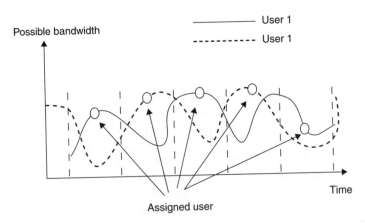

Figure 12.21 Dynamic scheduling of resources in HSPA network.

so that HSPA can be added to a range of handsets at different costs. For example, not all handsets might support the higher order modulation scheme, which is called 16QAM. Also, the number of code variations might also differ between handsets. The new result is the data rate that the handset can support in both directions. Early devices in the market supported category 12, which meant a downlink rate of 1.8 Mbps. Data cards were then introduced with category 6 support, meaning 3.6 Mbps. Category 8 devices coming onto the market in 2007 are able to support 7.2 Mbps. The theoretical maximum data rate is category 10 device, which can operate at 14.4 Mbps in the downlink.

HSPA is an exciting development in the mobile RF networks story. It looks set to dominate the mobile broadband standards globally because it is an extension of UMTS, which is already the most widely deployed mobile standard in the world. It is anticipated that by about 2012 there will be a 70% penetration of mobile subscribers using broadband devices and services. Other standards, like EV-DO, will be popular in certain regions, most notably the Americas where this standard follows their region-specific version of wide-band CDMA services similar to, but not based upon, UMTS.

12.4.2 WiMAX

There's a lot of press coverage about WiMAX, so it seemed useful to include a brief discussion in this book, at least to enable readers to position the technology in our mobile ecosystem. WiMAX, or 802.16x, was *not* developed or intended as a mobile technology. Its origins are in finding wireless solutions for what telecoms companies called 'the last mile' connection. This is the final connection from the local telecoms exchange to the user's premises, such as houses and offices. In many deployment scenarios, especially the developing world, finding a wireless technology for the last mile is important because it enables cost-effective and rapid deployment. There is no need to dig roads and lay cables, all of which is very expensive. Simply install a WiMAX base station in an area and then ship (or install) WiMAX access points to the customer, in to which they plug a standard phone. In fact, WiMAX, because of its potentially high data rates (40 Mbps) can provide connectivity for data in addition to voice, so customers could buy telephony and internet services and access these via the same physical box.

The wide-area application of WiMAX is what makes it a lot different to WiFi, or 802.11x. It is not really an extension of WiFi. It is a different solution to a *different* problem. Not only is WiMAX a metropolitan-wide solution (3–10 km), as opposed to local area (few hundred metres), but it has a number of attributes that make it suited to the robust delivery of telecoms services. Two key attributes are QoS mechanisms and better security. QoS is vital to the delivery of telecoms services. With WiFi, the resources in an access area, such as a coffee shop or hotel lobby, are shared indiscriminately between the users without regard to their services and needs. For example, one user might be using an internet telephony service (e.g. Skype) and suddenly get a very poor quality service when a nearby user starts a video stream from a website (e.g. YouTube).

Despite all its excellent capabilities as a wireless access technology, WiMAX was not conceived for mobile applications, which is the focus of this book. WiMAX was designed for static devices. As we have discussed earlier in this chapter, the mobile environment is very harsh for RF signals. Moving devices are difficult to cater for in terms of getting a clean signal to and from the device. However, perhaps inevitably, the WiMAX standards group has created a version of the standard that allows it to work in a mobile fashion. It is easy to understand the logic behind such a development. Firstly, wherever WiMAX is deployed for blanket telecoms coverage in an area, it makes sense to attempt to use the same infrastructure to offer mobile services to the same (or new) customers. Also, the licensing and industry structure behind WiMAX is such that competition is possible with other broadband services, notably the mobile operators already committed to UMTS with its HSPA derivatives. In other words, for various service providers and vendors, WiMAX is an opportunity to enter into a market that was hitherto closed to them because mobile operators being supplied by well-entrenched vendors of mobile equipment and infrastructure already occupied it.

The various flavours of WiMAX are referred to by their standards naming convention, which is 802.16x, where the 'x' denotes a variant. IEEE 802.16a focused on fixed broadband access, as discussed above. IEEE 802.16-2004 enhanced the standard by providing support for indoor coverage. The IEEE 802.16e standard (also known as IEEE 802.16e-2005) is an extension to 802.16-2004 that adds data mobility. However, it is possible that the evolution of this standard in terms of deployment will see the mobile version more focused towards mobile computing devices, such as laptops and PDAs, rather than consumer mobile phones. 802.16e could also be used for other types of data-intensive device, like Mobile TV units and the so-called Personal Media Players (PMPs) such as the iPod Touch. This device incorporates a WiFi modem so that music files can be downloaded directly to the device via a WiFi access point, whether in the home or in a public place. However, such a device is still tethered to access point availability and coverage. A WiMAX, or HSPA, enabled device could allow music downloads to take place anywhere and at convenient speeds that would facilitate a compelling user experience. If we compare the download times for a 4-MB music file using different standards, then we can see that only mobile or wireless *broadband* technologies can really support such services:

Service	Typical data rate	Download time (4 MB music file)
GPRS	50 Kbps	$\sim$10 mins
EDGE	100 Kbps	$\sim$5 mins
UMTS 3G	256 Kbps	$\sim$2 mins
HSPA (Cat 12)	500 Kbps	$\sim$1 mins
HSPA (Cat 10)	4 Mbps	8s
HSPA (Cat 15)	10 Mbps	3.2s
802.16e	20 Mbps	1.6s

12.5 TECHNIQUES FOR ADAPTATION

We may decide to design our application to allow different levels of service depending on the data rates available to our devices. There are two approaches to the problem:

1. *Manual approach* – ask the user to make choices

2. *Automatic approach* – use network information to guide choices about content type and formats

The manual approach can include:

- Giving the user a choice to view different versions of the site ('low bandwith' and 'high bandwidth' version)

- Designing the site in a way that data-intensive information is separate from the main flow and can be optionally pulled in by the user (e.g. not showing images and asking the user to 'reveal' those images)

The manual approach entails the user making choices about what they want to see and then having to specify their preferences. The problem with this is that the user may be intolerant of such approaches, especially because offering more than one choice may often seem counter-intuitive. A user may get confused as to why they are presented with such options.

The manual approach can also end up masking the fact that data rates change, keeping in mind that the rate of change compared to the typical length of a session might be noticeable. Therefore, a user may start a session with a particularly slow link and elect to view the low bandwidth. The link speed may subsequently improve and the user is still using the current view, needlessly foregoing the benefits of the richer content. An automatic approach would be better (which is not to say that a manual override should not be present).

Designing the site to be tolerant of variable bandwidth access is clearly a good idea, although it possibly involves extra work to formulate a suitable design. However, the mechanism for handling more than one version of the content has already been discussed when we looked at the J2EE presentation layer and the general issue of device-based adaptation, so this scheme would nicely accommodate a solution to the rate adaptation problem.

An example of such an approach is to replace pictures in a page with links, or low bandwidth spacers[26]. On some pages, pictures can easily constitute a large proportion of the overall data size, so treating them in this way makes sense. Some browsers give the user the option not to display images. This means that the user agent does not go back to the web server to fetch any images. These two approaches differ and can produce different results. The main limitation of allowing a browser to exclude image fetches is that we cannot rely upon it as a solution. Not all browsers support such a mode; therefore, we need to implement the solution at the origin server, not the browser. Furthermore, browser-based omission of pictures can produce undesirable results, such as adversely affecting the page layout. For example, if the browser does not know the size of an image[27], then we will not know how to layout the page to allow the rest of the content to flow around the image space that remains. However, if we purposely built a page without the image, then the layout could be fashioned accordingly.

Whatever the design approach used to adjust the visual appearance of pages according to available data rates, the use of an automatic means to control the content stream is preferable. However, the issue remains to identify a suitable mechanism for this. Fortunately, there is an ideal opportunity available using the UAProf mechanism discussed earlier in the book

---

[26] A spacer is some replacement image with low resolution, or possibly just blank, that can be clicked on to reveal (fetch) the actual image.
[27] There are techniques to find out the image size. Ideally, it is included as a parameter within the mark-up itself, so the browser has the information. Failing that, the browser can attempt to fetch the image, but using HTTP *byte ranges* to fetch only a limited amount of the image. In some file formats, like PNG and GIF, the image size information is embedded at the start of the file.

when we looked at how to very content according to fixed-device capabilities such as screen size (refer to Chapter 9). In the UAProf file included in the browser's HTTP request header, it is possible to insert semantic information about the connection speed. This is an ideal solution because it enables a potentially high degree of control because each response from the server can adjust its output according to each request, as opposed to a blanket control across the entire session (i.e. opting to view the 'low bandwidth' version for the current session).

The challenge is how to insert the required semantic information, and what to insert. What we need is some kind of estimation of the bandwidth available. It is relatively easy for a mobile to estimate the downlink bandwidth to the device. This is available from the device itself by consulting the control software on the modem chipset (see Chapter 10). The browser is also able to estimate bandwidth itself by timing how long HTTP- or WSP-response streams take to arrive at the device. Dividing the stream size (page size, image size, etc.) by the time, we get the estimated bandwidth. If the browser estimates bandwidth, the measurement will include all network delays, which is a more accurate reflection of the user experience. If it takes five seconds before an HTTP request generates a response, then this delay should factor in the measurement in order to calculate the *effective bandwidth*. For example, a 100-kilobyte page that takes eight seconds to stream, plus five seconds delay, gives 100 divided by 13 kilobytes per second, which is 61.5 Kbps[28].

As discussed on numerous occasions throughout the book, usability is affected by delays. Generally, users do not have enough patience to wait around for a web document with long delay. One may expect to receive a web document in a reasonable waiting time t, say 10 seconds. In our adaptive system, the browser reports a bandwidth b to the server, say $b = 10$ kilobytes per second. Then the web server will use content-adaptation to send the web document with size not greater than $b \times t$, i.e. 100 KB in this case.

The browser can also take into account its current downloading activities when making a new request. For example, the user could be in the middle of downloading a large MP3 file to listen to a new music track. A browser request made during the download interval should take into account that bandwidth is already being used by the device, which possibly means that extra bandwidth allocated to the new request will be less than expected. For example, if the device is operating in a cell where it could sustain 120 Kbps, the MP3 download might use most of this bandwidth. Therefore, if 120 Kbps is the bandwidth estimate for a concurrent browser request, this might result in the server producing an adapted file that is much too large for the bandwidth that the browser assigns to downloading the page (assuming that the MP3 download is not to be affected).

The strategy to allow the browser to estimate available link speed seems a useful one. However, it does have some potential flaws. Firstly, if there is a significant delay between the last HTTP response stream and the current request, the network conditions may have changed compared with the last response. Perhaps there is more congestion and consequently the network is much slower, in which case the browser might be over-optimistic in its measurement report and consequently the content adaptation will be non-optimal. The other potential flaw is that the browser is not the only application on the device. Other applications are potentially vying for bandwidth. For example, video streaming or audio streaming applications may work independently of the browser. If the activities of such

---

[28] Remembering that bandwidth is measure in bits, not bytes (i.e. Kbps means kilobits per second).

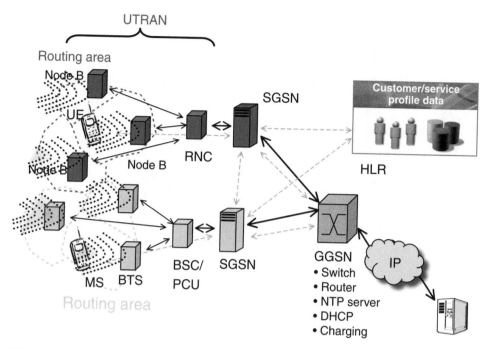

Figure 12.22 Simplified network architecture (GPRS/UMTS).

applications do not figure in the link speed estimate, then the browser report may not be realistic.

A solution to the problem of link speed estimation is to use transport-layer estimation techniques that take into account all the available resources end-to-end in real time. To achieve this, equipment in the RF network should be accessible to provide the required estimates. There are probably a number of strategies to achieve this, but the general principle is to ask the RF network itself what resources are available at a particular moment for streaming information to a particular mobile device in a particular cell. As Figure 12.22 shows, there are various network elements used to build a cellular network. We shall soon explain the elements in more depth, but for now observe that each cell is served by its own base station, and a group of base stations are concentrated into a controlling element. The controlling element, such as the Radio Network Controller (RNC) in a UMTS network, or the Base Station Controller (BSC) in a GSM/GPRS network are aware of how much RF network resource is available in each cell under their management. It is their job to keep track of the devices and allocate resources. To put it crudely, if a particular cell has a total capacity of X Mbps and our RNC is aware that there are N devices using Y Mbps between them at a particular moment in time, then to stream a response to our $(N + 1)$th mobile, the RNC could potentially report that we have $X - Y$ Mbps of capacity available. This is a somewhat simplified suggestion, but ultimately the potential to do something like it does exist within the network.

At the device end, instead of asking the browser to report measurements (which it still might do, as this gives us extra information on which to base decisions), we can involve the control software in the RF modem, as it has the best view of the RF link characteristics. It

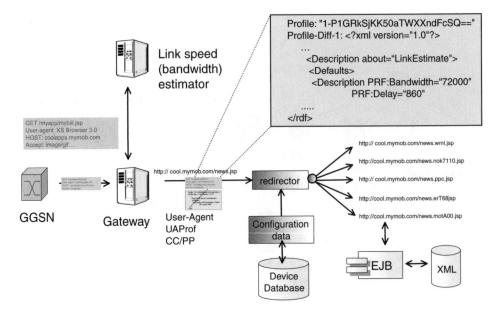

Figure 12.23 Inserting link-quality information via a gateway using UAProf.

will know how much data is passing through the modem at any one time, and it will know about the link quality measurements (e.g. power, signal quality, etc.) that it is required to make as part of its operation in the network.

The best thing about using UAProf as our means to convey the link estimate is that any entity in the network can add profile headers (*profile-diff*), not necessarily the user agent itself, although with UAProf, we have the potential to combine both. In terms of network-generated UAProf information, a gateway or proxy sitting somewhere between the device and our origin server could add the UAProf to the requests, as shown in Figure 12.23. This is very convenient, as it means that we have a way of inserting link capability information without requiring any special protocols or modified origin server solutions.

Having attained our link estimate, we can calculate how much data we should attempt to stream in the response. In deciding on the optimal page size to suit available bandwidth, it is debatable whether the size we are interested in should be the size of the document in question, or the aggregate size, which means the document plus all the resources it refers to and that also need to be fetched from the server. Recalling our discussion of HTTP 1.1 (including WSP and W-HTTP) the initial document can be streamed in and displayed before or whilst the remaining content is being streamed, not forgetting that such a process is pipelined (overlapping). Of course, this is a somewhat finer point that doesn't need much deliberation. Using an aggregate will suffice and any fine tuning can be applied in light of operational experience.

What are the content-sizing strategies that we can apply to take advantage of dynamic adjustment? We have already considered somewhat dramatic options of text-only sites, which could be the lowest common denominator. But how might the content be adapted otherwise? In addition to image removal, we can consider image resizing and using different levels of compression, especially for JPEG images, where the compression technique is

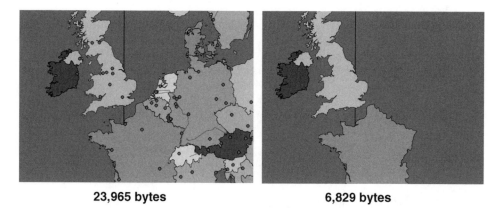

23,965 bytes   **6,829 bytes**

Figure 12.24 SVG files with different level of details.

lossy, which as we discussed means that we can sacrifice image quality for data size. In future, more images will be constructed using vector drawing techniques, using standards like Scalable Vector Graphics (SVG). Complex drawings can be quite verbose in their description, so we could consider dynamically adjusting the amount of detail. For example, in mapping application, if there is lots of bandwidth then we can send a very rich schematic with detailed map features, such as boundaries, rivers, roads, etc. If bandwidth is restricted, then we can pare down the schematic to reveal only the major tributaries and so on. An example of this is shown in Figure 12.24, where the area of interest (United Kingdom) is shown in the map with and without the neighbouring countries. By not showing the neighbouring countries and also removing the cities, we can show how the amount of data to be streamed is reduced.

12.6 CELLULAR NETWORK OPERATION

In the previous discussion, we introduced the simplified network architecture of a UMTS or GPRS cellular network, as shown in Figure 12.22. We are concerned in this section with the basic network architecture, particularly how we interface the IP network with the RF network, so that we can begin to see how we shall connect our J2EE content world with the device network.

Elaborating upon the architecture (see Figure 12.22), we already discussed that in cellular systems the cells are supported by base stations, which are the sources of RF connectivity in each cell (or sector). In a GSM/GPRS network, the name given to the base station is *Base Transceiver Station*, or BTS. In UMTS, the equivalent unit is a *Node-B*. These units actually contain the RF transmission systems and the required RF antennae to enable the RF waves to propagate in the required sector pattern. Some antennae disguise themselves as trees, or other objects, to avoid unpleasant eyesores on the landscape, such as the 'Scots Pine' shown in Figure 12.25.

Base stations are housed in equipment cabinets that can accommodate several radios, one per cell (sector). The number of radios depends on the nature of the site, which come in a variety of sizes, but are loosely categorised into macro-cells (up to 20 km radius for

Figure 12.25 Cellular antenna disguised as tree.

GSM), micro-cells (a little as a few hundred metres) and pico-cells (usually limited to indoor coverage, such as the floor of a large building). Installing smaller and smaller cells is the main way of achieving higher traffic capacity. This is the origin of the microcellular concept – small cells to fill 'hot spots' in dense urban areas, such as central business districts, main road intersects and so forth. Pico-cells extended the same concept indoors[29] (where other services are also feasible, such as a cellular Private Branch Exchange, or PBX[30]). In a cellular network, all the cells are uniquely identifiable by their cell ID, which is an address that the network uses to keep track of where cells are in the network.

Base stations transmit and receive the RF signals on one side (antenna) and connect back to the network via wire-line signals on the other. A base station needs to connect the information stream to the core of the RF network (e.g. switch), so a data connection is required, which is usually a leased line. Occasionally, it is difficult to route a suitable wire-line to the station, so microwave wireless connections (or short-range optical equivalents) are sometimes used. The signals from the base stations are concentrated into a controlling element higher up the network chain towards the eventual boundary with the outside (IP) world. In GPRS systems, the BTS connects back into a BSC. A BSC can handle as many base stations as it is designed to handle[31]. Similarly, in a UMTS network, the RNC manages a multitude of Node-Bs, the exact number being manufacturer dependent.

We are mostly concerned with the packet data support in the network, not voice, so we shall not dwell on the voice communications features. In both 2.5G and 3G networks (e.g. GPRS and UMTS), the rest of the network, behind the RF front-end elements, is very similar and is called the *core network*. The RF component in GPRS is collectively called the *Base station Subsystem* (BSS), whilst the RF component in the UMTS network is called the *Universal Terrestrial Radio Access Network*, or UTRAN.

In the voice network (*circuit-switched domain*), the signals work their way up the pyramid of network elements towards the public telephone network via a switch (*Mobile Switching Centre*, or MSC), in the data network (*packet-switched domain*), the data packets work their

---

[29] It should be noted that RF from cellular systems does propagate into buildings in any case, but here we are talking about the very specific idea of increasing capacity inside of a building by installing a dedicated base station.
[30] Cellular PABX (Private Automatic Branch Exchange) is where mobile phones can also act as internal phone extensions, allowing such features as short-code dialling and extension-to-extension calls free of charge.
[31] Meaning that the number of somewhat arbitrary and product dependent – there is no specified number.

way from the Node-Bs towards the IP interface of the network support via the *Gateway GPRS Service Node* (GGSN).

Beneath the GGSN sit the *Serving GRPS Service Nodes* (SGSN). These take control of the packet data support for a group of cells, each group known as a *routing area*. In terms of handling data coming into the network from the IP world, like perhaps a WAP-Push message, the underlying IP protocols only know IP addressing, they do not know about cell IDs or mobile IDs, or any other type of cellular routing information. The SGSN and the GGSN provide the necessary routing of IP packets to the correct cells where the target devices reside, and carry packets back again to the GGSN, shuttling between the internally routable packets to the externally routable IP packets with their IP addresses. Part of the shuttling involves dynamic routing due to the constant movement of the mobiles, thus requiring *mobility management*. The SGSN and GGSN also take care of various security features in the network, including encrypting data exchanged between these network entities.

12.6.1 Getting Data In and Out

The GGSN is the key network element as far as our IP layer is concerned. As its name suggests, it is the gateway to the RF network from the IP network. It is where IP data enters and leaves the RF network, going to and or coming from the mobile devices. It is where data physically passes from our RF network into the IP network, usually to a LAN run by the operator on which sits the GGSN as an IP-addressable entity. Internally, the cellular infrastructure itself is not IP-based. It has its own set of internal protocols optimised for the support of mobile devices in an RF environment, taking into account such requirements as security and mobility-aware data routing. We can think of the GGSN providing an IP overlay onto the cellular network, presenting the outside world with an 'IP-view' of the cellular network, such that devices have IP addresses and can receive and transmit IP traffic.

Such are the independent habits of users that they are not all using the infrastructure at the same time. Usage has a statistical flow that largely reflects the fact that humans and their patterns of behaviour (business and social habits) are erratic, or unpredictable. A user might access their email, check a few messages and then do nothing for a few hours. During that time, perhaps a few WAP-Push messages arrive. Meanwhile, another user is busy streaming MP3 files all day, listening to whatever takes their fancy. The consequence of diverse user habits is that different users require different network resources at different times. Therefore, our core network works in a way to assign resources on an as needed basis in order to achieve an efficient sharing of what is ultimately a limited resource (the RF spectrum itself). Setting up data connections for devices in each cell is part of the *connection management* role of the core network. We can think of this as allocation of physical resource – actual packets on the RF interface, assigning codes in the case of CDMA, or assigning slots in the case of TDMA.

The assignment of services at the RF level depends on the type of service requested by the user. For streaming applications that require a repetitive flow of packets with a fixed delay, the network takes a certain approach to assign RF resources, one that is suited to real-time services, such as videoconferencing. For non real-time applications, like the flow of packets in an HTTP request, the core network takes another approach to assignment of RF resources. These activities are all part of connection management.

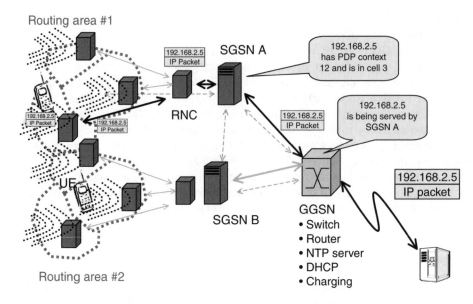

Figure 12.26 Routing packets via session and connection management.

Session management is the assignment of complete packet-switched circuits once the connection management has taken place. A critical aspect of session management is giving a mobile device an *active PDP context*, where PDP stands for Packet Data Protocol. We can think of this as the cellular equivalent of activating an IP device by giving it an IP address. Packets through the core network flow as *Packet Data Units* (PDUs), for which we need an active PDP context in order for our mobile to exchange PDUs. The user requests a PDP context, although their device software makes the request implicitly whenever an application on the device requires an IP connection, such as the WAP browser. We can think of the mobile requesting a PDP context as bringing it 'to life' on the cellular data network, ready for IP connectivity. As soon as the device has a PDP address, it is ready to send and receive IP data. Its routing and location information are known to the GGSN and relevant SGSN, which both collaborate to get data to and from the appropriate devices according to IP address allocation, as shown in Figure 12.26.

This has been a very simplified discussion of the cellular network, but the detail is beyond the scope of this book. We have omitted other features of the cellular network that lay in its *service domain*, and we shall come back to some of these, but otherwise we need to focus more on how we connect our IP network to the RF one. Throughout the book, we have been examined IP-based protocols in some depth, such as HTTP and WSP. We have assumed that the devices have IP addresses and are contactable via IP. We have seen how the cellular network can support IP-addressable devices that roam whilst maintaining a communications pathway to and from the IP network. We will now look in more detail at the interface between the IP network and the RF network.

12.6.2 Gateway GPRS Service Node

The GGSN is clearly an important node in our cellular network because it controls the IP view of the network. This book has dealt mainly with IP-related service architectures,

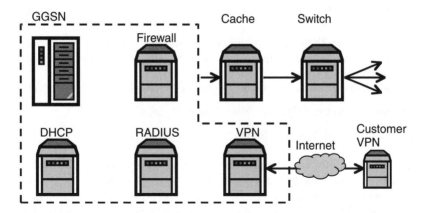

Figure 12.27 GGSN architecture.

therefore, out of all the RF network nodes, it seems most appropriate to understand the GGSN in more depth, especially in terms of how it might affect our considerations for application development.

Figure 12.27 shows a possible architecture for a GGSN. To an extent, we can think of a GGSN as a kind 'all-in-one' IP system, providing many of the crucial IP networking services that we need for an IP network to function, such as:

- Routing

- IP address allocation

- Authentication

- Firewall

- Caching

- IP switching

- Virtual Private Networking (VPN) support

The GGSN is essentially like a router in an ordinary IP network. Routing is concerned with taking IP packets and forwarding them from one router to another using the most efficient (or available) path to get the packets to their final destination. All kinds of criteria exist for routing, such as minimising congestion, minimising delay, and so on. However, in the case of the GGSN, the routing tables will not have any of these criteria when constructing input (IP address) versus output (mobile address) mappings, or vice versa. The routing algorithm is altogether more straightforward, it being simply to route the packet to whichever SGSN is serving the device with the destination IP address on the packet; in other words, it is 'location-based' routing[32]. Clearly, the SGSN and GGSN need an internal protocol to manage the routing information. The GGSN needs to know where to send packets to,

---

[32] Once could argue that ordinary IP routing is also location-based. After all, that is the whole point of routing – to send packets to a device *located* somewhere on the Net. However, unlike with the Internet, the core network does not possess a multitude of potential routes to forward packets, so the performance criteria that routers are designed for simply do not apply.

bearing in mind that devices move, with the distinct possibility that they may move from one SGSN to another.

In order to route an IP packet to a mobile, we first need to have an IP address assigned to a device. There are several strategies for doing this, all under the control of the GGSN:

- Assign an address from a configurable pool of addresses in the GGSN database

- Use the internet protocol Dynamic Host Control Protocol (DHCP)

- Use an address taken from the Home Location Register (HLR)

Within these methods, there are degrees of flexibility. For example, the IP-address pool in the GGSN configuration database can be set to any range of addresses. Different pools may exist that contain IP addresses reserved for different users. This allows segmentation of the user base into different IP address pools, something that might be useful for a variety of reasons. The allocation from the pool can be dynamic, based on a cyclical first-in first-out allocation of addresses, or static, if a fixed IP address is required on a per mobile basis. Static IP addresses can also be stored in the user's customer record found in the network's master directory, known as the Home Location Register, or HLR, and already holds other user information.

When using the internet protocol DHCP to assign IP addresses, the DHCP server can be internal to the GGSN, or externally hosted as part of the operator's local IP-network infrastructure. Alternatively, any DHCP server can supply IP addresses, on a user-by-user basis. Perhaps an Internet Service Provider (ISP) has its own IP requirements for users accessing its servers using the RF network, in which case the ISP's DHCP server can allocate its own IP addresses.

Regardless of the IP address allocation method, the management of IP addresses by the operator can take advantage of *Network Address Translation* (NAT), if required. This is another feature of the GGSN. For accessing the operator's services on internally hosted server farms, the operator will assign internally routable IP addresses from one of the private address ranges specified in RFC 1918[33]. However, these addresses do not route across the Internet[34], so when a device requires access to an externally hosted service, then the GGSN translates the packets to another IP address taken from a pool of public addresses assigned to the operator.

Using the VPN feature of the GGSN, it is possible to route traffic securely across the Internet to another network that is also using a VPN, a process called *secure tunnelling*. Most GGSNs support a variety of VPN configurations. This allows the support of secure services, such as mobile access to enterprise networks, including intranet access, CRM access and email. The core network also supports encryption internally; node to node, as well as across the RF interfaces, therefore securing the entire data path from the device to the enterprise, as shown in Figure 12.28. However, it is not true to say that this is end-to-end encryption unless the application itself supports encryption on the device and the application server, such as HTTP or W-HTTP using SSL, as discussed when we looked at IP protocols earlier in Chapter 7.

---

[33] http://www.faqs.org/rfcs/rfc1918.html
[34] Because they are allocated for internal usage only, routers are supposed to be programmed not to route these addresses onto the public internet.

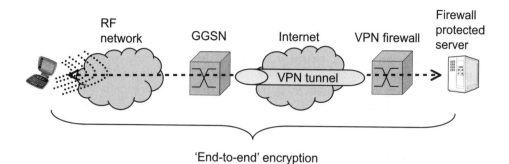

'End-to-end' encryption

Figure 12.28 VPN enables end-to-end encryption across the Internet. (Reproduced by permission of IXI Mobile.)

The GGSN can also support RADIUS[35] authentication. This has nothing to do with authenticating users onto the RF network or onto any downstream application, such as the basic or digest authentication schemes we discussed earlier in Chapter 9 when looking at how to secure J2EE applications. The RADIUS protocol and the support for it at the GGSN allows authentication onto another network, such as an IP network used for hosting applications, or a separate IP network, like one run by an ISP who might want to make sure that only its customers are accessing its resources. RADIUS is currently the de facto standard for remote authentication. It is very prevalent in both new and legacy systems, so its support at the GGSN is a useful provision.

In addition to these essential IP networking functions, some GGSN products provide other integrated functions like HTTP caching, which can be useful for accessing web pages. We have discussed earlier in the book how caching in the browser can provide improved performance when accessing web pages that have not changed since last viewed. This function can also be available by proxy in the network. Network-side caching in itself does not really improve performance that much for the end-user in terms of faster page loading across the RF connection (which was the motivation for browser-based caching), but is more useful for limiting the amount of web traffic coming into the core network, warding off congestion and so forth.

Other functions, such as IP switching and firewall protection are housekeeping functions, albeit important ones that may still have impact on our applications, as we shall now discuss in the sections that follow.

The core network is a tightly controlled network. Effectively, the operator of the network owns the connections to the mobile devices and is fully at liberty to manage the connections how they see fit. Sometimes, this might include apparently punitive network policies. For example, some operators have been known to limit the flow of IP traffic only to internally hosted servers, thus 'garden walling' their network, preventing others from utilising it for their own external applications. Other operators may provide access to external networks, but in a restrictive manner. For example, it may only be possible to access external WAP sites via the operator's WAP gateway. This has the effect of preventing certain applications operating properly that wish to use HTTP for non-web applications, which is not such a

---

[35] RADIUS stands for Remote Authentication Dial In User Service. RADIUS is a protocol specified by the IETF working group.

strange idea. The firewall may block inbound IP connections on certain port numbers, which again restricts flexibility in application design. Rightly or wrongly, many IP restrictions might apply in a cellular network. The implication for developing mobile services is to ensure that the operator can support the proposed design strategy, ensuring that no IP restrictions will adversely affect the service offering. The operator should provide details about their network configuration. Many operators have developer forums that developers can join and subsequently gain access to network configuration information.

12.7 ACCESSING NETWORK ASSETS

In addition to the RF network's support for packet communications, there is a range of other services that a cellular network can offer. Depending on the type of application we want to develop, some of these services could be essential and it would therefore be a good idea to gain programmatic access in order to allow integration of these services into the application under development.

The types of asset that an operator has are:

- Voicemail services

- Call-control (e.g. call forwarding, call conferencing, etc.)

- Charging and billing mechanisms

- Bi-directional text messaging, including sophisticated mechanisms for micro-charging, like reverse-billing (i.e. paying for receiving messages)

- Multimedia messaging systems

- WAP gateways

- WAP Push gateways

- Fine-grain IP management of mobile devices (e.g. GGSN capabilities)

- VPN and other security mechanisms (courtesy of GGSN again)

- Interactive voice response systems

- Location-finding infrastructure to locate mobiles

- Numbering management schemes (e.g. short codes, alias code, multiple-device numbering, etc.)

Firstly, we should examine the problem of exploiting these assets from the perspective of the operator, since they are the one who owns the network and has to find a way of making its assets available to third-party developers and other interested parties.

Figure 12.29 shows us a possible context of the mobile network operator's business. As the figure shows, the mobile operator view is one that places their network at the centre of the mobile cosmos. All other networks are tributaries to their network and business. The operator probably views other entities firstly in terms of commercial relationships, not technical ones, though clearly some technical interfacing has to occur, which is the problem we are going to look at.

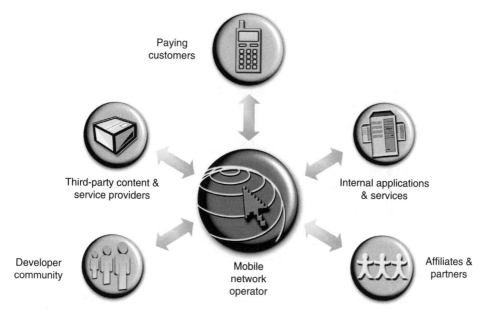

Figure 12.29 Operator context of supporting services and applications.

The primary external network is the network of customers. These are the paying customers. This is an obvious point, but its nakedness is worth examining for a moment. Historically, the development of mobile voice networks relied on customers paying for access to services. No aspect of the operators' offerings has ever been free of charge, thus setting the precedent that mobile services cost money. This is a key consideration for wanting to gain access to an operator's customer base: access to paying customers. Furthermore, the mechanisms for charging, billing and collecting dues are already in place in an operator's network. A partnership with the operator based on offering them a means to increase the *average revenue per user* (ARPU) might generate cash if the operator is willing to yield a share of the revenue to the partner. This is why the new world view of the operators includes all the relationships shown in Figure 12.29.

It is regarding this world view that we shall proceed to discuss how we exploit the technical possibilities and opportunities covered in this book within a revenue-share model. Let's recap the challenges facing the operator, in order that we have a context for proposing technical solutions.

Here is a list of some of the challenges facing the operator:

- *Give users something interesting to use and pay*
  - Not necessarily what they 'want' as they don't know what it is they want yet! Creativity is required!
  - Compelling enough to generate revenue

- *Balance service delivery and subscriber demand with capital expenditure on the service delivery platform*

- *Cope with uncertainty in the next generation*
  - Content and service types
  - Pricing models
  - Consumption patterns

- *Providing a rich 'open' end-to-end service package*
  - Building a third-party service platform that enables any application to be launched from the network meeting both the operator needs and the third party's.
  - Embrace a technological approach that the developer community will welcome
  - Reach the 'event horizon' of a leading IT company, not a utility company

- *Entering into 'networked business' models with:*
  - Affiliates
  - Third-party service providers
  - And even competitors

We first mentioned these challenges in Chapter 4, so we shall now proceed to elaborate on that discussion.

Uncertainty seems an appropriate theme for the current discussion. As the first point elucidates, the users probably *do not know* what it is they want until they see it, feel it, touch it, use it, abuse it or modify it. However, the 'build it and see' approach is almost regarded as taboo in the post dot-com crash; cast-iron revenue has to be demonstrable or else! A conservative financier's grip may strangulate many good ideas because they lack a proven business model, or are not a finished product. However, we should be honest that there are no proven business models for mobile services and the fantastical amount of 'expert' analyst reports has gotten us nowhere. However, no one likes to throw money at something with no apparent scope for success, so a means to mitigate risk would definitely be worth obtaining.

Lessening of uncertainty would come as a great relief to the operators, who already face too many variables, most of them with unfamiliar roots. Looking at the second point in our list, the impact of not knowing what services to launch causes a delicate balancing act to be entertained. On the one hand, investment in new infrastructure is required, such as building the 3G network. Current revenues from 2G have bottomed out, so we need to move somewhere. However, new infrastructure is expensive, so the operators would like it if demand for new services was at least evident, if not gathering pace, so that investment is balanced by new revenue, if not now, then at least at some foreseeable and hopefully predictable point in the future. The absence of knowing what types of services are going to prove popular compounds the lack of demand problem because of perpetuation of inexperience and the lack of precedents in pricing of new services.

Pricing uncertainty actually frightens operators and can even cause extreme adversity to risk, which is not what we want. They talk of not wanting to launch services that 'kill the network', which presumably means where something becomes overly popular at the 'wrong' price point (i.e. too cheap). The apparent need for novel and complex pricing schemes heightens this fear. There is a concern that this might lead to pricing anomalies that could pave the way for exploiting the operator's precious resources at the wrong price point. For example, if operators attempt to create a so-called *event-based pricing* structure, it has the potential to be exploited by users interested in minimising how much they pay per chunk of data (e.g. per kilobyte). Let's say we will get charged 10p per email, the operator having figured out that this is a pricing strategy that user's can relate to and thereby positively assess their willingness to spend money in this fashion. However, this might prove to be so popular, that the operator ends up setting a precedent for service price expectation that in the end is too demanding on network resources at the service price point. With a multitude of services, there is also the danger that pricing anomalies become vulnerabilities. For example, for simply transferring data, there may be a volume-based 'transfer tariff' offered to users. However, it is perfectly feasible that some users might design a system that uses email as its transport mechanism and subsequently exploits the better pricing model. More surreptitiously, an enterprising user could design a modified email client that concatenates email messages, so that they end up getting five for the price of one[36], or something like that.

These examples may not be very realistic, or even likely, but the point is that the operator has a finite resource (RF capacity), so they would not want to act in a way that undercuts their ability to exploit it. This is obvious, but the way to avoid it whist still launching many new services with different pricing schemes is not obvious. Paradoxically, to achieve success, the operator needs a vast and diverse set of services in order to attract users and their money in the first place. This will entail opening up their network assets for third-party access, which will present all kinds of pricing headaches due to the high number of service permutations possible in offering these assets.

This last point relates to the final points on the list of challenges facing the operators. These are to do with opening the network up to third parties. Given the anxiety that comes from uncertainty, it would seem natural for operators to baulk at letting third parties in the door to offer their wares whilst the operator has such a clouded understanding of mobile services. To an extent, such conservatism is justified, except for the fact that required service diversity is impossible without calling upon the creativity of as many third parties as possible to build applications. This is the most likely path to success and one that has a good deal of consensus. It seems the best approach towards finding the 'killer application' or 'killer cocktail' that will usher in the new era of mobile services.

Given the need to involve other parties in building the next-generation mobile services success story, strategies are needed for building a suitable partnership framework. Furthermore, it is not enough just to build relationships on paper, creating 'partnership schemes' that are not able to empower real change. Real mechanisms have to emerge that allow other players to trade their creativity and willingness to take risk in return for a share of the revenue. Invariable, these players are software experts. This is a new world where the giant money-raking billing machine of the operator has to court quick-minded software gurus who have at their disposal an insight into many of the technologies discussed throughout this book and, hopefully, an instinct for what users might want. They are people who instinctively know, or can find the unfettered time in figuring out how to get the best from technologies like J2EE, J2ME and the entire gamut of techniques that enable powerful services to arise from our network of networks, even including some of the newly emerging paradigms, like P2P.

---

[36] It is unlikely that they could get all their emails in one message, as presumably the operator will impose an upper limit in any case, such as 100 kilobytes per message.

There should be as few restrictions as possible placed upon the creativity of software engineers and their adroitness in exploiting new technologies and the assets of the operators. Ideally, they need 'full access' to the RF network. If we think of the RF network simply as a set of capabilities and assets, then surely the paramount approach is to offer these assets to the best minds who know how to create the right mix that leads to something spectacular in the hands of the end-users. To do this, we need to find a way to let these minds into the front and back doors of the network.

The exploitation of software techniques in the mobile world is leading to the following trends:

- *Componentisation of the Service Delivery Platform (SDP)*, with chunks of next generation mobile services being wholesale developed as platform components by third parties, such as content-management systems for downloadable content (i.e. ringtones, wall papers, screen savers, video clips, etc.).

- *Rationalising the platform* – transforming the entire RF network into a kind of 'cellular operating system' that is intimately accessible to external software processes.

- *Componentisation of the RF network assets* – realising that hitherto embedded network assets, like voice mail, call forwarding and so on, should become components in this 'cellular operating system'.

On various occasions throughout the book, we have discussed the idea of a Service Delivery Platform (SDP). There is no official definition of an SDP, but the idea it to build a software platform to host common mobile services, the platform itself providing the means to offer a range of ancillary software functions common to mobile services. The first point above is describing a trend towards third parties providing pre-built parts of this platform as 'service in a box' components.

A common target for platform components is development of a *download server*. This is a system to enable the management of downloadable content to devices, such as ringtones, games and screen savers. The point is that these platform products do not solely provide a narrow part of the solution, like the content database and download mechanism itself, but the entire gamut of functions that an operator would need to run a 'downloads business'. What's more, the functions are extensible or programmatically accessible so that other services can be built that need a 'downloads component', but are not pure content services in their own right.

In the case of a download server, we might anticipate 'out of the box' functions like:

- Content management

- Enabling content providers to submit their own content and suggest prices

- Enabling the operator to review the content and override pricing

- Enabling the content to be published to the discovery server

- Content control, including digital rights management

- Adding digital rights management to the content to prevent unlicensed usage

- Determining who is able to access content

- Classifying content into genres, irrespective of the owner's categorisation

- Content discovery

- Enabling the content to be discoverable via a WAP or web portal

- Enabling searching and ranking of content

- Enabling download rates to influence promotion of content on the portal

- Content billing

- Providing mechanisms for billing the users

- Charging mechanisms for per-play or one-off charges

- Token-based billing

- Time-limit-based operation

- Content analysis

- Tracking success of different content types

- Tracking popularity of each specific item

- Monitoring which devices request which content

In addition to these functions, some vendors might also deliver pre-packaged content with the platform, mostly through syndication deals already struck with content providers, such as games programmers, ring tone providers and so on.

There are some key challenges with the SDP approach. One of them is how to enable flexible and rapid deployment of new services at any point in the future. If we stick with our download example and focus on games for a moment, then we can examine the issues.

A service to provide first generation Java games is relatively easy to build and support because the games are mostly standalone games with simple pricing models. The main challenge for the operator is provisioning the user with the games. In most cases, this involves dedicating part of the operator's portal to the games, probably calling it an arcade, with some basic ranking of titles according to popularity, or maybe just a simple catalogue based on genre (e.g. puzzle games, platform games, retro games, etc.). Then a download mechanism is required, which is nothing more that the ability to point the phone's browser to downloadable files that are subsequently downloaded using WSP or HTTP (as described at the end of Chapter 11). Then, there's the billing aspect, which for most first generation games systems is based on a one-off charge implemented with a reverse-billed text message.

This is all relatively straightforward. However, what about providing games that offer greater levels of sophistication, requiring access to operator assets? For example, we want to allow content providers to supply games that take advantage of the location-finding capability of the RF network. Perhaps we want to allow multiplayer games where messages can be sent to each other in real time, possibly using the operator's IM infrastructure. Perhaps we want games to be able to send multimedia messages to other users, or even voice messages. In fact, the variety of ideas and options is quite vast if we ponder on what is possible by making the network assets available to developers. There is no doubt that this would significantly augment their ability to provide interesting services to the end-users.

In terms of accessing the network's resources, we can think of the network as just another software application that can respond to requests, as shown in Figure 12.30.

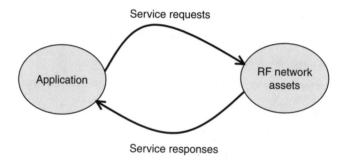

Figure 12.30 RF network is just another 'software' asset.

At the lowest physical level, this is made possible thanks to the GGSN in the core network, which facilitates an IP-based view of the RF network, which means we can apply any of our IP networking principles and software services to accessing the resources in the RF network; thus, the RF network 'plugs into' our IP network. From a software services perspective, we could even think of mobile devices as just another part of the entire RF network portfolio of software services and assets, enabling us potentially to map the entire RF network domain into the J2EE world, almost considering the RF network as just like any other EIS tier. The RF network is now a connectable service, accessible via APIs, Java messaging or even as EJBs. What makes this possible is the service deliver platform, the software layer that converts the raw network assets into forms that can integrate into our J2EE world.

It would require a whole book to discuss the assets of an RF network in any depth. Instead, we shall focus on the means of accessing and integrating with these, especially in relation to the J2EE and Web Services worlds. We shall look at trends in this area and see how 3G mobile networks can accommodate this idea as an integral part of their design, rather than an embarrassing afterthought. In terms of the assets themselves, then we shall look at just one of them in depth, in order to open our minds to the possibilities for next generation applications. Our focus will be the location-finding capabilities of a network, the basis for developing *location-based services* (LBS), which we shall describe in Chapters 13 and 15. In Chapter 15, we shall also attempt to put forward application ideas that bring together all of these concepts, if only as a thought experiment to encourage ambitious ideas for future mobile services.

12.7.1   J2EE Revisited

Figure 12.31 shows a possible J2EE system for a mobile video-on-demand application. Perhaps it would be useful for streaming old B-Movies[37] to commuters otherwise bored on the train journey home. The diagram is a very high-level view, but it is conceptually plausible. On the left, we have our servlets and JSPs, taking care of the presentation layer within a model-view-controller pattern, as discussed in Chapter 8. On the right, we have the guts of our system, not just the EJBs, but also the external network entities with which the EJBs interact using the powerful J2EE APIs. These entities include a database server (accessed via JDBC) that contains the catalogue of movies, a payment server (accessed

---

[37] B-Movie streaming services are becoming increasingly popular on the Internet.

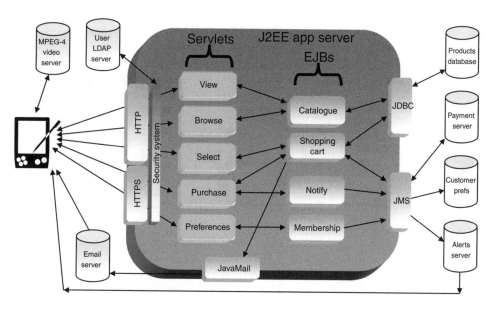

Figure 12.31 A possible J2EE mobile application architecture.

via JMS) to collect payment, a preferences database (JDBC) that stores user personalised information, such as a 'Wish List' of movies to see in the future. There's also an alerts server (JMS) to send an alert when a selected movie becomes available, or when payment clears, or to remind the user to view the film within the specified time limit, should they not view it straightaway. Finally, using the JavaMail API, we can send the customer a receipt for payment, if they request one, or we can send them a newsletter about the service.

This possible architecture for an application works well within the operator's environment, as presumably all the different resources are available within the reach of the LAN. However, what happens when we want to develop a similar system and subsequently host it outside of the operator's LAN?

Running the application platform outside of the operator's LAN presents us with the problem shown in Figure 12.32. On the right-hand side of the diagram, we have the J2EE system, but some of the required external entities (to the right) are not present on the LAN hosting the application server; these are the ghosted parts in the diagram. For example, the operator has a payment gateway that implements a micro-payment system. The operator also has a server for submitting text messages and another for sending WAP Push messages. However, all these assets are sitting behind in the operator's network and behind a firewall, as shown in the figure.

What we would like to accomplish is to find a way for the J2EE system in the figure to work by somehow accessing the missing resources that are present within the operator's network. We would like that in return for providing services to the end-users, our application owner receives a share of the revenue. In terms of accessing the operator's hidden assets, we do not want to have to implement private leased-line circuits to connect with the operator's network. The most obvious solution is to access them securely over the Internet, given its prevalence and all that we have said about the power of internet-based protocols and software solutions.

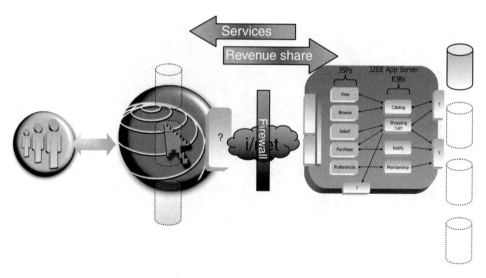

Figure 12.32 Problems with running a J2EE system outside of the operator's LAN.

The diagram indicates some of the problems with this approach. Firstly, no one connects to the Internet without placing a firewall at the boundary between the LAN and the Internet itself. The job of the firewall is to prevent any unauthorised internet traffic from penetrating the LAN. Techniques to button-down the boundary are numerous, but include, for example, closing all IP ports except for a very few that are subsequently closely guarded by the firewall in conjunction with the servers that sit on these ports on the inside (i.e. which have their own security mechanisms to add further levels of control to monitoring and defeating unwanted traffic).

A port that is typically open, at least for outbound traffic, is port 80, which is the default port for HTTP traffic. For organisations that host their own web servers, then port 80 is also open for inbound connections. Using HTTP and port 80 safely and securely is now a very well understood process and many software and hardware tools have become available to make it as secure and efficient as possible, mostly motivated by the obvious need to access the World Wide Web. Very often, companies will host their own web servers on their LAN, so they already understand the process of opening up port 80. Most web-related and firewall infrastructure has powerful tools to manage this process safely.

The popularity of HTTP traffic on the Internet has led to the emergence of Web Services, which, as we discussed in Chapter 7, proposes the use of HTTP for intersoftware traffic rather than human-machine traffic; thereby enabling software to talk across firewalls courtesy of port 80 being open. We can think of Web Services as the emergence of a parallel web space that is a network of machines, not people. Clearly, Web Services is not concerned with visualisation of information, such as web pages. With Web Services, there is no marriage between HTTP and visual markup languages like XHTML. The objective of Web Services is information interchange. Therefore, as we might expect, XML is the primary means of marking up the data flows, so where we have the HTTP/XHTML partnership for browsers, we have HTTP/XML for software services. Technically, there little difference between the two uses, especially as XHTML is in fact an XML vocabulary.

Figure 12.33 shows the principle of Web Services. The systems at either end exchange XML messages. We can see a sample message indicating a request from our application

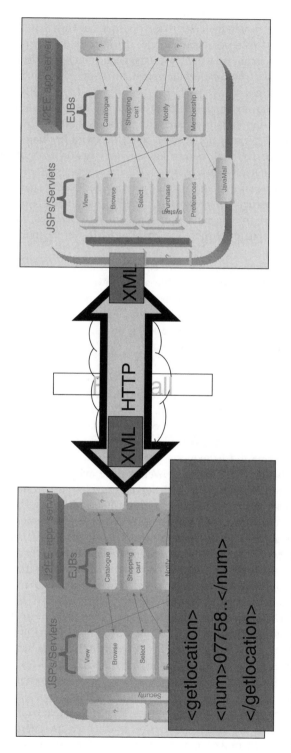

Figure 12.33 Principle of using Web Services to access operator assets.

to the location-finding platform in the operator's network. We can imagine an XML tag pair called <getlocation>, which indicates to the operator network that we want to get the location (i.e. coordinates) of a particular mobile, the identity of which we indicate in a nested tag pair <num>, wherein we would insert the phone number of the mobile device. In response to this request, we would expect the operator network to return the location results in the response HTTP stream.

The HTTP protocol allows our applications to enjoy the same fetch–response paradigm for software communication as is used for web pages. However, we are not concerned with fetching web pages. We may not even be concerned with fetching data per se. More specifically, we are concerned with asking software services on a remote machine to do work for us and return an outcome. In fact, with J2EE and programs like servlets, this is what fetching web pages is also really about – running a remote service, like a servlet on the J2EE web engine, but with a particular target result, namely generating a return content stream that articulates a web page interface in a browser. With Web Services, we perhaps should think of the process conceptually as more like enabling RMI over the Internet, although this is not what we actually do.

This latter point is an interesting one. We could have stuck to RMI, or perhaps any of the J2EE APIs, as they are all IP-compatible at the lowest level. However, they do not run very well over the Internet, especially as most of the ports they use are not typically open in default firewall configurations. Furthermore, Figure 12.33 also shows that the two ends in the Web Services dialogue could be black boxes. We do not need to know how the operator implements their system, as long as it can 'talk' Web Services over HTTP. The same applies to the other end, although we have focused on using J2EE throughout this book, for reasons justified elsewhere. Therefore, Web Services is software-technology agnostic – it cares not how the services and requests are fulfilled, just that the dialogue abides by the agreed protocols embedded within an XML stream running over HTTP.

Apart from the XML vocabulary that we need to implement Web Services, the remaining part of the puzzle from Figure 12.32 is the big question mark representing the entity that handles the inbound Web Services requests. Perhaps it is not obvious, but we cannot rely upon the operator network assets as having Web Service interfaces, mostly because these assets are legacy resources that never had Web Services as a design consideration. Moreover, it is likely that the design of most of the internal resources did not take into account the possibility of external (third-party) access, hence the mechanisms required to support third-party access are probably missing. In addition to APIs, in order to enable programmatic access to an asset plenty of other functions are required, such as charging mechanisms (e.g. per text message submitted, or per location request) and auditing, mechanisms that are essential to facilitate third-party access within a workable commercial framework.

Before we address the problem from the operator's point of view, we should consider implications for our J2EE system architecture. In our discussion of mobile services using J2EE, we have emphasised the usefulness of the powerful J2EE APIs for accessing the support tier. The use of JDBC or JMS, or other powerful J2EE APIs has been essential to enable dialogue with external entities. However, these APIs do not work over firewalls, so we must now turn to HTTP and XML and utilise Web Services. Fortunately, as we have already examined in some depth, the J2EE platform is more than capable of handling these interfaces, so we are not dispensing with its powers, but it is clear that we have to re-engineer our approach slightly. How we do this depends on whether we are repurposing an existing system to use Web Services, or designing one from scratch, although common approaches may also apply. It is not that difficult to think of implementing a 'Web Services engine'. This is a software layer on top of the J2EE web engine, which enables EJBs to make Web Services requests via an API. The function of the layer will be to take care of the HTTP dialogue, and the translation of messages to and from the XML domain. The exact nature of the API will depend on which of the proposed Web Services protocols is used. The two favourites are SOAP and XML-RPC, which we shall introduce briefly.

We have looked at the way that our J2EE system can interact with the operator network to request services over the Internet. We shall return to some of the features of the Web Services protocols (SOAP and XML-RPC), but first we should examine the higher-level system considerations, especially from the network-operator perspective in order to see if there is a unified or systematic approach that can be taken to 'web enable' their assets. Furthermore, although we may have a mechanism for making remote service requests, be it using SOAP or XML-RPC, we have not said anything about what those messages might sensibly contain and whether or not there are any emerging standards in this regard. It would seem an important area to standardise across the industry, otherwise the blossoming of applications and services being sought is unlikely to happen. In most markets, there is little incentive to develop mobile applications unless the entire market (i.e. all networks) is accessible in a consistent manner.

12.7.2 Service Delivery Platforms based on Web Services

Providing a Web Services interface to the operator's network is a good idea. It provides a unified and standardised interface for third parties to gain access to the 'cellular operating system'. The unification comes from enabling a common means to access any of the network assets – whether it be requesting the location of a mobile or submitting a text message, we use HTTP/XML. This means that developers can implement powerful applications that potentially can access a disparate range of powerful network assets but within a single software paradigm. This simplifies implementation and allows the application of system-wide software services where necessary, such as a common authentication apparatus. For example, we could insist that all requests come from the same registered source address, or we could apply a common ID token to all requests, masking it from others using secure sockets layer to encrypt the message transfers. The standardisation comes from the Web Services initiative itself, which is the subject to open standards agreement and itself built on the open standards of XML and HTTP. This means that increasingly, J2EE systems will be Web Services cognisant by default, either via existing applicable APIs or newly emerging Web Services ones.

Web Services dialogues are very loosely coupled. The two ends need make very few, if any, assumptions about each other. Operating characteristics and implementation details can remain hidden. This is due mainly to the fact that Web Service access is via a URI, like http://www.mymobileoperator.com/locationfinder. Any web-connected software application can therefore access what lies behind this URI. It is the ultimate in global and unfettered access to a programmatic resource. There is no inkling of what lies behind this URI. As long as something meaningful (and expected) returns from the URI call, then the Web Services client can go about its business and should be happy.

This loose coupling is a fantastic characteristic of Web Services. It means that the operator can chose to implement whatever they want to 'sit behind' the URI, and they can also change it whenever they feel like it, without any fear of upsetting clients, so long as the same results are returned (i.e. in their correct XML format). The network operator can take advantage of this fact because it enables the network assets to be available in a controllable and predictable manner without dangerously exposing the lower-level details and mechanics to potential third-party users. For example, a location-finding platform may have an expansive API capability, but only a subset of which makes sense to be publicly accessible. Moreover, perhaps the fine-grain control of its features needs repurposing for public access. For example, perhaps the default interface primitive for a location query returns a lot of information in its response, some of which should not be accessible by third parties. This is not a problem. The Web Services layer can tailor responses as the need arises, such as removing or adding information to the responses as required. The response to a query could contain a restricted set of fields in the XML stream, omitting much of the data that is inappropriate for inclusion in a third-party query response.

Access to the operator's network might not use Web Services standards all of the time. Some of the network assets may already have historical HTTP interfaces implemented by gateways, and these might remain in place for backwards compatibility, probably alongside a Web Services equivalent that will gradually take over from the legacy interface. Whether HTTP or Web Services, Figure 12.34 shows us a possible architecture for our application-centric RF network.

At the top of the figure, we have our application layer where the third party or operator-hosted applications sit. This is not limited to applications connected directly to the Internet. There is no reason why applications running on the devices themselves cannot access the

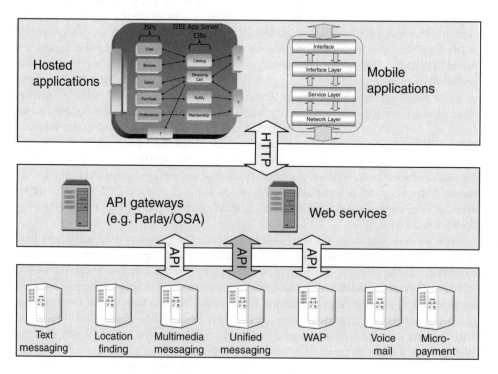

Figure 12.34 Service layer for mobile networks.

Web Services layer. For example, a J2ME application (MIDlet) running on a device may need to make a location query. In the absence of location-finding APIs local to the device, the MIDlet could access the Web Services layer, provided it is able to converse over a Web Services stack (HTTP/XML), which in some resource-limited cases may be an overreach, but otherwise entirely possible.

In the middle, we have the Web Services layer itself, which is responsible for terminating and servicing the Web Service requests. It acts as an interlocutor to the network resources that sit in the services layer below. A variety of APIs are in abundance for interfacing with different network elements, such as the text-messaging platform (Short Message Service Centre, or SMSC), for which we could internally use a protocol like Short Message Point-to-Point (SMPP) for example. The text-messaging platform is a notable example of a resource that probably already has a web interface via an existing gateway server. Text-messaging gateways are already in abundance for providing HTTP access, although historically not using any of the Web Services protocols (although the trend towards Web Services has accelerated).

One of the exciting things about Web Services is the URI approach. With web pages, we are comfortable about the idea that links (anchor tags) enable the reader to jump from one page to another. The same idea applies to Web Services and the application-centric web. It is a particularly relevant to the Semantic Web, a concept we discussed earlier in Chapter 5 and that we shall return to soon. We can make Web Services requests and embed further URIs within the requests, so that applications can talk to other applications as they go about their business. It is perfectly feasible for a Web Service response to tell the calling service where to go next in order to get further assistance with a particular request. For example, a location request could redirect the caller to a URI where mapping information is available for the requested location, tourist information, or some other geocoded data. These reference resources could sit anywhere on the Web, not restricted to the operator's network, so powerful additional Web Services can be easily incorporated into a mesh of services that enable a powerful mobile service to come to fruition.

12.7.3 Standards for the Service Layer APIs – Parlay/OSA

We have been looking at ways of providing external applications to gain access to the mobile operator network resources via the Internet. Having discussed how this could be achieved using Web Services (HTTP/XML), we have not addressed the issue of what messages we can send using any of the Web Services protocols (e.g. SOAP, XML-RPC, etc.)

Intuitively, the messages supported by the Web Services layer must surely reflect what the underlying APIs can support. Long before Web Services emerged in the rapidly developing world of internet solutions, the Parlay Group[38] was already addressing the issue of standardising on a set of universal APIs to access commonly found features in telephony networks (fixed or mobile). This would allow applications developers to develop applications that could run in any Parlay compliant environment. This would allow operators to host third-party applications, which is, after all, our objective, returning to the primary challenge of allowing a 'killer cocktail' of applications to flourish on the operator's network, which is a necessary step towards reaching the tipping point with mobile services. Hence, we can think of Parlay API calls as the API interface into our 'cellular operating system'.

[38] http://www.parlay.org/en/index.asp

OSA (Open Services Architecture) refers to the agreed architecture for mobile services developed by the Third Generation Partnership Program or 3GPP[39], and also adopted by 3GPP2. These are the industry forums responsible for developing the specifications for 3G cellular systems, like UMTS. These bodies selected Parlay as the basis for providing the API for OSA.

Parlay/OSA is a set of APIs. The functions covered by these APIs include the following:

- Mobility
- Location
- Presence and Availability Management
- Call Control

- User Interaction
- Messaging
- Content-based Charging
- Policy Management

These APIs are divided into common functional areas. Presence and Availability Management for example is to do with functions relating to detecting the state of the user. For example, we could have states such as:

- User is busy – in a call
- User if busy – out to lunch
- Use if available via IM
- User is available to receive phone calls

We discussed these ideas in action during the prelude to the book (Chapter 1), showing how a user can take advantage of such abilities to allow more effective management of their time and tasks using these various modes of state management. Presence and Availability Management (PAM) as an API is designed to assist the development of all kinds of interesting applications and services using potentially any of the common communication systems (instant messaging, email, text-messaging, voice, etc.), mobile or fixed[40]. It is probably true that currently most of the telecommunications networks do not provide such features for third-party access, so the Parlay APIs represent an exciting development for mobile services.

If we combine PAM with something like Call Control, then we get some very exciting possibilities. For example, we can use our presence information to control call handling, such as automatically diverting all voice calls to voicemail whilst in a meeting, allowing access to the recipient only via IM provided the user is logged in with an IM client. If not logged in with an IM client, then invitations could be sent to join an IM session, the invitation being sent via WAP Push. The possibilities and permutations seem vast, and here we are only talking about using two of the APIs in the Parlay/OSA set! Incorporating features across all the APIs, which is easily done thanks to the Web Services loose coupling, some potentially amazing services are surely possible!

In addition to the APIs mentioned above, there is the Framework API, as shown in Figure 12.35, which provides some of the housekeeping functions we mentioned earlier that any unified approach would need to have. For example, it enables applications to identify

---

[39] http://www.3gpp.org

[40] The 3GPP OSA is obviously concerned only with mobile services, but the Parlay group is concerned with all modes of communication, including wire-line (fixed).

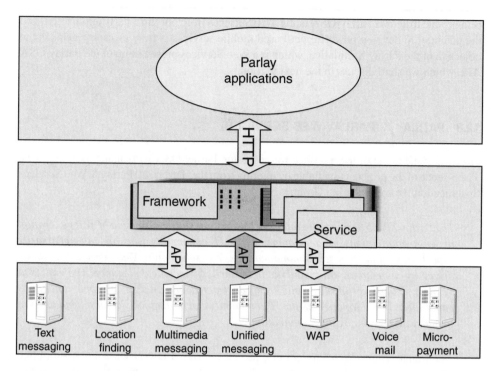

Figure 12.35 Parlay gateway.

themselves, to authenticate and to discover what capabilities (services or APIs) are available via the Parlay/OSA gateway, which is the name of the network element that implements the Parlay APIs. This is a layer of abstraction similar to the way the Web Services layer functioned in Figure 12.34. Telecoms products and platforms do not necessarily come with Parlay APIs built in. Therefore, we need to add them retrospectively, which means adding a new network element to translate Parlay/OSA API calls into ones natively understood by the underlying infrastructure.

This existence of the Parlay/OSA gateway means that applications are isolated from the specific protocols used within the network by the network assets. This allows networks to change without affecting existing applications and services. Networks can be upgraded without any impact on the applications accessing the Parlay/OSA APIs. The Parlay/OSA gateway is the element that implements the Parlay framework. Parlay/OSA gateways are generally not built by the operators, but by specialist software vendors, like Aepona[41], Telcordia Technologies, Inc.,[42] and BEA Systems[43].

The Parlay/OSA APIs are designed to enable the creation of advanced telephony applications. Formerly, we would have referred to some of this capability as Intelligent Networking (IN). However, perhaps more significantly for the future of mobile services, the way to think of this powerful capability is that Parlay/OSA allows for the 'telecom-enabling' of any IT application or service – not just 'telecom-enabling' but 'mobile enabling', too! All we need

[41] http://www.aepona.com/
[42] http://www.telcordia.com/
[43] http://www.bea.com/

is to be able to access Parlay/OSA using Web Services from our third-party application, and the potential to develop powerful net-based mobile services is truly awesome. This is the objective of the Parlay X initiative, which is a Web Services presentation of the Parlay/OSA API, which we shall discuss in the next section.

12.8 PARLAY X (PARLAY WEB SERVICES)

Parlay X is the name of the interface for accessing Parlay/OSA APIs using Web Services. The essence of the project is available by quoting from the 'Parlay 4.0 Parlay X Web Services Specification' from the Parlay Group:

> *The Parlay X Web Services are intended to stimulate the development of next generation network applications by developers in the IT community who are not necessarily experts in telephony or telecommunications. The selection of Web Services should be driven by commercial utility and not necessarily by technical elegance. The goal is to define a set of powerful yet simple, highly abstracted, imaginative, telecommunications capabilities that developers in the IT community can both quickly comprehend and use to generate new, innovative applications.*

In addition to being a Web Services interface, the Parlay X interface is a much-simplified presentation of the fully blown Parlay/OSA APIs. This is to help meet the objective of attracting widespread usage and facilitating quick development cycles. It should be relatively easy to utilise powerful features in the operator's network from our J2EE application (or handset application).

It is a truly exciting prospect that with just one request from a URI, it is possible to initiate a phone call between two parties (*Third-Party Calling*), or to forward an incoming call to another number if the user is unavailable (*Network-Initiated Third-Party Call Control*).

An example of how we could utilise Parlay X in a web-based context is to use the Call Control feature to indicate to our J2EE application (via a Web Service request) whenever a call is made to our device during office hours.

The types of features that will be available initially via Parlay X are as follows:

- *Third-Party Call* – this is the ability to initiate calls from one caller to another, such as putting a trader in touch with his stockbroker automatically when a price threshold is reached on a stock being watched.

- *Network-Initiated Third-Party Call* – this is the ability to control calls that are initiated by a caller (mobile phone). For example, to detect out-of-hours product-support calls to a particular number and forward to the appropriate support person by first looking up from a database who is on call that night.

- *SMS* – to submit and receive text messages, check for delivery, etc.

- *Multimedia Message* – to submit and receive MMS messages etc.

- *Payment* – to initiate payment sessions according to different charging methods such as one-off payments, regular payments, etc.

- *Account Management* – checking account status, such as account balance, account credit expiration, etc.

- *User Status* – this is like the presence management features of Parlay discussed above, but a very limited subset.

- *Terminal Location* – this subset of the API enables the location of a mobile to be ascertained.

Let us just look at one of these Web Services components in slightly more detail, in order to appreciate the power of simplicity of the Parlay X solution and just to examine briefly the anatomy of a Web Services message using SOAP.

The Parlay API is constructed from service calls. These are defined like the following example, which is taken from the Network-Initiated Call Control API:

```
handleCalledNumber(EndUserIdentifier callingParty,
   EndUserIdentifier calledParty,
   out Action action)  { XE
"handleCalledNumber(EndUserIdentifier callingParty,
   EndUserIdentifier calledParty,
   out Action action)" }
```

What this function specifies is a Web Services message that will be initiated by the Parlay/ OSA gateway when a call takes place to the *calledParty*.

The message *handleCalledNumber* requests the application (our application) to notify the gateway how to handle the call between two numbers, the *callingParty* and the *calledParty*. The method is invoked when the *callingParty* tries to call the *calledParty*, but before the network puts the call through to the *calledParty* (i.e. the called party will not know they are being called at this stage). It is interesting to reflect that the *calledParty* does not have to refer to a human receiver; it could be an auto-answering service (Interactive Voice Response, IVR). The application in the return HTTP stream (reply SOAP message) is expected to return the *action*, which tells the gateway to perform one of the following actions:

- **"Continue"**, resulting in normal handling in the network, i.e. the call will be routed to the *calledParty* number, as originally dialled

- **"EndCall"**, resulting in the call being terminated; the exact tone or announcement that will be played to the *callingParty* is operator-specific

- **"Route"**, resulting in the call being re-routed to a *calledParty* specified by the application

This is an extremely powerful facility, easily accessible to any competent web programmer. These days it is possible to implement Web Services with simple scripting languages, thus not even requiring any knowledge of a complex J2EE environment. Contrast this with the complex IN programming that would have been required otherwise, out of reach of most programmers.

We could use this feature as a simple filtering process on calls to a particular number. For example, we could check the recipient's activities via their online calendaring application (e.g. Lotus Notes or Microsoft Exchange). If the calendar API is flagging a meeting for that particular moment, then we could route calls to a voicemail number. Alternatively, we could look in the recipient's online employee profile to extract the number of their personal assistant or department secretary and route the call to their number.

We cannot overestimate the power of this type of service being made available via a simple API. Ordinarily, with or without Parlay/OSA (but not Parlay X), the complexity of determining and handling a call in mobile network is well beyond the ability of an average software programmer unfamiliar with a telecommunications environment. The standard IN approach requires a high degree of network expertise, not just programming knowledge. Using the Parlay X Web Services approach, even someone with rudimentary programming skills and practically no telecoms knowledge can rapidly create a powerful mobile service. Someone perhaps skilled in programming for groupware applications, like Lotus Notes, suddenly is able to 'telecoms-enable' their wares with flare. This is perhaps one of the most exciting aspects of the emerging trends in mobile applications construction and is an oft-overlooked aspect of the '3G platform'.

12.8.1   What does a Parlay X Message Look like?

Here, we briefly touch upon the anatomy of SOAP, which stands for Simple Object Access Protocol, or at least it used to. Being as SOAP actually has nothing to do with *objects* in the Object-Orientated Programming (OOP) sense; the current owners[44] of the specification abandoned the acronym, or at least its meaning. It now, apparently, stands for nothing and is just a name for a particular Web Services protocol that we shall now examine in the context of its usage in Parlay X.

By now, we are familiar with the HTTP methods, such as GET, the one most often used to request a resource from the origin server. However, with Web Services, we are interested in passing information back and forth, so we use the POST method in the request so that we can include the SOAP message in the request stream.

It is easier to demonstrate a SOAP message by repeating the example from the SOAP specification:

```
POST /StockQuote HTTP/1.1

Host: www.stockquoteserver.com

Content-Type: text/xml; charset="utf-8"

Content-Length: nnnn

SOAPAction: "Some-URI"

<SOAP-ENV:Envelope
  xmlns:SOAP-ENV="http://schemas.xmlsoap.org/soap/
     envelope/"
  SOAP-ENV:encodingStyle="http://schemas.xmlsoap.org/
     soap/encoding/">
```

---

[44] SOAP is under the auspices of the World Wide Web Consortium (http://www.w3.org/TR/SOAP/)

```
    <SOAP-ENV:Body>
        <m:GetLastTradePrice xmlns:m="Some-URI">
            <symbol>DIS</symbol>
        </m:GetLastTradePrice>
    </SOAP-ENV:Body>
</SOAP-ENV:Envelope>
```

In the SOAP request, we can see the familiar HTTP headers, including the POST method and the path ('/StockQuote'). We can also see the host header: www. stockquoteserver.com. It is the responsibility of the resource at the implied URI (http://www.stockquoteserver.com/StockQuote) to extract the SOAP body ('envelope') and process it. This could easily be a Java servlet using its built-in HTTP API to extract the body from the request. It could then use one of the Java XML processing APIs (JAXP) to parse the XML and extract the fields.

There is an additional HTTP header in the request; one we have not seen before, called 'SOAPAction' header. This header is a server-specific URI to indicate the intended nature of the SOAP request. This allows the server to determine what the request is all about without having to open the XML message to get the details. This might be useful for filtering SOAP requests, for example to ensure that certain requests are unable to get through, or to enable routing of requests in a distributed implementation (more than likely[45]).

The extra header here is peculiar to sending SOAP messages using HTTP, which, as we have discussed, is an attractive solution. However, just like many other XML-payload messaging systems, SOAP is not tied to a single transport protocol (i.e. HTTP). It could just as easily work over SMTP, or JMS, or one of the proposed modern alternatives to HTTP, like Blocks Extensible Exchange Protocol[46] (BEEP). Recognising that HTTP was indeed designed for web pages and not things like Web Services, and given the prevalence of the concept, based on globally addressable resources (i.e. via URIs), alternatives have been proposed that are not so closely tied to web pages. We can think of asking, 'How would we design something like HTTP, if we had to do it all over again, knowing what we know now?' Of course, that's an obvious question to ask about anything, but in the case of HTTP, something like BEEP is what the answer might be. To be a bit more precise, according to the IETF's task force on BEEP[47], it is:

> [a] standards-track application protocol framework for connection-oriented, asynchronous request/response interactions.

Getting back to the SOAP message itself, it comes in two main parts, the header (i.e. not the HTTP header, but the SOAP header) and the body, as shown in Figure 12.36. The body may optionally include information about reporting fault conditions.

The critical component is of course the body, this telling the Parlay X gateway what we require from it. In the example above, we can see the crux of the request, ignoring the XML paraphernalia, which is a service-request called 'GetLastTradePrice'. The place where the

---

[45] We have not discussed the architecture of a Parlay gateway, as it is beyond the scope of this book. However, it is a networked software system just like any other in the way that it has to be able to scale and access all kinds of network resources, etc. Therefore, not surprisingly, it is common for Parlay gateways to use the J2EE platform, as it is the ideal candidate for supporting such a product. Big J2EE vendors, like BEA Inc., are not surprisingly very heavily involved in supporting Parlay initiatives.

[46] http://www.ietf.org/rfc/rfc3080.txt

[47] http://www.beepcore.org/

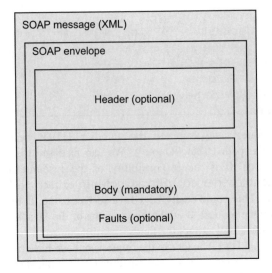

Figure 12.36 Anatomy of a SOAP message.

gateway can go find the XML vocabulary definition for this request is the 'Some-URI' reference, which is a parameter in the request tag itself. Parlay X defines Web Service calls using a language called Web Services Description Language (WSDL), which itself is formulated using XML. This is similar to the DTDs we met earlier in the book when looking at XHTML and its cousins.

For the 'GetLastTradePrice' request, the WDSL would tell us that there is a field called 'Symbol', so the Parlay gateway will be searching for this field (<Symbol> tags) in the body of the SOAP message.

Having extracted the fields, the Parlay gateway routes the request to the appropriate handler in the gateway platform, which could be any part of a scalable software system; there are no restriction on how the Web Service requests are processed.

The results from the request are returned in the HTTP response, which itself contains a SOAP message, just like a HTTP stream from a web server contains a XHTML payload.

```
HTTP/1.1 200 OK

Content-Type: text/xml; charset="utf-8"

Content-Length: nnnn

<SOAP-ENV:Envelope
  xmlns:SOAP-ENV="http://schemas.xmlsoap.org/soap/
    envelope/"
  SOAP-ENV:encodingStyle="http://schemas.xmlsoap.org/
    soap/encoding/"/>
```

```
<SOAP-ENV:Body>
    <m:GetLastTradePriceResponse xmlns:
        m="Some-URI">
        <Price>34.5</Price>
    </m:GetLastTradePriceResponse>
</SOAP-ENV:Body>
</SOAP-ENV:Envelope>
```

If the mechanics of the HTTP process itself runs smoothly, then we get the familiar '200 OK' response status header. However, the bit we are most interested in is the SOAP message itself, which has the same structure as the request message (i.e. header, body, etc.) As we can see, a result for the stock-price query has returned with a price – <Price>34.5</Price>, so the process has been successful. Any type of software could be behind servicing this request – it really is of no concern to the requesting application, only the result is of interest.

If we consider sending an SMS via the Parlay X interface, the request has the following structure:

```
sendSms (EndUserIdentifier[] destinationAddressSet,
        String senderName, String charging,
        String message,
        out  String requestIdentifier) { XE
        "sendSms(EndUserIdentifier[]
        destinationAddressSet,  String senderName,
        String charging, String message,
        out String requestIdentifier)" }
```

The corresponding XML within the SOAP body would look something like:

```
<sendSMS>
    <destinationAddressSet>4487718776351
      </destinationAddressSet>
    <senderName>    EmailAlerts</senderName>
    <charging>Bulk5000 01283940</charging>
    <message>You have a new email from Joe Somebody
      </message>
</sendSMS>
```

The SOAP message *sendSms* requests the Parlay X gateway to send an SMS, specified by the *String* message to the specified address (or address set, which is an array of addresses), specified by *destinationAddressSet*. The application can also indicate the sender name (*senderName*), which the user's terminal displays as the originator of the message.

Charging arrangements (*charging*) can also be specified, which is the charging scheme defining how the SMS is going to be charged. The application can receive the notification of the status of the SMS delivery. To do this, it must make a separate SOAP request using the *getSmsDeliveryStatus* message. In this case, the *requestIdentifier* (see above definition), that was returned by SOAP response to the above message, can be used to identify the SMS delivery request.

We have only touched briefly on SOAP and its mechanics, but the principles are straightforward. The detail is really in understanding the message formats. However, once we have a software function that can take care of Parlay X messaging on our J2EE platform, the greater understanding is then figuring out what the Parlay X APIs consist of and how best to use them in creating interesting mobile services, which is our real goal.

13

Mobile Location Services

13.1 'I'VE JUST RUN SOMEONE OVER'

There were nearly 6,420,000 car accidents in the United States in 2005[1] – that's a lot of car crashes! Approximately 2.9 million people were injured and 42,636 people were killed in auto accidents in 2005, based on data collected by the Federal Highway Administration.

Using a cell phone to call for emergency assistance can be an awkward affair. The biggest problem is pinpointing the location of the accident. 'It's near a lamp post near a post box near a turning' is not quite good enough for the emergency services to find where the accident has taken place. Because of the number of deaths and serious injuries that are aggravated by lack of a timely response from the services, the Federal Government (Federal Communications Commission – FCC) in the United States introduced a law, called the wireless *Enhanced 911* edict (E911[2]). This stipulated that a mobile phone operator must be able to locate physically (geographically) a mobile making an emergency call.

According to the E911 website[3]:

> *The wireless E911 program is divided into two parts – Phase I and Phase II. Phase I requires carriers, upon appropriate request by a local Public Safety Answering Point (PSAP), to report the telephone number of a wireless 911 caller and the location of the antenna that received the call. Phase II requires wireless carriers to provide far more precise location information, within 50 to 100 meters in most cases.*

[1] http://www.car-accidents.com/pages/stats.html
[2] http://www.fcc.gov/911/enhanced/
[3] http://www.fcc.gov/911/enhanced/Welcome.html

Next Generation Wireless Applications, Second Edition   Paul Golding
© 2008 Paul Golding

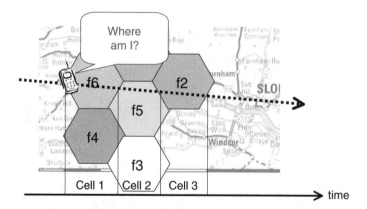

Figure 13.1 By default, a cellular network knows our whereabouts.

What reason made the FCC ask a mobile operator to carry out this location task and not some other agency? Figure 13.1 reveals the answer, which shows that by virtue of its cellular nature, the mobile network already knows the 'whereabouts' of each subscriber; this is the *mobility management* function inherent in a cellular network.

As Figure 13.1 also shows, this information may well be useful to the subscriber, not just the emergency services. The user might ask, 'Where am I?' This question turns out to be an interesting question that most of us ask frequently for one reason or another.

13.2 'WHERE AM I?'

At the edge or centre of each cell or sector in a cellular network, there is a base station. Operators know where these base stations are, because they were installed at fixed known locations. The mobility management function of a cellular network constantly keeps track of mobiles that are *switched on* (they don't need to be in a call). In fact, mobiles are in regular contact with the nearest base station just to keep tabs on parameters such as signal strength to ensure that the best station is being accessed, and to periodically tell the network which of the reachable base stations seems best to 'camp onto' in readiness for making or receiving calls, or exchanging data. It is imperative that the network knows where a mobile is so that any inbound calls can be routed to the appropriate cell. The same applies for inbound data, such as a WAP Push or text message. Hence, if we know which base station a mobile is currently camped on, then we know its rough location with respect to its proximity to the serving cell site (base station). At the very least, we can have the network report 'mobile X is currently attached to base station Y which is at our Maidenhead site'. Hence, mobile X is in Maidenhead. This is the basis of the *Cell ID* approach to *Location-Based Services* (LBS).

With the ability to determine the location of a mobile known, all kinds of applications and services become possible. The UMTS Forum suggested a classification of LBS into four service areas:

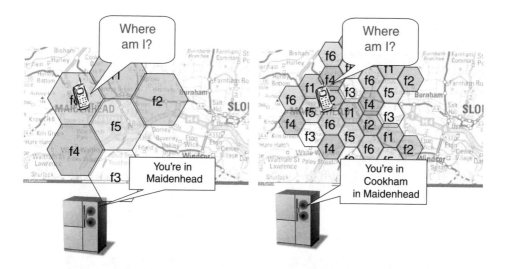

Figure 13.2 Cell ID is only as good as the cell size.

1. *The Commercial LBS (or Value-Added Services)–* Value-added services that typically the user will pay money for 'Where is the nearest . . .?', 'Where's my buddy?', Maps, directions, location-enhanced games and entertainment, etc.

2. *The Internal LBS –* The use of the location information for internal operations, e.g. location-assisted billing, usage patterns, network measurement, operations and maintenance (O&M) tasks, supplementary services, etc.

3. *The Emergency LBS –* Assist subscribers who place emergency calls. This service may be mandatory in some jurisdictions (e.g. E911 in the United States)

4. *The Lawful Intercept LBS –* The use the location information to support various legally required or sanctioned services (e.g. covert security force operations).

After the initial grievances[4] of having to deliver a tracking system to suit FCC ideals, the belated excitement about the potential of location services in the commercial domain took over. It seems there is no end of ideas for location-enabling services. Some services are possible because of location finding, whilst it is a means of augmenting others, leading to a rich potential for location enabling of many applications. We shall discuss accuracy and its implications later, but a range of applications are possible, some better than others given a certain resolution. As Figure 13.2 shows, the accuracy of Cell ID is susceptible to cell size. Smaller cells are clearly better, as it means the mobile can be more accurately located.

---

[4] The operators protested that the costs of fulfilling E911 should be met by the FCC, not by the operators themselves.

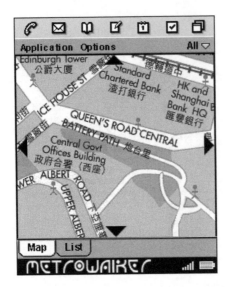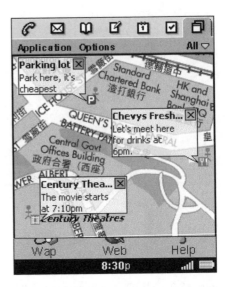

Figure 13.3 Buddy finding and instant messaging. (Reproduced by permission of Magic E.)

Once we have a location-finding system in the mobile network, then subscribers can be placed on a map. An exciting possibility is 'buddy finding', as shown in Figure 13.3.

The mapping application shown here (showing Hong Kong) appears to be locating users with high accuracy. In a dense urban area, the cell sizes can be quite small, such as 100 metres, so even with Cell ID, this type of accuracy might be possible. However, the requirement for accuracy depends on the application. It may be sufficient in many cases just to know the general proximity of someone with respect to either a fixed point or another subscriber. Two applications that use this approach are 'child minding' and 'mobile tag'.

The concept of a child-minding application is monitoring the whereabouts of a child in relation to a home zone or 'free-to-roam' zone. If the child leaves the zone, then someone is notified, perhaps the parents who have gone out for the evening. The parents could specify the 'curfew' zone and the action to be taken if the curfew is broken, such as sending both parties a text message indicating the breach of trust, as shown in Figure 13.4.

Here we can see that Johnny is supposed to remain within the 'Happy Valley 3' region. The map here shows each of the districts on Hong Kong Island, but we can imagine that each one has its own cell and so this resolution is achievable using Cell ID (we can discuss other methods of location finding later). If the LBS system in operation has a resolution limitation, then it is better to indicate this visually to the users so that they properly specify a reasonable curfew zone. There would be little point in letting the parents draw their own zone boundary, which may end up being smaller than the resolution of the system. This would be misleading and would contravene good usability practice.

There are a variety of ways that the application might operate, all configurable by the parents, who probably have password access to the application management console, which might be a desktop application or a mobile one. If the child strays from the zone, they could be warned of their erring, giving them a chance (time limit) to return to the zone before the parents get notified. The exciting thing is that with Parlay X Web Services (as discussed in

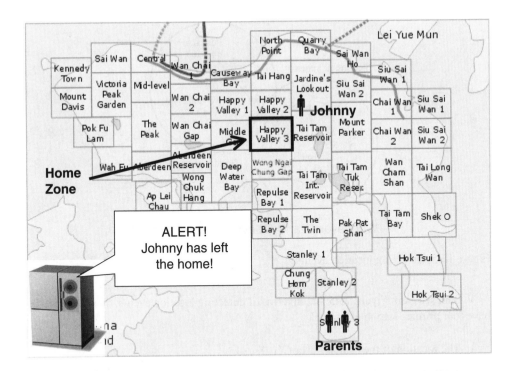

Figure 13.4 Child minding using LBS.

Chapter 12); the application could first detect the location violation using the location API and then use the text-messaging API to send the warning message. In other words, it is easy to build a system like this using the power of Parlay X.

If the child persists in roaming outside the zone, then a while later, after another warning has been fired off, the parents are notified. This could be followed by a third-party call being initiated by the Parlay gateway using the Third Party Call API. An alert could also be sent to a friend, such as a neighbour.

This all sounds wonderful stuff (except to the would-be errant child of course), but such a service is contingent on several things:

- The child has to keep the mobile with them at all times.

- The mobile has to be switched on (an obvious point, but an obvious limitation, too!).

- The child has to consent to being tracked (a possible and likely legal requirement in many jurisdictions, but we shall get to this issue later).

- The resolution has to be within a usable limit.

- The monitoring of location has to be sufficiently frequent to achieve the desired result (no use if the child can zip off out of the zone for 15 minutes and come back before the next location query).

In terms of location querying, there are a number of approaches, but Figure 13.5 illustrates the general problem.

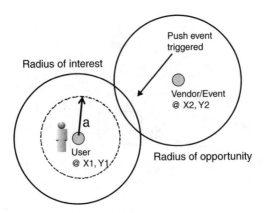

Whenever *Ri* and *Ro* overlap, send alert

Figure 13.5 Detecting location events.

We can see from Figure 13.5 that in terms of detecting location events, then we actually have four primary variables to track:

1. The location of the primary user (X_1, Y_1, a)

2. The location of the event or place (X_2, Y_2) that we wish to compare with the location of the user

3. The radius of user's interest (R_i)

4. The radius of opportunity (R_o)

These variables need some elaboration. They are variables that underline concepts to help model the location-finding problem, but should not be taken as a definitive articulation of the location problem.

The location of the user is the most obvious measurement or variable. The primary coordinates are the x, y position (X_1, Y_1), but it is necessary to take into account that location-based systems, like all measurement systems, have a degree of error or uncertainty, which here is articulated as a radial quantity (a). So we can think of the user as being represented by a virtual circle that roams across the map's surface, bearing in mind that the size of the circle may well change as the user roams from place to place, depending on the available LBS resolution in their locality.

The location of the place of interest is also an understandable variable, and presumably most of the time we know its whereabouts with a decent level of accuracy, so we don't really expect a margin of uncertainty in real terms, even though there might be one in actuality (due to human reasons).

The radius of interest specifies a zone around the user that we may want to use in order to understand the geographical extent of the user's interests. For example, if we have a tourist application and the user is on foot, then perhaps a good assumption is that they are willing to walk no more than two kilometres to get to any attraction of possible interest. This radius can be set to any value of course. If the user is driving along a major road, then for the purposes of notifying the user of service stations, the radius of interest might extend to tens of kilometres, perhaps even as much as a hundred, all depending on the circumstances.

The radius of opportunity really represents the extent to which a particular place should be of interest to a user. For example, if a particular eatery were running a buy-one-get-one-free deal on a soft drink, or some other low-value product, then probably only people within the immediate vicinity would respond to that information, were it 'beamed out' from the eatery. However, for a science-fiction buff, a science-fiction book fair might be worth beaming out over a few kilometres, or even over a whole town.

In general, it would seem that most location-finding problems can be articulated with these four parameters. A key challenge for large LBS systems is to constantly monitor all four of these parameters for each user and for each potential other user, place or event of interest; all in real time. This presents a significant real-time data collection and sifting problem.

The objective of tracking these radii is being able to monitor when they intersect. The coalescence of radii is the cause of location events. Being able to detect and respond to these events is at the heart of many location-based systems. The more fine-grain the location information becomes (or higher resolution the LBS gets), the higher the frequency of events, compounded by the increasing numbers of users of LBS applications. The processing requirement can become a formidable task.

To keep track of events in real time is a major computing and software challenge that falls outside of what we have looked at in this book thus far, such as the link–fetch–response paradigm of the Web (HTTP/XHTML-MP). Figure 13.6 illustrates the essence of a system for tracking location events.

Figure 13.6 shows a generic architecture of an events-driven software platform to facilitate LBS applications. We have our now familiar J2EE platform at the bottom, which could be hosting the actual location-enabled application itself, or could be a gateway, like a Parlay X gateway; it doesn't really matter which at this stage of the discussion.

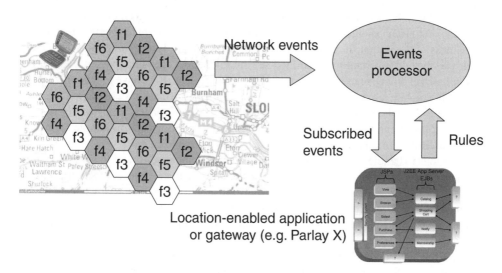

Figure 13.6 Processing of location events.

Two primary problems need addressing. Firstly, what is the nature of events processor itself? Secondly, what is the means of receiving events into the J2EE platform?

The first question is not something we shall address deeply, but it is worth dwelling on for just a short while, especially as this type of real-time processing is not something we may be used, especially if approaching the software design from a web background. An essential thing to point out is the nature of the events coming from the network, or, more crucially, the volume of events. If we are planning to monitor continually all the mobiles in our network, the frequency and volume of location updates will be vast, especially in a fine-resolution system. A user walking along a street in a major urban area could walk from one cell to another every few minutes. However, if we are using satellite tracking (Assisted GPS, which we shall discuss shortly), we may well be tracking updates much more frequently than this. Let's consider a user with a walking speed of something like 4 km/h, which means 4000 m every hour. If we assume something like 25 m accuracy (even less is possible), then we potentially have 160 location updates per hour for a constantly walking user. This has to be multiplied by the number of users, modulated of course by their own speed and resolution parameters. We can see that we are talking about large volumes of events, especially if they are concentrated into a single application (platform) for monitoring.

The problem of monitoring large volumes of events is not just the raw data flow, but the number of comparisons needed in order to interpret the data. At its crudest, this depends on how many opportunities are in the database. If we imagine a database with 100 opportunities, we have to compare the location of these opportunities against all the incoming location update events to see if we get a match (depending on what the condition is, but at its simplest, it could be the overlap of the two circles we looked in Figure 13.5, the circle of interest and the circle of opportunity). Of course, we can envisage all kinds of optimisation of the events monitoring solution. We wouldn't necessarily have to compare all opportunities against all events if we pre-arrange the data into geographical zones.

No matter the means of comparing location, the crux of the matter is that a hefty processing task is at hand. The whole process needs to take place in real time. It is a difficult design problem to achieve the minimum level of real-time response across a large volume of transactions. For example, a surge in the volume of events (e.g. as more and more users start moving faster) should not slow the performance of the system to the detriment of the timeliness of the output. In other words, if our sci-fi buff is approaching the sci-fi fair in their car, they do not want to be notified several kilometres after they passed it because a whole batch of other events crowded out the processing queue. If we have a location-enabled game, like mobile tag (see Figure 13.7), then real-time performance is critical to the game. We have to be able to detect constantly the whereabouts of players with respect to each other, so players can tag an opponent. There are many variations on this simple theme, all of them sensitive to timely processing of the location variables.

Solutions to handle vast amounts of input data and comparisons tend to use in-memory database technology. In-memory applications run entirely in the dynamic memory (RAM) of the computer, thus avoiding hard disks, which are orders of magnitude slower than RAM. Companies like TimesTen (which is now part of Oracle) produce solutions based on this approach that are specifically aimed at events processing, originally for financial transaction applications and more recently targeted at mobile applications.

Real-time events handling has been a challenge in the finance industry for a long time, just like many other problems from the finance industry that prompted solutions within the J2EE specification. Therefore, not surprisingly, the designers of J2EE included features that make J2EE suited to asynchronous events processing.

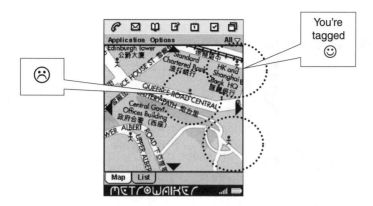

Figure 13.7 Mobile tag. (Reproduced by permission of Magic E.)

When examining how J2EE mobile applications can interface with the events processor, throughout the book we have focused on a design pattern for J2EE mobile applications that divides the logic into two parts; presentation logic and business logic. It is possible to feed events into a J2EE platform via HTTP, such as attaching event messages to HTTP requests using POST methods. To an extent, this is a viable solution. Indeed, it has to be viable if Web Services is going to work, especially in a Parlay X context where we can expect the Parlay gateway to send event-based SOAP messages to our application in just such a manner. However, we shall return to this consideration after we examine a feature of J2EE that is more ideally suited to message handling, namely JMS and a particular type of EJB (called a *Message-Driven Bean* (MDB), which we shall discuss shortly).

13.3 MESSAGE HANDLING USING J2EE

The concept of software collaboration via message passing is sometimes confused with collaborating via Remote Procedure Calls (RPCs), which is something different. RPC is more like what we discussed in depth when we looked at Remote Method Invocation, or RMI. Unlike RMI, messaging is a very loosely coupled process. The sender and receiver remain totally disassociated from each other. In fact, the sender of a message may not even know who consumed it. It could be a general message produced, like 'this just happened in cell 12' and any software process listening (subscribed) to events about cell 12 can pick it up. The fire-and-forget nature of message sending is a useful communications paradigm that we call *asynchronous messaging*. An example of the usefulness of asynchronous processing in general may serve us well, but still thinking about location services.

As Figure 13.8 shows, a user has unexpectedly arrived in a foreign place (perhaps for a last minute business meeting) and wishes to find and arrange accommodation, food and entertainment. Using the 'My Area Search' application, the user selects the required search fields and then specifies some particulars: four-star, non-smoking hotel room, somewhere to eat Indian food, and somewhere to watch a comedy film that evening. This type of request is a common requirement for travellers of all kinds. Location-based services have a lot to offer the traveller, particularly if the information service is available in the user's language and uses familiar terminology.

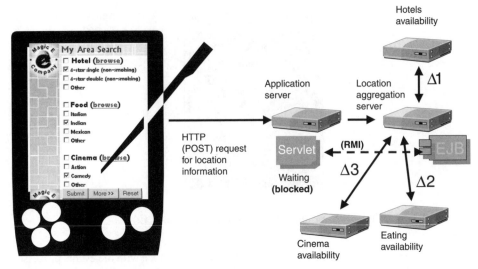

Δ1 or Δ2 or Δ3 could be long delays (minutes)

Figure 13.8 Traditional communications can get bogged down waiting for responses. (Reproduced by permission of Magic E.)

The emphasis in this example is on the query being a real-time concern for actual venue information, not just a static directory of venues (such as the *Yellow Pages*). For example, the user wants to know where the *available* rooms, not just hotels in general. The user would also ideally like to book a room, book the restaurant (unlikely[5]) and book the cinema ticket. The back end required to fulfil a request would probably involve interfacing with a variety of systems. For example, to understand hotel room availability, we need to find access to a hotel information system, perhaps several of them, covering a variety of hotelier chains. To access cinema seating availability, we would certainly need to interact with a cinema back end, perhaps for several cinema chains, too. The problem is that each of these queries could take a long time to process, especially if we are reliant upon the gathering of results from several sources. It is conceivable that parts of the process could even take minutes to complete, depending on the back-end processes, many of which could be on legacy machines[6].

Without examining the details of our system design, we can well imagine that at some point in the process a software call is needed to gather one element of the required information set. Subsequently, the calling software may have to sit idle waiting for a response (whilst a process somewhere on the end of the command chain gathers its data, such as on the hotel information system). In Figure 13.8, we show that a servlet on our application server is making an RMI call to a location aggregation server, which is subsequently being blocked; which means it is waiting for a response and cannot continue until it gets it. In other words, the servlet is blocked from completing its task, which is to hand the result back to a JSP for rendering the user interface (XHTML) via a reply HTTP stream. The consequence

---

[5] It is unlikely that restaurants would manage information to this degree. It is conceivable, and at some point likely, that major eateries would advertise some real-time data, like special offers and menus, possibly daily, but enabling seating availability to be placed online is perhaps too much to expect.

[6] We often forget that a large proportion of information systems continue to run reliably on legacy platforms that emerged long before the heady world of J2EE and Web Services, etc.

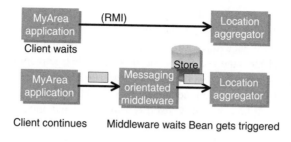

Figure 13.9 Messaging acts like an 'interlocutor' for the software call.

of this blocking is that the user, having submitted their web-based query (i.e. a form via HTTP POST) has to endure the agony of an irksome egg timer, or other animated 'progress' icon, whilst the whole process waits to complete. The user has no clue what is going on and would probably assume that the entire process has died and they may be tempted to abort (press 'stop') and start again, or get frustrated, possibly considering the entire service as dysfunctional.

Figure 13.9 presents an alternative to the blocked RMI call (top of diagram). Instead of making an RMI call to an EJB that makes the query and waits, the EJB fires off a message to all the information sources asking for the queries to be answered. The EJB then returns control back to the servlet, which can hand off to a JSP to generate a courtesy web page for the user. This page might state that the query is underway and that the user may have to wait for a certain time period to receive a result, which will be notified to them. At least this means that the user knows what is happening and they can then use their browser for other tasks while waiting, perhaps to check email, local tourist guides, or just finding a taxi from the airport to the town (where the hotel is hopefully going to be available). Back at the application server, reply messages will eventually arrive from the information brokers. Once enough information gathers, or a timeout interval elapses, the software can push out a WAP Push message to the user, steering them to a URL where the results are available (having been stored somewhere in a local database after the message gathering process).

As Figure 13.10 shows, J2EE has an API, called JMS, for sending software messages. Where do these messages go? Just as email messaging has an infrastructure of email servers,

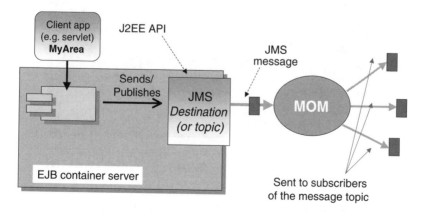

Figure 13.10 Sending messages using Java Messaging Service (JMS).

messaging has an infrastructure of *Message-Orientated Middleware* (MOM). Among other functions, the MOM provides the critical *store-and-forward* function of any messaging system that requires resilience to message loss. Messages are labelled with topics and then broadcast to potential message recipients and it is up to the recipient to take its message according to the message topics it is interest in (*subscribed* to).

We shall not discuss the detailed mechanics of a MOM and messaging in general, only mention the basic principles. The MOM itself is provided by a number of vendors. Popular products on the market include:

- Rendevouz by Tibco

- MSMQ by Microsoft (now part of .NET infrastructure)

- MQSeries by IBM

- Tuxedo by BEA Incorporated

The method for receiving a message into a J2EE application is via a special type of EJB, called a *Message-Driven Bean* (MDB), as shown in Figure 13.11. An MDB does not interact with any software entity (e.g. another EJB, or a servlet) to receive and process requests for its services. It can only 'come alive' by receiving a message with a subject (topic) that correlates with the topics it has subscribed to on the J2EE server. For example, if it subscribes to all messages with a moniker of 'Message about hotels in Munich', then in response to a message with the same topic passing through the MOM from some source, the J2EE server will pass the message to the bean and then cause the bean to execute. The bean has only one method (or function) and this runs when the bean runs in response to an inbound message of interest. The MDB is a J2EE program, so it can access any of the J2EE APIs it likes, including RMI, so it is able to call on the help of other beans to achieve any programmable task possible on the J2EE server(s). In our earlier example, an MDB would subscribe to several topics in relation to an initial enquiry from the user, such as hotel availability, cinema tickets, etc. Each time the bean 'wakes up' to a messaging event, it could amass the information (e.g. in a database) until all expected messages have been

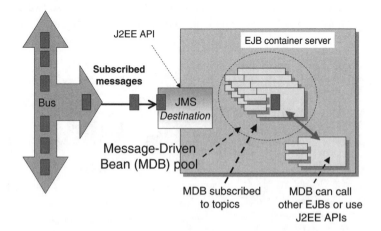

Figure 13.11 Intercepting messages using Message Driven Beans (MDB).

received to enable the user's initial query to complete. It could then use HTTP or JDBC to submit a WAP Push message to the user, informing the user that the query is finished. The message would contain a URL to reach an application (e.g. servlet) that can present the MDB's aggregated messages (from the database) in a meaningful manner.

It has become increasingly popular for mobile service delivery platforms to incorporate support for messaging systems, largely because their asynchronous 'fire-and-forget' nature enables systems to be constructed that are scalable in the context of high-volume, real-time event processing.

When thinking about location-events, it is easy to see how a message-driven software infrastructure maps neatly onto the event-driven world of location-based services. We can envisage a generalised mechanism for location-based applications that uses a messaging bus to pass round location events. These events could be a mixture of general events, of wide interest to many subscribers, and specific events, of interest to a particular user or group of users. The message topic can be used to discern intended recipients (if there are any). In addition, MDBs have a message selection logic that enables conditional execution, managed by the J2EE application server. For example, messages could be in constant circulation regarding shopping offers from a chain of supermarkets. The participating J2EE server could bring alive a particular MDB using a condition like, 'If the offers involves audio goods and is in the Maidenhead area'. If this condition becomes true, then the relevant message arrives at the MDB, which then runs, consumes the message and does something useful in response.

13.4 ACCURACY OF LOCATION-BASED SERVICES (LBS)

Returning to the methods of location finding, a major concern for any service or application is the accuracy of the location-finding process. Perhaps more important, is not just the accuracy, but also the consistent quality of location finding to a certain minimum level of guarantee, otherwise it is difficult to envisage a service where the user has to suffer hit-and-miss performance.

Figure 13.12 shows some of the problems associated with cell-based ID. Cell boundaries do not follow the neat honeycomb structure suggested by an idealised explanation of cellular systems. The edges suffer distortion due to the irregular physical topology of the cell and its physical influence on the strength of the RF signal throughout the area. Cells also overlap,

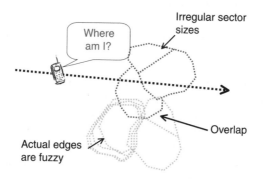

Figure 13.12 Uncertainty of location based on cell boundaries.

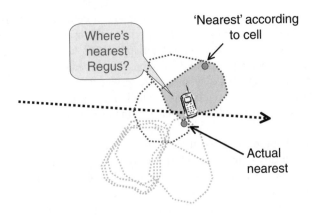

Figure 13.13 Uncertainties in cell size can distort location finding.

as they are required to so to a certain degree in order for mobiles to smoothly handover from one cell to another as they move around the network area.

Figure 13.13 gives us an indication of the affect that cell uncertainty could have. Let's say our user urgently needs to use a temporary office facility (such as the Regus chain in the United Kingdom), probably to host a last-minute meeting. The user consults with their location-finding application to ask, 'Where is the nearest Regus?' The figure indicates that the system determines the nearest location using a simple cell-overlap technique. In other words, the Regus in the same cell serving the mobile device is the one that is determined as being nearest. However, as the figure shows, the actual nearest Regus office is in a neighbouring cell, so it doesn't get picked up by the system.

This example of inaccuracy seems contrived and we might feel able to posit various ideas for attempting to lessen the problem. This is true. However, the use of cell-based finding techniques remains problematic and the above example is still a representative illustration of the issues. Cellular systems are notoriously unpredictable and difficult to tune – the RF goes wherever it pleases and it is not possible to tightly control cell site locations, never mind the obvious impossibility of controlling variations in the terrain. The vagaries of cellular communications remain apparent to all those who continue to experience less than satisfactory service from time to time, such as dropped calls or garbled speech.

However, perhaps the greatest problem with the Cell ID approach is variation in cell size. We acknowledge that the small cell sizes in dense urban areas are probably sufficient for many types of application, notwithstanding the above problems. However, the existence of small cells is incidental to the needs of location finding. They are small because there are many people in the areas they need to cover. This coincidentally allows accurate location services in those areas. However, the accuracy needs of an application will often have nothing to do with the density of people. For example, if we want to get the detailed whereabouts of another user (with their permission), then the small cells may be sufficient to support this if the user is in a small cell. However, if a user wants to use the same application, but in a more sparsely populated area with larger cells, then it no longer works well, or possibly not at all.

Figure 13.14 shows some actual cell site locations (triangles) near some of the water parks in Cotswolds hills in England. Returning to our E911 type of service requirement, if a tourist needed emergency assistance at one of the lakes, then trying to locate them using

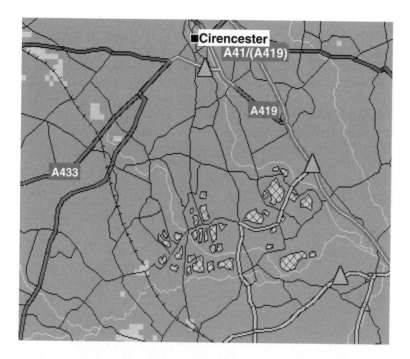

Figure 13.14 Cell locations near Cotswold water parks.

Cell ID is going to be problematic. The same would be true of any service that needed to locate a particular spot within the mesh of lakes.

Clearly, if we need high levels of accuracy, then we need to look at alternative solutions to compliment Cell ID.

Figure 13.15 shows us a possible improvement. The concentric rings emitting from the base stations indicate intervals of propagation distance at regular intervals of time, reminding us that RF waves travel at a finite speed, albeit very fast. If we can measure the time that signals take to get from the mobile device to the nearest base stations, then we have an idea of how far the mobile is from the cell site. A simple geometric calculation (*triangulation*, or similar) based on the Observed Time Difference (OTD) of transmissions from each base station will enable the calculation of the whereabouts of the mobile device, as shown in Figure 13.15 by the shaded area on the map close to where the darker rings overlap.

For this system to work, we need the mobile device to report the propagation measurements to the triangulation system to perform its calculations. However, by taking other measurements in the network, the triangulation accuracy can make better use of these reports. In GSM or UMTS a *Location Measurement Unit*, or LMU, makes radio measurements to support positioning of mobiles. In the case of OTD, the LMU measures the exact timing of transmissions in order to provide an accurate reference for making the comparative measurements. This enhancement to the measuring process, as well as enhancements to the handsets to better measure the timing differences, is an improvement over standard OTD, called *Enhanced-OTD*, or E-OTD.

The measurements from the mobile and the LMU are made available to a higher entity in the network that can perform location interpolation and decide where the mobile is based

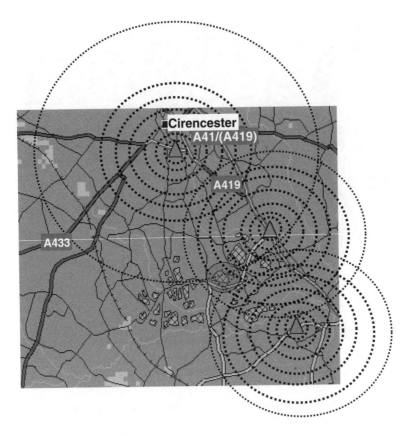

Figure 13.15 Using timing delays to improve accuracy.

on whatever coordinate system is used (e.g. Cartesian). This is called the *Serving Mobile Location Centre*, or SMLC – see Figure 13.16.

There are many variations on E-OTD, most of them using a variety of possible measurements and other metrics in the network to achieve the some basic triangulation approach. The accuracy of most of these systems improves when more fixed references are available to give us an indication of any distortion in the measurements made by the mobile. We can think of the LMU as being a fixed timing reference that enables reduction of uncertainty in the mobile measurements. The corrective techniques are mostly attempting to overcome the fact that because the RF seldom reaches the mobile directly (*Line-of-Sight*, LOS), then any measurement by the mobile cannot be relied upon to be a true indication of its distance from the cell site. The reported distance and the actual distance are seldom the same and some gross distortions are possible under certain landscape topologies, both in urban or rural environments.

The most accurate method of location finding is the *Global Positioning System* (GPS) network of satellites (see Figure 13.17). This is essentially still a triangulation technique, but with a direct LOS from the mobile to the satellites, plus a superior ability to measure timing offsets. The offsets allow measurement to a fine degree, translating into an on-the-earth resolution of a few metres. Due to its military origins and uses, the publicly accessible signals are deliberately conditioned to introduce measurement errors, a process

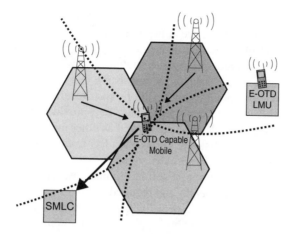

Figure 13.16 Observed time difference used to improve location-finding accuracy. (Reproduced by permission of Magic E.)

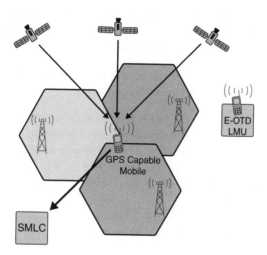

Figure 13.17 Global Positioning System (GPS) uses satellites. (Reproduced by permission of Magic E.)

called *Selective Availability*. However, after September 2000, this was switch off and so the full accuracy of GPS is available for civilian use, including mobile applications.

With a basic GPS receiver, we may expect better than 20-metre horizontal accuracy on the ground. The performance will vary depending upon the particular receiver and the level of solar disturbance of the ionosphere. However, there are several techniques to improve accuracy, one of them called *Differential GPS*, or D-GPS. This uses receivers in known locations to measure their location and apply various models of the GPS network parameters that would work to correct the reported measurement, bringing it back to the actual location. A well-tuned GPS system could reach accuracies of 3–5 metres, whereas a well-tuned D-GPS could reach accuracies of 1 metre.

The system proposed for location finding in cellular networks is *Assisted GPS*, or A-GPS, which sends GPS measurements made at the base stations to the mobiles so that they can better perform GPS calculations. Demodulation of signals from the GPS satellites occurs at a relatively slow rate and requires that the satellite signals be relatively strong; this can be perturbed by difficult terrain, such as the user being in a dense urban area with lots of buildings. To address this limitation, an A-GPS receiver receives context data from an A-GPS unit in the network (e.g. at a cell site). Context data provides the receiver information that it would normally have to demodulate as well as other information, which decreases start-up time (i.e. to make the first satellite location fix) from possibly two minutes down to somewhere in the region of 5–10 seconds.

Although A-GPS sounds fantastic, it still has limitations, especially coverage because it is subject to usage patterns whereby the required satellites are in direct LOS from the mobile. This is not always possible, especially in built-up areas and when indoors.

A possible solution is to adopt a hybrid approach using a combination of location solutions in effective unison, to gain the best of both worlds, such as the coverage of Cell ID combined with the accuracy of A-GPS. Working with the Daimler-Chrysler Corporation, Dr. Elizabeth Cannon of Calgary University (Canada) has been exploring ways[7] of combining GPS with existing, commercially available, inertial navigation systems – sensors inside a vehicle that record its rate of acceleration and direction. Her work focuses on developing algorithms and error modelling that will provide the best mathematical ways for merging the two different types of information provided by the GPS and inertial systems. Dr. Cannon says:

> We want to develop a car-based system that would allow it to continuously position the vehicle to centimetre-level accuracy in real-time...

Whilst this extremely precise positioning is perhaps excessive for buddy finding or E911, it's the kind of ability that opens whole new realms of possibilities, such as autonomous driving and ultra-accurate location services. 'Accuracy is addictive', Dr. Cannon notes. 'People start to think, I could do this or that.' Perhaps in the near future, jet-powered camera drones will be able to fly straight to the scene of an accident to assess the situation ahead of emergency personnel arriving.

This is a nice segue into raising the question, 'So what accuracy do we actually need from a location-finding solution? There is no single answer of course. It depends on the application. However, the comment from Dr. Cannon seems an important observation and perhaps a motivation for pursuing the most accurate systems possible from the outset. We may well find that accuracy is a key driver to the success of LBS. At the end of this chapter, we shall explore possible service ideas that examine many of the technologies we have investigated thus far in the book. We shall see in the example that very accurate location finding is a necessity.

13.5 INTERFACING LBS APPLICATIONS WITH THE CELLULAR NETWORK

The remaining issue is how to get the location information into our location-enabled application. We have previously discussed accessing location information via Web Services APIs

---

[7] http://www.eng.ucalgary.ca/Press_Release/2002/SteacieFellowship2002.htm

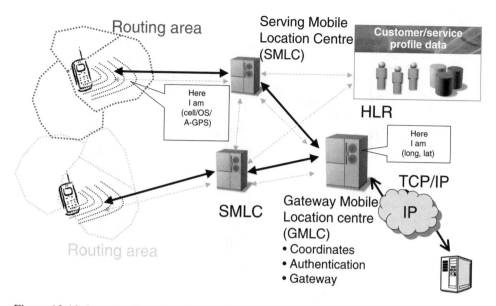

Figure 13.18 Location-based services architecture in the cellular network.

like OSA of 3G, which is an implementation of Parlay (Parlay X being the Web Services presentation).

Figure 13.18 shows the basic architecture of the location infrastructure in a cellular network. Just as we saw that clusters of adjacent cells formed routing areas for SGSN data concentration, there are similar routing areas of the network, each served by a SMLC. This is responsible for gathering the actual location measurements from devices and for the presentation of this information in a useful and agreeable form to the higher entities in the network, including, eventually, our location-enabled applications, J2EE or not.

The information from the SMLCs is concentrated into a *Gateway Mobile Location Centre* (GMLC), which is a secure boundary point for external entities to attain mobile location information about a particular subscriber. Apart from providing the main interface with the outside world, the GMLC makes location information available in terms of subscriber ID and not device ID; hence, why it has a relationship with the Home Location Register (HLR), which stores user information that enables device ID to be associated with user ID. This is necessary because within a cellular network, a phone number does not identify a device, but an International Mobile Subscriber Identity (IMSI) does. Therefore, when asking for a mobile's location, an application probably expects to enquire using a phone number and so translation needs to take place to the IMSI.

The GMLC can present location information in certain formats. It is up to the operator how they want to make location services available to third parties, but clearly a good deal of housekeeping functions are also required, such as authentication, account management, charging management and so on. These could all well be taken care of via a Parlay or Parlay X gateway that sits in front of the GMLC and becomes the main interface point for location applications, as shown in Figure 13.19. It is unlikely that a location-enabled application would ever interface directly with the GMLC.

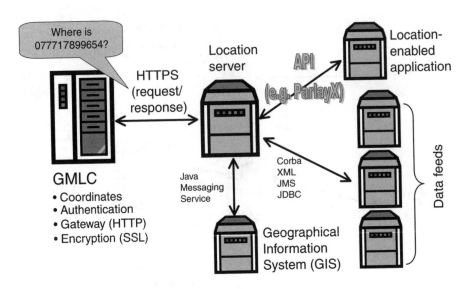

Figure 13.19 Possible LBS application support architecture.

Figure 13.19 shows several features of a possible LBS application environment, or service delivery platform. The GMLC itself is the main point of interface to the cellular network, it providing:

- HTTP access to coordinate information

- Authentication to protect unauthorised use

- Encryption (SSL) to protect location queries

As the figure shows, it is unlikely that the GMLC is ever directly available to the applications that need to use its facilities. Intermediate 'middleware' servers will provide additional services and present them via suitable APIs to the applications that require location information, be they internal to the mobile operator or external (i.e. third-party applications).

Typically, we would expect a Web Services API for location-based services, although whether or not this is Parlay X is another matter. Perhaps, eventually most Web Services service points in mobile networks will become Parlay-compatible and the Parlay X API will grow in its sophistication. However, it has to be said that a large number of more specialised 'non-telecoms' services will emerge from operator gateways, and these will have a Web Services presentation, but probably not within the Parlay X framework initially. For example, multimedia content management systems, photoblogs[8] and other such services will probably become standard to the point of becoming utility functions that other services could utilise; hence, why they shall eventually become available for third-party access. Later on, an extensible framework may emerge whereby the Parlay framework may allow for the addition of extended features. This is the beauty of the loosely coupled nature of Web Services in general – it hinges on accessing services via URIs and this makes it possible for

---

[8] Photoblogs are weblogs that enable a photo diary to be kept by sending pictures from a mobile camera phone.

service providers and application developers to easily aggregate a wide set of Web Service resources into a single service or application portfolio.

An example of a location service that an operator might want to offer as a utility service to third parties is a *Geographical Information System* (GIS). This provides many of the basic services associated with geography, such as maps and place locations, and this is a function beyond the Parlay X API set. A critical element of a GIS is the ability to perform *geocoding*, or the reverse process. Geocoding is the transformation of geographical data into polar or Cartesian coordinate format. For example, if we take a directory of restaurants, then more than likely their postal addresses will be in the directory, but not their geographical coordinates, as this would probably be an unusual parameter in a standard directory of restaurants. Therefore, the directory needs to be geocoded, which is to identify postal addresses in terms of coordinates. This will then enable a location-enabled application to process the whereabouts of restaurants.

Products already exist on the market, such as ESRI's ArcWeb, which is a hosted GIS solution that offers a Web Services (SOAP) interface, with features like:

- Accesses dynamic street, demographic, and topographic maps, shaded relief imagery and more

- Determines the locations of street addresses

- Generates routes and multilingual driving directions between multiple locations

- Provides place finder capabilities

- Performs census demographics mapping and reporting

- Uploads user-defined points of interest for geocoding – ESRI hosts this data for the client, so that the client does not have to host it

- Performs reverse geocoding (going back from coordinates to postal addresses).

- Provides thematic mapping

- Displays maps using a choice of projections

ArcWeb for Developers works with Internet standards including HTTP and XML. It uses XML-based SOAP to communicate, making it compatible with the majority of Web Service frameworks available today, including Parlay X. This means that our developer community using J2EE design patterns centred on SOAP Web Services could easily integrate GIS features into their Parlay X application; this should greatly enhance what can be done using Parlay X, as we shall examine shortly.

The Web Services from ArcWeb is constructed from components divided into the following functional areas:

- *Place Finder*– ranks a candidate list of place names and associated latitude/longitude coordinates for a given input place name.

- *Address Finder*–determines the latitude and longitude coordinates for street addresses.

- *Route Finder*–returns textual multilingual driving directions for a multipoint user-defined route.

- *Map Image*–provides access to a wide variety of dynamic maps.

- *Proximity*–returns all Points of Interest (POI) locations within a user-defined distance of a specified location (i.e. find all POIs within 5 miles of x, y) or determines the nearest specified number of POI locations to a specified location (i.e., find nearest 3 POIs to x, y).

- *POI Manager*–allows you to upload a custom set of points to ESRI, where they are geocoded and stored for later use.

- *Query*–determines the physical, environmental, or cultural characteristics of a specific location.

- *Utility Service*–allows you to change the projection of your map image.

- *Account Info*–lets you use a SOAP interface to access account information (such as usage statistics).

ArcWeb is part of a wider suite of products, called ArcLocation, which can be used to construct a service delivery platform for location-based services. One part of the suite is the Solutions Connector. This is an API suite that handles integration between ArcWeb Services and industry-standard mobile applications and Web servers. It also handles integration between ArcWeb Services and the GMLC. It currently supports the Location Interoperability Forum (LIF) Mobile Location Protocol (MLP).

As we might expect, MLP uses an XML-based vocabulary. The service level elements are XML messages and there is no specified transport, but it could be HTTP, SOAP, or any available method for exchanging XML messages. The baseline MLP specification suggests how to use HTTP, using the POST method to submit queries. Whilst this sounds similar to SOAP, recall that the XML messages in SOAP are themselves packaged in the SOAP envelopes (also XML) before appending to the HTTP stream, even though SOAP is also transport-agnostic, though commonly seen in conjunction with HTTP for obvious reasons to do with its prevalence.

What an operator may provide in their service delivery platform is a connection into various information services, such as place directories, events directories and so on. This information could be used by third parties or by operator applications, most likely on a per-enquiry charging structure.

A plethora of exciting mobile services can be envisaged thanks to the availability of a powerful service delivery platform for LBS, especially with a Web Services presentation; even more powerful if it is available alongside a Parlay X interface. The application developer can pull in resources very easily, seamlessly integrating features of many Web Services components into one application or service portfolio. Clearly, the Web Services paradigm works well with our application based on J2EE, but it is also likely that much of the platform itself is implemented using J2EE, such is its appeal to platform developers.

Some interesting possibilities arise when J2EE is used in conjunction with J2ME clients on the mobile devices. For example, ESRI have a J2ME API available that seamlessly integrates into their Web Services platform. The ArcLocation J2ME toolkit is used to build mobile client applications that effortlessly consume ArcWeb Services. It is based on CLDC and the MIDP (refer to Chapter 11). Applications built using the toolkit will run on all devices that support MIDP, so this is an exciting possibility for MIDP developers.

13.6 INTEGRATING LBS APPLICATIONS

In the introduction to the book (Chapter 2), we talked about becoming 'smart integrators', being able to take the new technologies and rapidly integrate them to provide exciting mobile services. It is highly likely that this is the future of mobile services; a constant stream of services that users selectively subscribe to according to their needs. It is even possible that users could construct their own services to meet their precise needs. Although the means to do this requires some further thought, there should be no doubt that such an eventuality is possible given the highly conducive nature of the underlying technologies we have been discussing throughout the book in terms of facilitating flexible and rapidly changeable solutions.

Earlier in this chapter we talked about the nature of location services being very much event-driven. There are so many possibilities for location services and many lead to the generation of events. For example, it has become increasingly popular to buy items via online auctions like eBay. For some items, the postage and packing costs are sometimes prohibitive. Also, there is a tendency to have a passing interest in certain goods, but not a compulsion to buy them. Under these circumstances, the use of location-based services may assist the buyer by introducing them to goods available within their locality, wherever that happens to be. Perhaps the user has a casual interest in acquiring a certain type of old car, just for fun, perhaps as a restoration project.

We can imagine a service whereby the user indicates their interest in a particular car, as shown in Figure 13.20. Here we have the buyer submitting their details about the car to an 'agent'. This is a fancy word for the concept of a software process doing work for the user, on their behalf, as an ongoing background task whilst the user does other things. The user might not be actively using any service on their device. It could be dormant. But the agent in the network carries on its task, which in this case is looking for the car, but within the context of the geographical movements of the user, searching for suitable cars within a reasonable proximity of the user at any one moment in time as they travel around. Of course, the use of the word 'agent' tends to indicate a process that is permanently engaged

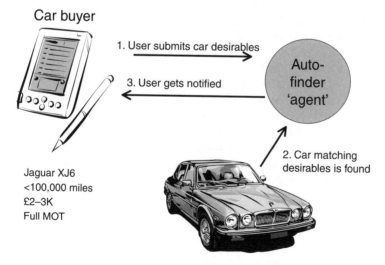

Figure 13.20 Using an 'agent' to find a product.

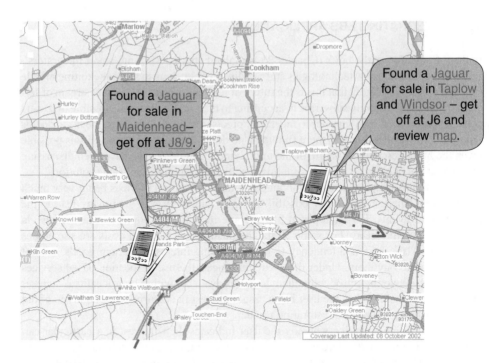

Figure 13.21 User being notified of nearby cars for sale that may interest them.

in hunting on the user's behalf. This is not quite the case. What is more likely is that a particular condition is set up in an events-processor platform, an idea that we discussed earlier when looking at detecting the overlap of the two radii: the radius of opportunity and radius of interest. In a moment, we shall look at possible manifestations of the application at the J2EE level.

What the agent does is to facilitate a service that alerts the user whenever he or she comes within a reasonable proximity of a car for sale, as shown in Figure 13.21.

As the user proceeds down a major route, shown on the map, the service might send alerts at various intersection points. Here we are able to combine many powerful features from the operator's service delivery platform to make this application work.

Firstly, we need to detect that the user is within range of a car for sale. This happens by setting an event condition. This part of the service relies heavily on asynchronous messaging across several loosely coupled platforms. We are interested in two types of event. First, the event that indicates a car has become available for sale that fits the user's criteria, irrespective of where it is located. The second event is simply the movements of the user moving around from one place to another; a stream of location updates. We can imagine the first type of event as causing messages to be sent from the car database to the application, causing a MDB to insert the item into the location-event processor. Now, if the processor works by comparing the proximity of coordinate pairs, whatever their format[9], we need to first geocode the car's position, because it will enter the system with a postal address, but we

[9] Coordinates in the MLP have different forms, such as Ellipsoidal and Universal Transverse Mercator (UTM).

need to calculate proximity using coordinate information. This conversion could be done via a Web Services call to the location service delivery platform, utilising a 'geocoding API'. The coordinates can then be submitted to the location-event processor, which is a real-time database like TimesTen. The location events of the user are subsequently fed into the location events processor and whenever a proximity match occurs, the processor fires out a software message via the MOM.

Part of the agent is implemented as a MDB that subscribes to messages' topics related to our user's requests. The exact construction of topics is a matter of system design, but it could be, for example, that we have instances of MDBs that listen for messages labelled with the user's phone number as the message topic. Various addressing techniques and design patterns need to be considered for implementing such a system efficiently, but even a rudimentary paper design will reveal a variety of possible approaches very quickly. The kind of trade-offs that the designs will have to tackle will be system resources versus real-time performance (as ever).

By designing our system in this way, the MDB will come alive when the user enters the proximity of an available car. What we could envisage is that the message itself contains the cars details and whereabouts. What the bean does is to encode the location and car details into a WAP Push message, using anchor tags (hyperlinks) to enable the user to access details about the car and details about its location.

Using the GIS platform, the WAP Push message can link to mapping information and also directions for getting to the car. However, it is likely that the user will first want to understand more about the car. This could be done in several ways. The user could link to a URI that has the car's details. The user could also be offered the chance to place a call with the car owner. This could be done via the browser itself, via the inbuilt dialler application on the device picking up the number for a link, or via a Parlay X third-party call API.

The J2EE components shown in Figure 13.22 are by now familiar to us. The presentation logic, constructed from servlets and JSPs (as per the model–view–controller pattern) takes care of the user interface. We have elements to configure the application, such as setting preferences and configuring the agent. We have elements to view the results, like selecting maps, getting direction and checking the availability of the car by viewing its online catalogue entry. On the back end, the EJBs are responsible mainly for integrating the services from the service delivery platforms into our application. We have a messaging interface (via JMS) into the location server to set up the agent events and a messaging bus to receive alerts from the location agent's server (events processor). There is also a JDBC database connector into the agents database itself, as we will need somewhere to store all the various agents' configurations, not just for each user, but for each product or service of interest, and there could potentially be hundreds per user, especially for a general-purpose location-enabled agent, not confined to just searching for cars.

It is possible that the user will not be able to, or want to, respond to a location alert. Perhaps our user didn't have sufficient time to go and investigate, or they were unable to contact the seller, for whatever reason. A variety of follow-up responses is possible in that scenario. In the current example in the figure, we use the JavaMail API to send an email to the user with all the event details included, so that a record of the event is then stored in the user's inbox for later reference. This can then be accessed by the user later on as a non real-time reminder about the event, so that they can return to it when more convenient, either from the mobile device or via desktop access.

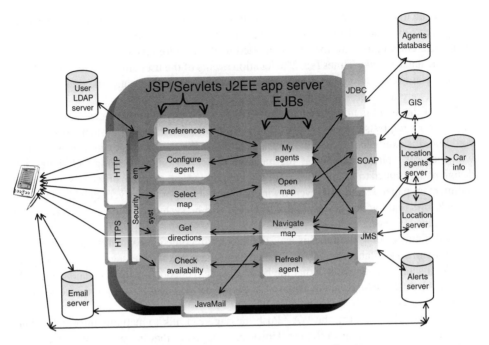

Figure 13.22 Possible J2EE architecture for the car agent application.

The alerts server shown in the figure is an aggregated platform that can support WAP Push, text messaging and even multimedia messaging. The latter is a possible option for including a photo of the car.

Although greatly simplified and contrived, we can appreciate from this brief application example how the various technologies and software services can be brought together and smartly integrated, especially taking advantage of the loose coupling of both the Web access combined with the loose coupling of event-driven software design. The ability to access some major, very powerful, APIs via Web Services and Java Messaging make for a much easier implementation path and a greatly enhanced end-user experience. If the Web Services and other service delivery platform aspects not available, it is difficult to see how an application like this could be successfully delivered, at least not without a significant investment of resources. Were this investment and specific low-level design to be required, in the absence of 'smart integration' technologies, the likelihood of lots of these sorts of applications arising would be very low.

Figure 13.23 reminds us that our location-based service is actually taking place across a very complex network of powerful entities. The application of J2EE technology and the service-delivery platform concept enables our application development to be free from knowledge or involvement with the inner complexities of the network of networks that make up the mobile network we have been touring throughout this book.

New possibilities are emerging all the time. For example, it is now possible, using products like iBus from Softwired AG[10], to send software messages across the

---

[10] http://www.softwired.ch/index.html

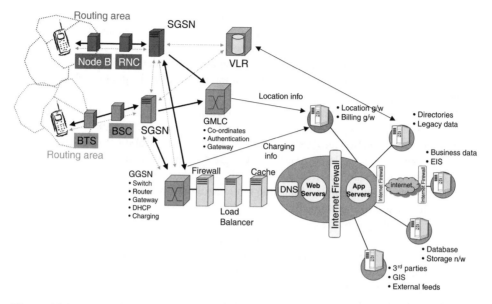

Figure 13.23 The entire system is a complex one.

message-orientated middleware ('bus') directly to MIDlets running on the mobile devices, which is an exciting possibility. It is an interesting idea that probably offers major benefits in enabling some powerful design patterns to emerge for event-driven applications, especially those that seem to arise in LBS service portfolios, due to their common nature. For location services in general, there is no single best-fit solution, so it is not possible to say that wireless Java messaging is in any way an ideal approach. Nevertheless, knowing that this is possible is a major guiding star to add to our constellation of possibilities within the mobile service technology cosmos.

Shortly, we shall look at another example of a location-based service, one that ties together even more technologies than the example we have just considered. However, before doing that, we should briefly investigate what the Multimedia Messaging Service is all about.

13.7 MULTIMEDIA MESSAGING (MM)

Here we shall examine a brief overview of the *Multimedia Messaging Service* (MMS). This is so we can consider alternative delivery mechanisms to the web-centric approach we have looked at so often in this book, even though MMS and web technologies are closely related, which is not surprising.

The push paradigm is clearly a powerful part of mobile service considerations, and it is almost unique to mobility, although we could pretend that email is effectively a push technology in one sense. However, the power of push technology in the mobile context is its immediacy, given that we carry our mobile devices with us most of the time. Thanks to the intimate relationship with a mobile device, it is possible to interrupt the user at any time with information that may be of use to them. In this chapter, we have been building both the 'business case' and technological framework for being interrupted by location-sensitive events, even making the claim that the ability to monitor and notify events in this context is probably a key enabling resource in a mobile network. Its importance cannot be overstated, which is why we have dedicated this entire chapter to examining this aspect of the RF network alone, in isolation, and its close ties to the other networks in our mobile network model.

The ability to send a text message is commonplace and barely needs explaining, other than our earlier glance at the ability for our application to interface with the text-messaging infrastructure via Web Services APIs, like Parlay X. However, the ability to send multimedia messages needs some explaining, largely because the content is more complex and the delivery mechanisms are different than for plain text messaging, so the possibilities need exploring. Our earlier example of pushing out location information, such as cars for sale, seems to lend itself to a multimedia message. The thumbnail photos of cars is a common feature of car selling activities, so to extend this to the mobile world seems a natural fit. Pushing the picture of the car, as opposed to asking the user to browser to it, seems to have immediacy and an appeal that enhances the user experience. This is probably true of many similar event-driven scenarios. We now need to see what type of content can be embedded in an MMS message and how to do it. We also need to look at how this facility relates to the web design patterns we have been considering thus far throughout the book.

The easiest model for think about MMS messages is to think of pushing simplified web pages to the user. In fact, this is rather an apt description. The way that MMS delivery works is by sending an initial *Service Indication* (SI) message to the device, which contains a URL embedded in a text message, as shown in Figure 13.24. The MMS client on the device then automatically pulls down the contents from the URL and via HTTP or WSP (or W-HTTP). The SI message is a WBXML tokenised message in order that its contents are compressed enough to fit within the 160-character text-messaging limit.

The MMS client fetches the content from a server that is storing the message that was sent to it via HTTP or SMTP. As the figure shows, the encoding of the contents of the message is in one of two formats: WML or SMIL. We have met WML before in our discussion on mark up languages and WAP. We have not yet met SMIL (pronounced 'smile') and we shall examine this shortly.

According to the UMTS specification [3G TS 23.140 V3.0.1] for MMS, there is no definitive presentation format for a Multimedia Message, or MM. The visual rendering

Figure 13.24 Mechanism for sending an MMS message.

on the device of an MM is known as *presentation*. Various types of data may be used to drive the presentation. For example, the MM presentation may be based on a WML deck or *Synchronised Multimedia Integration Language* (SMIL)[11]. Other presentation models may include a simple text body with image attachments. WAP has not specified any specific requirements for MMS presentations. UAProf content negotiation methods should be used for presentation method selection and earlier in the book we described how this can be done when pushing data to a device[12]. However, clearly for MMS to be successful, a degree of conformance is required between devices, but we shall discuss this in the next section.

Since its adoption for MMS, the use of SMIL has gone through several interpretations, just as we saw with XHTML, XHTML Basic and then XHTML-MP. However, before we examine those, let's briefly look at SMIL and what it can do.

13.7.1 Composing MMS Messages

SMIL is an XML vocabulary. It enables multimedia elements to be presented within a compatible client (SMIL browser or SMIL player). We are already used to multimedia elements being possible within a web environment. We can view images, read text, play videos (via players) and listen to sounds, so why the need for yet another markup language? Multimedia authors, especially from the world of animated and sequenced multimedia presentations, have long been used to the concept of a time base. This is the idea that within the presentation of information, there is a need to stipulate what happens when. This is the most critical aspect of SMIL. In short, SMIL allows authors to write interactive multimedia presentations. Using SMIL, an author can control the timing of events within a multimedia presentation, associate hyperlinks with multimedia elements and control the visual layout of the presentation on the viewer's screen (both in 2D and across time).

In moving from version 1.0 to 2.0 SMIL can now be used to instigate a time-based control to XHTML elements and also another language called *Scalable Vector Graphics*, or SVG. This combination enables nearly any kind of visual presentation to be constructed and delivered across the Internet using open standards.

With SVG, instead of using graphics files to represent images, XML commands can be used to indicate drawing primitives, such as rectangles and circles.

With SVG, a rectangle would be rendered by the XML description as follows:

```
<rect fill="#69ABFF" y="200px" x="16px" width="200px"
      height="100px" style="stroke: #69ABFF" />
```

This will render a rectangle that is sky blue in colour (as indicated by the RGB values #69ABFF) and would be 200 pixels wide, 100 high, placed on the screen 16 pixels across and 200 pixels down (from the top left hand side, which is the origin) – see Figure 13.25; you'll have to imagine the rectangles are blue.

---

[11] http://www.w3.org/TR/SMIL2/

[12] The point being that UAProf information is normally only available when a user agent requests web resources, not when pushing to a hitherto unknown user agent with hitherto unknown display capabilities.

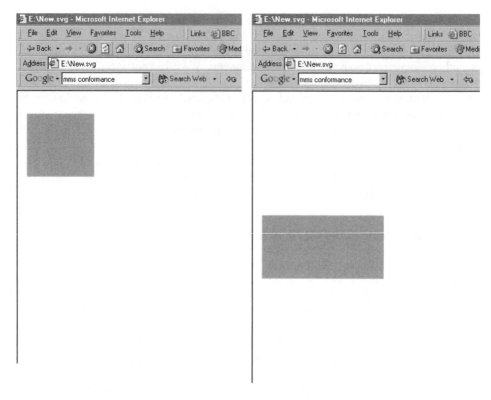

Figure 13.25 Animation of a rectangle using SVG and SMIL.

Using the animation module of the SMIL language, it is possible to add animation to the rectangle by controlling its attributes over time.

```
<rect fill="#69ABFF" y="200px" x="16px" width="200px"
      height="100px" style="stroke: #69ABFF">
      <animate attributeName="width" from="100px"
      to="200px" begin="0s" dur="3s" />
      <animate attributeName="y" from="15px"
      to="200px" begin="0s" dur="3s" />
</rect>
```

The <animate> tags stipulate that the attribute named within the tag should be ranged from one value to another over time, the duration being specified by the 'dur' attribute.

The ability to describe any number of primitives with SVG and to animate them gives us a very powerful graphical presentation technique. To create this animated sequence using XHTML and standard images is difficult and consumes a lot of memory. Clearly, with SVG, is takes just a few lines of XML in a compatible viewer.

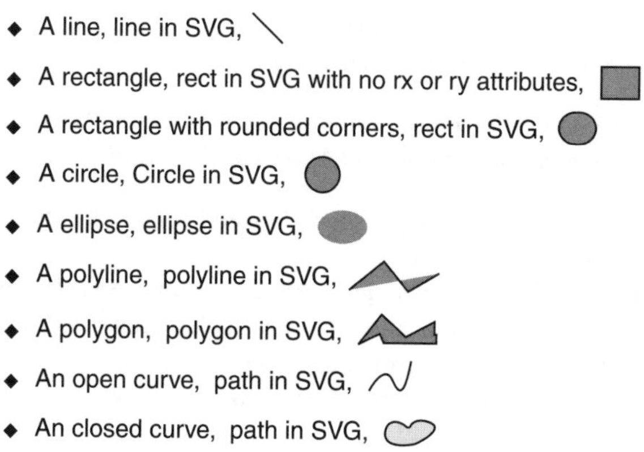

- A line, line in SVG,
- A rectangle, rect in SVG with no rx or ry attributes,
- A rectangle with rounded corners, rect in SVG,
- A circle, Circle in SVG,
- A ellipse, ellipse in SVG,
- A polyline, polyline in SVG,
- A polygon, polygon in SVG,
- An open curve, path in SVG,
- An closed curve, path in SVG,

Figure 13.26 Different drawing primitives supported by SVG.

There are many drawing primitives available in SVG, as shown in Figure 13.26. It is also possible to have layers in the rendering, similar to SMIL, so primitives can appear on top of each other as required.

SVG has some powerful possibilities for mobile services. SVG 1.1 is modular in the same way that XHTML is modular. This makes it possible to pick-and-chose the modules in forming new profiles. Naturally, lightweight profiles of SVG have emerged for wireless devices (and devices other than the desktop computer). The central effort in this regard is via the W3C project *SVG Mobile*. From SVG Mobile we have two new SVG profiles[13]: *SVG Basic*, which is a subset of SVG targeting PDA devices, and *SVG Tiny*, a subset of SVG Basic targeting mobile phones.

The 3GPP, in their definition[14] of requirements for codecs and standards for mobile phone presentation formats, state that SVG Tiny should be supported by next generation phones, and preferably have full SVG Basic compliance.

SVG Basic is rather close to full SVG in terms of its features, but leaves off some of the more complex elements that require a lot of CPU resource. SVG Tiny goes further in stripping scripting, filters, gradients, patterns, opacity, and style sheet (CSS) support. However, perhaps crucially for mobile applications, the *Animation Module* is preserved in SVG Tiny. Naturally, this is supported in keeping with the trend towards animated messaging content, as supported by SMIL in MMS.

As we previously learnt when looking at XHTML, and XML as its foundation, the modularised versions of XML-based languages enable any one of them to act as a host language. This means, for example, that an XHTML document could include SVG elements. Similarly, SMIL, which is also XHTML based, could contain SVG elements. SMIL itself could also contain XHTML elements, although that possibility is not yet supported (or suggested) by 3GPP.

Whilst SVG Tiny is a mandatory media type stipulated by 3GPP, this does not mean that all phones will support its inclusion in MMS messages. The basic minimum standard for

---

[13] W3C Recommendation: 'Mobile SVG Profiles: SVG Tiny and SVG Basic', http://www.w3.org/TR/SVGMobile/, January 2003.
[14] 3GPP TS 26.234 V5.5.0 (2003–06)

MMS is defined by the MMS Conformance document[15] maintained by the WAP Forum (now Open Mobile Alliance, OMA), which states:

> *The MMS conformance document defines the minimum set of requirements and guidelines for end-to-end interoperability of MMS handsets and servers.*

Let us first look at MMS in more depth to understand its capabilities and to see how to interpret the conformance requirements.

As already mentioned, the primary language used to compose MMS messages is SMIL[16]. This language takes the approach of allowing the display area to be divided into regions, which is called a *layout*.[17] Having defined a layout, the various regions within the layout can be used to display multimedia elements in a time-controlled manner. If we take the following code as an example:

```
<smil>
  <head>
    <layout>
      <root-layout width="160" height="140"/>
      <region id="Image" width="160" height="120"
      left="0" top="0"/>
      <region id="Text" width="160" height="20"
      left="0" top="120"/>
    </layout>
  </head>
  <body>
    <par dur="6s">
      <img src="pghead.jpg" region="Image" />
      <text begin="2s" dur="2s" src="greetings.txt"
      region="Text" />
      <text begin="4s" src="TheEnd.txt"
      region="Text" />
    </par>
    <par dur="10s">
      <img src="message.jpg" region="Image" />
    </par>
  </body>
</smil>
```

The code clearly shows two blocks encased by the <head> and <body> tags, very similar to XHTML, with the <layout> tags nested in the <head> tags and defining the screen regions we referred to a moment ago. In the XML, we can find two <region> tags that define areas we label "Image" and "Text", where we are eventually going to place image

---

[15] OMA-IOP-MMSCONF-2_0_0-20020206C: MMS Conformance Document available from http://www.openmobilealliance.org/

[16] Synchronized Multimedia Integration Language (SMIL 2.0) W3C Recommendation 13 December 2005 http://www.w3.org/TR/SMIL/

[17] A *layout* would be similar in concept to a *frameset* in HTML.

Figure 13.27 MMS message is compartmentalised into regions using SMIL layout tags.

and text *media objects*, as shown in Figure 13.27. On the top we see the "Image" region and on the bottom we see the "Text" region. These are arbitrary regions as far as SMIL is concerned, and we could define many regions all over the viewing space, to be used by whatever media objects we like. However, the MMS conformance document indicates a limited structure, like the one shown here. This is in order that messages get correctly rendered on as many devices as possible in a way that is consistent with the original message composition, thereby facilitating interoperability between devices within what is otherwise a wide range of capabilities.

If the layout is divided into four quadrants, like the flag of England, then this may prove problematic for some devices. It may be that many devices can support the layout in principle, but may have to display it in a modified form if the media objects are larger than the screen size apportioned to each quadrant, as shown in Figure 13.28.

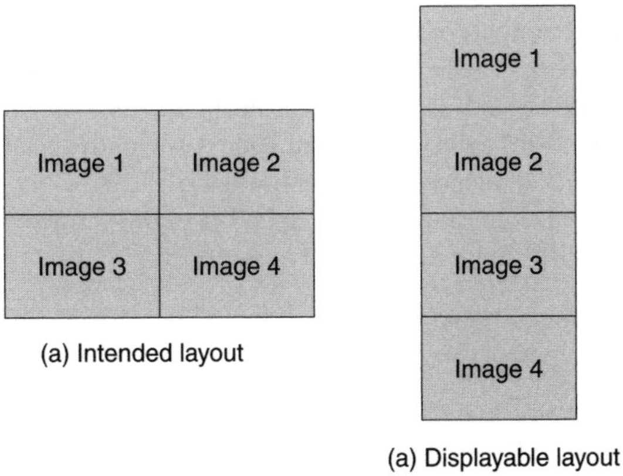

Image 1	Image 2
Image 3	Image 4

(a) Intended layout

Image 1
Image 2
Image 3
Image 4

(a) Displayable layout

Figure 13.28 Intended layout modified by device with narrow display.

Figure 13.29 Using SMIL, MMS messages can cycle media objects in a 'slide show'.

Examining the SMIL, we can see that there are two sub-blocks in the body with <par> tags. We can think of these as being like slides in a slide show, except each slide is a multimedia container rather than a static image.

For the first 'slide', the duration (dur) is 6 seconds. Upon loading, the first media object is loaded, which is an image (<img> tag) which is an external image file addressed by its URI (relative to the document). The <img> tag specifies that the image is going to be loaded into the "Image" region (aptly named, but the name is in fact arbitrary in SMIL, as long as it conforms to the XML rules of being well formed). Hence, we can see that the first image ("pghead.jpg") gets loaded into the region labeled "Image" as seen in Figure 13.27.

We would like to add some text to the first image, so this is done using a <text> tag to refer to an external file, which in this case is a plain text file containing the string 'Totally Distracted by . . .'. In the 3GPP specifications, the <text> tag can refer to an external XHTML file that contains text with XHTML formatting, but we have not shown this here. Referencing XHTML is not part of the MMS Conformance recommendation. The critical parameters in most of the tags are the 'begin' and 'dur' tags, which have their fields populated with time values in seconds, such as "3s" to indicate 3 seconds. Using these parameters, we can step through a series of media objects and slides. The above SMIL snippet produces four primary events in our sequence, as shown in Figure 13.29.

Displaying images and text is not the only capability of SMIL. We can also insert sounds and movie clips. The 3GPP supported media types are:

- *Speech/Audio*

  - AMR Speech Encoder (narrow and wideband)
  - MPEG-4[18] AAC Low Complexity (AAC-LC)
  - MPEG4 AAC Long Term Prediction (AAC-LTP)

---

[18] ISO/IEC 14496-3:2001: "Information technology – Coding of audio-visual objects – Part 3: Audio".

- *Synthetic Sound*

  - Scalable Polyphony MIDI (SP-MIDI) content format[19]

- *Video*

  - H.263 Profile 3 Level 10 decoder[20]
  - MPEG-4 Visual Simple Profile Level 0 decoder[21]

- *Bitmap Images*

  - GIF87a, GIF89a[22]
  - PNG[23]

- *Still Images*

  - JPEG, baseline DCT, non-differential, Huffman coding, as defined in table B.1, symbol 'SOF0'[24]
  - JPEG, progressive DCT, non-differential, Huffman coding, as defined in table B.1, symbol 'SOF2'[25]

- *Scalable Graphics*

  - SVG Tiny profile
  - SVG Basic profile

- *Text*

  - XHTML Mobile Profile

- *Character coding formats supported*:

  - UTF-8[26]
  - UCS-2[27]

In terms of sending the MMS, the complete message has to be encapsulated into a *Protocol Data Unit* (PDU) to be sent to the MMS messaging centre (MMSC). From the mobile, a composed message is usually submitted using HTTP POST or its semantic equivalent in WSP. From a non-mobile client, a message is usually submitted via HTTP or SMTP, but could also be submitted using FTP.

---

[19] Scalable Polyphony MIDI Specification Version 1.0, RP-34, MIDI Manufacturers Association, Los Angeles, CA, February 2002.

[20] ITU-T Recommendation H.263 – Annex X (2001): 'Annex X: Profiles and levels definition'.

[21] ISO/IEC 14496-2:2001: 'Information technology – Coding of audio-visual objects – Part 2: Visual'. ISO/IEC 14496-2:2001/Amd 2:2002: 'Streaming video profile'.

[22] CompuServe Incorporated: 'GIF Graphics Interchange Format: A Standard defining a mechanism for the storage and transmission of raster-based graphics information', Columbus, OH, USA, 1987. CompuServe Incorporated: 'Graphics Interchange Format: Version 89a', Columbus, OH, USA, 1990.

[23] IETF RFC 2083: 'PNG (Portable Networks Graphics) Specification Version 1.0', Boutell, T., *et al.*, March 1997.

[24] ITU-T Recommendation T.81 (1992) | ISO/IEC 10918-1:1993: 'Information technology – Digital compression and coding of continuous-tone still images – Requirements and guidelines'.

[25] ITU-T Recommendation T.81 (1992) | ISO/IEC 10918-1:1993: 'Information technology – Digital compression and coding of continuous-tone still images – Requirements and guidelines'.

[26] The Unicode Consortium: 'The Unicode Standard', Version 3.0 Reading, MA, Addison-Wesley Developers Press, 2000, ISBN 0-201-61633-5.

[27] ISO/IEC 10646-1:2000: 'Information technology – Universal Multiple-Octet Coded Character Set (UCS) – Part 1: Architecture and Basic Multilingual Plane'.

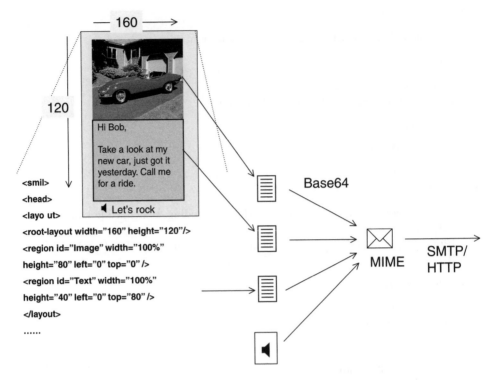

Figure 13.30 Encapsulating MMS components into MIME message.

When being submitted and transferred through the messaging infrastructure, the MMS components are encapsulated[28] into one entity, as shown in Figure 13.30. This aggregated structure is based on the well-known message structure of Internet email which is defined in [RFC 822][29], [RFC 2045][30] and [RFC 2387][31]. For transport over WSP, then the WSP specification provides mechanisms for binary encoding of such messages and serves as a basis for the binary encoding of MMS PDUs.

The PDU is one complete data structure and suitable for transmitting over WSP or HTTP, as shown in Figure 13.31. It has a header section which contains header fields. These are what give the message an identity and enable it to be transported meaningfully through the messaging network; otherwise the MMS message itself is just a collection of media objects. We need to know who is sending the message, to whom, what it is about and other such parameters that we would ordinarily expect in an email message (upon which the PDU

---

[28] OMA-MMS-ENC-v1_1-20021030-C: 'Multimedia Messaging Service Encapsulation Protocol', Version 1.1, 30-October-2002, Open Mobile Alliance.
[29] 'Standard for the Format of ARPA Internet Text Messages', Crocker, D., August 1982. URL: http://www.ietf.org/rfc/rfc822.txt
[30] 'Multipurpose Internet Mail Extensions (MIME) Part One: Format of Internet Message Bodies', Freed, N., November 1996. URL: http://www.ietf.org/rfc/rfc2045.txt
[31] 'The MIME Multipart/related content type', Levinson, E., August 1998. URL: http://www.ietf.org/rfc/rfc2387.txt

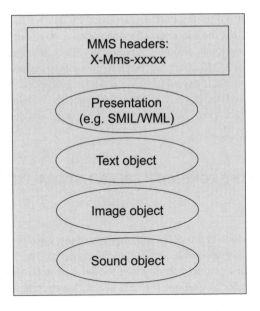

Figure 13.31 MMS PDU structure.

transport method is based). Examples of the headers are:

- *X-Mms-Message-Type* – this specifies the PDU type.

- *X-Mms-Transaction-ID* – a unique identifier for the PDU.

- *X-Mms-MMS-Version* – the MMS version number.

- *Date* – date and time of submission of the PDU. If the field was not provided by the sending MMS client, the MMS Proxy-Relay inserts the time of arrival of the PDU at the MMS Proxy-Relay.

- *From* – address of the originator MMS client. The originator MMS client sends either its address or an *Insert-address-token*. In case of a token being used, this indicates to the MMS Proxy–Relay that it is required to insert the correct address of the originator MMS client.

Within the message itself, each of the referenced objects is first converted to a format that makes it suitable for sending as character-encoded text. This is done using Base64 encoding, which is something we came across before (refer to Chapter 9) when we looked at obfuscating the username and password tokens used in basic authentication in HTTP. Unlike with basic authentication, in the case of MMS, Base64 is being used for its originally intended purpose, which is to make binary files available as characters so that they will pass character-orientated communications pathways, such as HTTP. It is not used here for obfuscation, although it has that benefit.

Each encoded object within the PDU is identified by its own header. The MMS specification actually states that the presence of a presentation object (e.g. a SMIL file to bind
everything together) is optional, although it would normally be expected. In the absence of
a presentation object, the MMS client decides how it wants to layout the remaining objects,
so clearly this is an undesirable aspect of MMS if we are interested in controlling the visual
appearance of the received messages. Each encoded object is delineated with its own header
(separator), so that the receiving client can extract the objects and decode them in readiness
for rendering.

13.8 GETTING IN THE ZONE WITH SPLASH (SPATIAL) MESSAGING

13.8.1 Introduction

Finally, we turn now to look at possible combinations of the various technologies discussed
in this chapter, especially to enable new types of services. The ideas explored include
some of the technologies explored throughout the book. I hope that it provokes thought
and leaves you with higher aspirations for next generation mobile services than when you
started reading this book.

13.8.2 Connectedness of Things

Location finding, P2P technology and social networking – what do they have in common?
I'm not sure that I have the answer(s), but I instinctively feel there are some connections and
possible emergent themes for mobile services. I am going to attempt to identify some here.

Leaving messages in space containing pictures and video clips that are searchable according to who's in them might be a powerful concept worth enacting. To do this, we need
a few current technologies and ideas to converge, as is probably going to be the case for
most of the exciting apps that will emerge in the coming 3G era.

13.8.3 Making a Splash

Imagine walking down the road and as you turn a street corner, up pops a message on your
mobile phone. Using fine-resolution location sensing, a real-time application has detected
you entering an area where messages have been pinned to a virtual notice board on that very
street corner. Using software on your mobile device you can read the messages, including
viewing pictures and watching video clips, all deposited by previous occupants of the same
physical space who left messages there at some bygone time.

This is the concept of *Spatial Messaging* (SM), the ability to post a message on a
virtual notice board anywhere on the Earth for someone else to pick up. The idea has
many interesting variants and possibilities and these shall be examined here as a vehicle for
exploring some of the myriad opportunities for next generation mobile technologies.

Sometimes called *air-graffiti*[32], it may also be useful to think of, and refer to, this
phenomenon as 'splash messaging', as in 'making a splash', wherever we go. (If we were

---

[32] This was a commonly used term a few years ago for the concept of pinning messages in space, but I have been
unable to find its origins.

trying to be clever, we could also intimate Pollock as a source of inspiration, as in the order that comes from the chaos of splashing paint, though it isn't obvious that any order would necessarily emerge from spatial messaging.)

We should not limit our view of this messaging paradigm just to text – obsessed as we currently seem to be with texting – judging by the ongoing large volumes of messages and the increasingly attractive deals to tempt us further into the messaging habit. Leaving text messages pinned in space was an early idea, one that was originally discussed when location-finding technologies began to emerge and well before MMS had arrived on the scene. Strictly speaking, 3GPP[33] includes video clips as a supported media type in MM, so video is a possible format for spatial messaging, although text, photo and sound elements seem more likely for now.

Interestingly, the concept of associating synthetic information with the real world is, in its general form, much older than the text message, dating back to those early days when some of us got excited about virtual reality, or VR (seems a very departed era). The idea of VR now oddly feels dated, like something from a very old run of Tomorrow's World on the BBC, even though it remains a futuristic idea yet to happen (in any mainstream sense, never mind mobile[34]).

However, VR may be about to make a come back in another guise, which is *Augmented Reality*; the overlay of real world images (what we see) with virtual ones (what we project). This is how the idea of spatial messaging had once surfaced, because in order to augment the real-world view with synthetic information, a spatial coalescence between the two worlds is desirable (though not necessary). The likelihood of cheap wearable displays[35] is now just around the corner and with location-technology and ubiquitous high-speed mobile networks, who knows what's possible?

Being as comfortable as we are now becoming with picture messaging, the suggestion of leaving a picture message in space, be it a logotype, manufactured image or an actual photo, is probably not so hard to imagine taking hold. Even simpler, leaving a voice message for the next passer-by is even more plausible, perhaps an addendum to PTT services. We can easily imagine leaving voice messages when leaving the front door of the home, perhaps to let the spouse know where we've gone. Upon his or her return, the generation of an alert when nearing the home would prompt the listening of the message.

With the location-finding technology that exists now in mobile networks and with picture messaging infrastructure in place, this concept is entirely possible to implement today. However, there are several challenges. One of them is the available accuracy of the location-finding system. With the most rudimentary forms of system, such as associating location with the cell the user is in, the best accuracy is somewhere around 150 metres in the high density of cells found in crowded urban areas. Elsewhere, this can deteriorate significantly, as cells begin to span kilometres, not metres.

However, with A-GPS and other techniques, including hybrid solutions, the accuracy tightens to less than 10 metres. Some futuristic ideas are also under consideration, such as the use of inertial sensors to track movement from known fixed points[36]; the idea of pinpoint

---

[33] 3GPP stands for 3G Partnership Project, the collection of interests to develop standards for the next generation of mobile device technology, such as UMTS.

[34] I attempted to research mobile VR concepts for a PhD back in the early 1990s. The sponsor discouraged the work, considering it too esoteric.

[35] For example, see Xybernaut, URL: http://www.xybernaut.com/

[36] Refer to the earlier discussion regarding Dr Elizabeth Cannon.

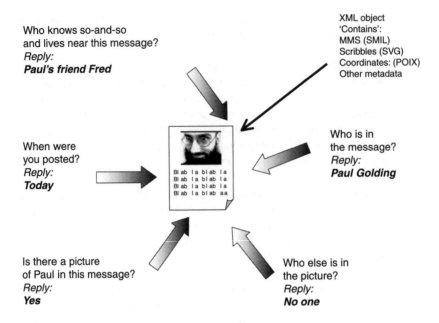

Who knows so-and-so
and lives near this message?
Reply:
Paul's friend Fred

XML object
'Contains':
MMS (SMIL)
Scribbles (SVG)
Coordinates: (POIX)
Other metadata

When were
you posted?
Reply:
Today

Bl ab l a bl ab l a
Bl ab l a bl ab l a
Bl ab l a bl ab l a
Bl ab l a bl ab aa

Who is in
the message?
Reply:
Paul Golding

Is there a picture
of Paul in this message?
Reply:
Yes

Who else is in
the picture?
Reply:
No one

Figure 13.32 Concept of active multimedia messages.

accuracy is no longer a pipe dream. Other inertial technologies, or the use of 3D visualisation technologies, also enable a directional element to the augmented reality experience. For example, standing on a street corner, we could be alerted differently according to which direction we are facing (or holding the mobile device).

Another challenge is leaving a video messaging pinned 'in the air', primarily because we would need some heavy-duty wireless networking to sustain the video streams. However, with 3G networks or supplemental hotspot technologies (e.g. WiFi or 3G TDD[37]), this is not that incredulous. Using streaming, both up and downlink, this seems possible.

In a moment, we shall look at things we could do to make multimedia splash messaging more interesting and more personalised. However, let's look first at the potential social connectivity we can build around a network of splash messages, their contributors and consumers. If we are going to leave messages pinned in space, then we might start thinking about ways to connect these messages with people, software agents, and possibly machines. How do we engage usefully with an emerging fabric of messages in the virtual world that also have a spatial existence, splashed across the physical world, on street corners, in shopping malls[38] and just about anywhere we may care to roam, locally, nationally, or globally[39]?

---

[37] TDD stands for Time-Division Duplexing, which is a technique for alternating the transmission and reception of signals by assigning time slots. Used in conjunction with wide-band CDMA, it has some interesting performance attributes that make it suitable for high capacity 'hot zones' of RF coverage in contradistinction to WiFi, which has a much smaller coverage potential (hence 'hot spots'). For example, see IPWireless: http://www.ipwireless.com/
[38] Indoor positioning is an area of intense research, carefully scrutinised by retailers. For a useful summary, see 'Retrieving Position from Indoor WLANS through GWLC', by G. Papazafeiropoulos *et al*. Can be found at http://www.polos.org/reports.htm
[39] It is conceivable that splash messaging can be accessible anywhere on the planet, whether the local network operator supports it or not. This should provoke some interesting thoughts in all respects of the socio-techno and even legal-politic implications.

We can think of the splash messages themselves as being active objects that we can send software messages to and expect replies from, as illustrated in Figure 13.32. The idea that we can ask about messages, about their contents, their owners and recipients brings an important dimension to the spatial messaging concept that puts it into the cyberspace realm, rather than a huge collection of inert media objects, like a draw stuffed with old photo albums in need of explanation.

Proxies, acting on behalf of real people, could generate the software messages (queries), or they could be from software agents, or even machines; a navigation application in a car could follow a trail of messages to find something interesting at their source (at the discretion of the trail-setter).

To encourage and facilitate an open spatial messaging network, we need to use open standards as much as possible, as well as commonly established methods of data interchange (e.g. HTTP, SOAP, etc).

XML is a useful way to describe a splash message. We could think in terms of spatial envelopes formed by XML documents that can describe the nature of the spatial posting (e.g. recipients, topics, etc.) and also host the presentation part of the multimedia message, which itself is XML (such as SMIL[40]). If the Mobile Location Protocol[41] XML vocabulary were modular, we could use this to describe location. We then need other XML vocabularies to describe information such as the message owner. The proposed XML format for vCards might do for this.

The spatial messaging world is different from the text-messaging world, which is altogether more ethereal. Splash messages could potentially have a very long lifetime, a lifetime presumably controllable by the messenger, perhaps even retrospectively so, and this might make for interesting real-time dynamics in various messaging scenarios. Leaving information pinned at a location for a limited time may be a useful means to attract people to a place within a certain period. This has obvious uses to bring people to venues, such as shops. The idea can be used to tempt people to convene at a certain place and time, for whatever reason. Taking this a stage further, using a string of linked messages with limited life spans, we can encourage people to move from one place to another, like an old-fashioned paper chase.

How might we start exploiting social connectivity in the spatial messaging world? One attractive feature of mobile phones is that they each contain an address book that is constantly available, probably updated frequently, and a potentially valuable source of information, not just to its 'owner', but also to others, were the contents made accessible for sharing.

What if address book X talks to address book Y, which talks to so-and-so's (Z) address book, and so on at every node in the mobile network? We could find out who X and Y both know, whom X knows that Z might want to know, and so on. P2P technologies (e.g. JXTA[42]) could facilitate the unhindered sharing of address books, allowing software agents to crawl intersecting peer groups to find an interesting person to call or intercept in cyberspace. Permissions-based protocols would help to avoid fears of surreptitious Trojans accessing anyone's precious 'black book' of numbers.

Clearly, phone numbers offer us a unique identifier in the mobile phone network, but to establish connections with the Internet world, email addresses are probably a more useful

---

[40] Refer to the earlier discussion in Section 13.7.
[41] See Open Mobile Alliance Location Forum: http://www.openmobilealliance.org/tech/affiliates/lif/lifindex .html
[42] JXTA is an open initiative to develop P2P protocols and technologies, URL: https://jxta.dev.java.net/

identifier. Increasingly, we shall find email addresses appearing in mobile phone books anyhow, thus readily enabling connections between the two worlds. This option is also eligible in a P2P application, thus avoiding interrogation of centralised directories to match numbers with email addresses.

One of the great things about this approach is that we can use text messaging for P2P communication, even with devices that aren't running a client. We could simply get humans to respond to P2P text messages!

However, how could we make social connections more meaningful, other than by simple association of address book entries? In the Internet world, an emerging standard for identifying relationship semantics between net users is the Friend-Of-A-Friend (FOAF[43]) file, which is an XML file describing relationships of one person to their declared contacts. We would probably want our mobile peers to formulate a FOAF file from our phone book. We can then use this as our semantic bridge to link people whenever we need to.

Currently, in MIDP[44] 1.0 of the shrunk-to-fit Java version for phones, the Java runtime environment is sand-boxed, preventing any access to the phone book from a MIDlet Java program[45]. With MIDP 2.0, this is no longer the case, and since MIDP2.0 readily supports HTTP (as did its predecessor), we have the potential foundation for a wide-area relationship-mining network running on a grid of mobile phones. FOAF could provide the method of describing our connections that we find in address books.

In itself, FOAF and its mobile embodiment, whether it be facilitated by P2P or not, is a powerful networking tool. It is worth noting that MIDP2.0 has policy-based control over how MIDlets access phone resources, like the address book, thereby providing us with a privacy control mechanism to protect unwitting users from those Trojan activities we feared earlier. In addition to its own networking potential (for social and business networking activities in a trusted network), this wide-area relationship mining can be used as an enabling technology in its own right, and we are entertaining such ideas here.

One idea then is to connect relationships with the people in photos left hanging in our spatial messages. Another is simply to connect relationships with spatial messaging senders and recipients, and another is to connect all these monikers in a buzzing network of social interaction, where we get people, place, time, opinion and serendipity all in one package.

The ability of X and Y to conjure up interesting uses from this available cocktail is surely going to be limited, but in the hands of millions of users, exciting things will unquestionably emerge. Let's familiarise ourselves with what splash messaging can consist of. We can leave 'splash' messages hanging at a particular place – anywhere where we get network coverage and anywhere we visit. We can:

- Leave video clips
- Leave text messages
- Leave pictures

[43] 'FOAF Vocabulary Specification: RDFWeb Namespace Document', 24 May 2007, URL: http://xmlns.com/foaf/0.1/
[44] The Mobile Information Device Profile (MIDP), combined with the Connected Limited Device Configuration (CLDC), is the Java Runtime Environment (JRE) for today's mobile information devices (MIDs) such as phones and entry level PDAs, URL: http://java.sun.com/products/midp/
[45] Address book access is possible if the phone vendor provides an API to enable access, but this would not be a standard MIDP 1.0 feature.

Figure 13.33 An annotated picture message.

- Leave audio (recorded there and then or pre-recorded)

- Leave links to any other resource on the Internet, including clickable phone numbers

- Leave any combination of the above

- Leave annotated videos and pictures, as shown in Figure 13.33 (annotations can be via hand sketching, using a stylus)

- Use post labels on images showing whom they depict, linking the depiction to an address book entry if necessary

- Store annotations and labels as SVGs[46] to maintain our desire to use open standards in order to ease widespread processing of our messages (SVG is a media format supported by 3GPP)

Lest we forget to think 'out of the box', let's ponder for a second on the thought that we don't need to go anywhere to splash a message. There is no reason to assume we have to visit a place to send a message there. We could do it remotely. We could send a video clip to a particular location and wait for some passer-by to pick it up. It could be anything, like a goodwill message for a friend as they arrive in a certain place.

   This does raise some interesting legal issues, among many that probably frighten me if I were honest. Perhaps we have to think about spatial authentication as a means to verify that we actually went to a place to leave a message, although the possible separation of the device from the person is an obvious obstacle. That needs some thought, but it is worth pursuing, as it is probably important to some applications. For example, mobile ticketing and coupons using pictures is now a distinct possibility, but open to various types of fraud.

---

[46] Refer to earlier discussion in Section 13.7.1 on SVGs.

The possibility of verifying message receipt according to location has some interesting applications.

Another legal issue that comes to mind is the possibility that spatial 'publishing' in public places actually constitutes broadcasting in a legal sense and would be subject to broadcasting laws, or publishing ones, including due care against libel. Is this the case? The answer is not clear, nor is it clear that existing legal frameworks can cover what may turn out to be some very strange social and commercial activities hitherto never contemplated by law. Even if existing laws do apply, such as broadcasting laws, one immediately suspects that legal loopholes will exist and their existence will emerge in unexpected ways.

The legal issue is clearly a serious one, an obvious concern being offensive content, such as adult material posted where kids can pick it up. There may be nothing new about the nature of the threat, but the Internet is notoriously difficult to police, and its invisible connection with space may bring undesirable consequences. The daubing of public places with potentially anonymous messages has obvious negative connotations: yet more ramifications to ponder in this brave new connected world. It would seem that digital identification is required coupled with a reliable method of establishing trust. This is so much easier to state than to implement.

To avoid information overload (or lessen it), we can leave news of our splashes using a suitable form of message summaries: such as RSS for example. It's certainly possible. Thus, we are able to add the ability to subscribe to spatial content. That may be useful for some applications. Perhaps we would like to receive messages from a particular shop that we regularly pass, but only about a particular topic if there's something new to say about it.

With all the semantic information in place, there are so many ways of connecting. To name just two obvious ones, we have:

- Co-depiction[47] (friends or FOAF shown in the same picture)

- Coalescence (friends of FOAF in the same place)

With co-depiction, a nomadic user would be alerted to multimedia messages with pictures of people they know, or who their friends know. With coalescence, just the coincidence of any metadata to do with friends (or FOAF) could cause an alert. The nomad can access the MMS messages associated with the alerts, add replies, make contact, etc. They would also be free to make their own annotations of the messages.

13.8.4 Splash-Messaging Summary

The location-finding capabilities of a mobile network enable messages to be location-stamped, not just time-stamped. If the messages are multimedia messages, then we have the added dimensions of a shared visual and audio experience. Allowing the sender to add annotations in a suitable format (e.g. XML), powerful semantic references become possible and we move from the flat messaging realm into a multidimensional geosocial realm. With a suitable infrastructure, messages are no longer inert, but can become active participants in cyber-conversations, able to respond to queries about their contents, authors and recipients.

Within the global collection of mobile address books, there is a powerful network of social connections. Using P2P technologies, users can tap deep veins of social capital. Conjoined

---

[47] 'Photo metadata: the co-depiction experiment', Dan Brickley, URL: http://rdfweb.org/2002/01/photo/

with the geosocial aspects of the spatial messaging and additional FOAF semantics, powerful possibilities seem likely to emerge.

In many parts of the world, text messaging has proven to be a cultural phenomenon[48] with many unexpected characteristics. Among other things, social scientists have noted its effect on how people meet and gather[49]. It seems obvious that the ability to add spatial awareness and involve the extra senses of sight and sound will surely lead to hitherto untold uses with increasing profundity.

Whilst the technology to achieve this extension to our communications reach is not all that sophisticated, the possible challenges to our sense of privacy and judicious use of new technologies probably surmounts to a disproportionate plethora of legal and social implications.

[48] Rheingold, H., *Smart Mobs – The Next Social Revolution*. Perseus Publishing, Cambridge, MA, USA (2002).
[49] Katz, J.E. and Aakhus, M. (eds), *Perpetual Contact. Mobile Communication, Private Talk, Public Performance*. Cambridge University Press, Cambridge, UK (2002).

$$14$$

Mobile 2.0 and IMS

14.1 INTRODUCTION

IP Multimedia System (IMS) is a network that enables connectedness within both the
telephony and computing realms, fixed and mobile. Whatever the future of mobile telephony
and mobile computing holds, one can confidently say that connectedness is a vital theme.
We have discussed this to an extent in Chapter 3 when we looked at becoming an 'Operator
2.0'. This chapter will guide you through the world of IMS.

14.2 MOBILE TRANSFORMATION

The mobile world is going to transform. Eventually, we shall move from mobile telephony
to mobile computing. In Chapter 3, we looked at the trends and discussed this transition
more in terms of evolving from a paradigm of 'dial-to-talk' (or 'text-to-message') towards a
more general 'click-to-connect'. IMS is a cause for optimism in making this change because
it enables an architecture that supports generic connectivity between two devices, no longer
bound solely to voice and texting protocols and services.

Telephony has a long history and is well understood by its users. It is also one-
dimensional – with telephony, all you can do is make a call. On the other hand, mobile
computing doesn't have much of a history, nor a user base. What is it and what is it good
for? There are precious few ideas and no consensus. Advocates of the Web 2.0 idea are
convinced that mobile computing means unfettered mobile access to the Web, including
the latest and greatest technologies, such as AJAX. Certainly the public has almost zero

This chapter is based on a paper first written by invitation from Wiley for a special delegates meeting at Osney
Media's Fixed Mobile Convergence conference, Paris, September 2005. It has been revised for this book.

Next Generation Wireless Applications, Second Edition   Paul Golding
© 2008 Paul Golding

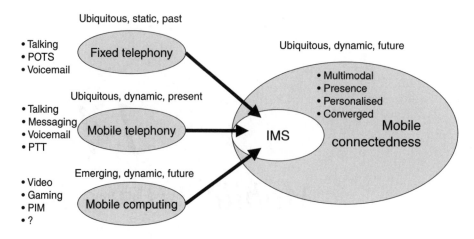

Figure 14.1 IMS as a catalyst for future mobile paradigm of connectedness.

conception of mobile computing, so they aren't about to rush out and pay for it. Never-theless, we know that we need to get there, especially with the growing pressures from convergence, with Mobile TV most likely acting as a catalyst.

IMS is an interesting development in the evolution of mobility. It has grown out of the mobile standards consortium 3rd Generation Partnership Project, or 3GPP. At its simplest, it replaces circuit-switched calls with packet-based ones whilst also allowing other media connections besides voice, such as video or an instant-messaging session. However, IMS is so much more. IMS can also support mobile computing connectivity paradigms. This is significant and worthy of exploration. Were it the case that telephony could be reinvented, or rationalised (i.e. cost-reduced), by a network upgrade that would also lead to viable mobile computing services, then IMS, as the enabler for all this, seems an important development. IMS is a potential catalyst for merging the disparate worlds of telephony (fixed and mobile) with the nascent world of mobile computing, as shown in Figure 14.1. A new paradigm emerges along with a new set of service potentialities: a new market perhaps. One could ask if IMS is to telecoms what the World Wide Web is to IT. Perhaps not – there are some key differences that we shall discuss later.

In this chapter, we shall explore IMS as both telephony enabler and mobile computing enabler and what that might ultimately mean to users and operators alike. Understanding both modalities is essential to grasping its future significance in the mobile world. We cannot do this without examining some of the wider concepts of communications and computing from a very high level. How we perceive things in a wider context matters. The humble text message makes this clear. To its inventors it was perhaps an engineering afterthought, largely incidental to the GSM plot. To its current users, especially the youth, it is an integral part of daily social life. These are two completely different worldviews of the same thing. It seems useful to attempt to understand how things might eventually be perceived, because that's where their fate lies and tells us where the future is, or isn't. IMS is on the one hand a bunch of protocols, interfaces and servers, but the world of IMS is what matters, or the worlds it will lead to. This is a journey through the world of IMS, not just its protocols.

14.3 IMS – WHAT IS IT REALLY?

In a room full of people, mostly techies, who were discussing IMS, the question arose – 'what, essentially, is IMS?' You can imagine such an arresting interruption. The IP guys in the room said, 'it's a type of router'. The applications guys said, 'it's a type of platform'. The marketing guy (you only need one of them per dozen or so techies) said, 'it doesn't really matter, as long as we can brand it' (I'm not joking).

The real question, perhaps, was 'Why do we need IMS?' That might seem a bit pointless having already committed to it in the standards bodies. Indeed, IMS is rapidly becoming the standard for all next generation networks, not just in the mobile telephony world. However, the question is still valid because of where it leads to in our understanding of this brave new world of 'next generation' communications.

If you want the short answer, then operators need IMS because of its potential to both save money and increase revenue, essentially as a technology upgrade to their core network. It can do this in ways that no other single technology investment can seemingly achieve and it is relatively future-proof. Users need IMS because of the new levels of connectedness that it can bring to their lives. If we take the world view that mobile users pay for connectedness, not texts or minutes – then if we can find new ways of connecting, we ought to be able to charge more money. This assumes that we don't reach a connectedness saturation point, but the experience on Web 2.0 shows that this is unlikely in the near future – offer users a new way to connect and they seem to have enough capacity to absorb them. Connectedness is a broad term. Connecting to friends and family is one level, whilst connecting to ideas, trends and personalities is another. Both are part of the mix of mobile services. Put simply, the more we can connect users, the more chances to charge them for it. Therefore, if IMS increases connectedness, then the opportunities for revenue ought to increase too. That's the hope anyway.

All of those opinion-givers in the room were probably correct in their own way, except perhaps the marketing guy (remember the WAP 'branding' fiasco?). IMS can save money because of what it is ('a type of router') and it can enable new services because of what it allows ('it's a type of platform'). It is a generic session-level router, which means it allows devices to find each other and services, and then to connect for whatever purpose they like – talking, viewing, gaming, sharing, etc. This capability is generic, which means that it can be used to implement a wide range of services. Therefore, IMS is a single infrastructure investment that has a multitude of potential uses and paybacks besides being a flexible IP upgrade for core voice services. Mobile circuit networks can only really support calls. IMS networks can support other services, too. Moreover, its architecture allows these services to interact, with potential multiplier effects, as we shall reveal.

The fact that IMS is IP-based and open-standard implies the usual cost-saving advantages of relatively low implementation and support costs, competitive vendor market, flexible and cost-effective upgrade paths, low-cost development tools, etc. Crucially, IMS is also a de-centralised network where much of the service logic resides at the end-points. These can connect via well-understood (i.e. IP), easy-to-implement (i.e. IP) and open interfaces (i.e. IP). This makes for a plug-and-play world of service add-ons to the core IMS network, just as widespread internet-aware operating systems have done for software ('it's an operating system'). Easy service introduction will inevitably lead to a plethora of new service possibilities, which in turn will lead to greater choice of services for the user.

By deploying IMS, operators can get all the benefits of an open IP-based paradigm, both as an overhaul to their telephony network and as a framework for delivering many new services using applications from a potentially large supply base. This sounds similar, perhaps, to the revolution seen on The Web. However, in case it slipped our attention, IMS is a subsystem of the operator network. IMS is a deliberate attempt to gain internet-like advantages, technical and commercial, but within the operator domain (i.e. control). This is a proactive step to avoid the operator-dreaded alternative, which is that their network simply becomes a pipe to and from the Internet domain. In theory, IMS-like architectures and functions could be part of the Web 2.0 ecosystem.

14.4   WHY IS IMS IMPORTANT?

Having just discussed what IMS is, the next question to ask is why is it important? Obviously, any new technology that promises simultaneously to lower costs whilst enhancing revenue potential is important. However, IMS isn't just any new technology. To understand why, we have to look more broadly at the evolution of modern communications.

What is The Web? You might like to answer that one yourself first. Again, we can resort to all kinds of definitions. An IP person might talk about HTTP. An applications person would probably talk about browsers and AJAX. What would the marketing person talk of? Well, I think they're still working on it. Traditional marketing narratives don't seem to apply online, although that is a controversial topic in itself. To users, the Web is really a place. The way to access that place is via the browser. As we discussed in Chapter 5, the browser is unlike any other piece of computer software previously deployed. It was a major milestone in IT. Almost without exception, all other pieces of software tie themselves to a particular function and set of information. However, the browser is almost entirely function and information agnostic. It is a *universal client*. No matter what the information is, the browser can handle it: train times, email, health care, hobbies, tax laws, and so on, ad infinitum (8 trillion pages and counting!). HTTP and HTML allowed this decoupling. Similarly, IMS, with its SIP core (see later) has the potential to decouple connectivity services from the underlying infrastructure. All that's missing to make it truly analogous to the Web is a universal client. However, it could be that one emerges (more on that later) and, therefore, we shall have all the ingredients for a web-like revolution in personal communications, at least in theory.

If the Web sits in our psyche as a phenomenon of 'placeness' – somewhere to visit – then the 'IMS-Web', if we can call it that, might introduce us to a realm of connectedness. The metaphor for the Web is still the printed book – we visit and read *pages* on the Web. With IMS, the expected metaphor is buddies, as in the ones we are used to finding in our IM clients, except extended to beyond just IM and to include things, not just people. The Web is a place we visit whereas the IMS-Web will be an environment that we are connected with. This perspective might sound odd, but we are trying to grasp the wider potential for IMS, which, just like the Web, is potentially something much bigger than a bunch of technical specifications.

14.5   START HERE: INTERNET TELEPHONY, OR VoIP

IMS is an IP-based communications environment standardised by 3GPP, but its technical roots are planted in the Internet. However, within the mobile *telephony* ecosystem, IMS is essentially an IP-based alternative to the conventional signalling architecture already used in current mobile networks to control the flow of calls. Therefore, we start our journey through the IMS network from a telephony perspective. We shall end up at more general

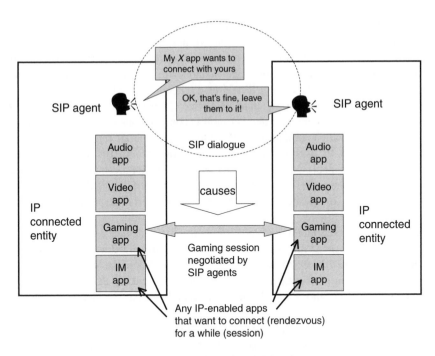

Figure 14.2 SIP agents negotiate IP-based service connections.

'connectedness' applications and return to explore the notion of an 'IMS-Web', or universal connectedness paradigm. Does such a thing exist, or could it?

The core of IMS is based on Session Initiation Protocol (SIP), which supports the necessary signalling between two IP-connected entities to enable them to establish a connection suitable for some kind of media-exchange. SIP is sometimes called a *rendezvous technology*, as it allows network entities to rendezvous for the purposes of exchanging data, but – and this is the important bit – it doesn't get involved in the actual exchange of the data itself, as shown in Figure 14.2. SIP is an IP-based signalling protocol that can facilitate IP-based connections, including voice. The use of SIP in this fashion gives rise to a form of Internet Telephony, more generically referred to as Voice over IP, or VoIP.

Seen from this perspective, IMS is SIP is VoIP. However, as its name suggests – IP Multimedia Subsystem – IMS is about more than just voice. What the 3GPP had in mind was multimedia. SIP enables any media connection to be established between two IP-connected entities. The SIP entities negotiate the connection (i.e. which sockets to use and which IP protocol) and negotiate the media type, voice or video, or both. Other sessions are possible too, like gaming and IM (see Figure 14.2) which we shall examine later. First, let's explore the core protocol for IMS, which is called SIP.

14.6 SESSION INITIATION PROTOCOL (SIP)

14.6.1 Making the Connection

SIP is a protocol at the heart of IMS. It comes from an Internet Engineering Task Force (IETF) standard, called RFC 3261. However, SIP is also a wider set of ideas. In order for

SIP to work, IP-networked entities are required. These entities, whatever they are, engage in a *SIP dialogue* using SIP *user agents*, as shown in Figure 14.2.

Two SIP user agents cannot communicate without outside help. They need help in finding each other the same way that two phones need a phone number (address) in order to connect. In the world of IP, entities talk using IP addresses. People don't have IP addresses, so we can't contact Aunt Sally[1] on her SIP-enabled device via her IP address. What we can do is contact her via a telephone number that her mobile company has assigned to her. However, SIP user agents address each other using IP addresses, not phone numbers. With a mobile network that supports IP, like GPRS or UMTS, it's not a problem to assign an IP address to Aunt Sally's SIP-capable device. However, we then need a mechanism to discover that IP address so that when we tell our SIP device to contact Aunt Sally's device via her phone number, it has a way to turn the number into an IP address in order to route SIP messages over IP packets.

There's an added complication to locating Aunt Sally's device via IP. In these silicon-abundant days, IP is easy to implement in software on nearly any device with a silicon chip in it. Therefore, unlike with GSM, we need not confine our SIP client to a phone-like device. A SIP agent could run on practically anything, such as a desktop PC, a PDA and so on – even a lawnmower![2] For those that like the comfort and familiarity of a desktop phone, SIP comes in that flavour, too.

Each IP device that we might connect with using SIP must have its own IP address. Thus, we require two steps towards discovering how to call Aunt Sally using SIP. Firstly, we need to go from an MSISDN (Mobile Station Integrated Services Digital Network) to an IP address (think of SIP user agents as dialling IP addresses whereas we dial phone numbers). Secondly, if Aunt Sally can be contacted on a range of SIP-enabled devices, we need to know which one to contact her on (or potentially all of them).

It is the job of SIP servers to solve this connectivity problem. In the first place, what these do is to allow Aunt Sally to register which device she's currently using, as shown in Figure 14.3. Aunt Sally does this by some menu on her device(s), but the underlying user agent deals with this at a protocol level by sending a SIP REGISTER message from the device to the proxy. This tells the SIP network at which device (i.e. which IP address) to reach Aunt Sally.

The user agent that wants to call Aunt Sally now has a means to locate her, and so we now have a means to establish an end-to-end SIP dialogue. Subsequently, SIP messages can be sent from one end to the other, such as the important SIP INVITE message. This is used to ask Aunt Sally is she wants to accept an invitation to connect with another SIP device.

SIP dialogues are always two-way. There is a *request* and a *response*, so Aunty Sally's device will respond to the SIP INVITE with some other message within the SIP vocabulary, even if it's just to say 'OK – I got it'. An important attribute of SIP to keep in mind for later discussions is that SIP is a two-way communication scheme. Unlike the synchronous HTTP for fetching web pages, either party in a SIP dialogue can initiate a SIP message to its counterpart. Contrariwise, only web browsers can initiate communication with the web server, not vice versa.

Now that we have introduced the concept of the SIP server, along with registration, we can look at its role in enabling a SIP dialogue to take place, as shown in Figure 14.4. Here

---

[1] I don't know who Aunt Sally is, but I'm fed up with Bob and Alice always sending messages to each other. XMPP's Romeo and Juliet are clearly more inspiring, but I dare not detract from Shakespeare's great works.
[2] I'm getting tired of the fridge as the archetypal networked home appliance.

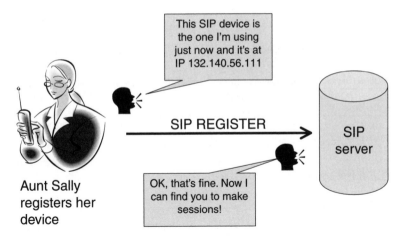

Figure 14.3 Registering a SIP device with a SIP server.

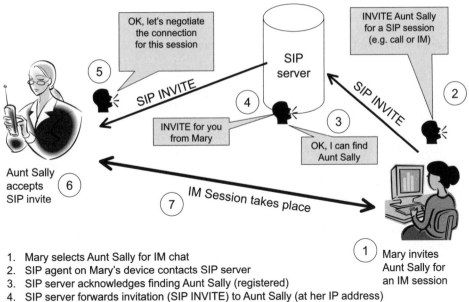

1. Mary selects Aunt Sally for IM chat
2. SIP agent on Mary's device contacts SIP server
3. SIP server acknowledges finding Aunt Sally (registered)
4. SIP server forwards invitation (SIP INVITE) to Aunt Sally (at her IP address)
5. SIP agent on Aunt Sally's handles invitation and session connection
6. Aunt Sally told of invite and accepts Mary's IM invitation
7. Mary and Aunt Sally conduct IM session and live happily ever after . . .

Figure 14.4 A SIP invite from one user to another.

we see a SIP invite from Mary to Aunt Sally, using the SIP server as a proxy that can route the invite to the right destination.

Figure 14.5 shows a view of the message flow for a SIP dialogue. The flows show setting up a telephony session *and* an Instant Messaging (IM) session, just to show that it is flexible enough to handle session management for different services and their endpoints.

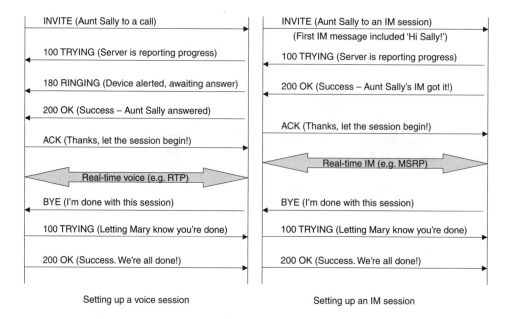

Figure 14.5 SIP call flows for telephony and Instant Messaging (IM).

In the Web world, resources on the Internet have human-friendly URL names, not just IP addresses. Similarly, users in the SIP world have human-friendly names called SIP Uniform Resource Indicators, or URIs (pronounced 'yurees'). A SIP URI for Aunt Sally might look like aunt.sally@purpleprovider.com, which is a potentially more meaningful and recognisable name compared with an IP address. SIP servers and user agents will deal primarily with SIP URI identifiers for registering and connecting, although the usual translation to IP addresses is obviously going to take place underneath.

The resolution of SIP URIs to IP addresses is done using DNS, an obvious favourite for such transformations, already widely used for other address resolutions, including email and web addresses. It is possible for a SIP URI and an email address to be the same. We could email or call Aunt Sally on aunt.sally@purpleprovider.com. More-over, we could push-to-talk or IM at the same address, too. The notion of a common address for communications is a form of convergence, and one that is especially obvious to the user.

A SIP URI actually includes the 'sip:' prefix, just as Web URLs include the 'http:' prefix. Aunt Sally's fictitious SIP URI would be sip:aunt.sally@purpleprovider.com.

As we stated earlier, SIP is at the heart of the IMS network. IMS-enabled mobile devices will have SIP user agents running on them, as part of the software protocol stack on the device. The user agent clearly requires the presence of another piece of software to transport these messages using IP (usually known as an IP stack). In the IMS network itself, we need the SIP servers we have just been talking about that allow user registration and proxy connections. In IMS, these servers are called Call Session Control Functions (CSCF), as shown in Figure 14.6 and described in the following section.

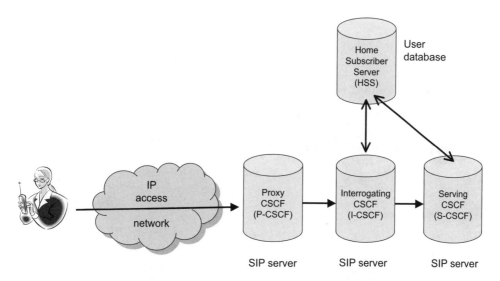

Figure 14.6 Simplified IMS architecture.

14.6.2 The CSCF Triad

As shown in Figure 14.6, the core of the IMS network is a set of SIP-server functions. There are three of them: Proxy CSCF (P-CSCF), Interrogating CSCF (I-CSCF) and the Serving CSCF (S-CSCF). In any given implementation, all three could well reside within one physical chassis, or even the same physical processor card in a rack. We shall now discuss each function:

1. *P-CSCF* – The P-CSCF is the first point of contact within the IMS for a SIP user agent residing on a User Element (UE), which is 3GPP parlance for saying end-user device. The P-CSCF could be located in the home or visited network and acts as an inbound/outbound proxy for all SIP traffic to and from the IMS clients. As the initial point of contact, the P-CSCF initiates user authentication and establishes a secure connection with the IMS client. Nothing gets by without the P-CSCF saying so. The P-CSCF then passes the SIP traffic to the appropriate S-CSCF via the I-CSCF if necessary (see I-CSCF later).

   The P-CSCF ensures that SIP messages are valid for each connected IMS client, including proper SIP message formation (e.g. syntax) and ongoing validity of authentication credentials. It also checks that media negotiations are appropriate for each IMS client, taking into account device capabilities and resource policies in place for the user. This is discussed later when we look at the Policy Decision Function (PDF), which controls resource allocation policies that the P-CSCF is expected to police.

   The P-CSCF also handles compression of SIP messages, which for mobile networks is principally a mechanism to reduce messaging delays rather than bandwidth. The P-CSCF can also generate charging information, which it routes towards a charge

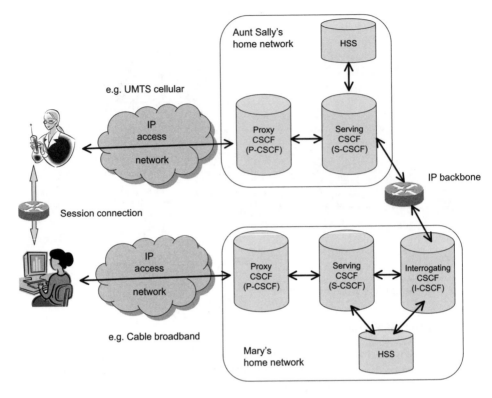

Figure 14.7 IMS connection end-to-end across two networks.

collection node via a standardised interface and protocol (called *Diameter*). The P-CSCF can also assist in the provision of emergency call handling and lawful intercept, and it can help prevent SIP signalling Denial of Service[3] (DoS) attacks.

2. *I-CSCF* – The I-CSCF is the boundary proxy for the home IMS network. The I-CSCF acts as a SIP Proxy to the home network, but one of its main functions is to find the target S-CSCF that is going to handle the SIP registration and subsequent dialogue for each IMS client. The I-CSCF retrieves the whereabouts if the S-CSCF assigned to a user via a request to the Home Subscriber Server (HSS) – see Figure 14.7. The protocol for the HSS interface is Diameter. A special function that the I-CSCF can perform is Topology Hiding Internetwork Gateway (THIG). This function hides the inner topology and details of the home IMS network by altering the contents of SIP messages to replace various SIP headers that might reveal network details that are pertinent to processing of SIP messages inside the home network. Once the I-CSCF has routed the initial SIP registration messages to the designated S-CSCF, it can remove itself from the signalling path, unless it is performing the THIG function.

---

[3] A DoS attack is where a rogue user agent attempts to swamp the SIP server with so many messages that it is too overwhelmed to offer service to legitimate users.

3. *S-CSCF* – The S-CSCF is the main server in the SIP signalling path. It provides many of the traditional SIP server functions described earlier, including registration. Similar to the I-CSCF, the S-CSCF also connects to the HSS via Diameter. A key function of this interface is to perform user authentication, as the users' security credentials are stored in the HSS database. In addition, if a user dials a SIP recipient via their telephone number instead of SIP URI, the S-CSCF provides the translation service (via DNS[4]).

Another key function implemented via the HSS interface is the downloading of *User Profiles*. These profiles tell the S-CSCF how to handle the SIP dialogue for each user, particularly how to route the SIP dialogue into the service layer of the IMS network (see 'IMS Service Concept' later). For example, if a user is a subscriber to a voice-messaging service, then the S-CSCF will know this from the user's profile and route traffic to the voice-messaging server accordingly. We shall return to this routing function later when we discuss application servers.

All the CSCF servers have the ability to generate charging information, as and when required by the network operator, depending on applicable charging policies.

14.6.3 Media Support

As shown in Figure 14.8, SIP devices, via the CSCF layer, have the ability to establish contact and conduct a SIP dialogue, which means they can exchange SIP messages such as a SIP INVITE. What happens next after they've initiated a connection? SIP user agents exchange SIP messages in order to negotiate a media connection. This means that the agents will agree to exchange a stream of packets between two identified endpoints (IP Sockets) attached to a particular application (e.g. voice, gaming, etc.). Moreover, the precise attributes of the media connection, such as the voice compression scheme in the case of a voice call, will also be agreed during the dialogue.

Negotiation of the media connection is essential because different devices will have different media capabilities. If we want to call Aunt Sally from our desktop video phone, we need to check that the device she is currently using (the one she registered with the SIP server or the S-CSCF) is video-capable. Ideally, we wouldn't attempt to make a video call if we already knew in advance that Aunt Sally's current device doesn't support video. This prior indication is possible via the power of *presence*, a very closely related and strategically important topic that we shall return to later.

SIP itself doesn't support media negotiation, although it can support caller preferences[5] to identify to the called party that we prefer a certain connection type. SIP messages, such as the INVITE message, can be used to carry embedded messages in their payload. There isn't really any restriction of the payload content[6], except that it should be rendered into plain text and be in a recognisable format (called a MIME type). For establishing media sessions, a language is used called Session Description Protocol (SDP – RFC 2327). Looking at just a snippet of an SDP message, we can appreciate how it works. When our INVITE message

---

[4] DNS E.164 Number Translation, as specified in RFC 2916.
[5] Caller preferences are defined in RFC 3841 'Caller Preferences for the Session Initiation Protocol'.
[6] Although there is a size restriction for SIP messages of 1300 bytes total.

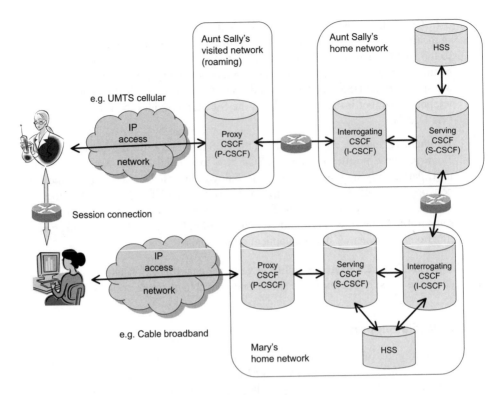

Figure 14.8 IMS connection from a visited network.

reaches Aunt Sally, the tail end of that message will contain the SDP message, a snippet of which might look like the following:

```
m=audio 20000 RTP/AVP 0
a=sendrcv
m=video 20002 RTP/AVP 31
a=sendrcv
```

This looks rather cryptic, so what does it mean? The 'm' stands for media. The first m-line is requesting a connection to be established for sending audio packets using Real-Time Protocol (RTP) as the IP transport protocol. The second m-line is to establish a video connection, also over RTP. The sockets that the sender wants to receive media on are 20000 for audio and 20002 for video. The last digit is the codec required for the session (0 means the audio codec G.711 μ-Law and 31 means video codec H.261). The a-lines in this case are indicating that the media connections should be bidirectional.

This SDP request is just a wish list. It constitutes an offer to Aunt Sally's device, which shall need to come back with a response. SIP and SDP allow SIP clients to negotiate compatible media connections. If the entities agree on the media parameters for the session, then the SIP dialogue will conclude with a successful acknowledgement of the offer–response cycle and then the devices can establish their media streams. It's worth keeping in mind that a given service provider will ensure (as a matter of commercial sense) that device capabilities are fairly standardised and consistent on their network, so that a given SDP offer is likely to get a successful response.

It is possible for the SIP dialogue to become very chatty, which means that lots of SIP messages are exchanged before the session is finally established. This could cause considerable delay before the media session gets a green light to start. For some services, this could be disruptive, such as PTT where the user would like to start talking the moment they press the button. In these cases, the SIP entities can start sending the media early with the assumption that the negotiation and connection is going to succeed. Alternatively, media can be buffered locally on the device in readiness for transport. These methods are particularly effective for PTT where the delay between pushing the talk button and actually talking would otherwise be too long, causing a poor user experience. Don't forget that in mobile services, a good user experience is vitally important.

With SIP and SDP, we have the means to establish multimedia communications between any two IP devices. This is the essence of IMS. Furthermore, the SIP messaging vocabulary is rich and flexible enough that it allows call-handling logic to be added by a server that can sit in the SDP signalling path. Such a server could provide call forwarding, divert, voicemail and so on. Of particular note is the capability of SIP to support call forking, which is where the initial SIP INVITE message can be routed, via a SIP-forking proxy, to several devices at once (see Figure 14.9). If Aunt Sally wanted to, she could register multiple devices with the IMS network and ask for calls to be routed to all of them. Whichever device she takes the call with will finalise the SIP session with the calling device. The alerting on the other devices (forks) will then cease.

14.6.4 Media out of IMS control

The IMS network is only concerned with the signalling layer of communications. The media layer is handled outside of the IMS network and in the mobile network via its access network nodes (e.g. GGSN) and IP backbone. Therefore, it is left to the mobile network to police the media connection. For example, whilst the IMS network might negotiate a certain media connection, it cannot police it, as it takes place outside of the IMS network. Potentially, the IMS network could negotiate one type of media connection (i.e. audio) whilst the devices could then establish another (i.e. video). Thus, there is a potential loophole to bypass correct charging or management of the mobile network resources. To address this issue, the IMS network includes a policy adjunct to support harmonisation between media authorisation in the access network and media authorisation in the SIP network (during SIP dialogues). This is described later when we look at QoS and resource policing.

14.6.5 Telephony Gateway Support

Given the obvious use of SIP/IMS to support telephony applications, IMS includes support for interconnection with non-IMS voice networks, such as the PSTN (as specified in 3GPP TS 29.163). There are distinct IMS entities that support internetworking. These entities connect to the IMS SIP core (CSCF servers). The telephony gateway support in IMS is similar to what we would expect in any SIP network, mobile or fixed, trying to interface with the PSTN (see Figure 14.10).

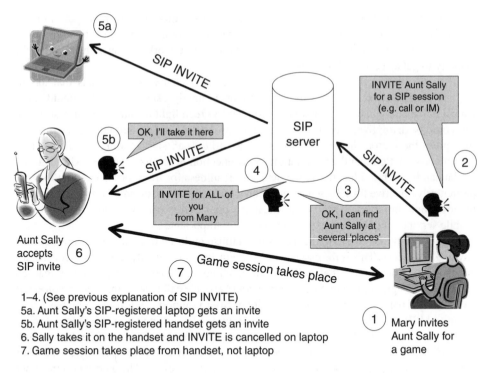

1–4. (See previous explanation of SIP INVITE)
5a. Aunt Sally's SIP-registered laptop gets an invite
5b. Aunt Sally's SIP-registered handset gets an invite
6. Sally takes it on the handset and INVITE is cancelled on laptop
7. Game session takes place from handset, not laptop

Figure 14.9 SIP forking of session invites.

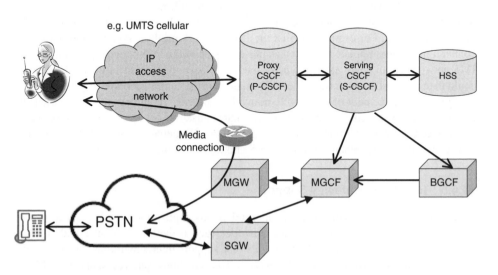

Figure 14.10 IMS gateway to legacy phone networks.

The functions of the IMS telephony gateways elements are:

- *Media Gateway (MGW)* – this provides the necessary adaptation of one media encoding and transport scheme on the IMS-side (i.e. packet-based) to whichever scheme exists on the non-IMS side (circuit switched).

- *Signalling Gateway (SGN)* – this supports the necessary signalling dialogue with the PSTN, such as ISUP (ISDN User Part).

- *Media Gateway Control Function (MGCF)* – clearly, the media gateway needs to be controlled in order to assign inbound and outbound ports and to set up the required adaptation scheme between them. It is the job of the MGCF to control the media gateway. Therefore, the MGCF needs to be connected to the core IMS network to handle SIP-based call legs, and to the SGW to handle the ISUP-based call legs; we can clearly see these two interfaces in shown in Figure 14.10.

- *Border Gateway Control Function (BGCF)* – for calls made from the IMS network to the PSTN, there might be more than one MGCF to handle them. A network could have connections with a number of PSTN networks at different border points on their network. Alternatively, the IMS user could be roaming onto a visited network and their call to a PSTN might best be handled via the MGCF in their home network. For these reasons, and potentially others, the BGCF exists to select which MGCF should handle the call before the SIP traffic is routed to that MGCF.

14.6.6 More than just SIP

The IMS network consists of more than just a few SIP proxies grafted onto the mobile network. Precisely because it is in the core of a mobile network, it has features peculiar to that environment. As one might imagine, most of these features support the operational aspects of using SIP in a mobile network, such as charging and billing, roaming, authentication, QoS and resource policing, which we will now take a look at in the sections that follow.

Charging and Billing   Of critical interest to any mobile service is charging and billing. Hence, as we might expect, the integration of IMS into the core of a 3G network comes with the necessary hooks for supporting charging. Some extensions to SIP were necessary in order to facilitate charging in the 3GPP context.

Roaming   This is essential to the cellular network concept and so support for roaming had to be included in the IMS architecture. This, as explained above, is the reason for the tripartite nature of the core IMS network.

Authentication   An IMS user needs to be authenticated (and authorised) before using the IMS network to access IMS services. In mobile networks, the smart card based Subscriber Identity Module (SIM) is integral to overall user authentication onto the access network. In 3GPP, the UMTS SIM (USIM) is an additional security apparatus on the user device. For IMS, the 3GPP has specified an IMS SIM (ISIM), which supports information specific to IMS registration. This includes public and private user identities for general user identification and shared secrets to support ciphering.

We shall skip the details of the authentication process here, but note that the ISIM (or USIM) participates directly in the IMS authentication process during registration (see earlier discussion). The authentication process, although using the IETF-standard digest authentication, uses a 3GPP-specific authentication algorithm (AKA). For backwards compatibility, IMS authentication is still possible with a USIM, but not a SIM. Either way, the mechanism for authentication is still mobile specific.

Quality of Service (QoS) and Resource Policing   SIP is used to negotiate a session between two SIP user agents. The sessions, whether they support media or not, will be IP-based and require IP networking resources to support the required data flows. The network provider may wish to exercise control over resource allocation to support these flows, either to enforce usage policies on a per-user basis (i.e. limiting an audio-only customer to audio-only connections), or to ensure appropriate resource allocation to support the desired QoS on a per-user basis. Therefore, IMS includes mechanisms for QoS and resource policing that work in tandem with the core SIP network. The key agent of resource management is the P-CSCF, as this is the gateway in and out of the IMS network.

An adjunct to the P-CSCF is the PDF (Policy Decision Function), which implements the COPS (Common Open Policy Service) protocol (IETF RFC 2748). The PDF can be used to inform the P-CSCF of specific user-level or network-level policies prior to session negotiation in the SDP messages. This can provide both QoS and resource-allocation control. Moreover, the PDF, in conjunction with the P-CSCF, can implement an authorisation token scheme to ensure that device requests for IP resources are in line with prior, and PDF-endorsed, SIP-level negotiations, as shown in Figure 14.11.

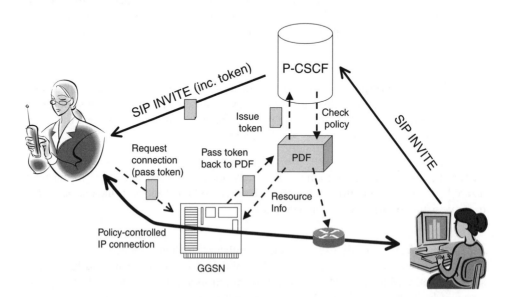

Figure 14.11 Managing resource allocation policies in the IMS network.

In the case of a mobile network, the P-CSCF uses COPS to communicate policy requirements to the GGSN. This is done on the Go interface. However, it should be understood that the COPS protocol is generic and can be used to implement policy with any COPS-compatible network element, such as an RSVP-capable router. In non-mobile IMS networks, the policy control is implemented at the border of the IMS network via a Border Control Function (BCF). We shall return to this topic shortly.

14.7 THE PROMISE OF A COMMON SERVICES ENVIRONMENT

With SIP, IMS has a core set of protocols for establishing an IP-based future for mobile telephony. Moreover, with the additional features just described, IMS presents a common environment for billing, roaming and QoS within the IP world. All of the above features needed to build an IMS core and to allow inter-operator working and user roaming were specified in Release 5 of the 3GPP specifications for UMTS. IMS-R5 was a distinctly mobile vision for an 'all IP' future. However, as we shall now discuss, IMS and its future lies beyond a purely mobile vision.

14.7.1 Seamless Mobility and Convergence

Any device that can handle IP packets is potentially capable of engaging in a SIP dialogue. It doesn't have to be a telephone device, nor wireless. SIP is access-network agnostic. Therefore, inasmuch as IMS is based on SIP, IMS is potentially access-network agnostic. 3GPP Release 6 of IMS fully embraces this potential by ensuring that IMS has true access independence. The gives rise to the potential of seamless mobility, and network convergence, particularly motivated by Fixed-Mobile Convergence (FMC).

Seamless mobility is the ability for a user to access the IMS network from any type of wireless access technology, as long as it supports IP and is capable of supporting the desired session that results from the SIP dialogue. This doesn't have to be a media session. A SIP dialogue could establish an IM session, a PTT session, or some other session; media and telephony is just one potentiality of IMS, which is what makes it so exciting.

Purists might argue that seamless mobility is only truly seamless when it is possible to move from one access network to another whilst maintaining an ongoing SIP session. Such a requirement is not so easy to meet. More work is required to define standards. Intersession seamlessness is still desirable in the sense that the user can move between networks without any manual network selection via a user interface. Again, an intuitive and easy user experience is important.

Some seamless services are currently receiving greater attention than others, most notably the case of VoIP for FMC convergence. Some possibilities, such as the widely discussed WiFi-Cellular combination, have been proposed as an ideal delivery mechanism for VoIP, given the relative ubiquity of these two networks and the promise of piggybacking on already paid-for broadband connections to the user's premises. It is worth stating that for WiFi-Cellular, IMS isn't the only solution. There are others and these vary in architecture according to whether the use case is consumer or enterprise orientated. Non-IMS solutions are outside the scope of our IMS journey, but of particular interest, mostly because of its early presence in the market, is the Universal Mobile Access (UMA) solution. Originally a proprietary solution from Kineto Wireless (in the United States), UMA has been adopted by the 3GPP under the rubric of Generic Access (GA). The GA approach is to extend the cellular network along the fixed network and WiFi pipe, so that, in effect, a new transport layer is introduced in the handset to network interface. There is now considerable excitement about the use of Femtocells in the home, which are low-cost mini 3G base stations. These use a similar principle to UMA in terms of extending the core cellular network interface all the way to the access point in the home, but don't require the use of an alternative RF network to UMTS.

Motivations for converged services can be found at all levels of the value chain:

- *Consumer interest* – 'Seamless connectedness': accessing multiple services consistently and cost effectively irrespective of device and location

- *Content provider interest* – 'Liquid connectedness': information constantly available to users irrespective of access method

- *Service provider interest* – 'Sticky connectedness': keeping the user connected to the network, to any service, from any device, over any access mechanism

14.8 IMS AS A CONVERGENCE CATALYST

14.8.1 Mobile Roots, Fixed Branches

The first release of IMS specifications in the 3GPP standards body (Release 5) took the view that IMS is a mobile network subsystem. Therefore, various assumptions and implementation decisions were made with the mobile context in mind. For example, the IMS core is assumed to be IPv6 only, in line with the general 3GPP vision. This is problematic for networks where IPv4 is still prevalent, like many existing IP networks. A number of other assumptions were made about the access network being essentially cellular, such as the existence of a smart card (SIM) to facilitate authentication.

For IMS Release 6, the 3GPP included a work item to give IMS greater network flexibility, allowing access from WiFi and last-mile networks (including WiMAX). What we now have is a well-defined IP-based network core that could potentially be used by any telephony service provider to realise an entirely IP-based telephony and connected-services vision. In a short space of time, IMS has grown up and out of its cellular origins. This gives mobile operators an egress into fixed-line environments. The significance of this should be understood within the wider context of IMS potentialities. Not only does IMS allow multimedia services, but it can carry a multitude of other services, such as IM and PTT, which could equally work within a fixed-line environment, abound with cheap PCs, converged devices and ever broadening access (see Figure 14.12).

The interest in IMS is significant amongst the mobile operator community and numerous equipment providers, both traditional mobile infrastructure suppliers and telco suppliers. Then there's the mushrooming applications market, not only serviced by established software companies like Microsoft, but by the usual plethora of start-up companies, too numerous to mention. IMS is potentially a plug-and-play world with relatively low entry costs.

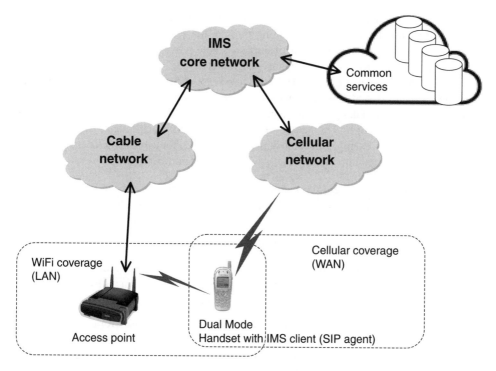

Figure 14.12 Converged access using IMS.

14.8.2   Spanning the mobile-fixed divide – TISPAN

With IMS receiving lots of support, it is no surprise that the ITU decided to adopt IMS for its Next Generation Network (NGN) vision and architecture. *Telecoms & Internet Converged Services & Protocols for Advanced Networks* (TISPAN) is a standardisation effort by the originators of GSM – ETSI (European Telecommunication Standards Institute) – to adopt IMS into the ITU's NGN vision.

TISPAN is an effort by the fixed operators to bring their networks into the 'all IP' world for all the same reasons that IMS makes sense for mobile operators: lower costs, greater flexibility and potentially new service opportunities. Indeed, many large carriers are facing a perplexing business challenge. Due to regulatory pressures, they are forced to open their networks to smaller and more agile providers who have the agility to bring interesting new services to users at competitive prices without worrying about the baggage of clumsy legacy networks. IMS provides a better basis for competing in the future.

14.8.3   A Winding Path to Convergence

It is a mistake to think that unified networks automatically lead to unified services. Not all service providers will want to provide the same services, or be able to, especially if they need to evolve their service offering from an existing service domain to the new converged services domain. Legacy will undoubtedly affect the route taken towards convergence. It is tempting to assume that an IMS-centric FMC architecture is the only solution, where all traffic for all calls utilise SIP signalling through the session layer of IMS. That's not necessarily the optimal solution. For example, a cable provider would already provide a legacy telephony service for residential users. Rollout of WiFi access points to enable dual-mode cellular/WiFi telephony should take this into account. Users won't dump their existing landlines without a good reason. Where legacy networks exist, there are various ways to achieve convergence, taking into account new services, old services and combined (not converged) services that bridge the two service domains until such time as a single service domain emerges. For example, in cable networks with telephony infrastructure, the existing landlines can be folded into the new service domain in a number of ways:

- *Internal extensions* – where the landline and converged device appear as extensions to each other, allowing call transfer, call pick-up and other extension-like services (all free of charge).

- *SIP forking* – where the landline is brought into the call path of the converged number.

- *Virtual numbering* – where the landline number is ported to the converged 'line', allowing a single service whilst maintaining the old number (and the new), but with distinct service profiles to allow separation of 'home' calls from 'personal' calls, etc.

- *Converged messaging* – where a single voicemail platform is utilised for the landline and the converged 'line', allowing single-point access and voicemail management, as well as interesting new service possibilities (e.g. via the converged handset, the user could manage the voicemail profile for the landline number, such as the number of rings before announcement, etc.).

There are also commercial realities to consider too, such as least-cost routing possibilities. For example, a call made from the WiFi access point to another landline could potentially be routed more cost effectively through existing cable gateways to the PSTN rather than via a partner mobile network.

Here we have only discussed one legacy issue and only for residential scenarios. The possibilities for other legacy scenarios and for enterprise networks are numerous. No doubt, IMS can provide the means to build a converged network, but there are stepping-stones to get there depending on the starting point and commercial strategy of the operator. These stepping-stones might well influence the IMS evolution and will determine whether we can stay on track for a single standardised IMS core for all operators.

We have also yet to see whether the grander IMS commercial vision can be realised. Let's not overlook the idea behind IMS, which is to bring Internet-style flexibility and benefits into the mobile operator world, reaping all the benefits whilst keeping a relatively tight grip of the customers! IMS might be built using Internet technologies, but it lacks The Internet's central, and seemingly crucial, theme, which is *openness*. Anyone is allowed to come and play in the Internet sandpit and dig for gold. With IMS, operators are retaining control over who gets to play. As a means to damn the already leaking dyke around the leaking mobile ARPU oasis, it remains to be seen if IMS is up to the job. Nevertheless, as a means towards a new services-led future for digital communications, it makes a lot of sense and has vast potential, especially for a services landscape that reaches beyond telephony. It is here where we now focus our attention – the world of applications and services – and take the final part of our journey in the world of IMS.

14.9   END HERE: BEYOND VoIP – APPLICATION SERVERS

SIP is largely a peer-to-peer (P2P) signalling method. This means that the logic (software) responsible for processing the call flow sits at the end-points (edge) of the network connection (i.e. it resides in the devices). It is true, SIP and IMS do rely on intervening servers, but keep in mind that these servers provide the necessary assistance for two (or more) SIP agents to find each other. Afterwards, the two agents are left to negotiate and collaborate at their discretion.

The P2P architecture of SIP is very conducive to supporting powerful new services, simply because there's no messy integration into some behemoth centralised core (usually with archaic protocols and programming models). More or less, just granting access to the IMS network is enough to allow a powerful new service to be introduced to its users. This is indeed an exciting prospect for both IMS network providers and third-party application providers. Moreover, because SIP and IMS are open standards, third-party providers are encouraged by the prospect of developing applications that can run on many networks. Developing SIP-aware applications is becoming as easy as developing web applications. Furthermore, the growing sophistication and power of supporting software technologies, like J2EE, make it easy and low cost to develop scalable carrier-class applications on a relatively low budget. In theory, the path to producing scalable IMS applications is an increasingly familiar and easy one. As VoIP/SIP guru J. Rosenberg says:

> *It should be as easy to create a new phone service as building a Web page.*

14.10   IMS SERVICE CONCEPT

The IMS architecture envisions three layers, as shown in Figure 14.13. For most of this chapter, we have been discussing the session layer, which is the essential core of the IMS network – the SIP and CSCF servers and all the bells and whistles for billing, QoS, security and resource management. Then there's the access layer, which is mostly (and increasingly) outside the scope of IMS. Telephony sits within the session layer, including the required gateways to non-IMS networks, but we have already indicated that IMS is much more than just telephony. What remains is the applications layer.

The core concept of the applications layer is the Application Server (AS). SIP, as previously discussed, is a 'rendezvous technology', allowing companion applications to connect. The concept of an AS is that it sits in the SIP dialogue pathway and either involves itself in a dialogue directly with a SIP user agent or intercedes in a dialogue between two other agents. A classic example would be a voicemail AS. The SIP call is eventually routed via an S-CSCF, which attempts to forward the SIP Invite message to the called party's SIP device. However, if the receiving party doesn't answer, then the S-CSCF can route the SIP invite to a SIP-capable voicemail server sitting on the ISC interface.

The voicemail server accepts the SIP invite on behalf of the user (using the user's SIP URI address to associate the call with the appropriate mailbox[7]) and negotiates a media connection between the calling device and a media port on the voicemail platform that can

---

[7] In theory, the SIP URI should be sufficient to identify the user and therefore the associated mailbox. However, for various reasons, the SIP URI might not be one that the voicemail server recognises. Therefore, there have been suggestions for adding extra information to the SIP URI specifically for the purpose of identifying mailbox attributes, as suggested in RFC 4458 (http://www.ietf.org/rfc/rfc4458.tex).

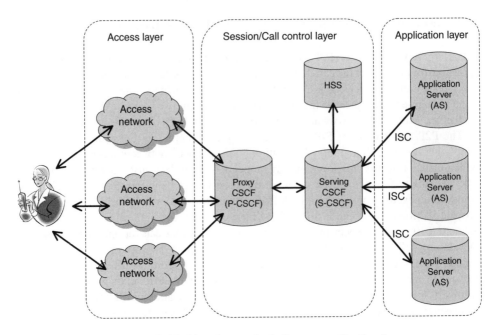

Figure 14.13 IMS can be divided into layers, including an application layer.

receive the incoming voice packets. The voicemail server can now use SIP, such as the *SIP Notify* message, to inform the called party that they have voicemail waiting for them. Of course, the called party will eventually use SIP to connect to the voicemail server and listen to the voice message via a media connection from a media port back to the device itself.

It is the job of the S-CSCF to extract user profiles from the HSS during the registration of each user. These user profiles describe which SIP servers (AS) are to be involved in the SIP signalling path for each particular connection that might be attempted across the CSCF core. This is done using Initial Filter Criteria, which are 'triggers' that fire when certain conditions are met in the SIP messages for a particular user.

3GPP defines the following set of roles that application servers can take:

- *Terminating User Agent* – the AS receives and terminates the SIP pathway.

- *Redirect Server* – the AS can ask the originating user agent to redirect the SIP dialogue to another endpoint.

- *Originating User Agent* – the AS is able to initiate a SIP dialogue.

- *SIP Proxy Server* – sits in between two user agents and proxies the dialogue.

- *Third Party Call Control Entity* – able to set up a dialogue between two other end points, e.g. to play an announcement. In this mode, the AS sets up two dialogues with each end point and then acts as an intelligent relay in the middle, called a Back-To-Back User Agent (B2BUA). A PTT server is a prime example.

The idea of well-defined AS roles is to ensure a workable collaborative model for the service layer, bearing in mind that there might be several application servers available to participate in any given IMS session, which will be discussed fully in the following examples.

14.11 SERVICE EXAMPLES

We shall now look at possible collaborative service examples in order to show the diverse and powerful capabilities of the IMS network.

14.11.1 Multimodal Chat

In this example, the incoming SIP messages are routed through all of the AS entities shown, which are a Presence server, an Instant Messaging (IM) server and a Push-To-Talk (PTT) server, as shown in Figure 14.14. Each server manages a separate service domain in its own right. However, let's examine a possible collaborative service example, in order to demonstrate the power of IMS and how AS entities can behave as service enablers.

Let us imagine an IM session with three users, A, B and C (dull names, but easy to reference). Users A and B are accessing the IMS framework via IM clients running on their PCs. A is on a WiFi-connected laptop in a coffee shop and B is at home on a cable modem. User C is connected via 3G, using a smart-phone device with integrated qwerty keyboard. Clearly, in this figure, we have multiple access technologies. Let us now examine some elements of the session. This is a complex collaboration at the SIP dialogue level, but we shall skip most of the detail in order to illustrate the concept (there are several ways to implement multimodal chat).

- *Presence* – Here, the presence state of each presentity is visible in each of the IMS clients. Let's call this generic presence-viewing client a 'buddy browser' (a topic we shall return to shortly). The buddy browser reveals:

  – User A – Online
  – User B – Online
  – User C – Online/PTT

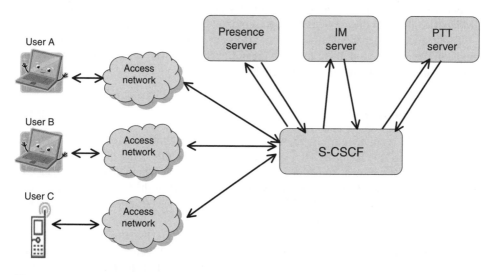

Figure 14.14 Collaborating application servers in IMS.

- *IM* – Currently, all the users are chatting via IM and can see each other's IM messages in their IM clients. This is a three-way IM session. The IM messages are sent via SIP using the SIMPLE standard. These messages are handled via SIMPLE clients on User A and User B devices. User C is using an IMPS client to connect with an IMPS server, which is using an IMPS-SIMPLE gateway to connect with the IM server.

- *PTT* – The presence state of User C is indicating 'online' and available for the IM session. It also indicates being contactable via the PTT service. Currently, User C cannot communicate via PTT to the other participants: C's presence state indicates non-contactable via PTT.

- *Presence/PTT* – User A wants to enter the conversation using PTT, but doesn't have a laptop PTT client. User A therefore connects to the PTT service via a GPRS handset (in addition to the ongoing laptop session). When User A connects to the PTT server, the Presence server receives an update of User A's PTT presence status. This is reflected in the buddy browser, which now reveals:

  - User A – Online/PTT
  - User B – Online
  - User C – Online/PTT

- *PTT/IM* – User A and C can now communicate via IM and PTT. User A will use their laptop to type IM messages and can use their handset to interact via PTT as and when desired. User A and C can communicate privately via PTT outside of the current IM session, or they can communicate via PTT within the current session (via a 'group buddy' representing the ongoing group session). User B does not have a PTT client, but can still hear the PTT dialogue via a media connection to the desktop. It is also feasible for the PTT dialogue to be routed via SIP to the IM server itself where speech-to-text transformation could take place and its output inserted into the IM dialogue.

The above example shows the mix-and-match capability of IMS-connected service enablers, combining to create a new service. Such meta services will become commonplace in the IMS world, although they potentially represent interesting charging challenges.

To illustrate further the flexibility of the IMS environment, let us consider another example in the next section, this time with greater integration with the mobile core network using location information as a service enabler.

14.11.2 Push-To-Taxi

In this example, seen in Figure 14.15, we are going to look at how to catch a taxi (cab) using a mobile phone. This is not new. Of course, we can dial a ride using plain old-fashioned telephony, but let's look at something a bit more fancy. This example is also a chance to introduce some further features of the IMS environment, such as the Open Service Access – Service Capability Server (OSA-SCS). This is one of my favourite examples because it shows just how versatile the IMS architecture can be.

With the use of presence information, we can do some interesting things. Imagine in our buddy client that we have a buddy (presentity) called 'taxi'. When we want to hail a cab, we simply click on the 'taxi' buddy icon, with its presence set to 'for hire' and, assuming

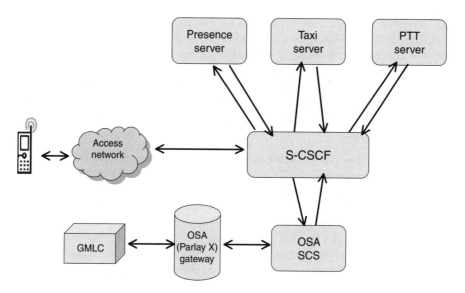

Figure 14.15 Collaborating application servers demonstrating server-enablers.

we have a PTT client, we start talking, e.g. 'Hi – I'd like a cab'. Looking at Figure 14.15, we will now go through the process:

- *Initial Filter Criteria/Taxi* – Using filters that identified the destination address as a 'taxi', the S-CSCF routed to SIP message to the taxi server via the ISC interface.

- *Presence* – The presence server is maintaining the presence information of individual taxi drivers from a particular taxi company. The taxi server acts as a *Watcher* to the taxi drivers' presence information and is notified via SIP of any changes. Each driver uses presence to indicate his or her 'for hire' or 'busy' state.

- *Taxi server (AS)* –The taxi server looks up the location of the inviting party, having extracted the user id from the inviting SIP message. The taxi server also looks up the nearest tax driver whose presence state is 'for hire'. Having done that, the server can now redirect the invite towards the PTT server, having inserted the user id (SIP URI) of the nearest cab driver.

- *Location via the OSA-SCS* – For this example, the mobile network operator has already deployed an OSA Parlay-X gateway in order to provide a unified (Parlay X[8]) API into some of the operator's assets, including the Gateway Mobile Location Centre (GMLC), which provides location information for each subscriber in the network. Here, our service user is on the mobile network and so their location information is accessible via the OSA gateway. The OSA-SCS allows the taxi server to gain location information via SIP Subscribe messages.

---

[8] Parlay X is a Web Services presentation of the Parlay API, which is a standardised telecoms API (see http://www.parlay.org) that was adopted by 3GPP for its Open Services Architecture (OSA). OSA is concerned with providing a single open API to gain programmatic access to many of the assets in the operator's network, including messaging platforms (e.g. SMSC) and location platforms (e.g. GMLC).

- *PTT/Presence* – The cab driver receives the passenger hail 'Hi, I'd like a cab' via a barge call in their PTT handheld (or hands free!) and responds. Now, more than likely, the cab driver will respond with a message announcing that he's picking up the fair and asking the exact whereabouts of the passenger. However, this last bit is not entirely necessary (although likely) because the presence information for the 'passenger' buddy automatically includes the location of the passenger. The presence server inserted this information after receiving a presence update from the taxi server, which knows the location of 'passenger'.

From this example, we can see how easy it is to construct new services from key service enablers, such as presence and PTT. Furthermore, in the case of a mobile network, we have shown how 'legacy' services or assets can be utilised via an appropriate SIP gateway (OSA-SCS). The S-CSCF played a crucial role in making sure that the SIP dialogue for PTT sessions with a 'taxi' buddy were routed via the taxi server, which sits on the ISC interface.

The above two examples have illustrated the power of the buddy/presence interface for the presentation of services to the end user. In the next sections, we shall now explore the potential of this buddy/presence paradigm as a universal client for future mobile connectedness.

14.11.3 Avatar Chat

IMS video sharing allows a user to add two-way video to an existing voice call. It is sometimes described as 'see what I see' by marketing folk. The benefit is that a user could augment their conversation with video in the spur of the moment. It is often the case that during a phone call a user might describe something that relates to where they are or what they are doing. However, such remarks or thoughts are unlikely to have led to a video call being made to start with. Let's face it; most people don't ever think to make a video call, even if they know how. If a user is already in a voice call and there's an attractive button glaring on the screen to add video to the call, then it might lead to a greater use of P2P video than we see currently in most mobile networks. This is the benefit of using SIP. The session can be negotiated at any point. Thus, having established a voice call, the data bearer can be used to carry SIP signalling to initiate an accompanying video connection.

Avatar chat uses a similar principle to augment, and hopefully enrich, an existing call. Instead of seeing video, each user sees an avatar representation of the other person, as shown in Figure 14.16, and this is used to communicate via messaging, gestures and even lip-syncing. Gestures can include backslapping, high fives, handshaking, or even throwing a punch (ouch!). The avatar session is established using SIP. The movements, messages and other avatar control commands are also sent via SIP messages. Of course, this application requires an avatar client on each handset. It is feasible to break out the avatar chat to non-IMS devices using a video gateway to convert the avatars to video streams. The benefit is that the service will be compatible with any video-capable device. However, this is very inefficient way to implement avatar chat.

14.12 THE UNIVERSAL CLIENT AND WEB 2.0

On the Web, the web browser is the universal client that makes the web so successful. A plethora of services can be delivered through this one client. It would seem that the 'buddy

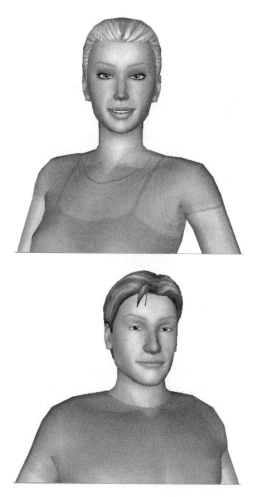

Figure 14.16 Avatars for mobile avatar-chat applications. (Reproduced by permission of
GoFigure Mobile.)

browser' is a key client for IMS-related services, not just person-to-person, but, as we
have just seen, presentity-to-presentity, which is a more general class of communications
possibilities. Presentities can be human, automaton or avatar (i.e. an automaton pretending
to be human).

It is an increasingly common view that the presentity-based connectedness paradigm is
a key enabler for next generation services. These could range from simple telephony, if we
can still call it simple, to sophisticated multimodal communications, moving more towards
the realm of mobile computing and away from mobile telephony, although always with
the telephony potential lying somewhere underneath; after all, it will always be possible
in the IMS world to make a call. However, more generally, calling is just another mode of
connecting in the IMS world.

The universal client possibility leads easily and naturally to multimodal services. Besides
the huge potential in the services domain, a universal client would seem to give IMS-based

FMC solutions a distinct advantage over alternatives, like UMA, which seem stuck with the relatively inflexible and limited feature sets, such as those of the GSM mobile services layer. Compared with the IMS services layer, there should be little doubt about the final winner.

Mobile has traditionally be voice and texting centric. The evolution to IMS, allowing a more generic connectedness architecture and usage paradigm, will clearly be something new for mobile users. We can increasingly expect users to connect to a number of services via presence-enabled 'address books'. In this sense, mobile communications will enter a new and important phase, which is why I had chosen in various workshops with operators to call this 'Mobile 2.0' which is not the same thing as saying Mobile Web 2.0, which is how one could use Web 2.0 paradigms and concepts within a mobile network. However, I think that there is a connection between IMS and Web 2.0, which is quite simply the potential to build SIP-based services using Web 2.0 as the user interface. This is still a largely unexplored area, despite many attempts I have made to initiate such thinking with various handset software companies.

14.13 CONCLUSION

IMS is a powerful enabler for next generation services. We have established how its SIP core, incorporated into a collaborative set of proxy functions, allows flexible session-based connections to be established end-to-end. Moreover, given its IP origins, so long as an IP-access network exists at either end, the IMS core can establish a session for the end points to utilise however they like. The potential therefore exists for multiple-access convergence, which might translate into device convergence, especially in some attractive cases, like WiFi-Cellular combinations. This gives rise to fixed-mobile convergence (FMC) possibilities.

Admittedly, with FMC, more than one implementation path is possible, especially taking into account legacy services and legacy networks. An 'all IP' network is attractive due to its cost savings and ability to utilise IP-enabled software with all its low-cost, service-enabling potential, but might not be reachable within a single step. The devil is in the detail of these steps. However, as a central platform for both transitioning to, and allowing the exploitation of, an IP universe, IMS seems to have what it takes. Moreover, IMS already has wide support from the mobile community, giving sufficient weight to propel the idea into the fixed community via ITU's next generation network ambitions.

The convergence potential for IMS services is an exciting possibility. As we have demonstrated, IMS seamlessly allows the integration of 'connections' (of any type) with IP-enabled services. The peer-to-peer nature of SIP allows the services to be built at the edges of the network, away from an entwined and centralised quagmire of service nodes that historically has stifled service creation and adoption.

For many IMS services, it is likely that they will be visible to the users as presentities in the 'buddy browser'. The potential for a universal client for next generation mobile services is an exciting prospect and might well be the missing link (we dare not say 'killer app', but perhaps 'killer paradigm') from mobile telephony to a future paradigm of universal connectedness, of which mobility is just an attribute rather than an essential core theme.

We have journeyed from our starting point of IMS is SIP is VoIP to a much broader and exciting destination, namely that *IMS is universal connectedness*. The service potential seems vast, which will surely bring new revenue opportunities.

15

Mobilising Media and TV

15.1 INTRODUCTION

Having looked at the various ways that mobile services can be delivered to the device and then rendered into a user interface, let's consider the options for *media companies* who want to make their content and services accessible via mobile. In this chapter, we shall look not only at the technological means to bring media services to mobiles, but also the commercial options in terms of possible revenue models and how to collect the money from the users (i.e. billing). We shall examine all of the major options for mobilising the *media experience*.

The ability to reach very large audiences through a single broadcast medium and with a single story or message per audience is becoming increasingly difficult. Media experiences are becoming more diverse and fractured. Young people are tuning out of sitting in front of the TV set to watch broadcast TV and spending more of their time surfing the Web. At the same time, we know that nearly everyone carries a mobile with them at all times and that these devices are increasingly capable of delivering some type of media content to the user. The potential for mobile as a new mass media channel is luring many media companies into the mobile world. However, mobilising the media experience isn't easy. There are many unique challenges to the mobilisation of content. At the same time, the media experience itself is evolving, mostly due to new trends on the Web, including user-generated content, social networking and citizen journalism. This chapter looks at the options for mobilising the media experience. We shall also examine the all-important question of how to make money in the process. This is in order to understand the impacts on the technology and service design.

15.2 WHY 'EXPERIENCE'?

I hope by now that one of the key messages in this book is clear, which is the importance of the user experience. That's why our mobile ecosystem model starts from the user network and works outwards. But why talk of mobilising the media experience and not just content? The media business is about the production and distribution of content ranging from hard news to soft entertainment, although the boundaries are becoming increasingly blurred with categories such as 'infotainment' and 'edutainment'. The concept of content has evolved and continues to evolve. For example, a film is no longer just a film. It is a merchandising platform, an online game, a ringtone and so on. These days, a film could include a free pizza, or a pizza could include a free film, using the mobile to link the two merchandising opportunities into one experience. The modes of engagement have also grown. Users are no longer simply watching and listening, they are playing, interacting, networking, commenting, sharing and even creating. In other words, they are clearly experiencing a media theme, not just consuming content. Thus, the bar on engaging the audience has been raised. Newer means of engaging are constantly being sought and mobile is the newest frontier in engaging the audience.

Is the media industry moving into the mobile space or are mobile companies moving into the media space? The answer is both. Thus, mobile companies and technology providers need to adapt their businesses to this trend. Some of them already are. No longer in the phone business, various service providers are moving into the experience business. Witness the evolution of handsets from portable phone-call devices to Personal Media Players (PMP), games platforms and beyond. Look at companies like Three (or 3), who are now calling themselves a media company, not a mobile operator.

The ability to capture users and engross them in a wider set of thematic experiences is a design goal that is very different to just connecting people. It might cause us to rethink our definition of mobile services. It might even cause us to revise our view or model of the mobile ecosystem. After all, the media experience extends across a variety of devices and access technologies, more than just mobile, as shown in Figure 15.1.

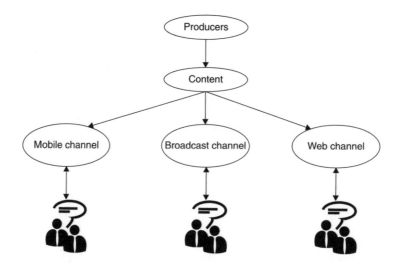

Figure 15.1 Multiple media channels.

It is tempting to contend that the ability to involve users successfully in the next big movie theme might well yield more revenue than investing in the next messaging service to replace texting. Okay, that might be an exaggeration or an unrealistic comparison. We shouldn't lose perspective. Plain old telephony and texting on mobiles is a massive global industry that dwarfs global film revenues or global advertising revenues. People still want to talk and talking makes money, but there's only so much talking and texting to be done. User expectations and requirements evolve. Mobile has already proven its power in the world of content, a world where remarkably, ringtone revenues can outstrip conventional music sales of the same title. This is why so many media companies are eager to capitalise fully on the potential of mobile.

15.3 UNIQUE MOBILISATION CHARACTERISTICS

Before looking at the specific options for offering media services on mobiles, let's first consider some wider issues relating to the nature of mobilisation. The key ones to focus on here are pervasiveness, personalisation and merchandising.

15.3.1 Pervasiveness – Always On

Unlike any other means for accessing media services and content, mobiles are always with us. We seldom put them down. Even at night, users are known to sleep with their mobiles next to their bed, even under their pillow; the alarm clock is a popular use for mobile phones! The mobile has become an integral part of our daily routine, as familiar as clothing and personal accessories. Indeed, it has become a fashion item.

Pervasiveness means that content providers need to think differently in terms of how they involve the audience in the media experience. It isn't necessary, nor useful, to think in terms of discrete time slots and 'programming'. This is 'old school' although mimicking of existing TV channels is often perceived to be an entry point into mobile. The key point is that audience availability is not a design issue – it is always available. It is probably better to think in terms of user context. What is the user doing? Where is the user? Who is the user with? Where is the user headed? These questions are more to allow personalisation of the user experience. We mould the service around the user. We no longer expect the user to mould their life and habits around the service.

Parallel to pervasiveness, there is also the opportunity provided by mobile to think of services in terms of multiple modes of engagement within one experience. The most obvious example is allowing users to use their mobiles to vote during TV broadcasts. The integrated mobile TV experience takes this to a new level. Users could be listening to a pop video and simultaneously purchase the accompanying ringtone or even tickets to a concert. The mindset here should be that *Mobile TV is not just TV on the mobile*.

15.3.2 Personalising the Experience

Mobiles are by their nature very personal items. Users seldom, if ever, share their mobiles. They hold on to them dearly and like to customise the experience, such as changing the ringtones, wallpapers and so on. Thus, media providers need to allow users to personalise the experience to suit their tastes and preferences. This is different to the mass broadcast channels like TV and print. Indeed, this potential for a highly personalised experience is what led Nicolas Negroponte[1] to coin the term 'narrow casting'. Interestingly, because of the necessity to adapt the services to suit a wide variety of handset types, personalisation (of a sort) is more advanced in mobile than any other medium, including the Web.

Interactivity is part of the mobile experience. Making this more personal is essential to a great user experience. Users are no longer passive consumers of content. They expect to interact, share and communicate, all in a manner that is meaningful to them. The more that a user feels as if the service is 'talking' to them, the more captivating it becomes. Service stickiness is vital in the fickle world of digital consumption. There are a number of ways to exploit this for mobile, but the key is for media providers to recognise that interaction needs to be a default part of the storyboard for any mobile media service. The powerful communications potential of a mobile can be exploited in ways not present in other media. For example, in a mobile TV service, it is perfectly possible to allow users to exchange messages with each other as part of the experience.

15.3.3 Merchandising – Paying is a Familiar Experience

Unlike many web-based services, it is hard to find any mobile service that doesn't require payment from the outset. This is good news for media companies generally, who expect to get paid for their content. Micro-payments for content are now well established, typically via a premium-rate text message. A number of third parties aggregate these payment services, so it is no longer necessary for a content provider to establish relationships directly with mobile operators in the target markets. Indeed, a good service provider who can host a content platform, like Motorola, will already have the necessary aggregator relationships in place to ensure that any service launched can be charged for.

15.4 THE CONTENT EXPERIENCE

Before getting into the specific options for mobilising media content, let's examine the types of experience that a user is likely to want. Here we can generalise to a set of use cases that are common to experiencing content on mobiles (and even desktops), as shown in Figure 15.2.

The boxes in the figure are self-explanatory, but they are shown here in order to act as a guide to the steps in the process of mobilising a media experience. Each box actually provides its own set of challenges, but the important first step is to ensure that there is a set of mechanisms in the mobile media offering to facilitate each of the boxes.

15.5 MOBILISATION OPTIONS

What are the options for mobilising the media experience? The options are becoming increasingly diverse. They can be categorised in terms of distinct technological approaches in the same way that physical media distribution is limited by the available media formats,

---

[1] http://web.media.mit.edu/~nicholas/

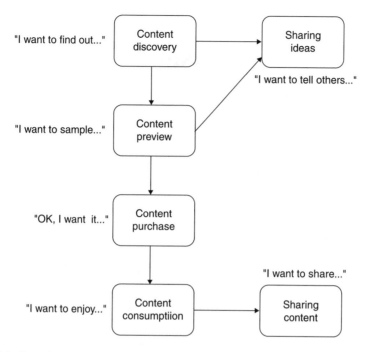

Figure 15.2 Generic media content use cases.

like DVD and VHS. The mobile clearly imposes constraints on the user interface, such as small screens and limited processing power, so the options for mobilisation are still largely constrained by technology. Take Mobile TV for example. Some handsets can support the data rates necessary to stream live video, whilst others can't. However, those that can't might still be able to play offline video files with reasonable resolution via an MMC card; hence, we can divide the possibilities here into online versus offline and so on. This is just one example of categorisation based on handset capabilities. Let's look at the most prevalent options.

Figure 15.3 shows a simplified map of how we can get content into the hands of users. Mostly we are concerned here with the 'downstreaming' of content from the producers to the users, not the other way around, which we shall discuss later. The first foray into mobile is often constrained to one of the downstream routes before getting in to more innovative solutions. However, some of the technologies used for downstreaming can be utilised for upstreaming too, as we shall see.

15.5.1   Client versus Clientless: to WAP or not to WAP

The most basic categorisation of the technology for implementing a mobile media experience is *client* versus *clientless*. A client is a piece of software installed onto the handset specifically to support the display of media content in a mobile-friendly fashion. For example, a magazine provider might issue a client that enables their magazine to be read on the mobile, but obviously in a condensed and mobile-friendly format.

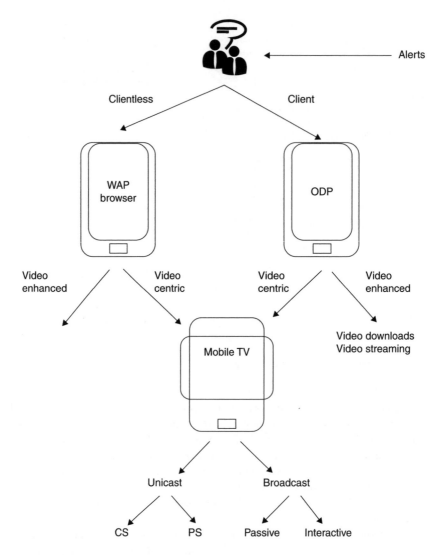

Figure 15.3 Options for mobilising media services.

There is no standard client solution on the market. Some web technologies, like Flash, have appeared in mobile form, but not as a standard. Standardisation is at a much lower level, which is the software technologies used to support client programs, the most widespread being the Java platform. Most phones on the market today support Java. This means that client solutions can be utilised with the knowledge that most users will be able to download and use them. In this category, the most common type of solution is the ODP solution that we discussed in Chapter 11.

Not all phones support the same features, so a client solution will either need to adapt to the available features on the phone or be designed to the lowest common denominator. In practice, a kind of hybrid approach is often used where families of devices are targeted with different versions of the client and each client supports the lowest common denominator within that family. For example, think of having three families: 'Entry' experience, 'Good' experience and 'Great' experience. Users aren't asked to make such a choice, which would obviously be clumsy. After all, who would deliberately choose 'Entry' over 'Great'? The type of client is selected automatically when the user first accesses the service and requests the client. What they get is the 'Cool Media Experience' and it is adjusted automatically to be 'Entry', 'Good' or 'Great' according to the user's phone.

As users become aware of a better experience available on better handsets, this can be exploited to up-sell a more expensive solution to the user. However, the up-selling process isn't necessarily one that the media company can influence if they are not involved in the retailing of handsets.

The overriding advantage of a client is that the user interface is the best possible. This is because the software approach used to implement clients can support better access to the native screen and interface capabilities of the underlying device. This might equate to smoother graphics, greater sound and so on.

On the other hand, the clientless solution offers a more limited interface, but has a significant advantage over any client, which is the much wider device support. This is what makes it such an attractive option for media companies and an obvious starting point. The so-called clientless approach still uses a piece of software, but it is the WAP browser that we know well. The advantage is that it is included in nearly every phone on the market. WAP is more flexible in terms of enabling content to be mobilised. The similarity to the Web makes it easy to understand and implement. Moreover, because the WAP browser is already installed on phones by default, there is no additional hurdle of downloading, installing and configuring clients onto handsets – this process remains alien to a large proportion of the audience and can therefore be a serious impediment to service adoption.

WAP is already widely used by most operators to support their own mobile content portals. Hence, it is a realistic and a well-understood means to offer content to users. Furthermore, any media company can set up a WAP site today with very little expertise. WAP is so commonplace, that it is no longer necessary or sensible to ponder a 'mobile strategy' before committing to a WAP site. Just get on and build one! This message has been slow to propagate generally, not just within the media world, so there are still very few sites designed for mobile.

One reason for the slow rollout of WAP sites has been the lack of tools to construct and customise pages for WAP. A unique challenge for mobile that no longer really exists in the desktop world is the need to ensure that the site will display properly in the target browser. However, these technical considerations can be hidden with the use of smart tools. There are now WAP-site creation tools that put the process back into the hands of the marketing and creative personnel and, as much as possible, integrate the process into existing content handling workflows. Integrated content management is a topic that we shall return to because of its growing importance in a converged media world.

Unlike WAP, a client requires provisioning on a handset, which means it has to be installed and configured. This is a precondition for experiencing any of the content use cases shown in Figure 15.2. It is increasingly easy to provision client software 'over the air', which is the process of downloading and installing already familiar to the Web environment. However, thus far many mobile users have been reluctant to download programs onto their mobile, mostly because they are unfamiliar with the process and may also have fears of large download times and costs. However, as users increasingly download content and games for their phones, the idea of downloading an application is becoming increasingly familiar.

Moreover, the newer devices make this an easier process, due to faster download speeds and more intelligently designed user interfaces that hide most of the superfluous technical dialogue from the users. One thing about mobiles is that the wider user population expects a relatively 'tech-free' environment due to expectations already set by the simple press-button nature of mobile telephony, quite unlike the more technically involved process of operating a PC (despite ongoing improvements).

15.5.2 On-Device Portals: Using Clients to Engage the User

The decision to go client versus clientless is not necessarily an either/or option. As discussed above, a WAP site should be part of the standard offering and with new tools it is now relatively easy to mobilise content for the 'WAP channel'. However, media companies will usually aspire to give the best brand experience possible and so the advantage of a client, with its superior user experience, is a very real attractor. When extending the service to a client solution, there are a number of possibilities in terms of how to package services via the client. As mentioned above, the trend already underway in the operator world is to use an ODP solution.

There are two options in terms of client solutions: transactional and informational. A transactional client allows financial transactions to take place via the client, such as the purchasing of music tracks. This will usually include the ability to try-before-you-buy. This is a useful merchandising approach because it makes it easy for the user to 'play with' the content and then very easily click a button to buy.

The pre-loading of content on the device means that a user doesn't have to endure download delays when browsing the content catalogue. The responsiveness of a pre-populated catalogue is a major contributor to the improved user experience. Obviously, the catalogue has to be downloaded at some point, but the usual approach is for the client automatically to initiate the download at some dormant period when the user isn't using the service (e.g. at night time). Thus, the download is hidden from the user and all they see is a refreshed catalogue that is convenient to browse. Background downloading has only been possible on those devices that support a program running in background mode. On most J2ME devices, this is not possible with a MIDlet, although it is possible to wake up a MIDlet using the push mechanism discussed earlier in this chapter. MIDP 3.0 overcomes this limitation through the support of concurrent MIDlets. Hence, an ODP client could remain active in the background and handle idle-period downloads of fresh content with limiting the user's ability to run other MIDlets, such as IM programs and games. Keeping the content fresh is a key part of any media-rich experience.

A preloaded catalogue has to be small enough for economical download, in terms of time, device memory size (often at a premium) and cost (someone has to pay the data charges to download the catalogue – see later in this section). Hard decisions have to be made about what to include in the limited size catalogue. The usual and obvious choice is to include the most popular content, like the top ten music tracks or top ten news stories. Thus, the 'fat head' of the distribution curve becomes important again, although accessing the long tail is still supported via search mechanisms. Search is used to find content not in the device catalogue. To facilitate access to a back catalogue, it is often necessary to jump out of the client to a WAP site. On most current mobile phones, it is easy to click from a client into a WAP browser without the user manually launching the browser.

There is a balancing act between refreshing the content often enough to keep users buying and not too often that it drains resources (e.g. download costs and battery life). The on-device catalogue is made from popular items and the content is usually rotated daily. This approach, combined with making the back catalogue available via WAP, provides a robust mobile content solution for most media experiences. An ODP usually preloads the content, at least the previews anyway. This necessitates 'short head' selection, as opposed to long tail, although this is still supported via search engines. As network speeds improve, preloading of content becomes less of an issue. For example, with a category 12 HSPA device (i.e. the 'slow' category), a sample 500 KB music file can be downloaded in seconds. Even so, I still believe that some preloading is useful because the instant responsiveness will be compelling and much more likely to increase content sales overall.

There are a couple of problems with the ODP approach. Firstly, the preloading isn't necessarily done in the background, thereby totally and utterly transparent to the user. In its most crude form a user might even have to press a 'refresh' button to see the updated catalogue. Moreover, if the ODP is in Java, then true background execution is problematic. The other fly in the ointment is accessing the long tail. Searching can be tedious on a mobile, especially now that the user has to keep popping back to the network to get responses and download previews. Again, HSPA will speed things up on the download side, but many ODP solutions push the user out to a WAP browser to access the long tail. Once the user leaves the ODP client, the user experience can quickly become fractured and the user may give up. With mobile data services, we should always work to the assumption that our archetypal user has an extremely limited attention span. That may well be the case, but the point is that tolerance to delays with mobile interfaces is generally very low.

MIDP 3.0 offers us a couple of advantages here. Firstly, it supports MIDlet concurrency, so that means that our ODP client can remain running all the time and thus stand a chance of offering true background downloading of new content. Secondly, and this is my hope, we can expect to see better integration between MIDlets and the browser, thus giving the user the impression of one continuous experience even if two programs are used to achieve it.

The alternative to the transactional client is a client that is used only to convey content to be read within the client, not purchased. This is useful for news and magazines. The client technology isn't that different from the transactional client, but there is a lot less sophistication, which usually means lower costs. Moreover, a transactional service requires the maintenance of a content server to manage the catalogue and related merchandising solution, which isn't necessary for something like a magazine service. This doesn't mean that a magazine, news or other type of informational service has to be offered for free – far from it. It is still possible to implement subscription schemes and this is the most common model for informational services offering premium content.

15.5.3 Offering Video Services

The devices suitable for handling video are a subset of the devices that can support clients and WAP. Also, video requires additional components back in the network to handle the streaming, which is a specialised format for mobiles called 3GPP.

Some client solutions include video capabilities, but these might well support proprietary video formats (not 3GPP). Most video-capable devices already have a media player capable of playing videos in the 3GPP format. Using a standards-based video solution to support video has the advantage of wider handset support and avoiding vendor lock-in. In addition, using a standards-based solution means that video can be supported without the need of a client at all, thus widening the prospective audience for the service. Video content can be accessed via a WAP session and the handset will launch the handset's built-in media player. Similarly, with most ODP solutions, it is possible to access the media player on the device via an API in order to play multimedia files.

There are essentially two types of video service: Video on Demand (VOD) and live streaming. VOD means that the video file is sitting on a content server waiting to be accessed via a mobile media player and it can be downloaded (or streamed) at any time. A typical example is a music video. It's the same piece of content accessed repeatedly by different users on demand. Live streaming is more akin to a broadcast where content is coming in to the content server as it happens (such as a news channel) and can be accessed at any point via a media player. Once it is played, it is gone. It can't be requested again like the VOD file, although previously live content could be packaged in this format, such as news digests available every hour.

The major challenge with streaming of video content is that a mobile must establish a connection with the streaming server that can support the continuous data stream, which is typically flowing at a high rate (compared with the rates needed to support digitised speech in a cellular phone call). Mobile networks were not designed originally for data streams, so only the newer phones can support video. 2.5G phones can support video services, but have limited video quality. It is widely recognised in the industry that video services aren't really viable without 3G phones, which are fully capable of supporting video streaming services, both VOD and live. Of course, with HSPA, things only get better.

However, the capacity of a mobile operator's network is not like the capacity of a broadband provider. The wireless resources are more limited. Hence, 3G services that are data intensive can end up competing for resources in the network. For this reason, many operators restrict the availability of video services on their network, particularly those services offered independently by third parties. Consequently, to ensure the availability of their video content, media companies may have to enter into direct agreements with the mobile operators in order to ensure that the operator will allow network resources to be allocated to the media company's video streams. Some operators might even insist on hosting the streaming services for the media company, at which point a revenue sharing agreement usually becomes necessary. On the plus side, the operator will probably include the content on its portal and spend some marketing effort on promoting the content.

15.6   MOBILE TV

For video-centric services, the media provider can elect to provide a fully blown Mobile TV experience. This is where a large part of the experience is intended to follow a programming concept similar to TV. The distinguishing factor of a Mobile TV experience over other types of mobile video services is usually the availability of an Electronic Program Guide (EPG) to support a continuous diet of TV-like programmes, just like a TV services at home. The EPG is how a user can tell which programmes are available at what times and is interactive to allow channel switching.

For a decent Mobile TV experience, a client is ideally required on the handset that will support the EPG interaction in a convenient format and also the channel zapping. EPG and switching support is not usually a standard feature in any media player on a phone, which is why a separate client is usually required for TV, even if it uses the media player on the phone (which is a typical approach). Standards are evolving all the time and eventually we can expect media players to also act as TV clients.

As we said earlier, Mobile TV should not be thought of as simply 'TV on the mobile'. What does this mean? Most importantly, simply making standard existing TV channels and TV content available on mobiles isn't the best recipe for success. Most obviously, mobiles are small and attention spans are limited. Mobile TV is often considered to be a 'killing time' service. For example, a user stuck in a bus queue will kill time by watching some TV. That's not the only way to think of mobile TV though. An important consideration is immediacy, which is why we emphasised pervasiveness from the outset. The audience for certain types of 'TV content' will want to get it straightaway, such as a live football match (assuming that the user is on the move and not in front of their 42-inch LCD screen at home or in the bar).

Other types of programming have been proposed for mobiles, although it is still early days. For example, allowing the audience to follow plot lines in-between episodes has been discussed (and proposed by the author). This can be made to feel like real-time unfolding of the story whereas the TV broadcasts are synchronised with certain fixed intervals in the story timeline. This tethered-TV approach is similar to how we expect mobile gaming to unfold for certain types of deeply immersive console games and also virtual worlds, like SecondLife. 'Real-time' events can be engaged with via mobile, giving a different type of experience to the richer episodes offered when sitting in front of a large screen. Indeed, some games will crossover into TV dramatisations and vice versa, as has already happened, but with the mobile element providing a uniquely non-stop involvement.

There are two ways to support Mobile TV – unicast or broadcast. Let's now consider the differences.

15.6.1 Unicast (and Multicast) TV and Video

Unicast simply means sending content over an IP connection that is set up from a single source (i.e. server) to a single end-point (i.e. client device). Importantly, the resources used for that connection cannot be used for other users or services. The resources remain dedicated to the user until he or she frees them up by terminating their use of the service. For example, if I connect my media player on my 3G phone to a streaming server in the network, the service is unicast.

The advantage of a unicast connection is that the user gets to access whatever video they want and when they want it. If they want to use it to access the Number 1 hit video for that week, then that's also fine. It doesn't matter what the user does, so long as the content is available. Furthermore, the access of the content is also totally under the user's control. If he or she wants to pause it, rewind it, start it again, keep watching it over and over, then that's fine too. The disadvantage is that the user exclusively ties up the resources used to support the unicast connection. As we learnt in Chapter 12, in a mobile network the physical network resources are limited. In a typical 3G cell, we have only enough resources to support about 10 concurrent video connections of moderate quality. In fact, 10 is probably a good number. Even if we manage to find ourselves in an HSPA network, we can still only expect about 30 concurrent video users[2].

What if we want to send the same content out to many users? First, why would we want to do this? Well, we might want to offer a blanket service that is more akin to a continuous TV transmission. With a technique called *multicast*, this is possible. With multicast, many users can access the same IP connection and receive the same content at the same time. Clearly, what we lose in this arrangement is the ability for each user to control their viewing experience, so this type of technology is definitely only good for TV-like services. Multicasting has always been possible with IP connections. However, with 3G networks, they are unable to assign the same physical RF network connections to multiple users. Therefore, multicast doesn't save us any network resources. This is being addressed with an extension to UMTS called MBMS (Multimedia Broadcast Multicast Service). This allows the same physical RF connection on the downlink (i.e. from the network to the devices) to be shared, potentially among an unlimited number of devices.

There are in fact two ways to establish an IP connection with a mobile device over 3G. The first is the regular packet-switched (PS) connections that we have just been discussing. The second is to make a dial-up video call into the network via the mobile switch, which is called a circuit-switched (CS) call. The advantage of making a video call to access video content is that it is very easy for the user to do. All he or she has to do is dial a number. It couldn't get simpler. Moreover, the operator or media company doesn't have to worry about whether or not the user has the right client software, browser or settings programmed into his or her phone. Video calling just works. There is no configuration necessary. It requires a piece of equipment in the network called a video gateway to enable the video call to connect to a video-streaming platform via IP. The gateway might be available from the operator or via a third-party service provider. It is increasingly common for providers of premium-text short codes to provide video gateway services. This is because the video content can be accessed via short code, even one shared with texting services. The business model is charging a premium to access the video short code and then sharing the revenue made on the call. The disadvantages of CS video access are that it is usually lower quality than PS and can only be unicast in operation.

15.6.2 Broadcast TV and Video

For truly mass consumed TV services, Mobile TV can be supported using broadcast technology. The evolution of Mobile TV standards has been influenced heavily by existing digital TV standards for terrestrial TV, hence the wide interest in the DVB-H standard. DVB-H comes from the Digital Video Broadcast forum that first developed a standard for digital terrestrial broadcast TV, called DVB-T to replace the old analogue PAL standard. DVB-T uses an RF transmission scheme called Orthogonal Frequency Division Multiplexing (OFDM) to pack lots of bandwidth into the frequency spectrum assigned to TV transmission, which is typically in the UHF band. We don't go into how OFDM works here, except to say that in order to extract the OFDM signal at the receiver, a lot of heavy-duty processing is required. That's fine for a box of electronics sitting on top of the TV where there's a relative

---

[2] These are mostly finger-in-the-air estimates based on typical network conditions and video qualities seen the field.

abundance of continuous power, space and cooling. However, for mobile devices, it's a killer. Hence, DVB-H was proposed as a variant of DVB-T that allows a much lower processing overhead suitable for handheld battery-powered devices. Of course, the overriding factor in reducing complexity is in the fact that a mobile device has a small screen. Therefore, less information is required to encode a TV picture that is only a few inches across as opposed to a few feet. Moreover, as video compression continues to improve, the use of the newer MPEG4 video compression schemes also allows a significant reduction in information to be sent across the air compared to the use of MPEG2.

The thing to keep in mind about DVB-H is that it really is a mobile TV technology. In other words, it is designed to support mobile TV. This is different to any of the generic mobile network technologies, like UMTS, that are designed for point-to-point data services of any description. When an entire communications network is designed with one purpose in mind, it is always possible to achieve greater efficiency than with a general-purpose network. All the components in a purpose-built network can be ideally selected and tuned to work together in harmony for one purpose. This is mostly the case with DVB-H, except that its origins are not mobile TV, but terrestrial TV, so it is not as optimised as it could be. This is not the case with MediaFLO, a competing standard, which was designed with one purpose in mind that being the support of mobile TV services. MediaFLO uses a similar transmission technique (OFDM) but we shouldn't jump to the conclusion that it is therefore the same as DVB-H. There are differences in the design and there is more to sending a TV signal than just the RF technology. There is a variety of signal processing and network design parameters to take into account. Therefore, it is claimed by Qualcomm, the inventors of MediaFLO, that MediaFLO is a superior mobile TV technology, which in practical terms means more efficiency, which means greater battery life and lower overall network costs.

There are yet other technologies for broadcast TV and only time will tell which standards will prevail. As with other standards globally, there will usually end up a clear winner, but there is always room in the market for more than one technology. However, with mobile services it is important to understand how critical handset availability and choice is to the success of the service. Hence, for a media company who is in the position to be involved with the launch of a mobile TV network and service, the practicalities of handset choice might well be the dominant consideration. In Europe, the backing of DVB-H by the European Commission and its Nokia-influenced roots are likely to have a strong impact on the success of DVB-H over other standards. We don't go into broadcast technologies in this book, but there are a few key concepts that are useful to understand.

The first is that it is a broadcast technology. This is an obvious thing to say and almost too dumb to mention. However, if you are used to mobile technology for the past decade or more, as I am, then it takes a bit of getting used to the fact that a broadcast network is fixed in terms of transmission and very oblivious to the receiving devices. All the content is broadcast at once! The transmitter doesn't have a concept of who is accessing the content – they tune it and take it or leave it. To make all this work, there is a set of tables broadcast on a 'hidden' channel (i.e. a control channel) which tells the receiver where to find each channel in terms of which 'frequency' to tune in on (except with OFDM, it isn't a frequency at all). To allow the user to figure out which channel they want, an Electronic Service Guide (ESG) is also transmitted. This is what says which programmess are on and when they are on. There is scheduling and programme information used to provide an Electronic Program Guide (EPG) for the user and there is hidden data used for such things as telling the DVB-H decoder where to look in the broadcast stream to extract and decode the selected content.

There is only one other flow of data to consider besides the channels and the ESG, which are the *key streams*. In TV services that are paid for, it is necessary to implement a Conditional Access System (CAS). All the channel content is encrypted, which means it can only be viewed if the user device has a key to decrypt the required channel. The user gets this key from the key stream. The key stream is used to transmit a mass of keys to all the users who have paid for content. If they paid for content, then they will find their key whizzing past on the key stream at some point and this can be downloaded to unlock the content. The keys are matched to the user device by another key on the smart card (e.g. SIM). This key is used to unlock the key in the key stream, which is used to unlock the content. It is even more complicated than that, but a full explanation is outside the scope of this book.

15.7 MOBILE TV IS NOT TV ON THE MOBILE

The ultimate TV experience on the mobile will almost certainly combine unicast and broadcast technologies, along with all other media services in one integrated user interface and experience, as shown in Figure 15.4.

The user will be able to enjoy very high quality channels on the broadcast network and then switch whenever they like to more personalised content via VOD and live streaming on the unicast network. The channel capacity of a mobile broadcast TV network is typically 20 to 30 channels. On a unicast network, the concept of channels begins to blur or fade and it isn't that useful to talk of numbers of channels. In effect, the number is as many as there are pieces of viable content to keep the user engaged. It seems almost certain that

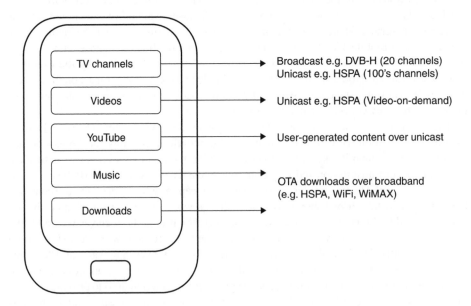

Figure 15.4 Integrated user experience.

most mobile TV services will combine broadcast with unicast, although it is important to understand that mobile TV services will evolve from different backgrounds, some from existing TV franchises (whoever they are and however they're configured in a particular region) and some from mobile operators. It is conceivable that in some countries and on some networks, especially from an existing broadcast franchise, that the service could be limited to broadcast only without the need for a mobile network for unicast services or to support interactive services. It is also possible that in some networks, the content on the mobile could be very similar to the content on the existing TV networks just because of a particular appetite for a certain type of programming that won't need reconfiguring for mobile in order to generate revenue.

Where TV has already been available on mobile devices, mostly over unicast networks, it is already clear that a different kind of approach is required other than trying to mimic what is already done on existing TV networks. Clearly, if mobile TV is, as is often thought, a 'killing time' service, then it needs to be adapted to suit short bursts of irregular length viewing. Generally, with mobile data services, we often think in terms of 'grazing' or 'snacking' on bite-sized chunks of content. A user isn't going to be that thrilled with the prospect of only being able to watch intervals of lengthy television programmes where the viewing is terminated at some awkward point in the programme's flow. If you've ever experience watching a film on an airplane and then missing the end due to landing, you'll know the feeling.

There are three key approaches emerging for tailoring the TV experience to the mobile user:

1. Interactivity

2. 'Made-for-mobile' production

3. Time and place shifted viewing

15.7.1 Interactivity

It was clear from the start of the activities with DVB-H that interactivity would be a key requirement. Clearly, if the DVB-H receiver is built into a mobile phone, then the device is capable of establishing a data connection back towards the network in addition to receiving broadcast transmissions. Therefore, interaction with the programming is possible. More-over, audience participation via mobile is well established in many markets, such as voting via premium-rate texting. Therefore, in the ESG it is possible to transmit web addressed alongside programmess so that links can be overlaid on the screen for the user to press. They will then be taken to a WAP site where they can interact via the usual means available in WAP pages (e.g. forms, buttons, etc.). Of course, interaction can be taken in a number of directions beside just voting. Participation can include voting, quizzes, feedback to the studio and even users sending in their own content.

With DVB-H on a Java phone, there is the possibility to access the DVB-H tuner via a J2ME API and thereby facilitate a much richer experience within an ODP solution instead of asking the user to jump out of the built-in TV client into a separate browser session. Once within the ODP (or browser) users can also be tempted to buy associated content, such as ringtones, wallpapers and other downloadable items.

15.7.2 Made-for-Mobile Production

Gradually, as media companies begin to better understand the mobile viewing experience, content is being produced specifically for mobile viewing, such as the short *mobisodes* that have accompanied some mainstream TV programmes. This trend will undoubtedly continue. It seems likely that for some productions, the producers will think of multichannel viewing from the outset and seek to spread an entire viewing and interacting experience across a range of viewing devices and scenarios.

Another key reason for considering an alternative approach to mobile viewing is that the mobile screen is small. This inevitably limits the viewing experience, especially for certain types of content. Even basic issues, such as displaying graphics on the screen, will need to be addressed. For example, some news channels display ongoing ticker information across the bottom of the screen in a variety of sizes and speeds. Some of this is often difficult of impossible to view on a small screen. Other problems occur with fast-moving scenes, which might be a problem for certain types of snapshot-edited films (i.e. 'MTV-style').

15.7.3 Time and Place Shifted Viewing

If users are going to dip in and out of TV on their mobile, they will inevitable find themselves out of sync with the programming. They might miss the start or end of a show. For some types of programming this might not be a problem, but for it will generally be an irritation that may even put users off of viewing TV on their mobiles. One solution is to allow users to catch up with the programme later, or to 'rewind' to the start. On a broadcast transmission, like DVB-H, this isn't generally possible, except where a group of channels could be used to transmit the same content staggered in time. However, given the lower overall channel capacity of DVB-H network compared with fixed TV networks, this seems an unlikely use of DVB-H channels.

Time-shifting of content can take place in other ways. Firstly, the broadcast back-office could support Time-Shifted TV (TSTV). This is where the broadcast TV content is constantly fed into a VOD system and is thereby available on-demand. In the mobile setting, this could then be made available over the unicast resources, such as UMTS. For certain types of live event, such as sports, it seems likely that TSTV will accompany broadcast TV by default. For example, whilst watching a live football matches, part of the screen display will gradually build up links to allow the viewers to step back in time to watch action replays, like goal shots.

TSTV can also be used across viewing devices so that the user can start watching a programme on their mobile and then later catch-up via their set-top box at exactly the point they left off. This is part of a wider set of possibilities with convergence, as discussed in the following section.

15.7.4 TV-Centric Convergence

TV is already something well known by viewers. It has largely been an experience based on the home, where it is still evolving. The introduction of IPTV has enabled new modalities of experience, such as VOD and the Personal Video Recorder (PVR). To an extent, users expect Mobile TV to be an extension of the home TV experience. There are a multitude of use cases for converging the two experiences. Some examples are as follows:

- *Aggregated EPG data* – here the user is able to view the EPG data for both fixed and mobile programming on either device (and probably also the Web).

- *Remote PVR operation* – this is where the user can issue a remote control command from the mobile to their PVR. The PVR could be located in their STB as a built-in hard drive or could be a networked PVR feature (which is a giant VOD server in the network).

- *Place-shifting* – as discussed in the previous section, a user could finish watching a programme on their STB that they had been unable to watch any further on their mobile.

- *Bundled content* – a user could buy a package of content from one device (e.g. a movie-theme package on the STB) and access related content on another device (e.g. related ringtones, wallpapers and audio tracks).

There is a lot of excitement about TV-centric convergence because it combines one hugely popular service or technology – TV – with another hugely popular service or technology – mobile. Furthermore, as TV is about viewing content, there is no reason why the entire TV platform couldn't be extended to include user-generated content, such as digital photos, videos and podcasts. There are numerous additional use cases that could include user content, including the ability to access and share the content via the STB and the mobile.

15.8 COMMERCIAL CONSIDERATIONS

Whilst mobile services might offer many exciting possibilities technically, it is important that they offer commercial possibilities to match. The good news is that the mobile channel is already a merchandising channel and has been for some time. It is already possible to use the mobile channel to distribute content and receive payment. Of course, the value chain is invariably occupied somewhere by the mobile operators and they will want to take their cut. The costs here will appear as a revenue share that the operator will take from the micro-payments made for content. Nearly all micro-payment schemes possible, such as reverse-charged (premium rate) texts are implemented via the operator's billing platforms and utilise a direct financial relationship with the operator's customers. Money flows out of the micro-payment infrastructure as payouts to the billing aggregators who exist to buffer service providers from having to deal directly with all the operators (which often they won't accept anyhow). Payouts from the aggregators then flow back to the media companies less the aggregator's margin.

Micro-payment margins are only one element of the cost base associated with service delivery. Users have to access the services in the first place and this requires data transfers across the network. All data flows have to be paid for and the operator will not subsidise these costs unless part of a wider agreement with the media company. In the absence of such strategic partnerships, the users themselves usually pick up the data charges. This can get ugly because there are still a wide variety of service packages in place where the user can end up paying high rates for data transfers (unless they are directly to the operator's own services).

The market trend is slowly towards fixed data tariffs where the user will pay a fixed monthly fee for unlimited data transfers (or some high chunk of data each month). Meanwhile, the cost of data to the user is a real consideration and might thwart the use of more data intensive services, such as video or on-device portals (with preloading of data) without some sort of relationship with the operator. There is an indication that reverse billing of data is going to be offered on some networks. This means that the content provider and not the user, which under the right conditions will facilitate service usage that is still profitable to the provider, can meet the cost of data transfer.

The other element of cost is the platform used to deliver the services. Media companies can choose to license software and host their own platform or they can partner with a service provider to do it all for them, which is often the preferred route.

The reason that service providers are an advantage is the ongoing complexity of handling content distribution and fulfilment to a very wide range of device types over a range of networks, especially for multinational rollout. Hence, complexity exists and needs to be managed. Additionally, the lack of standards in some areas, such as client solutions, and the need to have a wide range of service offerings (e.g. client and clientless) to maximise service usage only adds to the complexity. It is possible to find excellent point solutions for video, Mobile TV, WAP content management, client solutions and other components, but the challenge is how to utilise these components to make an easy-to-use and compelling end-user experience. That is where the skill of a service integrator becomes essential to the design, management and fulfilment of an integrated solution that hides technical complexities from both the media company and, more importantly, the users.

Clearly, managed service provision comes at a cost, but this is often offset in terms of risk by active-user pricing schemes or revenue sharing arrangements that allow costs to be spread and pro-rata to the usage of the service. Hence, media companies will often only end up paying for a relatively low set-up fee and then for service usage only as the service is taken up by users. Other ways of offsetting the costs exist, such as the user of advertising, which is discussed in the next section.

15.9 MONETISATION

Having discussed the costs above, what about the revenue opportunities? Clearly, media companies have content that can be sold to users. There is an initial cost of mobilizing the content, but as we have hopefully explored above, the opportunities for accessing a large user base in a variety of interesting ways, including innovative new services bodes well for a potentially very lucrative revenue stream. Also, not only is there an opportunity for direct revenue from mobile services but also the indirect revenue facilitate by mobile access to bundled services, such as an online gaming community that has a mobile element.

15.9.1 Subscription Models

In terms of monetising the services, then the content provider can choose from a number of possibilities:

- Pay per item of content ('pay per bite')
- Pay for time-based access to a set of content that is kept ('subscription')

- Pay for bundled services ('Bundling')

- Pay for time-based access to a set of content that is borrowed ('rental')

To implement any of these subscription models, some server logic is required to keep track of users, their subscriptions and to request payment from the appropriate payment mechanism, no matter how this is done (e.g. via a billing platform). In a converged services environment, some of the bundling mechanisms can get complicated very quickly and may necessitate a single platform to handle subscriptions across all services. This is where service delivery platforms (SDP) can be helpful.

15.9.2 Advertising Models

There has been a lot of talk about mobile advertising, including a lot of hype. Some mobile operators have become interested in using advertising as a new revenue stream. Again, some perspective is required. The global advertising revenues are less than 10% of operator revenues, so advertising doesn't have the potential to subsidise or pay for mobile services, except in certain niche cases perhaps. However, most media companies are firmly entrenched in the advertising business and it is a familiar and healthy contributor to revenue. Extending this to the mobile space is an obvious step.

There is no clear 'unit' of mobile advertising. In the web world, advertisers can talk of cost-per-click (CPC) models, per-impression (CPM) and so on. No equivalent metric has emerged yet in the mobile world, although there are numerous attempts to mimic the web approach. It is perhaps still too early to verify the effectiveness of this approach. Within the media experience, there are various ways to introduce advertising on the mobile, as follows:

- *CPC and CPM models* – with visible ads within browsers and browsing spaces (e.g. ODP client).

- *Ambi-adverts* – continuous brand presence during use of a particular client (e.g. TV client) for a period of time, such as the length of a sponsored TV show. The ads can appear somewhere in the screen space and may or may not be actionable (clickable). This approach could also include video overlays in the form of text straps of logos.

- *Video-ad insertion* – placing adverts in the video stream, before, during or after content. Not too dissimilar to a standard TV advertisement approach. This approach can also include video overlays.

- *Text-ad insertion* – placing ads in various text streams that the user may access, such as IM, email and texts. These approaches are considered more intrusive and can also be perceived as spam, even if permission-based.

Advertising requires some kind of management process to deliver and most likely personalise the ads. Again, certain types of SDP are taking on this role, or at least putting in place the hooks so that various services can call a common set of services to drive the advertising profiles for each user.

Index

accept header (HTTP), 164
accept-encoding (HTTP header), 182
accuracy (location), 497–498, 502, 507–511
ACK, 184–186
active PDP context, 468
actor (UML), 54
ad hoc (RF network), 61
adaptation with JSP tags, 263
adaptation with XSL, 266
Adaptive Differential Pulse Code Modulation (ADPCM), 340
Adaptive Multirate (AMR), 455
address book, 360, 535
Ad-insertion (text), 587
Ad-insertion (video), 587
adjacent channel interference, 442
Aepona™, 487
agent (software), 53, 517
agility, 38
air graffiti, 59
AKA authentication, 556
alpha channelling, 417–418
Amazon™, 82, 87
analog baseband chip, 334
analog-to-digital conversion, 334
anonymous user, 107–108
antenna, 334
Apache Jakarta, 148
API security in MIDP 2.0, 419
application discovery, 394
application loader, 349

Application Programming Interface (API), 51, 56, 83, 139, 349
application server (J2EE), 147
application topologies, 365–366
applications (component), 28, 39
applications (infrastructure), 28
applications (user), 28
ArcLocation J2ME toolkit, 516
Arithmetic Logic Unit (ALU), 400
ARM V5E DSP extensions, 348
ARM V5T Thumb instruction set, 348
assembly language, 400
Assisted GPS (A-GPS), 502, 512, 533
asymmetric encryption, 309–310
asynchronous communication, 21
Asynchronous JavaScript Technology and XML (AJAX), 12, 42, 43, 111–113, 192, 541
asynchronous messaging (software), 133, 503
augmented reality, 533
aural interfaces, 323
authentication (anonymous), (*see* anonymous user)
authentication (device), 294
authentication (HTTP basic), 295
authentication (HTTP digest), 295, 298
authentication (HTTP), (*see* HTTP authentication)
authentication (web), 294
authentication with JAAS, 298
availability (application), 36
avatar, 4–6, 40, 566